ENERGY AND LAND USE

ENERGY AND LAND USE

Edited by

Robert W. Burchell
David Listokin

Center for Urban Policy Research
Rutgers, the State University of New Jersey
P. O. Box 489
Piscataway, New Jersey 08854

Published in the United States of America
by the Center for Urban Policy Research
Building 4051 - Kilmer Campus
New Brunswick, New Jersey 08903

The preparation of *Energy and Land Use* was aided by a grant
from the U.S. Department of Housing and Urban Development,
Office of Policy Development and Research

The statements and conclusions contained herein are those of the authors and do not necessarily
reflect the views of the U.S. Government in general nor particularly the U.S. Department of
Housing and Urban Development. Neither the Federal Government nor the Department of
Housing and Urban Development makes any warranty, expressed or implied, or assumes
responsibility for the accuracy or completeness of the information herein.

Library of Congress Cataloging in Publication Data

Main entry under title:

Energy and land use.

    Bibliography: p.
    Includes index.
    1. Land use and energy conservation—Addresses,
essays, lectures. I. Burchell, Robert W.
II. Listokin, David.
HD108.2.E53   333.79'16   80-22134
ISBN 0-88285-069-5

ABOUT THE CONTRIBUTORS

George Benda is Manager, Energy Development Bond Program, Illinois Institute of Natural Resources.

Raymond J. Burby is Assistant Director for Research, Center for Urban and Regional Studies, The University of North Carolina at Chapel Hill.

Robert W. Burchell is Research Professor, Center for Urban Policy Research, Rutgers University.

Robert M. Byrne is Associate Director, Urban Land Institute.

T. Owen Carroll is Associate Research Professor, The Institute for Energy Research, State University of New York at Stony Brook.

John A. Carver, Jr. is Professor of Law, University of Denver College of Law. He is also Counsel, Head, Moye, Carver and Ray, Denver, Colorado.

Paul P. Craig is Professor, Department of Applied Science, University of California at Davis.

David B. Crandall is President, Environmental Design and Research, Skaneateles, New York.

Anthony Downs is Senior Fellow, The Brookings Institution.

Edward Duensing is Librarian, Center for Urban Policy Research, Rutgers University.

David Engel is Director of Market Development, Residential Solar Demonstration Program, U.S. Department of Housing and Urban Development.

Duncan Erley is Senior Research Associate, American Planning Association.

Robert H. Freilich is Hulen Professor of Law in Urban Affairs, University of Missouri-Kansas City School of Law.

Bruce Hannon is Director of the Energy Research Group, University of Illinois at Urbana-Champaign.

Lawrence O. Houston Jr. is Assistant to the Secretary, U.S. Department of Housing and Urban Development.

Libby Howland is Research Analyst, Urban Land Institute.

James W. Hughes is Professor of Urban Planning and Policy Development, Livingston College, Rutgers University.

Martin S. Jaffe is Senior Research Associate, American Planning Association.

John F. Kain is Chairman, Department of City and Regional Planning, Harvard University.

Edward J. Kaiser is Professor, Department of City and Regional Planning, The University of North Carolina at Chapel Hill.

Dale L. Keyes is Project Manager, Energy and Environmental Analysis Inc., Arlington, Virginia.

David Listokin is Associate Research Professor, Center for Urban Policy Research, Rutgers University.

Stephen A. Mallard is Vice President-System Planning, Public Service Electric and Gas Company.

Daniel R. Mandelker is Stamper Professor of Law, Washington University (St. Louis).

Shri Manohar is Professor, Department of Architecture and Planning, Western Australian Institute of Technology.

Gerald Mara is Demonstration Data Manager, National Solar Heating and Cooling Information Center.

Mary Ellen Marsden is Research Associate, Institute for Research in Social Science, The University of North Carolina at Chapel Hill.

The Morgan Guaranty Survey is a monthly publication put out by the Economic Analysis Department of the Morgan Guaranty Trust Co. of New York.

David Mosena is Director of Research, American Planning Association.

Peter Lewis Pollock is Staff Urban Planner, Solar Research Institute.

Donald E. Priest is Director of Research, Urban Land Institute.

Arthur W. Quade is Principal Engineer, Public Service Electric and Gas Company.

Carl N. Schuster, Jr. is Ecological Systems Analyst, U.S. Federal Energy Regulatory Commission.

Robert H. Socolow is Director, Center for Energy and Environmental Studies, Princeton University.

George Sternlieb is Director, Center for Urban Policy Research, Rutgers University and Professor of Urban Planning and Policy Development.

Grant P. Thompson is Senior Associate, The Conservation Foundation.

Elissa B. Udell is Graduate Research Assistant, W. Averell Harriman College for Urban and Policy Science, State University of New York at Stony Brook.

Jon H. Weyland is Senior Consultant, Peat, Marwick, Mitchell & Co.

William Wood is Manager-Power Supply Planning, Public Service Electric and Gas Company.

Contents

BIBLIOGRAPHY

INTRODUCTION

Introduction: The Energy-Land Use Interface

ROBERT W. BURCHELL
DAVID LISTOKIN

ENERGY and land use are planning concerns that may appear similar conceptually, yet operationally are really quite far apart. In large measure this operational disparity is due to the governmental level at which these concerns were addressed: energy policy at the federal and state levels; land use decisions by each of over 40,000 local jurisdictions throughout the United States. The inability to address energy and land use coincidentally also is traceable to a development pattern in this country that had throughout its gestation period unrestrained energy resources. Land use decisions could be made without regard to energy considerations with only slight economic penalties.

Historically, the principal concern of planning was to ensure that developments falling within a community's scheme of land use had the necessary public services to enable them to reach fruition. Land use development in this climate was clearly related to the physical characteristics of the site, local transportation access, and the availability of traditional public services. Rarely, except possibly in terms of consumer preference for one fuel or another, were energy considerations given other than cursory attention. Energy to heat and provide hot water to a building envelope was assumed to be available at a reasonable cost. Fuel sources were thought of as interchangeable by land use category, and while developments may have had to face legal and environmental hurdles, energy considerations were never a source of concern.

Often in the development of new housing, aesthetics outweighed energy efficiency and in used housing it was far cheaper to provide more heat to compensate for heat loss than to properly seal and insulate the structure. In the case of nonresidential uses, glass office buildings and "garden factories" were developed in suburban locations with a clear emphasis of form over energy function. Further, the idea that our metropolitan mobility should or could be restricted by necessary sensitivity to energy conservation was unheard of. In fact, it was always cheaper to live, shop, and work farther out. For the most part, the market for these economic entities through their capital appreciation reinforced this individual locational decision.

Given this situation it was not unlikely that the planning field was totally unprepared for what was to come. Just as it was the American automobile industry's decision not to downsize its new automobile fleet in the wake of the 1973 oil embargo, the planning literature initially paid only passing attention to energy and land use. First generation, this "passing attention" consisted largely of statements in planning journals and occasionally in actual development plans that land use be sensitive to the increasing *shortages and costs* of energy. Second generation literature contributions expressed increasing concern about the *costs* of energy, but again specific linkages of energy and land use remained unstated and unexamined.

It is the purpose of this monograph—through its twenty-seven essays—to build a foundation for the clear explication of the relationships between energy considerations and land use planning.

Toward this end, energy and land use may be defined as: *land planning requirements to enhance energy supply and reduce energy demand*. These requirements recognize both: (1) the influence and pervasiveness of past development trends, and (2) the necessity, in some instances, to redirect past trends. Further, energy-sensitive land planning will most enduringly be influenced through comprehensive state and local land use controls that include energy considerations as a key element of their basic fabric.

Energy and land use is an initial realistic determination of the "100 percent" or "Triple A" energy-savings development location within the metropolitan area. This nucleus or node determination, based on the realities of transportation costs as well as the location of planned and existing capital investment, is an essential first step to subsequent land planning efforts. Just as zoning sought to determine and ensure the future viability of the R-1 residential neighborhood within each city, the most energy efficient, geographical locus within the metropolitan area must be isolated as a key development center. The basic question to be asked here is whether this location is the older central city, residual center of mass communication and most fixed rail transit systems, or has it shifted to the interstates and beltways of metropolitan America served by telecommunications, soft wheel mass transit, fuel efficient automobiles, and diesel trucking? Obviously, this is a topic for future research, however, there is an increasing body of thought that identifies the latter rather than the former as the emerging path of energy-sensitive land development.

Once basic locational decisions have been made, energy-sensitive land use must also be concerned with the structure and land manipulation measures that have to be undertaken to *foster* space heating /cooling *supply* and *limit* similar structure-related energy *demand*. This has at least two aspects. First, the *substantive planning evaluations/analyses* that must be invoked to determine the range of development accommodation effort required to curb current demand and, as important, the necessary steps to enhance energy supply. Second, *legal instruments* that will institutionalize these measures must be drawn, tested, and incorporated within existing land use controls.

In the first case, energy savings through structure sealing and construction alterations, site positioning, shielding through land berms and landscaping must be tallied and compared against: one another, the energy consumed (if any) in their development, and the energy lost due to negative locational factors (unique to a specific technology) that would cause them to be located farther away from energy efficient nodes. For instance: (1) trees saved for summer cooling may conflict with winter solar heating objectives; (2) the space required for land solar panels may reduce the optimal density of a development; (3) the benefits of cogeneration may be best realized only in built-up central city locations; and (4) the scale of large, energy-sensitive land sharing developments may limit their siting to peripheral metropolitan locations. The tabulation of potential energy conservation by specific form of attack, and the potential neutralizing of energy savings by conflicting energy-saving strategies is critical to this new dimension of land planning.

Further, measures to accommodate energy supply must also be evaluated for their impact on energy savings. Will nuclear plants and electric generation stations be developed in such remote locations that significant transmission loss occurs before this energy can be usefully transformed? Will the soot from close-in storage facilities for coal drive residential neighborhoods to more peripheral locations? Is the energy and labor consumed in saving, gathering, and transporting materials for recycling in excess of its true energy saving value? Will site specific solar installations (roof tops of single family homes) reduce immediate energy cost pressures and thus allow movement away from the most energy efficient metropolitan locations? These and other interactions must be considered by land use planning sensitive to the potential impacts of energy supply.

Finally, energy-sensitive land use planning is the transformation of these efforts to ordinance language, such that practices may be continued into the future in pioneering locations and shared with other jurisdictions that are just becoming sensitive to this issue. This is accomplished through federal, state, and local building and housing codes, regional development controls, and specific provisions in local ordinances relating to solar access, structure siting, and street direction, length, width, etc. Model and local ordinances must be carefully thought through to ensure that energy conservation in one facet of land development is not neutralized or exceeded by energy expenditures in another.

Energy and Land Use provides the beginnings of a base of reference on the scope of issues and necessary considerations to influence the development of metropolitan areas, cities, and neighborhoods with energy conservation and supply concerns as one of several motivating forces. The next section of this introduction traces the historical progress that has allowed us to conceive of the issues of energy and land use as they have been outlined.

Background

PRECURSORS TO THE CURRENT STUDY

In 1973, the Arab oil embargo shattered America's energy complacency. The new reality of expensive energy, whose supply was not assured, prompted many symposia, articles, and monographs considering the consequences to the American economy and life style (102, 114, 123, 127).* One such analysis was conducted in 1975 by the Rutgers University Center for Urban Policy Research (CUPR) (103). CUPR compiled invited essays from energy experts, attorneys, and environmentalists examining the short- and long-term energy situation, and discussing how traditional land use patterns/controls would be affected by changes in energy availability and cost. The authors also considered other emerging influences on land use such as new environmental constraints/review procedures and the growing legal scrutiny of exclusionary zoning. The invited papers were published in a

---

*Numbers refer to the citations found in the bibliography section of this monograph.

monograph entitled *Future Land Use: Energy, Environmental and Legal Constraints* (103).

Future Land Use predicted that energy *consumption* would continue to increase in the future, albeit at a slower pace than the 3.5 percent annual growth rate of 1965–1970. Energy *supply* would be met by both foreign and domestic sources. *OPEC* oil imports would climb to a peak of about 12-million barrels a day in 1980. *OPEC* oil supply would be reasonably continuous, but the price would rise to approximately $11 per barrel by 1980. Such price escalation would spur production of *domestic* energy—coal as an interim step and nuclear, especially in the form of breeder reactors, as a long term strategy.

Future Land Use considered the consequences of these trends on development controls and the built environment. It predicted that the new energy realities would occasion for the most part only intermittent and relatively minor change in land use and adjustment in life style.

1.   *Builders would provide energy conservation measures*—in part due to anticipated legislation, in part due to consumer demand—but these changes would be slight and expensive conservation devices would be avoided.

2.   *The intrusion of smaller, single-family units as well as garden apartments and townhouses would continue in suburban areas.* These changes, however, would be modest, for the most part continuing shifts already occurring. Moreover, they were not solely the outgrowth of a new energy reality, but reflected demographic and economic influences: changes in lifestyle, declining household size, the rise of dual worker families, and a leveling of real increases in household income encouraged many families to seek smaller, less expensive, and easier to maintain forms of housing as opposed to the traditional single-family unit. The challenge to the hegemony of the suburban single-family house would also be supported by a new environmental regulatory process altering traditional subdivision practices and a growing legal assault against exclusion of lower cost housing in the suburbs.

3.   In addition to evoking these suburban changes, the new energy conditions would also affect *urban* areas by *encouraging some return of population and economic functions to cities.* As the case for the suburbs, energy's influence would be furthering a shift already underway before the 1973 embargo, and would be adding one addi-

tional factor to already multiple influences—demographic, economic, legal, and so on—fostering the urban return.

In hindsight, many of the predictions of *Future Land Use* were clearly erroneous. Energy costs were underestimated—the predicted $7 to $11 price per barrel of oil as of 1980 is at least 200 percent under the current rate. The 1980 energy consumption levels were overestimated by over one-third. Projected 1980 OPEC oil imports were overestimated by about 50 percent. These inaccuracies point to the difficulty of trying to "crystal ball" the energy future by extrapolating into the future past trendlines when basic influences—including technology, price controls, and the economy of consuming nations—are themselves in flux. A similar vulnerability is evident in other 1970s energy projections including the Project Independence study as well as the Ford Foundation Energy Policy Project (116, 485).

While one can point to individual over-and underestimates in *Future Land Use* concerning energy supply, consumption, and cost, the monograph's overall land use vision has stood the test of time. In the first half decade following the Arab oil embargo, builders are more energy conscious, there is a greater mix of housing in the suburbs, and there has been some move back to the cities. These changes have not, however, been dramatic nor solely the end result of a new energy atmosphere. In short, energy as a singular force, has *not* significantly altered the land use status quo. History, however, is not always a clear blueprint for the future. The energy/land use trendline of the 1970s may *not* continue into the 1980s because of certain significant energy changes.

THE CHANGING ENERGY PICTURE

Concern over energy *supply*—a fundamental fear after 1973—has receded somewhat as an issue. The possibility of an embargo by all major oil producers seems slim. Stoppages in oil production may occur in individual countries but such cutbacks will occasion only minor disruption. The Iranian-Iraquian conflict is a case in point. Contributing to the optimism on energy supply, is the softening of energy consumption due to conservation measures in homes and factories, the introduction of fuel efficient automobiles, and the economic downturn (thus reducing demand) in many national economies.

While energy *supply* has retreated as an issue, the *price* of energy has moved to the forefront. In 1975, OPEC oil was $8 a barrel. In 1980 it is almost $40 and there have been continuously increasing escalations in the magnitude of each price change, i.e. from $14 to $18 a barrel between 1978 and 1979 and from $18 to $38 between 1979 and 1980. In addition and related to the OPEC price jumps, *domestic* energy is also becoming more expensive as federal price controls are lifted. Tentative steps in this direction were taken by former President Carter. Gradual decontrol was included in his administration's major energy legislation—The National Energy Act of 1978. Accelerated price decontrol is likely under the Reagan administration. One of President Reagan's first actions was to issue an Executive Order eliminating the remaining provisions of a nine-year-old controls program on domestic crude oil. Domestic oil can now be raised in price to OPEC levels—a move expected to increase gasoline/oil costs by ten to twenty cents a gallon in 1981 alone. The price of domestic natural gas will also spiral if the Reagan administration modifies the Natural Gas Policy Act of 1978. This Act established a price decontrol schedule that stretched until 1985; the Reagan administration is considering an accelerated phase-out, a move which would compound the natural gas price increases already underway.

In the aftermath of these foreign and domestic actions affecting the cost of energy, American consumers are responding to calls for energy conservation. Gasoline consumption has leveled and is beginning to drop. Utilization of mass transit is up dramatically. There have been increasing purchases of insulation and other home energy conservation materials/systems. For the first time, energy is making a dent in the American lifestyle. With this change, it is opportune to reevaluate how energy will affect land use patterns/controls in the decade of the 1980s and beyond.

This volume, *Energy and Land Use,* continues and updates the dialogue begun in the predecessor volume *Future Land Use.* It similarly consists of invited essays by energy experts, urbanologists, economists, land use attorneys, and government officials. These individuals discuss the implications of the new energy realities to metropolitan development; examine the capability of land use measures to limit energy consumption by fostering energy efficient construction; analyze the effect of land use controls on conventional and new energy technologies such as nuclear, electricity, solar, coal-gasification and cogeneration; and consider specific federal, state and local measures to ensure an energy efficient land use system.

The introduction which follows is divided into two parts. Part I overviews the energy and land use literature and discusses its thematic emphases. Part II structures and summarizes the essays in the volume according to the identified themes.

.

PART I
ENERGY AND LAND USE LITERATURE:
ISSUES AND DIRECTIONS

Background

EVOLVING ENERGY-LAND USE RELATIONSHIP

Energy and land use have shared a dramatic relationship in America's development.[1] In colonial times, water energy was preeminent and consequently the placement of the nation's first settlements was often dictated by the presence of an appropriate water system—a stream, river, rapids, or falls. There were further energy influences:[2]

> In the pre-steam energy era . . . because there were no machines for transportation . . . settlements were [compact] and characterized by the juxtaposition of the various types of land uses. Commercial and residential activities were located either next to or within residential structures and these uses were in turn intermingled with small scale agricultural, educational, religious and other uses.

The advent of the steam age and coal energy in the 18th and 19th centuries lifted many of these restraints. There was more flexibility regarding the placement of communities—siting near flowing water bodies was no longer necessary. In addition, communities could grow larger, often expanding along the axial patterns of the new railroads. The small scale blending of land uses was no longer appropriate as the demand for land by larger and larger industries and businesses became voracious so as to disallow the prior mixing of modest residential and non-residential activities. In addition, many individuals chose not to live close to work but opted for commutation by railroad. In short, the intimacy of land uses was discontinued because of business necessity and worker preference.

While the new energy reality allowed greater flexibility in land use development, the energy delivery and production system itself made intense land use demands. In 1915, for example, coal had the following impact on Chicago:[3]

> Land use in Chicago was completely driven by the requirements of coal and coal-derived fuels. Transportation and coal-handling facilities dominated the landscape. The 23 million tons of coal passing through Chicago each year handled by barge, train and wagon required an extremely large materials handling system.

These factors add up to one conclusion. Large tracts of land had to be given over to coal, the fuel that made the city work, that made the city possible.

In the 20th century, the advent of inexpensive and/or new energy sources such as domestic petroleum/electrical energy which were more efficient and less demanding than coal, coupled with the introduction of the automobile and its cheap and unrestricted travel options, served to remove most of the vestigial restraints of energy on land use. Population and industry could now be diffused, first to the outer limits of the city, then to the near and far suburbs and finally, to the exurbs and rural fringes. These expanded and scattered metropolitan forms, in turn, grew to the point where they joined their neighbors as a megalopolis. Structural sectors of the metropolitan area which did not coincide with the new, "free" energy-land use relationships lost population and economic functions. This decline was most prominent with respect to urban centers, characterized by a bygone imprint of intensity and mixing of land uses. Railroad and other transportation systems which had flourished under axial patterns of growth similarly suffered in a new era of cheap energy and metropolitan diffusion.

EARLY ENERGY-LAND USE LITERATURE

American land use regulations date from the 1920s. Even at its beginnings, some of the early land use planners were concerned with the relationship of land use and energy. To illustrate, in the 1930s, the Tennessee Valley Authority (TVA) grappled with the strategy of using cheap energy to revitalize construction in and to stem population outflow from, a depressed region.[4] The TVA planners also considered the problem of "energy-boom" communities—cities springing up almost overnight as a result of hydroelectric construction. Other examples also bear mentioning. In the 1930s and 1940s studies prepared under the auspices of the National Resources Planning Board as well as numerous regional planning associations recommended land use strategies which would reduce the traffic, energy and other demands of spread city.[5]

These early linkages, though notable, were clearly the exceptions. American land use controls originated just at the time when the traditional restraints of energy on land use were being lifted and land use planners tended to disregard energy as a force shaping the metropolitan area. This attitude is evident in the effort to formulate basic zoning and subdivision practices. In the 1940s and 1950s, engineering criteria regarding traffic, sewer, drainage, and so on were the major yardsticks. In the 1960s and 1970s the substantive inputs to land use were broadened: environmental, social, economic and aesthetic concerns were deemed of prime concern. The broadening of the dimensions for reviewing and formulating land use controls did not, however, include *energy* implications. Basic land use texts treated energy largely as an engineering requirement to be satisfied, e.g. utility lines would have to be planned for, rather than a key element influencing land use itself.[6] Such treatment is understandable given the seemingly unlimited supply of cheap domestic and overseas energy in the 1950s and 1960s.

RENEWED INTEREST IN ENERGY AND LAND USE

In 1973 the Arab oil embargo changed the complacent energy perspective. In place of energy which was plentiful and inexpensive—the price per barrel of crude oil had *fallen* in real dollars from $2.50 in 1948 to $1.85 in 1972—energy supply was reduced by the embargo and price more than tripled over 1973-1974. The embargo was temporary but continued energy availability was problematic and further cost increases almost certain. With this change, energy became a key issue to those concerned with land use. By the late 1970s, the energy-land use relationship was considered in major studies conducted by the Urban Institute (126,127), the Real Estate Research Corporation (136,272), Booz Allen and Hamilton (207), RAND (264), National Bureau of the Standards (487), Brookhaven National Laboratory (131) and other research organizations/consulting firms. Energy and land use was further: (1) examined in hearings before federal, state and local goverments (137,146,315,413,422,484); (2) considered in articles in planning (101,125,149), law review (318,406,410) and other journals; and (3) reviewed as a central theme in many planning and other professional conferences (112,120,151,260).

The profusion of literature almost defies categorization. It covers many aspects of the energy-land use interplay from the different professional perspectives of engineering, planning, law, and architecture. There are, however, a number of underlying energy/land use themes which have received major emphasis across disciplines. These include:

1.   *Energy and metropolitan development.*
2.   *Land use and energy consumption.*
3.   *Land use and energy supply.*
4.   *Implementation of energy efficient land use controls.*

Energy and Metropolitan Development (See 100 Series Bibliography)

The historical, ready-availability of inexpensive energy contributed to the centrifugal diffusion of development away from the central city. With the 1970s change of energy supply first curtailed and then increased in price, numerous observers discussed the energy incentive to discourage continued sprawl and encourage *compact* development in cities and older suburbs. A considerable literature has developed on this subject (101,104, 105,111,119,126,127,128,137,142) focusing on the *existence/magnitude of an urban energy advantage and the ability of this alleged energy bonus to modify the outward development patterns of the past half century.* The themes of the studies considering energy and metropolitan development are highlighted below.

THE URBAN ENERGY ADVANTAGE

The argument that energy will encourage a return to the central city is based on the premise that cities are significantly more energy efficient for numerous reasons:

1. *The urban housing stock, because of its density and compactness, consumes less energy.*
2. *Cities benefit from advantageous transportation and commutation characteristics.*
3. *Cities can readily capitalize from emerging more efficient, energy systems.*

Many studies assert that the urban housing stock possesses an energy advantage (102,106,109). In part, the energy bonus is due to the *higher density* of the residential sector, i.e. comparable single-family units in the cities are located on smaller lots than their suburban counterparts, and since density is higher, energy consumption for transportation and other purposes is lower (109,110,111,118). The *type* of urban housing, consisting of a relatively high share of low and high rise multifamily units, in contrast to the preponderance of the single-family unit outside of cities, is an additional and important energy saving feature (113,129,137). As discussed in the Congressional report, *Compact Cities: Energy Saving Strategies for the Eighties:*

> In central cities, over half of the residential units are multifamily and townhouse types. In the remainder of metropolitan areas, over two thirds of the housing stock are single-family, detached dwellings. The multifamily homes are more energy efficient as a rule . . . New York City's per capita energy consumption is among the lowest in the nation because of the dominance of multifamily housing.

The transportation and commutation characteristics of cities are also alluded to as offering a further energy advantage (109,143,144). Because of the city's compactness, residents do not have to travel long distances for work or leisure activities. In addition, urban households have better access to energy-efficient, mass transit. This situation stands in contrast to the suburban resident who has to travel long distances, often to the CBD of the central city, and typically has no alternative to the automobile. The argument is therefore made that as energy costs escalate, suburban families will be induced to move back to the city.[8]

> One of the most popular assertions is that relating to the effect of higher energy costs on urban housing markets. Noting the upward swing in gasoline prices and the American dependence on the automobile for daily travel, many persons have asserted that higher energy costs will result in increased demand for housing in U.S. cities. The basic premise of the energy assertion is relatively easy to understand: suburban residents now commuting to the city for employment will move to the city to have a shorter, cheaper, and more convenient journey to work.

In addition to the *current* energy advantages discussed above, many studies assert that cities are in an advantageous position to benefit from *future* efficient energy technologies (130,143,148). To illustrate, cogeneration and district energy systems are two energy saving approaches discussed later in this paper. Both technologies require a reasonably high density and a mixing of land uses—characteristics most readily found in cities.

In short, a considerable literature argues that cities are more energy efficient for many reasons. Reference is often made to cross national studies such as the Doernberg (110) comparison of the United States and Sweden which indicated that while the two countries shared an equivalent per capita GNP, Sweden consumed 60 percent less energy per person because of its higher density of uses and more intimate mixing of residential, commercial and industrial activities. Cities as an entity are viewed as the "Swedens of the United States"—more energy efficient because of their land use characteristics and therefore much more competitive under the new energy conditions to attract residents and businesses.

ENERGY AND CITIES: A REEVALUATION

In recent years several urbanists have *begun to question whether cities, in reality, benefit from a current or future energy advantage significant enough to reorient the pattern of land use development.* This literature directs itself to the city's energy-land use characteristics discussed above.

Is **urban** *housing more energy efficient?* There is no question that the city housing stock has a relatively high density and contains a larger share of attached housing—two energy saving characteristics. At the same time, the housing is *old* and is often energy inefficient because of masonry construction, ceiling height, room layout, lack of weatherization, outdated heating/cooling plant and other characteristics.[9] These drawbacks can be corrected by installing insulation, storm windows and new energy systems, but such melioration requires a financial capacity absent to the urban poor living in the most energy inefficient dwellings. These households do not have the cash reserves to pay for the needed improvemtns, and because of their low income, benefit only marginally from federal tax incentives for energy conservation. Thus, the urban housing stock may actually imply an *energy debit* rather than credit (148). These factors stand in contract to suburban housing/households where significant energy savings are realizable given the physical characteristics of the suburban housing stock and the relative affluence of the suburban population. As discussed in the 1980 *President's National Urban Policy Report,*[10]

Rising (suburban) household incomes, coupled with the widespread availability of alternative strategies for energy conservation, will enable the average household—and especially those households that can afford new homes—to avoid compromising their desire for the amenities associated with the single family dwelling and the private automobile.

Do cities have significant transportation/commutation benefits? Several studies also question whether most suburbanities suffer a long and expensive (relative to household income) journey to work. While the suburbs at one time served an exclusive bedroom function, the decentralization of jobs to suburbs in recent years meant that many suburban residents no longer would commute to the CBD but rather would drive to locations near their residences (111,142,148,216). Data from the *Annual Housing Survey* reveal that most suburbanites have a *similar* journey to work as their urban counterparts because many "suburbanites do not work in the city." Furthermore, even if suburban residence demanded travel by automobile to an urban workplace, the practical effect is minimal as far as impact on the household budget is concerned even after the large increase in energy costs of the 1970s. As summarized in a recent publication, *Energy in the Cities*:[11]

> The average U.S. household spends about 20 percent of its gross income on transportation. About 19 percent of this amount goes towards gasoline. The percentage of gross income going to gasoline is therefore .038. The net difference of 4 percent between the increase in gasoline cost (1973-1979) and median income (1973-1979) when compounded over six years totals 26.5 percent. Multiplying this 26.5 percent compound total by the average household expenditure on gasoline of .038 of gross income yields 1.007 percent. The total impact, then, of the gain in gasoline prices from 1973 to 1979 is 1 percent of the average household median income.

Are cities in an advantageous position to benefit from new energy technologies? Cities may have an advantage with reference to certain new energy approaches, such as cogeneration but it is problematical how they fare with respect to others. The potential for solar energy is one such example. The high ratio of building surface (for solar collection) to open land, the prevalence of flat roofs so that collectors can be freely oriented, and the general absence of foliage causing shadow problems are all features found in the urban landscape which lend themselves to solar energy generation. There are, however, many *impediments* to solar applications in cities (224,239). The height and orientation of city buildings can create solar access problems. There is also a penalty due to the compactness of cities whereby there is a high energy demand relative to the space (surface) available for solar collection. Covering every square yard of Manhattan with solar collectors would supply only about 15 percent of that borough's intense demand for energy.[12] In addition, applying solar to cities would require extensive *retrofitting* on high-rise structures—a more difficult and expensive hurdle than utilizing solar in new, low-rise construction.

Other energy systems may also not be appropriate for urban areas. The combustion of wood and other renewable resources is often precluded by stringent urban air pollution safeguards designed to prevent further air quality degradation. Utilizing coal may similarly be prevented. Wind generated energy is typically unfeasible because of blockage from existing buildings. The possibility of packaged nuclear plants in urban areas is also slim because of the potential threat to large masses of population.

The literature reevaluating the alleged city energy advantage also emphasizes the importance of *social* forces in influencing metropolitan development (111). Many white, middle-class families have sought a suburban refuge in order to segregate themselves from the urban poor and minorities. This desire is so strong that it often outweighs any energy arguments for urban return. The factual uncertainty of whether cities have a decided energy advantage makes the energy argument for urban return even less compelling.

In summary, the literature on energy and metropolitan development largely concerns itself with an assessment of the evidence supporting/denying an urban energy advantage. It also attempts to place these energy implications in perspective with social and other forces influencing the return to the central city. These analyses, in turn, involve consideration of the changing economic function of the central city; comparison of the journey to work of urban versus suburban residents and the role/cost of the automobile as a means of transportation; review of the urban housing stock and tenant/owner characteristics; and evaluation of the spatial requirements of the new energy systems.

Land Use and Energy Consumption
(See 200 Series Bibliography)

The literature in this area has focused on the related topics of the *land use determinants of energy consumption and the ability of land use to foster energy conservation. In both cases there has been a parallel evolution of general, sometimes theoretical, studies followed by a second generation level of analysis which leans to the empirical and pays more attention to the detailed subsets of the energy-land use relationship as it concerns energy consumption and conservation.*

LAND USE DETERMINANTS OF ENERGY CONSUMPTION

In the wake of the 1973 Arab oil embargo, attention turned to the land use variables affecting America's energy use. Initial studies typically evaluated the energy implications of alternative, *hypothetical* land use systems. The most well known of this genre of studies is the *Costs of Sprawl*, by the Real Estate Research Corporation (272). This investigation considered different categories of communities: "low-density sprawl" consisting of conventional single family development; "high-density planned" composed of multifamily/attached units with only 10 percent single-family; and "combination-mix," a blending of different categories of units and development modes, both conventional and planned. The *Costs of Sprawl* concluded that the "high density planned" scenario offered an energy savings of over 40 percent as compared to the "low-density sprawl" case. The study's methodology was subsequently questioned by Altshuler, Windsor and others who felt that many of its assumptions and conclusions were oversimplified and as such, the development scenarios did not adequately mirror real world development.[13] Despite these criticisms, the *Costs of Sprawl* is the most widely cited analysis of its kind (and is even more so the case out of the planning field) and typifies the methodological approach of the early energy consumption/land use studies.

This first round of literature has been followed by numerous *empirical* investigations examining energy consumption in *actual*, not hypothetical, development situations. One example is *Interaction of Land Use Patterns and Residential Energy Conservation* by Booz, Allen and Hamilton (207). This analysis examined energy consumption in twelve neighborhoods in three SMSAs, Tuscon, Washington and Chicago. It concluded that energy use is most strongly influenced by *household income*—consumption goes up with wealth—but that certain land use variables, such as those controlling home size, also bear on energy consumption. Other studies of this type include the Hittman Associates analysis of energy use in the Baltimore-Washington area as influenced by variations in the type/density of development and the A. D. Little investigation of energy use in four geographic areas as affected by dwelling/size (square footage) and type (i.e., low-rise versus high-rise) of construction.[14] The investigations of actual energy use were sometimes taken to a very *detailed* level. In the multiyear study (103,104) of the Twin Rivers (New Jersey) Planned Unit Development, engineers from Princeton University went so far as to measure energy use in different models/types of townhouses, (i.e. end versus attached units) and different areas of the same townhouse, (i.e. roof, basement, bedrooms, and so on).

In addition to their examination of actual energy consumption, the second generation studies also were more *sensitive to the interactive dynamic between energy and land use*. This interest is evident by the attempt at *modeling* the reiterative energy-land use relationshps. One of the most prominent efforts in this direction was T. Owen Carroll's formulation of an energy variant of the Lowry housing-land use model (214). There was also interest in tracing the intervening actions influenced by land use which resulted in greater or lesser energy use. This type of analysis characterizes the series of studies considering the land use/transportation/energy relationship. Fels and Munson examined the transportation and energy implications of nine alternative development patterns in the Trenton (New Jersey) metropolitan area (118). They considered how land use affected the choice of transportation mode—automobile versus mass transit—and how this selection translated into different levels of metropolitan energy demand. Edwards and Schofer, in an Illinois study, similarly compared the transportation implications of alternative development patterns such as sprawl versus pure linear growth (228). Other examples are contained in the bibliography (227,240).

THE POTENTIAL OF LAND USE
TO FOSTER ENERGY CONSERVATION

Most of the major energy studies in the last decade including: the Ford Foundation Energy Policy Project, the Project Independence-sponsored investigations, the National Academy of Sciences' interdisciplinary projections, the *Report of the Energy Project at the Harvard Business School* (116,236) and the recently issued (1981) Sawhill study conducted by the Solar Energy Research Institute stressed the importance of *energy conservation*. The *Harvard* Report, for example, concluded that energy conservation was the *key* energy strategy because it could reduce energy consumption by 30 to 40 percent. And unlike other energy options, the benefit of conservation did not hinge on a technological breakthrough and could be realized without the possible safety and environmental hazards associated with turning to nuclear, coal or other energy sources. Similar energy conservation recommendations are found in influential professional and policy studies such as the AIA's *A Nation of Energy Efficient Buildings*[15] and Lovins' *Soft Energy Paths* (328).

All of these analyses stressed the importance of *using land use/building controls as a mechanism to foster energy conservation*. In the *transportation* sector, for example, a shift from reliance on the automobile to mass transit would reduce energy consumption. Mass transit users can be economically gathered at specific collector points. This requirement, in turn, necessitates land use controls fostering compact development as opposed to sprawl (109,108,216).

In the *building sector*, specifically the construction of residential/nonresidential structures, land use/building controls are an especially potent means for encouraging energy conservation. These regulations affect: *(1) the energy efficiency of the building envelope; (2) site design requirements;* and *(3) the scale of development.*

1. *Energy efficiency of the building envelope.* Several civil engineering studies have pointed to the significant savings in energy possible by improving the insulation barrier of the building envelope. The Federal Energy Administration, after examining the effect of a test house in the District of

Columbia, concluded that the use of standard insulation devices would yield a 25 percent energy savings per structure; improved devices increased the savings to 35 percent. The most detailed engineering report was conducted by Princeton University at the Twin Rivers PUD (103,104). The researchers found that a 67 percent reduction in annual energy consumption was possible with relatively simple modifications such as adding window, basement and attic insulation, and plugging air leaks between rooms and the basement/ceiling.

Insulation standards and the general energy efficiency of new construction historically were not governed by public regulations. This hands-off approach ended in the mid 1970s as the federal and state governments imposed mandatory energy requirements. While there has been debate concerning the restrictiveness/usefulness of the new standards, a point considered later in this paper, there is no argument that they can foster energy conservation in new construction.

2. Site design. Land use controls affect site design which in turn influences energy consumption. Energy use for *heating* purposes can be reduced by: orienting structures on a southeast to southwest axis so as to maximize the potential for active and passive solar heating; emphasizing a linear building arrangement which can reduce energy losses by providing protection from the wind; and taking advantage of natural and added vegetation shelterbelts which similarly can also reduce energy wastage. Energy consumption for *cooling* purposes can be lessened by taking advantage of vegetation shelterbelts and by using a staggered building arrangement which accentuates the "natural" cooling provided by breezes. The potential and implementation of energy sensitive site design controls have been studied by Victor Olgyay in *Design With Climate* as well as by Conklin and Rossant, Knowles and others[16] (219,245). The recent SAND (Site and Neighborhood Design) research by the Argonne National laboratory documented actual energy savings in five case study communities.[17]

The case studies, done in Radisson, N.Y.; Burke Centre and Greenbrier, Va.; Shenandoah, Ga.; and the Woodlands, Texas, represent a variety of climatic settings, development programs, and schedules. The studies reflect a diversity of approaches to energy conservation at the scale of site and neighborhood design (SAND) . . . This work has established that substantial energy savings can be had by modifying site designs during the development process. Estimated annual energy consumption reductions were as high as 60% in Shenandoah, the Woodlands, and Greenbrier.

3. Development scale. Large scale development can abet energy conservation. Clustered, attached and other types of energy efficient housing are more feasible and likely to be found in such projects. The same is true with reference to energy-sensitive site planning discussed above. Large scale projects often mix land uses and in so doing, reduce energy consumption involved in shopping trips as well as the commute to work. There is also the added factor that developers responsible for large scale projects tend to be relatively sophisticated and better capitalized and thus more willing and financially capable to supply or experiment with energy saving devices.

The energy conservation possible from large scale development has been examined by the Urban Land Institute and others (271,291). These investigations have discussed the theoretical energy savings of PUDs and New Towns and have examined the actual energy experience in prototypical projects such as Reston, Virginia. They have also considered the financial and regulatory hurdles which discourage large scale development and the commensurate changes needed to remove these barriers.

This discussion illustrates the close relationship between the literature considering how land use affects energy consumption and that examining the potential of land use to foster conservation. The *Costs of Sprawl* and other studies pointing to the higher energy consumption of conventional single-family tract development have evoked recommendations for planned, large scale, multi-use communities. The analysis of the land use-transportation-energy-relationship has culminated in proposals for a land use system encouraging mass transit. The micro-scale investigations of energy use, personified by Princeton's Twin Rivers study, result in less encompassing recommendations to conserve energy, such as the reintroduction of vestibules and exterior awnings as well as the use of curtains to reduce energy loss.

Land Use and Energy Supply
(See 300 Series Bibliography)

Most of the 1970s attention on energy supply focused on the issue of federal controls and *pricing*, namely that artificial price ceilings discouraged energy exploration and thus stateside additions to the nation's energy supply. To expand energy supply, tentative steps were taken at deregulation. The National Energy Act of 1978—the most significant energy legislation of the Carter administration—established a schedule for eliminating natural gas price controls and "artificial" price distinctions between gas transported interstate versus intrastate. The first energy-supply initiatives of the Reagan administration also have been in the area of pricing, as discussed earlier in this summary.

While pricing is central, other influences on energy supply have entered the national discussion.Land use controls are one such factor. These regulations impact energy supply by *(1) influencing the production/cost of energy sources and energy facilities, and (2) affecting the feasibility of innovative energy technologies.*

LAND USE AND ENERGY SOURCES/FACILITIES

Land use regulations bear on the extraction/cost of certain energy sources. One example is coal production. Coal is often looked to as "the" American energy resource because of this country's vast coal reserves recoverable through surface mining. Surface mining is subject to both federal and state land use controls. At the federal level is the Surface Mining Control and Reclamation Act of 1977, requiring surface mine operators to restore disturbed land to its appropriate original contour, replant trees and grass and prevent siltation and erosion. Strict reclamation laws have also passed in Kentucky, West Virginia, Pennsylvania and other states. These federal and state land use controls have been considered in the energy literature (303,390,331,334,356,358) for reclamation affects not only costs but the feasibility of surface mining. The uncertainty over *future* reclamation requirements adds to the transportation, environmental and other unsettled issues which retard large scale coal production.[18]

> In surface mining, the major problem is repairing the surface deterioration . . . comprehensive reclamation programs include restoring the surface topography, replacing the topsoil, fertilizing, vegetating and returning the land to productive use, whether agricultural, commercial, residential or recreational.

> The uncertainty about future Federal and state regulatory legislation has tended to inhibit the industry's initiative in developing new mines.

Land use controls also affect energy *facilities* (311,313,307). In some cases these uses are prohibited by local regulations. In this regard, the rejection of a major needed oil refinery by Durham, New Hampshire and St. Mary's County, Maryland received national publicity.[19] Even where an energy facility is not expressly prohibited, it can be impeded and delayed by *local* zoning, *state* wetlands and coastal zone controls, and *federal* "wilderness" and "wild and scenic river" standards and the review procedures inherent to these regulations. The literature in this area includes many case studies documenting problems of this type (343).

This regulatory situation has evoked several reform proposals. Two examples were the Energy Facility Planning Act proposed by the Carter administration in 1975 and the Land Resource Planning Assistance Act introduced by Senator Henry Jackson in the same year.[20] The former legislation encouraged comprehensive energy planning for the siting of energy facilities and under certain circumstances allowed for federal preemption of all state and local laws affecting energy facilities. The latter did not permit federal override but did call for federal support for multipurpose state land use planning of which energy facility siting would be an integral part. Both measures were not enacted and subsequent similar legislation met a comparable fate. The Reagan administration, however, is in the process of introducing specific mechanisms for dealing with the land use and other regulatory constraints to energy facility siting. While it is hazardous to guess, these mechanisms will likely not include federal preemption but instead will emphasize state government's lead in "rationalizing" the regulatory framework.

LAND USE AND ENERGY SYSTEMS/TECHNOLOGIES

Land use affects energy supply in yet another manner. By modulating the density and other characteristics of the built environment, land use controls bear on the feasibility of certain energy systems. Some energy technologies require or are most amenable to higher density situations or cases where different land uses are mixed. An example is cogeneration. Cogeneration lends itself to instances where residential and nonresidential development are in *close proximity* because in such cases the waste heat from the latter is available and can be efficiently delivered to the former. *Higher density* is also necessary because attendant capital costs can be spread over many users. In addition, high density reduces certain technical difficulties of the system such as the problem of heat loss when waste heat is delivered to scattered consumers. The order of magnitude of the necessary compaction is suggested by the analytical work conducted in conjunction with Modular Integrated Utility System (MIUS)—a form of cogeneration considered by HUD. The optimal implementation of the MIUS is at a development density *impeded* by some existing local land use controls.

Other energy systems require entirely opposite base conditions. Windpower, for example, is not feasible in instances where high density, high-rise development obstructs the flow of wind. Solar energy may also not be suitable in high density situations. The previous discussion in this paper concerning energy and cities indicated that several studies indicate that solar's physical requirement of access to sunlight and for collectors of a reasonable size implies that this energy system is best suited to moderate density suburbs or low density exurbs as opposed to higher density urban places.

The literature considering the effect of land use on energy systems is frequently technical in nature (323,335,346). Examples, in the case of solar, include: the NSF/NASA analysis, *Solar Energy as a National Resource*, examining the physical demands of solar collectors and the consequent applicability of solar in different land use situations;[21] the Knowles and Berry consideration of solar under different density scenarios;[22] and the solar futures work conducted under the auspices of the Technology Assessment of Solar Energy Project at the University of California. (An excellent compilation of many of the early investigations in this area is found in *An Overview and Critical Evaluation of the Relationship Between Land Use and Energy Conservation* [285]). Given their engineering orientation, these studies are particularly sensitive to changes in technology which can alter the land use restraints/opportunities for different energy systems. To illustrate, development of low cost insulation materials could allow cogeneration systems to be placed in low density areas. Conversely, production of more efficient/smaller solar collectors may lessen the restraints to solar energy in cities.

Implementation of Energy Sensitive Land Use Measures: Federal, State and Local Controls
(See 400 Series Bibliography)

OVERVIEW OF THE PUBLIC INITIATIVES

In the 1970s, the federal, state and local governments began to formulate land use/buildings/housing controls to reduce energy consumption and foster energy conservation.

At the *federal* level the direction has been to apply "carrot and stick" measures to prod states to adopt energy efficient building requirements. In 1975, as an interim measure, states were encouraged to adopt the Standard 90–75 Energy Performance Code developed by the American Society of Heating, Refrigeration and Air Conditioning Engineers Inc. (ASHRAE). More recently, the federal government has formulated Building Energy Performance Standards (BEPS) and is moving to encourage state adoption of BEPS.

In addition to accepting the ASHRAE 90–75 and BEPS construction guidelines, *states* have also acted to promulgate other energy-sensitive controls. The states which have moved most stridently in this area are those jurisdictions—Oregon, Vermont, Florida and Hawaii—which have assumed the most assertive role in the land use process. In Oregon, for example, a state Land Conservation and Development Commission (LCDC) is empowered to establish statewide land use goals that must be adhered to by local communities. One of the LCDC guidelines is that land should be managed to "encourage the conservation of all forms of energy." Similarly, land use should foster transportation corridors to "achieve greater energy efficiency." In addition, the LCDC has direct review power on projects of statewide significance. Such developments require an LCDC permit whose approval is contingent on the project satisfying energy conservation and other state established goals. A similar permit process, with energy usage as one of the state review criteria, is found in both Vermont (Act 250) and Florida (Review Process for Developments of Regional Impact).

At the *local* level, numerous communities have effected energy-sensitive land use regulations. In many cases, an energy element or equivalent is first adopted in the local master plan. This is followed by revision in the zoning/subdivision ordinances aimed at realizing the energy conservation goal through such means as increasing allowable density, reducing the minimum size of dwelling units, protecting solar access and so on.

THEMES OF THE LITERATURE

The emergence of these new public regulations has sparked several studies considering and comparing those approaches which are most comprehensive and innovative. The Gil annotation of energy and land use policy statements as well as energy sensitive zoning-subdivision controls is one such example (239). The Harwood study, *Using Land to Save Energy* (246) analyzed the land use directions taken by Florida, Oregon and Vermont as well as multiple local jurisdictions ranging from Davis, California to Baltimore, Maryland. The most comprehensive survey has been undertaken by Erley and Mosena of the American Planning Association (433). In their study, over 1,400 local, regional and state planning agencies were contacted. From this pool, they identified thirteen jurisdictions with particularly innovative energy conserving development regulations which were examined at length. Further compilations are found in the bibiography (422, 425, 438, 464).

The movement toward encouraging energy-sensitive regulations has also evoked attention to *practical aspects of implementation*. Legal issues must be researched and at times new legal ground broken. Solar energy, for one, has raised questions concerning property rights and solar access. This issue is fundamental to the feasibility of solar energy:[23]

> Another obstacle is legal uncertainty. Access to the sun is not a transferable property right in most of the United States . . .Yet such guarantee is essential to eliminate one of the main uncertainties associated with installing solar energy—loss of access to the sun.

Legal questions concerning solar access, including the provision/absence in common law property rights and alternative strategies to secure solar access via zoning, easements or other means has been the subject of scores of law reviews and other legal research. (See 401, 402, 406, 407, 409, 411, 412, 425, 445, 449, 453).

There are many other practical concerns to implementing energy-sensitive development via land use controls. Erley and Mosena identified scattered problems of administration—the new land controls pose an additional responsibility to planning staffs and also require an energy expertise which may be lacking (433). There further appears to be instances of the regulations being ad hoc and implemented as such, rather than comprehensive and integrated into the ongoing, planning process. Finally, implementation concerns often go beyond the land use regulations themselves. Allowing innovative energy sensitive development by zoning or other means does not address the many restraints to altering the status quo in development. Barrett, Epstein and Harr[24] have documented the hesitation of lenders to finance innovative housing/land use patterns. There is also conservatism on the part of some builders as well as consumers to depart from established patterns of product delivery or purchase. Such hesitation was documented in the 1977 ULI survey of its members and the more current 1979 survey of builders and homebuyers conducted by the University of North Carolina.

In summary, the energy sensitive land use controls are in their infancy; the federal, state and local measures discussed above all date from the late 1970s. As implementation moves towards a more widespread and mature stage, we can expect a growing literature to record the steps taken and more importantly dissect what is being accomplished by the new regulatory framework and to recommend how these controls can be made effective.

Summary

From a topic of relative obscurity, energy and land use has become a major concern to planners, land use attorneys, transportation engineers, public officials, builders and housing consumers. With this change, the energy and land use literature has expanded in number and broadened in scope. This section has attempted to identify some of the central thematic emphases of the major studies. (See the bibliography in this volume for a more comprehensive listing.) The discussion reveals the breadth of topics considered by the literature and a growing depth and maturity as indicated by a turning to a factual investigation, greater sensitivity to reiterative energy-land relationships and a closer integration of research and discussion of policy.

The following section structures and summarizes the papers in this volume according to the themes identified in the review of the literature. The papers are thus organized into the following four sections:

1. *Energy and the city—future land use patterns*
2. *Alternative land use measures to limit energy consumption*
3. *Alternative land use measures to assure adequate energy supply*
4. *Implementation of energy-sensitve land use measures: local, state and federal controls*

Each section contains five to six papers which summarize the major points of contention and agreement in each substantive area and which highlight the progress made and the efforts which still must be taken to develop a land use system sensitive to the new energy realities.

PART II
SUMMARY OF *ENERGY AND LAND USE* ACCORDING TO IDENTIFIED THEMES

Energy and the City—Future Land Use Patterns
(See 100 Series Bibliography)

This section of invited essays concerns the conceptual and functional relationships between energy and land use. For what purposes in the United States is energy used? How does energy use and purpose compare in urban and the suburban settings? Between transportation and heating/cooling needs? Between the demand of the residential and non-residential sectors?

Further, does energy demand affect or is it affected by land use patterns? What are the causes of American decentralization? What role in the decentralization of land uses is played by energy price/availability? Is the node of the city the most energy-efficient location in the United States? Has a new energy-efficient, locational "mid-ground" been established by the Interstate road system and metropolitan beltways?

Finally, do new energy supply conservation "answers" have inherent land use implications of their own: in terms of either competition for space or as accelerating agents for continued metropolitan decentralization? These are the questions addressed in this section.

Shri Manohar in his article, *Energy Efficient Planning*, begins this section of essays by providing a basic foundation of energy statistics—a stepdown analysis of the world, developed countries, North American, and United States energy consumption. He notes, for instance, that 80 percent of the world energy is consumed by 30 percent of the world's population and further, that the United States with 6 percent of the population is responsible for over one-third of the energy consumption. In terms of end-use distribution, Manohar indicates that, in the United States, 28 percent of the energy utilized is for space or water heating/cooling, an additional 25 percent for transportation, 37 percent for industry, and the remainder for utility consumption of various types and for miscellaneous uses. The efficiency associated with the various end uses to which energy is directed varies from 5 to 30 percent. Applying the inverse of the efficiency rating to the end use consumption percentage, Manohar indicates that significant energy savings (50 to 60 percent) are available in the United States through conservation measures related to the upgrading of the efficiencies associated with end uses.

Manohar reaffirms the appropriateness of the Ford Foundation's (Energy Policy Project) three alternative energy futures for the United States: (1) *Historical* (3.4% growth in energy demand); (2) *Technical Fix* (1.9% growth achieved through conservation related to more efficient space heating and smaller cars); (3) *Zero Energy Growth* (via post-industrial priorities related to a service oriented society). He attempts to develop the land use priorities of each growth scenario as well as the types of energy resources which must be drawn upon to support these scenarios. Manohar's prescription for the future realistically selects *Technical Fix* as the most likely solution for the United States and views *conservation* as the single greatest energy *production* weapon.

Lawrence O. Houston's article on *Pro-Development Policies for the Eighties* moves from Manohar's world perspective on energy to issues specific to the United States. Houston indicates that metropolitan growth and energy consumption increases need not be on a collision course but rather significant growth can take place with only modest increases in energy utilization. He states that this can be accomplished via "pro-developement" planning policies which emphasize:

1. reduction of per capita energy consumption for space heating/ cooling

2. reduction of per capita energy consumption for transportation purposes
3. reduction of per capita government services and facilities costs
4. reduction of per capita cropland and open space loss

These policies can be achieved through private development which emphasizes:

1. increased scale
2. heightened density
3. a greater variety of land uses within the same development
4. improved non-auto access to activities within developments
5. improved structure siting and weather protection

The public sector can play a role, at the *federal* level, through support of new communities and mixed-use development initiatives; at the *state* level, by identifying energy efficient development locations and supporting public transportation systems; and at the *local* level, by directing zoning and subdivision regulations to encourage developments which minimize trip origins/lengths and enhance energy-sensitive siting and building techniques.

George Sternlieb and James W. Hughes draw on Shri Manohar's description of North American urban form in their article *Energy Constraints and Development Patterns in the 1980s*. Manohar described U.S. metropolitan areas as polynucleated—i.e. having many small transportation-linked centers rather than an single dominant node. Sternlieb and Hughes note that this description characterizes the Interstate system and beltways of North America which have grown as spokes and circumferentials around the once-dominant, central city. These areas have had sustained capital investment over the past 20 years. Further, they have become linked to each other, to the central city and increasingly, to exurban areas such that they are now the most efficient locational nodes in the metropolitan structure. Beltway development is the new locus of the 100 percent or AAA location of the metropolitan area and most recent nonresidential growth has been keyed there.

Sternlieb and Hughes indicate that over recent history the business sector of the United States has invested a staggering amount of capital in nonresidential facilities (business, industrial, office space, etc.) in centrifugal fashion around the central cities of metropolitan areas. To declare this investment obsolete because of a future energy policy that would attempt to reinstate the central city as the dominant metropolitan node would bankrupt the nation.

Anthony Downs in his article *Squeezing Spread City—A Second Look* describes the forces that shaped the structure of metropolitan areas in the United States over the previous 50 years. In so doing, he makes the point that cheap energy, while a contributing factor to metropolitan suburbanization, is only one of several forces which have encouraged such a trend. The desire for lower density living, to be separated spatially from the poor, and finally, the growing obsolescence of the central city were also considerations for families to leave the city. Thus, given these multiple "pull" and "push" forces encouraging outmigration from the central city, strong pressures for continued decentralization exist despite the prospect of higher energy costs.

Downs further concludes that:

1. Households will continue to decentralize as long as the cost saving in homeownership exceeds the increased costs of commutation.
2. *Travel* patterns will be altered in response to increasing fuel costs before *settlement* patterns are.
3. There will be increased demands on public transit particularly those forms that are soft-wheel, highway oriented.
4. Select, older neighborhoods of central cities may experience some revival related to energy price but this will not lead to any general, broad-base revival of central cities for energy considerations.
5. A density increase is more likely to take place in the middle band of suburban locations than in the core areas of second-order or major central cities.

The essay by John F. Kain, *Metropolitan Form and Increasing Fuel Prices*, further rejects the "revitalized city due to excessive suburban commuting costs" scenario by initially detailing the actual consumer effects of fuel increases at the source of supply. He indicates that if we doubled the real price of fuel (at the barrel head) by the year 2000, it would only raise the price at the pump by 30 percent (transport costs, refinement costs, etc. diluting the pump price increase). Further, if the price at the pump is increased by 30 percent, this causes only a 10 percent increase in the cost of automobile ownership (maintenance, amortized purchase price, etc. diluting the ownership increase).

Kain also notes that the *real* cost of gasoline in 1980 is 5 percent lower than it was in 1960 and 12 percent lower than was the case in 1940. Thus, popular opinon to the contrary, there may be no *real* price increase in gasoline by the year 2000. Alteration of metropolitan development patterns to reestablish the central city as focus may be premature in light of the relatively slow changes taking place in the real growth of fuel prices.

Kain concludes his essay by noting that we need not recreate the wheel in attempting to estimate the impact of increased fuel prices on development patterns. The Western European example serves us well here. Western Europeans have reacted to severe fuel prices by reducing the size and increasing the efficiency of their automobiles and not by restricting the outward flow of their settlement patterns. Why shouldn't this scenario also hold true for the United States? According to Kain, it does— and Ford Escorts, Chrysler "K" cars, General Motors "J" cars, Volkswagon diesels and Japanese subcompacts will continue to inexpensively take us to our favored outer reaches of most metropolitan areas.

Two final articles in this section deal with the specific effects of an energy solution, i.e., increased use of solar energy, on the metropolitan development patterns. T. Owen Carroll and Elissa B. Udell in *Solar Energy, Land Use and Urban Form* discuss the potential continuance of "spread-city" through solar energy conversion in the suburbs. They note that solar energy collector requirements are significant for other than hot water systems. These systems often require significant land as well as roof space and further, the orientation of the structure under conventional development may not favor solar employment. Urban areas' vacant lots or cleared lands awaiting renewal are usually not large enough, contiguous, or available in sufficient supply to provide space suitable for solar development. Approximately one acre of vacant land is necessary to support the residential energy needs of 10 people and associated levels of commercial and industrial development.

Solar applications seem best suited to new, usually peripheral developments, where land area is available and the relationship between structures and land can be altered to maximize the site utilization of solar technology. According to Carroll and Udell, the land use implications of the various emerging energy technologies have not been seriously considered by planners. Yet, the configuration of the energy system that evolves in ten years or so will have vigorous and lasting effects on future urban form. Land use demands of energy systems will have to compete in the political arena with those of housing, business, industry, recreation, etc. As such, decisions which affect energy planning will have to be coordinated with those that affect land use. If this is accomplished renewable-based energy systems may well be a significant part of the land use priorities of a still dominant "spread city."

In *Direct Use of Solar Energy in the Compact City*, Peter Lewis Pollock indicates that solar energy need not be a centrifugal metropolitan force. There are applications for solar energy even in the most dense, inner portions of metropolitan areas. Although building placement and height variation are indeed problems in dense areas, densities in excess of 50 units per acre and adequate solar access are still compatible objectives. Pollock further cites evidence that it is definitely possible for most *existing* cities to meet the often espoused goal of 25 percent solar contribution by the year

2000. This can be done by: (1) increasing end use efficiencies (appliances, water/space heating systems, etc.); (2) increasing efficiencies in solar technologies; (3) employing energy supplies from outside the city; (4) relying on surfaces other than buildings for solar collectors; and (5) actively competing for surplus urban land sites as potential solar collector areas. For *new* cities: (1) residential development can be completely self-sufficient (80% of roof area consumed); (2) commercial development can be 67 percent self-sufficient (100% of roof area; 50% of parking area); and (3) industrial development can be 18 percent self-sufficient, using all available roof area and on-site facilities. A city which borrows excess capacity from residential uses, exhibits an average industrial/commercial base, and has no more than 7.4 persons per acre, can be totally self-sufficient. The density of 7.4 persons per acre quoted by Pollock is more dense than the current density of the average city in the United States. Thus, solar self-sufficiency is definitely possible for new cities and significant conventional energy support is available from solar sources for existing cities.

Alternative Land Use Measures To Limit Energy Consumption
(See 200 Series Bibliography)

This section describes the procedures that may be employed to limit energy consumption by the residential sector. These include: (1) development practices to emphasize the revitalization of close-in or further development of ring-oriented locations as well as development scale and type; (2) transportation planning, to minimize convenience, shopping and work trip distances; (3) site design to emphasize position of the building structure relative to external energy consuming/production forces (i.e., landscape planning to shield the building from these same forces); (4) structure design to maximize conservation advantages and application of natural space heating systems; and (5) space heating/cooling system upgrading to take advantage of recent efficiency improvements available through the revamping of traditional systems.

To this end, the section deals with: development location, scale, concentration, orientation and construction contributions to energy conservation. It points to the unanticipated yet demonstrable contributions of energy conservation as a growing source of energy supply. The section reviews specific recommendations for conserving energy in the residential sector and evaluates the dollar benefits to the consumer of alternative energy conservation schemes.

The first article by Bruce Hannon, *Energy and Land Use*, discusses the potential of energy conservation as a mechanism to enhance energy supply. Hannon indicates that energy conservation is possible in three overlapping but definable stages: (1) increased efficiency in *direct* energy *use*—i.e., more efficient space heating and cooling systems, increased auto fuel efficiency, etc.; (2) more *product use* efficiency—i.e., the substitution of energy consuming devices based on their differing abilities to minimize fuel consumption—bus or train versus automobile travel, gas versus oil residential heating, selected manual labor versus low-efficiency, mechanical loading, hauling and lifting devices; and (3) limitations on national economic activity with emphasis on reduced spending and equitably distributed lower cost living possibilities. Economic limitations appear necessary in a society of rising expectations, where energy has been cheaper than labor, capital and land, and thus any growth in consumer dollars (per capita aggregate income, disposable income, etc.) will usually translate directly into a demand for increased energy. The implications of the translation of greater household income into increased energy demand is the identical point made by both Downs and Kain in the previous section.

According to Hannon, both energy and product use efficiency can be encouraged through the formation of high density, ring/beltway oriented, suburban population nodes that will shorten work and shopping trips, create tighter living arrangements, and foster alternative product use choices.

Self-imposed, higher energy costs in these high density environments would further prevent family unbundling and the desire for individual space and more automobiles, usually associated with such family unbundling. Replacing low-density, residential frame structures with mid-density, mixed-use masonry structures would enhance product efficiency by substituting forest acre demand with increased demand for sand and water which exist in more plentiful supply. Small electric generators serving 5,000–10,000 residential users, whose waste heat could be used for industrial space heating/cooling, have twice the thermal efficiency of traditional systems serving much fewer residential users and not having residential application.

Economic growth can be maintained by moving to a more labor-versus-technology-oriented society—a shift which would simultaneously lower both future labor productivity/real incomes and resultant future energy consumption almost proportionately.

The second paper of this section by Dale L. Keyes, *Reducing Travel and Fuel Use Through Urban Planning*, discusses the variables which appear to affect fuel consumption through urban travel. These are: (1) the spatial arrangements of desired destinations; (2) opportunity costs of travel; (3) income of the traveler; and (4) the supply of travel opportunities. Keyes indicates that the dominant factors affecting trip generation are household income and automobile ownership characteristics. If these factors are held constant, very little relationship exists between metropolitan structure and travel patterns. Thus, if income is high, travel will be extensive as will be fuel consumption. This will be true in a Los Angeles setting (low density) or one resembling New Orleans (high density).

Though far from ideal, given the nature and associative strengths of the variables, a more energy-efficient metropolitan structure may be possible. This might take the form, as Sternlieb and Hughes suggest in their paper, of a smaller, multi-node metropolitan area, linked by a well-developed road network, with a large proportion of the population living in mid-density neighborhoods with a relatively uniform distribution pattern of jobs and residences. This metropolitan structure might foster shorter trips and, conditions permitting, greater use of "other-than-automobile" transportation modes.

If this type of metropolitan area could be achieved by the end of this century, we might be able to save 10–15 percent of the energy currently consumed for urban travel—an amount equal to over a quadrillion (1 quad) BTUs per year. This figure compares to a two quad annual saving by the year 2000 due to 20mpg efficiency of all new cars from 1980 onward, or a four quad saving by doubling the 1980 real price of gasoline. Thus, it is Keyes' conclusion that there are slight differences in the level and type of travel by residents of differently structured metropolitan areas. However, differences in transportation energy use in varying metropolitan areas are modest when compared to the differences that would be visible in sharply differing income settings. To control travel one must control income. Increasing gasoline prices are the strongest support measure for transportation energy conservation.

Another issue affecting energy conservation is development *scale*—the number of residential units and the proportion of nonresidential space which comprises a development. This factor is discussed by Donald E. Priest, Libby Howland and Robert M. Byrne in *Energy Conservation Through Large Scale Development*. The authors review how development scale affects space heating/cooling options. Mixing uses within a single development affords additional potential savings relative to the time sharing or system interplay made possible by the energy needs of more extensive and differing land uses. Further, larger, mixed-use developments tend to promote reduced recreational trips and convenience-oriented shopping trips which would add to the travel-related fuel savings discussed in the previous paper.

Large scale development historically has been a laboratory for residential energy experiments. It has been the scene of increased weatherization experiments—both retrofit and new construction, solar heating installations, community energy systems, cogeneration demonstrations, and so on. Energy conservation experiments and large scale development have been largely synonymous terms for over a decade.

In addition, large scale development has consistently demonstrated its ability to encompass creativity in structure design and offer variation in both site orientation and site development. This, for the most part, has been available from the underwriting, by large scale developers, of more extensive front-end development costs. Ideas learned here may be passed directly to smaller, less encompassing construction efforts.

Yet, the large scale development is in trouble. The costs of construction loans are high, and the financial track record of previous developments so poor, that seed money, public or private, is nonexistent. Further, the larger the development, the more it attracts the public's attention, and the more it is subject to multiple layers of intergovernmental development regulations.

If we are to further test community energy systems or residential/nonresidential energy cogeneration, development grants must come from other than the developers or the initial home purchaser's pockets. It is ironic that at the very time in the country's history when the federal government's involvement in local development is being reduced, there has never been a greater need for large scale (public?) experiments in energy-saving development techniques.

In *Energy Conservation Through Efficient Site Design,* David B. Crandall argues that, if basic design and planning concepts are pooled, it is possible to conserve energy use while upgrading the quality of the local environment. Crandall, like Hannon, distinguishes between two forms of expended energy—(1) *embodied* or that energy involved in the manufacture of a product or in its installation, and (2) *end use* or the energy consumed during the operation of a completed project. *Embodied* energies may be conserved during development by: (1) using natural drainage swales, thus eliminating asphalt, piping and grates/manholes; (2) minimizing grading and cut/fill by pairing intended use to existing slope conditions and maximizing the shielding and absorption capacities of on-site vegetation; (3) reducing the length and widths of roadways by clustering development and major streets not publicly dedicated; and (4) selecting a type of structure compatible with goals of development—achieving middle density aims employing more low-rise (less energy intensive) than high-rise (more energy intensive) structures.

End use energy may be saved through site selection, unit orientation and roadway layout. This is possible by: (1) selecting a place for development which will take advantage of natural site advantages (subsurface conditions, elevation, vegetation); (2) positioning housing units to minimize the effects of wind and sun; and (3) aligning roadways to maximize east/west directions and minimize their lengths and widths.

Local building, zoning and subdivision regulations, especially those in communities in the Northeast and North Central regions of the county, currently tend to restrict creative site development and site layouts which are energy-efficient. *Existing* controls need loosening to allow site plan flexibility; new controls (according to Crandall) should actively mandate site plans and building configurations which maximize energy conservation.

Robert H. Socolow in *Energy Efficiency in Tract Housing* looks at the possibilities of energy conservation specifically related to the building envelope. He notes that a productive society not only produces new goods of increased quality but regularly reviews and updates previously generated products. The analogy in the housing sector is to the standing stock of residential structures. According to Socolow, we are building more energy-efficient new housing today than has ever been the case; the real challenge (and potential for return) lies in the retooling of the standing stock. Socolow, repeating a phrase heard elsewhere in the conservation literature, notes that we must go back to the 1950 housing tracts and cut energy use in half by "drilling for oil and gas in old buildings." The energy-savings potential of such action is evident from cross national comparisons. According to data compiled by the U.S. Department of Energy, housing in the United States standardized for climate, consumes twice as much energy per square foot as housing built in Sweden. If the consumption rate per square foot could be made more equivalent, the United States might possibly save over two million barrels of oil per day or a saving equal to one-half of the current daily consumption of the U.S. automobile fleet.

Socolow, who has directed a nationally-famous housing energy conservation experiment in Twin Rivers, New Jersey, indicates that the new frontier in energy conservation is getting conservation information to consumers in a simple, standardized way. The most pressing need is to express the fuel efficiency of new and older housing in a form equivalent to the automobile gas mileage ratings. It is important, according to Socolow, to know the difference between a 30mpg and 15mpg house. His equivalent measure would be in gallons per year (fuel oil equivalent at 140,000 BTU) and the comparison above would be stated as the difference between a 1000 gallon and a 500 gallon house.

The key to saving energy in existing housing is to plug leaks—window sashes, door frames, electrical outlets, pipe holes, etc. According to Socolow the key is not to just insulate, but rather to trace the source of the energy leakage and plug it. This would be accomplished by a labor force with new skills and new equipment—"house doctors/technicians" capable of providing owners with diagnoses as to structural energy saving alternatives.

The last article of this section, *The Adoption of Energy Conservation Features in New Homes* by Edward J. Kaiser, Mary Ellen Marsden and Raymond J. Burby deals with the building industry's response to what they believe is consumer demand for energy conservation in housing. Obviously, for the housing which is to be built in the future, the active participation of the building industry in energy conservation measures is critical.

In a survey of North Carolina *developers*, Kaiser, Marsden and Burby found that more developers are including traditional conservation features such as storm window/doors, caulking, weatherstripping, additional insulation, etc. than has ever been the case before. Even in the near-temperate climate of North Carolina, there was a dramatic trend towards energy conservation measures in new home construction. In addition to basic "structure sealing" strategies, developers were attempting to: (1) more accurately pair heating system capacity with heating system demand; (2) use heat pumps to replace electrical heating systems; and (3) reduce the overall glass area of the building envelope. Almost no solar energy devices for either hot water or main heating system use were being introduced in new developments at this time.

The developers were also asked to identify their *future actions if* they sensed even further demand for energy conservation in the homes that they were selling. They responded that they would, among other things, add additional insulation to the exterior shell and introduce more energy-efficient fireplaces. What was immediately apparent was that no singular new energy feature prevailed nor was there any indication that additional solar applications would be tried.

In terms of subdivision design, developers appeared willing to add landscaping and provide weather shields through the use of berms, etc. They were much less willing to fight to orient all lots on a north/south axis, to institute solar access covenants or to attempt to go for a cluster/mixed use variance or rezoning.

Consumers surveyed by Kaiser, Marsden and Burby have a somewhat different view of measures to enhance energy conservation. They agree with developers on necessary sealing measures to improve thermal efficiency, they are not receptive to installation of heat pumps or more efficient fireplaces, yet they appear to be significant solar heating enthusiasts. Thus, energy conservation consumer demand and producer supply relationships are slightly out of sync. In this case, consumers appear more positive than developers about the future market and potential of solar equipment; developers are more positive than consumers about the market impacts of heat pumps and energy-efficient fireplaces.

The differing behavior of energy conservation supply and demand is related to where consumers and developers get information about energy conservation. Consumers obtain it from magazines, other advertising media and/or public/consumer information services; developers obtain information from their building, plumbing, electrical and heating suppliers.

Developers and consumers do agree, however, on requisite criteria for energy conservation mechanisms. These devices must (1) work, and (2) be accepted by the general public. Equipment which is too complex to repair by field technicians or which has an initial cost three to four times higher than less efficient, more conventional systems, will not meet the test of the market. While energy conservation is indeed being marketed within the structure as sold, there is still reasonable inconsistency as to what features are available and whose conservation objectives they satisfy.

Alternative Land Use Measures To Assure Adequate Energy Supply (See 300 Series Bibliography)

This portion of the monograph discusses the future for and potential land use ramifications of traditional/developing energy resources. The section begins with a projection of the demand and supply of *current* energy resources (fuel oil, natural gas, electricity) by land use sector (residential, commercial, industrial) and is followed with individual summaries on the *future* of: coal, nuclear, electricity, cogeneration and recycling. The section concludes with a discussion of the interface between various energy supply scenarios and standing goals for environmental protection.

Central to land use planning are the *space* requirements to accommodate:

(1) appropriate sites for solar generators and wind farms
(2) coal storage and transport, and
(3) security and safety of nuclear installations.

Aside from the issue of space are the issues of *technology status, production schedule and energy costs*. Specifically, what land use sector will receive the primary benefit (and thus planning responsibility) of increased coal production efforts? What sectors will receive the new influx of electricity supply fed both from coal and nuclear sources? What are the fuel-specific hurdles that must be faced by each industry to make their products more acceptable to the marketplace: nuclear—safety, coal—pollution, electricity—cost, etc.

This section discusses fuel supply and the land use consequences of alternative fuel sources: space, environmental degradation, transportation needs, locational preferences, etc. It also outlines the network of regulatory restraint which has developed to ensure that sensitive areas (open space, historical, recreational) are not encroached upon by the need for heightened domestic fuel production.

The article by Jon Weyland, *The Future of Traditional Energy Sources*, attempts to review the demand projections for traditional fuel sources compiled by the Energy Information Administration (EIA). With some data modification to produce comparable outputs by fuel sector, Weyland projects growth in overall energy demand by land use category as well as fuel source shifts within each category. His summary of the EIA Series II projections are displayed graphically below:

1980–1995 TRADITIONAL
FUEL SOURCE DEMAND CHANGES BY LAND USE SECTOR

	Residential	Commercial	Industrial
I. Overall:	Slight Growth (+12%)	No Growth (−1%)	Significant Growth (+78%)
II. Within Category:	Elect. ↑ Gas ↑ Oil ↓	Elect. ↑ Gas ↔ Elect. Oil ↓	Oil ↔ Coal ↑ Gas ↓

As is indicated by the diagram, *residential* energy demand will grow slightly (+12 percent) over the period 1980 to 1995. This is energy consumed by single-family, multi-family and mixed use residential structures. The share of residential demand borne by electricity will grow moderately as will that shouldered by natural gas; residential oil consumption will decrease significantly. *Commerical* energy demand, i.e. that consumed by finance, insurance and real estate firms, retail/wholesale trade, health/educational services, office buildings, and other commercial activities will remain essentially flat (-1 percent) in terms of demand in the near/mid-term future. This is due to relatively slow growth in commercial floor space over the period 1960 to 1977 and to substantial increases in efficiency of commercial-sector heating, cooling, water heating and lighting systems. Within the commercial sector, reliance on electricity will increase, the significance of oil will decrease, while gas consumption will remain flat.

Industrial growth—chemicals, primary metals, petroleum, paper, food, stone/glass products will occasion an increase in demand of close to 80 percent from 1980 to 1995 due largely to projected increases in value added in manufacturing for this period. Coal use in industry will increase significantly during the period, electricity will grow moderately; oil consumption will remain essentially even, while the use of natural gas will drop sharply.

Although an indirect source of supply, the demand for energy by electric utilities will increase by 4.2 percent from 1977–1990. Within the electric utility sector, use of coal and nuclear fuels will increase; natural gas and oil will decline.

Thus, on the whole, moderate growth in energy demand towards the year 2000 will be triggered primarily by the industrial sector. A shift to electricity will characterize the energy emphases of slower-growing residential and commercial land uses. An even more noticeable shift to coal as a basic source of energy will be displayed by industrial uses and by the electric utilities. The latter will also experience significant shifts to nuclear power over the period 1980 to 1995.

George Benda's article, *Coal and the Residential Energy Future*, describes the potential of coal as a residential space heating resource under four alternative utilization scenarios:

(1) *maximum decentralization*—the use of individual coal furnaces in each residence
(2) *maximum centralization*—coal as a feedstock for large, remotely-located energy generation plants
(3) *community-based secondary coal*—coal to be used as a feedstock for small, neighborhood-oriented gas and oil production
(4) *community-based primary coal*—coal as one of several elements of community-based energy resources

Scenario One would require a retrofit or new installation of a coal furnace in most residences. It would recreate the handling, storage and transportation inconveniences previously associated with coal, and further cause reasonably extensive environmental and solid waste disposal problems. This scenario is the least efficient use of the coal resource (heat arrives late and lasts long after initially needed) and poses significant cost and land consumption difficulties.

Scenario Two would involve the construction of coal-fueled, energy generation facilities at the outer reaches of the metropolitan areas. This would mean that rural areas would be dotted with coal conversion facilities (energy parks) producing electricity, and synthetic gas/oil. The efficiency of this strategy would be 1.5 times the first scenario, capital construction costs ($4-6 billion per park) would be high and rural environmental degradation, significant. According to Benda, this is the most probable of the four alternatives.

Scenario Three is the cogeneration alternative, using coal as a feedstock in moderate scale plants to produce gas oil, electricity, and heat. Each plant, of approximately 1000-5000 necessary nationally, would be tied to a specific neighborhood or large industry as its basic service area. Due to the planning and coordination necessary for the multipurpose functions of such facilities, this is probably one of the least likely of the four alternatives. Yet, the capital costs and environmental degradation for this scenario are relatively low and the heating efficiency, at 60-80 percent, is one of the highest.

The Fourth Scenario employs coal in conjunction with agricultural biomass and solar energy (broadly defined to include wind, hydropower and others) as the primary sources of a community's fuel. In this scenario coal would provide the basic resources of heat and chemicals for any manufacturing processes which are part of the community. This scenario has about the same costs as the previous alternative and less overall environmental impacts, with the exception of land consumption which is considerable. It does, however, require significant planning and integration of the processes of fuel and food production and recycling of community wastes. The likelihood of extensive implementation of such a system is not high.

Benda's concluding remarks concern the appropriateness of using coal either as a direct or indirect residential fuel source. Notwithstanding the balance of payments argument, he indicates that: (1) coal is a precious resource with multiple clinical and industrial applications, and (2) coal use can produce environmental degradation in the form of acid rain and CO_2/NO buildup. Finally, the direct use of coal as a fuel for centralized electric power production, the most likely use scenario, is the most wasteful strategy for the employment of this comprehensive energy resource.

The Morgan Guaranty Survey article on *The Future of Nuclear Power*, indicates that nuclear power is at a crossroads. While the demand is high for non-oil, power production and the energy cost of nuclear power per kilowatt hour (kwh) is one of the lowest in the industry, nuclear power production has an uncertain future. This is obvious in the relatively low number of nuclear power plants in planning, construction or operation in the United States versus other countries. According to Morgan Guaranty, the three major problems facing the nuclear industry are: (1) *safety*, (2) *waste disposal*, and (3) *proliferation*. *Safety* concerns a view by the public that a power plant will either explode or melt down and release its radioactivity. While the former is impossible, given the amount of plutonium necessary for an explosion, the latter is indeed possible, as was partially indicated in the Three Mile Island (TMI) incident. The nuclear industry points to enhanced safety precautions since TMI, and its safety record to date (no melt downs in 200 commercial reactor years) as indications that the major public concern is well in hand. They fully realize, however, that if a melt down ever did take place, evacuation would be difficult in some places and impossible in others. Evacuation planning is a major land use problem to be faced by the nuclear power industry over the next decade.

The second problem facing the nuclear industry is how and where to permanently *dispose* of nuclear wastes. Currently it is estimated that in temporary storage there is enough waste product to approximate the cubic footage of a 100 mile, four-lane highway, 10 inches deep. Spent nuclear fuel contains lethal wastes that must be protected against for the health of society. It is problematical whether the technology is at hand to permanently dispose of nuclear wastes. The nuclear industry feels that disposal is no longer a technical problem but rather a political one. Yet they realize that expanding *temporary* waste storage increases the potential for future limitations of *permanent* sites for nuclear disposal.

The final problem facing nuclear power is the problem of *nuclear proliferation*, the misuse of nuclear fuel to produce a bomb. There is a fear on the part of the general public that through sale, shipment or misappropriation, plutonium will fall into the hands of those who might use it indiscriminately. Or more generally, as the world gets more nuclear power, the chance for irrational application grows geometrically. While these apprehensions have been reacted to by the nuclear industry through shipment controls and multi-level plant security, these activites have not succeeded in allaying the general public's fears.

Given these problems and with the demand for electricity down during the latter part of the 1970s, capital construction costs increasing, and a general erosion of support on the part of the public, it is not difficult to surmise that the nuclear option faces a difficult road ahead.

John A. Carver's article, *The Challenge of Electric Generation*, overviews issues facing electric utilities. To better understand these issues it is first necessary to understand the electrical industry. Electricity currently constitutes approximately 20 percent of nonresidential and 10 percent of residential energy demands. The source of electrical supply is the electric utility industry which currently produces over 520 gigowatts (GW) of electricity or the equivalent of about 10 million barrels of oil per day. Utilities are fueled by coal (50 percent), oil/gas (40 percent) and nuclear sources (10 percent).

Early in its history, electrical generation was concentrated in small stations close to demand sources. As transformer technology improved, plants became larger and moved more distant from recipient users. Those plants served by coal moved closer to the fertile mine areas; those by nuclear power to remote geographical locations.

Thus, the scenario of the 1960s was increasing generation plant size serving an increasing consumer population. Carver indicates that during this period, the Federal Power Commission would regularly project electrical demand at 2-3 percent greater than the demand for all energy—anticipating a shift to a more electrical-based economy. However, as the United States economy has slowed, particularly the manufacturing sector, electrical demand has fallen off. In addition, technology has not been able to make more-competitive unit costs available to residential consumers. This has slackened demand in the short-run by rendering electrical heating to a fuel source of last resort, due to a high comparative unit price. Although electricity demand is projected to grow in both the residential and commercial sectors, those who are familiar with its economics indicate that demand could be several times current projections if technology could deliver cheaper power.

The challenge of future electrical generation is thus to: (1) reduce costs through technological advancement to make electricity a more competitive primary residential and commercial fuel source; (2) better anticipate the future growth of the U.S. economy and the demand for electricity within the various sectors of the economy; and (3) look for regulatory resolution which will indicate firmer trends to larger or smaller generation facilities. Given the uncertainty of the industry to rise to the challenge and the pluralistic nature of electrical utilities (over 2000 municipalities own and operate facilities for electrical generation), it is Carver's opinion that the next decade will witness a shift to smaller generation facilities, located closer to demand rather than supply sources and serving a dwindling residential consumer base. (It is interesting to note the noticeable differences of opinion between Carver and Weyland on the future scope of residential electric generation.)

The essay by Stephen A. Mallard, William Wood and Arthur W. Quade, *Cogeneration and Recycling as Energy Resources*, details the evolution, potential and limitations of these alternative energy generation methods. Mallard and his colleagues define *cogeneration* as the sequential generation of electrical or mechanical power from the same prime energy source or fuel. *Recycling* is the collection, processing and conversion of scrap into new raw materials. Both cogeneration and recycling are attractive energy alternatives because in each case additional energy is being garnered from essentially-spent energy resources.

For cogneration to be applicable in principal there must be a need for: (1) high boiler horsepower and (2) a large and continuous demand for energy. The most successful applications of cogeneration to date include the symbiotic relationships between the chemical/petroleum industries and the electrical utilities. They are geographically located on the fringe of dense, central city population and industrial concentrations. The chemical/petroleum industries provide fuel for the electrical utilities; the electrical utility produces steam and electricity for the chemical/petroleum industries. Even in these seemingly straightforward applications the economics of cogeneration is dubious. Since there must always be a back-up system, the use of cogeneration as a prime energy source cannot offer a one-for-one reduction in either equipment, operating personnel, or maintenance costs. Further, the basic principle of cogeneration, i.e. aggregation of land uses and sharing of power, may be slowly eroded in that there is a tendency for central city nonresidential land uses to disperse—a shift which can void any long-term energy sharing relationships. The urban application of the cogeneration-derivative, *district heating** suffers the same shortcomings. In addition, the costs to retrofit current oil consumers are often prohibitive. *ICES (Integrated Community Energy Systems**)* another cogeneration-based, energy sharing technique suffers from equal economic limitations and further demonstrates only moderately-efficient heat energy and electricity production.

Whereas cogeneration in one form or another has had close to one-half century of attention, *recycling* has only been in the eyes of the public since the mid 1960s. It grew in popularity for at a time of acute environmental awareness it was a procedure which both slowed the depletion of natural resources and reduced waste disposal problems. Recycling is an indirect

---

*The production and transmission of heat, hot water or steam from a central plant to residential, commercial or industrial users close to that plant.

**The production and transmission of energy for space heating/cooling and domestic water heating to adjacent facilities by a centrally-located refuse burning boiler.

energy resource in that less energy is expended to reuse a product than to create that product totally anew. Since the United States is the most prolific producer of refuse in the world, it would appear that recycling would be a natural strategy to enhance national energy supplies. Yet, it has not been for a variety of reasons. It is a costly process which saves very little in the production of similar or different* goods. As such, a premium cannot be paid for the recycled raw materials; donation and voluntary labor for collection and transportation to the recycling site frequently characterize most recycling operations.

According to Mallard, Wood and Quade, both cogeneration and recycling have intrinsic potential. They are ideas which are physically pleasing to conservationists and to the public at large, yet the technology which will allow energy from these indirect sources on a "pay-its-own-way" basis is still not at hand.

Roger L. Johnson's article, *The Status of Solar Energy Development*, provides basic information on the definition and components of solar energy and a view of the progress of the technologies associated with each component. Detailed are: (1) thermal applications; (2) fuels from biomass; and (3) the solar electric elements of solar energy. Within these respective categories differences are highlighted between: (1) active and passive applications, (2) wood and waste resources and (3) photovoltaic, wind, and ocean thermal electric alternatives.

According to Johnson, the *technology* for generating usable energy from most of the above-listed sources is currently available. Solar hot water and heating systems are being marketed, wood and other plant matter are being sold as sources of heat energy, methane is being tapped at dump sites for its energy applications, and wind energy farms and ocean thermal energy conversion experiments are revealing surprisingly-positive results. The author believes that solar energy (in its expanded definition) could eventually provide 20 percent of America's energy needs (by the 21st century) if its various technologies could be disseminated to the field. The inhibiting problem is not technology itself but rather the overcoming of high initial front-end investments, various regulatory constraints and product dissemination sluggishness. These factors suggest that in the short term, solar energy will be confined to traditional on-site installations (space and hot water heating) or that derived from wood and wood waste. Longer range applications in-

---

*Technically, *recycling* is the process in which the same product is made from the materials that are recycled; *conversion* is the process in which a different product is made from the materials that are recycled.

clude energy from methane tapping, wind farms, photovoltaics, and ocean thermal sources. Future exploitation of solar energy could be enhanced by vigorous commercialization of short range techniques and research/product development and subsequent commercialization of the longer range techniques. The federal government must be a catalyst for R&D; local governments must provide regulatory flexibility to accommodate these new energy formats.

The essay by Carl N. Shuster Jr., *Energy Siting in Critical Environmental Areas,* surveys the complexity of this issue by conveying the true difficulties of meaningful energy-environmental cooperation. Through an encompassing array of environmental protection legislation, the two goals of *energy development* and *environmental safety* have been often cast as competing and/or mutually threatening. As an example, energy facility development is referenced directly or indirectly for substantive/procedural environmental protection compliance in the following legislative acts:

1. *National Environmental Protection Act (1969)*
2. *Outer Continental Shelf Land Act (1953) (as amended)*
3. *Wilderness Act (1964) (as amended)*
4. *Wild and Scenic Rivers Act (1968)*
5. *Federal Water Pollution Control Act Amendments (1972)*
6. *Clean Air Act Amendments (1970)*
7. *Endangered Species Act (1973)*
8. *Coastal Zone Management Act (1972)*
9. *Reservoir Salvage Act (1960) (as amended)*
10. *Historic Sites Act (1935) (as amended)*
11. *Preservation of Historic and Archaelogical Data Act (1974)*
12. *Executive Orders for Floodplain Management (1977) Protection of Wetlands (1977)*

What appears to be lacking, according to Shuster, are the broad-base objectives for elements of land use which might be found in legislation invoking a national land use policy. The absence of such legislation means that the licensing of bulk (large scale) energy production facilities, both nuclear and fossil-fired, cannot be done in a systematic, interdisciplinary way that is sensitive to regional/national and long/short range environmental concerns.

Energy facility siting in environmentally sensitive areas requires a knowledge of: (1) the players involved; (2) specifics of the site or geographic location; and (3) the information which is available which may be drawn upon to simultaneously convey both energy supply objectives and areal environmental limitations. According to Shuster, there is good reason to avoid energy facility development in environmental climates where local habitat destruction might take place or there may be a potential for severe exploitation of renewable resources.

In most other environmentally sensitive areas, however, energy development could proceed providing there were assurances of: (1) multidisciplinary approaches; (2) adequate public participation; (3) third party impact assessments; and (4) pre-specified mitigation measures. These procedures would help to ensure the dual objectives of energy supply and environmental protection.

Implementation of Energy-Sensitive Land-Use Measures: Local, State and Federal Controls (See 400 Series Bibliography)

This final section of the invited essays examines the governmental land use measures which promote and support energy-sensitive development. It first considers the *framework* for such controls. How does the goal of energy conservation relate to and affect the other objectives of the land use system, such as environmental protection and public service efficiency? How probable is it that energy concerns will prompt significant land use changes? If change is likely, how should it be planned for so as to consider the engineering, economic, legal and other trade-offs associated with possible, alternative strategic formats? How could or should energy induced changes be incorporated within ongoing land use reforms, such as comprehensive growth management?

The section also considers the *direction, operation and consequences* of the energy-sensitive land use measures: What specific changes in land use and structure design controls have been made to foster energy savings? What are the emphases of these modifications? How innovative have they been? How have they or likely will fare, given legal, economic, and other institutional restraints? What are appropriate "carrot and stick" measures to overcome these barriers? Will the new generation of energy-sensitive controls affect the appearance and placement of the built environment? If so, to what extent?

The *framework* to the implementation of energy-sensitive land use measures is discussed in the first two essays by Daniel R. Mandelker and

Robert H. Freilich. In *Energy Conservation Implementation Through Comprehensive Land Use Controls*, Daniel R. Mandelker emphasizes the need for perspective and dispassionate analysis in considering: (1) the *merits* of emphasizing energy conservation as a key land use goal; and (2) the various *means* of achieving energy savings via the land use system.

Since the early 1970s, energy strategists have advocated a shift in American metropolitan development from a pattern of decentralizing functions away from the urban core ("low density dispersal") to one concentrating uses near the urban center ("containment-high density"). Mandelker does not quarrel with this recommendation but argues that the merits of "containment-high density" must be weighed against its costs and limitations. For instance, in emphasizing multifamily as opposed to traditional single-family construction, the "containment-high density strategy" implies a lower housing and recreation amenity. Concentrated development may also incur more significant environmental degradation. Mandelker mentions these impacts to emphasize the importance of airing the energy-housing-environmental *trade-offs* in selecting an appropriate land use system.

There are also trade-offs in the different *implementation* approaches to achieve energy-sensitive land use. Mandelker evaluates three alternative strategies: (1) "highly centralized, highly regulated system"; (2) "less centralized, less inclusive system"; and (3) "site specific review for energy impact." The first—the *"highly centralized, highly regulated system"*—involves comprehensive, unified planning/implementation powers wielded by a central body. The nearest examples in place, albeit less comprehensive than the idea depicted above, include the controls imposed by the Metropolitan Council in the Twin Cities and Hawaii's Land Use Commission. A centralized land use system offers the most stringent, comprehensive guidance to achieve energy savings. Its very inclusiveness—a marked departure from the traditional, decentralized and diffused land use controls—is likely, however, to engender strong political opposition.

Under the *"less centralized, less inclusive strategy,"* the existing land use system/implementing agencies are retained but special attention is paid to those decisions significantly affecting energy consumption, such as actions regarding the transportation sector. Montgomery County, Maryland, adopted such an approach in encouraging mixed-use zones near transit stations as a means to reduce energy expended for both job commutation and nonwork journeys. Mandelker observes that the "less centralized, less inclusive strategy" offers the advantage of relatively easy adaptation to the current land use system—it does *not* require the drastic changes of the "highly centralized, highly regulated strategy." In so doing, however, it is saddled with the disadvantages of the land use status quo, such as the absence of a comprehensive land use strategy and fragmentation of land use control agencies and the powers they wield.

A final strategy is *"site specific review for energy impact."* This approach expands environmental impact statement (EIS) evaluation of major projects to include analysis of energy consumption. California and several other states already require such an energy component within their EIS requirements. Mandelker notes that this modification would likely engender the least opposition because it merely expands the purview of the already well-established EIS system. The strategy, however, is also the least powerful in directing development to achieve energy efficiency because impact statements are *not* a land use control—the EIS process, at full much less reduced strength, is designed to *disclose* environmental impacts rather than *force* a particular reaction to the identified effects.

Mandelker's article is an important contribution in the conceptual evaluation of the energy-land use linkage. It emphasizes that energy conservation is only one of the many objectives of and influences on land use controls, although it has moved to forefront significance. In addition, the very goal of increasing energy efficiency can be achieved through different levels and types of modifications to the current land use system, each of which involves some mix of the trade-offs of: comprehensiveness, effectiveness, and political acceptability.

Mandelker's call for perspective and analysis in considering the energy-land use interrelationship is echoed in the section's second essay by Robert H. Frelich. In *Managing Energy Conservation Under Planned Growth*, Freilich emphasizes that energy is only one of the numerous forces affecting the pattern of metropolitan decentralization in the post-war era. Inexpensive, plentiful energy surely helped fuel the movement of jobs and households from the urban core, but other forces and actions were equally, if not more, influential. These include: FHA mortgage insurance policies in the 1930s and 1940s favoring suburban as opposed to urban housing construction; federal Interstate Highway policies in 1950s and 1960s facilitating outward movement into surburban and then exurban locations; and a federal income tax system providing substantial incentive to ownership tenure and new construction as opposed to rental tenure and rehabilitation.

Freilich also emphasizes that perspective is needed in considering the *effects* of metropolitan decentralization. While excessive energy consumption induced by sprawl is an important consideration, it is only *one* of *six* "crises." The others include: (1) central city decline; (2) environmental degradation; (3) fiscal insolvency; (4) excessive agricultural land consumption; and (5) increasing housing costs. Since energy is a single component of the set of "crises" affecting American society, land use planning must not limit itself to energy considerations. Instead, it must seek the complementary resolution of all six problem areas.

Freilich argues that such integrated planning is capsulized in *growth management* strategies. Growth management entails a comprehensive examination of areal development opportunities and restraints culminating in a blueprint establishing the optimal placement, timing, and sequence of growth. The technique was first experimented with and adopted in Ramapo, New York and subsequently implemented in several other jurisdictions throughtout the nation. In the Minneapolis-St. Paul area, for example, the regional planning body, the Metropolitan Council, adopted a series of controls to foster growth near the metropolitan center while discouraging development at the outer fringes—a pattern designed to maximize exploitation of existing public facilities/services while reducing energy consumption, environmental degradation and encroachment on agricultural lands.

Freilich's description of growth management places it as a control mechanism between Mandelker's "highly centralized, highly regulated system" and the "less centralized, less inclusive strategy." It is, however, more encompassing than both of these strategies because it addresses *other* concerns than energy conservation. In short, Freilich's essay provides a perspective for viewing energy-land use problems/opportunities and points the way—via growth management—to improve energy efficiency while achieving other planning goals.

The remaining essays in the section turn from a discussion of *framework* to analysis of the *direction, operation and consequences* of energy-sensitive land use controls. Included are essays which respectively: overview the energy-conserving legislative/statutory actions taken by numerous jurisdictions throughout the county; examine, in-depth, the efforts taken by one particularly energy innovative community, Davis, California; review the land use measures considered/effected to promote solar energy; and highlight federal activity to promote energy conservation by encouraging a new generation of building codes.

A national review of communities enacting energy-sensitive land use measures is provided by Duncan Erley and David Mosena in *Energy Conserving Development Regulations: Current Practice*. The authors surveyed 1,500 local, regional, and state planning agencies and from this group selected 13 communities with the most far-reaching energy-sensitive land use statutes in place. The initiatives taken by the 13 communities are examined at length to determine the types of actions underway and attendant planning-land use implications.

Erley and Mosena classify the land use statutes into the following four groups:

(1) The most common category moderates *heating/cooling needs* through such means as: encouraging/allowing energy efficient dwellings, including multifamily and earth sheltered housing units; and modifying site subdivision requirements to allow for solar energy use and natural cooling/wind protection.

(2) The second group reduces *transportation-related energy consumption* by encouraging compact development and a mixture of residential/nonresidential land uses—features permitting fewer trips in general and facilitating mass transit in particular.

(3) The third category lowers *energy consumption in construction materials and processes*. Permitting higher density development—with attendant reduction in the length of necessary utility and other infrastructure requirements—is one means of accomplishing this. Allowing less stringent subdivision extractions regarding cul-de-sac diameter, parking space provision and other improvements represents yet another.

(4) The final land use group turns from lowering energy demand—the focus of the first three groups—to *encouraging alternative energy sources and systems*, such as solar and wind energy and district heating cogeneration technologies.

The authors go beyond summarizing the innovative energy land use measures and review their *planning* implications. Like Mandelker, Erley and Mosena discuss the trade-offs of regulatory scope and acceptance. Comprehensive controls—covering all aspects of development—offer the greatest potential for energy saving, but in so drastically challenging the status quo, are least likely to be accepted. More politically palatable, although offering less drastic energy benefits, are incremental/partial energy conserving regulations. Erley and Mosena also stress the importance of planners in this process: developing a constituency for energy-sensitive controls through education and other means; experimenting with different approaches in order to increase the chances for energy savings in the face of changing energy technologies, costs, and availability; and developing requisite administrative/technical skills so that the emerging energy-land use review procedures can be effectively provided. Thus, the energy-conserving controls offer both new opportunities and responsibilities for planners in the 1980s.

Davis, California figured prominently in the roster of energy-innovative communities examined by Erley and Mosena. The Davis initiatives are examined in-depth by Paul P. Craig in *Energy, Land and Values: The Davis Experience*. It is instructive to examine Davis for this community has significantly moderated its *energy demand* while also increasing *energy supply*, mainly in the form of solar energy. Craig argues that these accomplishments, and the manner in which they were reached, stand in contast to the results and direction of federal energy initiatives as well as the experience of the nation as a whole.

Federal energy policy in the 1970s, stressed *centralized, technology-oriented* approaches to expanding domestic energy supply. This "hard energy" strategy included funding for coal gasification and liquification, shale conversion and light water and breeder reactors. Craig maintains that this approach, while having some merit, is costly, speculative and suffers from the drawback of imposing uniform *national* energy solutions instead of encouraging *local* responses as per local needs/priorities. More appealing is a "soft energy" strategy of *energy conservation* reached through a variety of *locally* adapted technologies. These include:

(1)   Imposition of stringent *energy conservation standards* in public building codes such as those adopted in Davis which reduced space heating needs to levels approaching hot water demands.

(2)   Development of *district heating and cogeneration* systems in those instances where land use densities permit an area-wide sharing of heat/electrical generation.

(3)   Improvement of *energy efficiency in transportation* through such means as: (a) mandating further automobile fuel efficiency standards beyond those currently imposed by the federal government; (b) encouraging mass transit by upgrading existing systems and adopting new approaches (i.e. light rail and computer directed rubber tired system): and (c) improving the efficiency of airplanes.

(4)   Development/utilization of *renewable energy sources* including solar, biomass, wind and wave systems.

Davis has successfully taken many of these steps. Its stringent energy-building code standards result in energy consumption in new construction of about 6 BTU/square foot per degree day compared to a national average of 10 BTU/square foot per degree day. The community encourages non-automobile transportation alternatives, most notably the bicycle. It also was one of the first jurisdictions to protect solar access and to take other actions to foster the utilization of solar energy. This attitude encouraged the construction of the nation's first solar subdivision—Village Homes. In the aggregate, Davis' actions have increased its *energy resiliance*—defined as the "ability to withstand future energy shock." One indication of this is that Davis consumes about 60 percent of the natural gas and about 80 percent of the electricity of comparable communities.

Craig argues that the Davis energy experience *can* be replicated successfully elsewhere. He further emphasizes that the very steps taken by this community to encourage energy conservation, such as turning to bicycling as a means of transportation and encouraging well-planned and energy efficient housing developments, have the added societal benefit of enchancing the *ambiance* of the built environment—an important element, if not the overriding objective of a comprehensive land use system.

Davis, California and many of the communities examined by Erley and Mosena made strenuous efforts to encourage solar power. This energy source, in turn, is feasible only if *access to the sun* can be ensured. Common law doctrine does not recognize "the right to the sun" as contained within the traditional bundle of property prerogatives—a gap evoking numerous proposals and legislative actions to guarantee solar access. These options and their consequences are examined by Martin S. Jaffe in *The Planning, Regulatory and Design Implications of Solar Access Protection.*

Solar access can be ensured by *private agreement* and/or two forms of *public guarantees:* (1) *lot-by-lot* and (2) *areawide protection.* The private approach consists of property owners establishing covenants or easements guaranteeing solar rights. It offers the benefit of flexibility since the private agreements can be tailored individually to each situation. At the same time, the private approach has numerous shortcomings. It is enforced by litigation or declaratory judgment—two time consuming and costly procedures. Moreover, the strategy is extremely vulnerable since it is based on *voluntary* agreement and thus can be easily thwarted by a recalcitrant property owner not wanting to limit/modify development options for the sake of a neighbor's solar benefit.

Public solar access protection in contrast, avoids these drawbacks. Enforcement is easier, not necessitating individual, private suits for compliance. Most important, the public controls are *compulsory* on all affected parties and thus are more effective and comprehensive. These advantages make the public guarantee of solar access a near pre-requisite for the meaningful development of solar energy.

One public guarantee approach—*lot-by-lot protection*—establishes a solar access protective zone around solar collectors in place. New Mexico's Solar Rights Act, for example, gives the individual *first* installing a solar system a prior right to direct sunlight. This grant protects against development activities which would shadow, or in other ways impede, operation of the solar energy system. Jaffe argues tht public lot-by-lot protection, while superior to privately guaranteed access, suffers from numerous drawbacks. The lot-by-lot option is evoked only when a solar system is installed—an event difficult to predict and thus plan for. In addition, the lot-by-lot system extends its protection only to the first individual applying for protection. Is such an exclusive solar franchise equitable?

More comprehensive and balanced treatment is made available by *areawide solar access protection*. Under such an approach, *all* owners would be offered a certain degree of solar access protection and *all* owners would have to yield some development options/rights. Areawide solar access would be afforded by, and contained within, traditional land use instruments. To reduce shading problems, the zoning code would modify height, setback, frontage and yard area requirements. Subdivision regulations regarding street/lot/building orientation, as well as landscaping vegetation requirements, would similarly reflect a new solar consciousness. To illustrate, to reduce shading difficulties, the zoning code might lower maximum building heights and allow zero-lot line siting on the north lot line while the subdivision ordinance might encourage/require east-to-west building orientation.

Jaffe notes that these and other changes would likely have a *marked impact* on the appearance of the built environment. Current, standard *suburban* subdivision placement—identically centering all homes on their lots—would be replaced by a much greater diversity. For solar access purposes, buildings positioned on a north - south orientation would have their *shorter* sides oriented towards the street, while their counterparts on east-west streets would have their *longer* sides oriented toward the front of the lot. The *city* landscape would also experience a lowering of maximum density in order to permit greater solar utilization. These revisions would evolve over time as planners and others involved in the land use process developed a sensitivity to solar potential and needs.

Solar access is a critical concern affecting the feasibility and utilization of solar energy. It is not, however, the only influence. The multiple factors bearing on solar energy utilization are considered by Gerald Mara and David Engel in *Legal and Institutional Barriers to Solar Energy*. Their discussion is based on the experience of the Solar Demonstration Program—a federal initiative responsible for funding the construction of almost 14,000 housing units using solar energy for space and/or domestic hot water heating. Mara and Engel consider a wide range of influences on the Solar Demonstration Program including: *zoning* and *building regulations, and financing, public utility surcharge* and other *economic factors*.

Public *land use* controls pose few hurdles to solar energy use. Zoning ordiances have traditionally *not* directly prohibited solar systems and such noninterference will likely continue in the future, given the heightened public sensitivity to energy concerns. Moreover, the direction in public land use controls, as examined by Jaffe and others, is to *encourage* solar through such means as protecting solar access.

Public *building codes* similarly pose few restraints to solar energy systems as these regulations typically allow solar systems, provided they satisfy certain performance levels. More troublesome, are time delays resulting from the building code administrative process. Code officials are typically unfamiliar with solar systems and thus the building code inspection/approval of solar installations is typically protracted. Mara and Engel note, however, that this problem may be transitory; as solar usage becomes more popular, building inspectors will become more familiar with and knowledgeable about solar and thus will be able to expedite processing.

Financial restraints to solar pose a much more significant barrier. In the absence of an established sales record for solar equipped homes, mortgage underwriters sometimes hestitate to finance such dwellings. The relationship of energy to public utilities and the rates they charge is another crucial concern. Some utility companies increase their charges to low volume energy users. If solar equipped houses, using substantially less electricity/gas than their conventional counterparts, must pay a penalty surcharge, then solar energy's cost savings—a strong argument for its installation—could be significantly eroded.

Mara and Engel state that these financial restraints are not fatal and moreover are likely to *diminish* over time. As mortgage underwriters become more familiar with the sales appeal of solar homes, they will likely discontinue conservative valuation practices. Utility companies will also, in time, recognize *their* benefit from widespread solar usage. Solar installation, for example, can decrease the threat of summer brownouts and blackouts, thereby precluding the need for utilities to expand peak load capacity. If utilities acknowledge the benefits of solar, they will likely reduce/eliminate penalty surcharges imposed on solar users.

Mara and Engel provide an important review of the barriers to solar energy. Based on the "hands-on" experience derived from the Solar Demonstration Program, they suggest that certain issues, such as solar access, which were believed to be paramount have, in practice, proved to be less troublesome. In contrast, other factors receiving little consideration in the literature, such as the solar/utility rate interface, have proved much more significant. As the implementation of solar energy grows, so will the clarification of issues and opportunities.

In their analysis of solar energy utilization, Mara and Engel touched upon the negative effect that building codes and their administration could have on innovative energy technologies. Building codes can also play a potent *positive* energy role by requiring that new construction meet specified standards of energy efficiency in such critical components as heating/cooling systems, building structural elements, and window areas. The energy conserving potential of building codes, and the effect of these regulations on land use and building design, is reviewed by Grant P. Thompson in *Federal Building Energy Performance Standards.*

Thompson considers the historical evolvement of building codes to energy concerns. Before the 1973 OPEC oil embargo, public building regulations did not specify required energy performance. The embargo led the American Society of Heating, Refrigeration and Air Conditioning Engineers, Inc. (ASHREA) to formulate, in 1974-1975, a code (ASHREA Standard 90-75) establishing energy performance levels for different building components. Under federal prodding, ASHREA Standard 90-75 was incorporated by many states in their official building codes.

State adoption of the ASHREA standard was a major breakthrough mandating energy efficiency in new construction. The ASHREA standard itself, however, was criticized since it established *component* energy performance guides rather than an *overall building* energy performance level. This shortcoming prompted Congress to mandate, in 1976, composite Building Energy Performance Standards (BEPS). The BEPS were developed over 1976-1979 through a complex procedure considering such factors as: (1) existing energy-saving technology; (2) familiarity of design professionals with these systems; (3) ability of recent (post-1973) construction to satisfy various, hypothetical levels of energy efficiency; and (4) varying local factors, such as energy cost/supply and climatic conditions.

Thompson evaluates the proposed BEPS as an important step forward but one which falls short of its potential, and Congressional mandate, to *induce* significant energy conservation. He traces BEPS conservatism to its underpinning of *existing* building practice and professional designer knowledge as opposed to *prodding* innovation and designer sensitivity to energy conservation. To illustrate, BEPS permitted "lenient" or "nominal" as opposed to "strict" energy standards in most forms of nonresidential construction because design professionals were unfamiliar with the "strict" level.

While conservative, BEPS constitutes an important control on new construction. Will it affect land use patterns and development practices? Thompson believes "it may or may not." He describes two contradictory energy-land use scenarios, *both* of which are equally plausible. Under one, BEPS *would markedly affect* development patterns by encouraging more energy efficient multifamily construction and changing traditional building orientation to maximize solar potential. These dramatic changes —at least those regulating solar—are similar to those envisioned in Jaffe's discussion of energy-induced land use modifications. Under a second scenario, BEPS would *minimally affect* land use patterns because its energy standards could be satisfied by changes to the building envelope itself, such as improving insulation in ceilings/walls, and utilizing insulated glazing. Thompson believes both scenarios, or a mixture of the two, are possible and thus only time will tell the precise impact of BEPS on land use.

Conclusion

The land use process in the United States reflects the complex interchange of forces within the overall society. Social exclusionary sentiments vie with the idea of equality. The historical exploitation of land as a limitless national birthright must come to grips with a new sense that resources are limited. The perspective that "new is better" similarly must contend with a growing appreciation of our heritage and the need to protect structures of historical quality. Laissez-faire sentiments compete with the need for societal control and if control must be wielded then at what level—Hamiltonian central guidance versus Jeffersonian decentralization. These are some of the underlying tensions in American society which are reflected in the land use arena in such themes as: anti-exclusionary zoning litigation versus local growth moratoriums, the building up and then scaling down of environmental review procedures, the debate concerning the appropriateness of local versus regional growth management efforts, and so on.

The 1970s saw energy leap to the forefront of societal attention. Energy quickly became a lightning rod for often acrimonious debate—perhaps because of the sudden change in its availability/cost, perhaps because these very changes were controlled by political forces outside of our shores. Not surprisingly, energy has become an important land use concern. It is a critical influence on, if not determinant of, how metropolitan growth is guided, land uses interrelated, transportation systems developed, growth managed in terms of sequence and timing, energy natural resources exploited and so on. This is not to say that energy has replaced the many other influences on land use—physical, social, or environmental—nor that there is full agreement concerning the energy supply/cost situation or the specific implications or different energy scenarios to land use. We are clearly feeling our way in a continuing national debate. It is hoped that this monograph will help guide and structure such debate—to define the critical energy-land use issues and to point the way to the development of a land use system reflecting both energy restraints and opportunities.

Notes

1. This discussion is based on the historical review in Technology and Economics Inc., *An Overview and Critical Evaluation of the Relationships Between Land Use and Energy Conservation* Vol. 1 (Washington, D.C.: National Technical Information Service, March 1976).

2. *Ibid.*

3. George Benda, "Coal and the Residential Energy Future" in Robert W. Burchell and David Listokin (eds.) *Energy and Land Use* (New Brunswick: N.J. Center for Urban Policy Research, 1981).

4. Mel Scott, *American City Planning Since 1890* (Berkeley: University of California Press, 1969).

5. Ibid.

6. F. Stuart Chapin, *Urban Land Use Planning* (Urbana: University of Illinois Press, 1966); Joseph De Chiara and Lee Koppelman, *Manual of Housing/Planning and Design Criteria* (Englewood Cliffs, N.J.: Prentice Hall, 1975).

7. U.S. Congress, House of Representatives, Committee on Banking Finance and Urban Affairs, *Compact Cities: Energy Saving Strategies for the Eighties* (Washington, D.C. Government Printing Office, 1980).

8. Steven A. Kursh, "Exploring the Relationship Between the Rebirth of Cities and the Journey to Work: Potential Implications of Higher Energy Costs" in Joel T. Werth, (Editor) *Energy in the Cities Symposium*, American Planning Association, Planning Advisory Service Report No. 349 (Chicago:1980).

9. *Ibid.*

10. U.S. Department of Housing and Urban Development, *The President's National Urban Policy Report* (Washington, D.C.: Government Printing Office, 1980). See also discussion in U.S. Department of Housing and Urban Development, "Metropolitan Development Patterns: What Difference Do They Make?" (Washington D.C.: HUD Office of Policy Development and Research, November 1980). This HUD publication is based on an Urban Institute study entitled *Urban Development Patterns*.

11. Kursh, "Exploring the Relationship Between the Rebirth of Cities and the Journey to Work."

12. See Steven Ferrey, "How Carter's Solar Program Could Eclipse the Cities," *Planning* (December 1978); John H. Giggons, Director, Office of Technology Assessment, Testimony before the Subcommittee on the City of the House Committee on Banking, Finance and Urban Affairs, October 16, 1979.

13. Alan Altshuler, "Review of the *Costs of Sprawl*," *Journal of the American Institute of Planners* Vol. 43 (April 1977), pp. 202-209; Duane Windsor, "A Critique of the *Costs of Sprawl*" *Journal of the American Institute of Planners*, Vol 45 (July 1979), pp. 279-292.

14. Hittman Associates, Inc., *Residential Energy Consumption, Single Family Housing Final Report* (Washington, D.C.: Government Printing Office, 1975), Hittman Associates, Inc., *Residential Energy Consumption, Multifamily Housing* (Washington, D.C.: Government Printing Office, 1974); Arthur D. Little, *Residential and Commercial Energy Use Patterns 1974-1990* (Washington, D.C.: Federal Energy Administration, November 1974).

15. American Institute of Architects, *A Nation of Energy Efficient Buildings By 1990* (Washington, D.C.: AIA, 1975).

16. Victor Olgyay, *Design with Climate: Bioclimatic Approach to Architectural Regionalism* (Princeton, New Jersey: Princeton University Press, 1963); Ralph Knowles, *Energy and Form: An Ecological Approach to Urban Growth* (Cambridge, Mass: MIT Press, 1974).

17. Argonne National Laboratory, *Community Systems Energy Saving Programs for Communities* (Argonne, Ill.: Argonne National Laboratory, 1979).

18. Tera Tech Inc., *Energy From Coal: A State of the Art Review* (Washington, D.C.: Government Printing Office, 1976).

19. See U.S. Congress, Senate, Committee on Interior and Insular Affairs, *Land Resource Planning Act and the Energy Facilities Planning and Development Act* (Washington, D.C.: Government Printing Office, 1975).

20. *Ibid.*

21. NSF/NASA, *Solar Energy Panel, Solar Energy as a National Resource* (Washington, D.C.: Government Printing Office, 1974).

22. Ralph Knowles and Richard Berry, *Solar Envelope Concepts: Moderate Density Applications* (Golden, Colorado: Solar Energy Research Institute, 1980).

23. Robert Stobaugh and Daniel Yergin, *Energy Future, Report of the Energy Project at the Harvard Business School* (New York: Random House, 1979).

24. David Barrett, Peter Espstein and Charles M. Haar, *Financing The Solar House* (Lexington: D.C. Heath, 1977).

SECTION I

ENERGY AND THE CITY— FUTURE LAND USE PATTERNS

Energy Efficient Planning

SHRI MANOHAR

Introduction

THE "energy crisis" of today is due to many complex reasons related not only to our way of life, but also to the lack of adequate distribution systems, and to our failure to utilize renewable alternative energy forms such as wind, sun, water and power. The difficulties are not caused by a lack of technology, but by inertia, by social, economic and political systems which react not to the needs of humanity, but to the considerations of the market place.

Our dependence on a high rate of energy flow and on wasted energy raises urgent ecological, economic, political and moral questions. Energy is the essential underpinning of almost all our society. Growth in industrial production, transportation, communication, population, knowledge for improving the quality of life and many other aspects of civilization are all related to and supported by energy growth.

After a fifty-year period of generally steady growth in demand and supply of energy, the developed regions of the world are now entering what is likely to be a long period of uncertainty and perhaps abrupt changes.

In its simple form, energy is the capacity for doing work. All energy harnessed by man, bought and sold to satisfy energy demands is considered to originate in one of the five fuels: natural gas, oil, coal, nuclear fission or others, (others includes geothermal energy, hydroenergy, solar energy, nuclear fusion, etc.). Until 1975, about 90 percent of the world-wide demand for energy was met by natural gas, oil, coal and nuclear fission, but it is expected that the fuel category "other" will have a rising share in world energy production in the next fifty years.

Real energy needs applicable to the entire world and not just a few are food and efficient shelter. Technological development of the recent past has attempted to improve the standard of living—a necessary and vital part of social and cultural evolution. But in the process of cultural evolution, man's dependency on energy also has increased. In fact, energy has created a cultural norm in society; an imbalance between the standard of living in the developed and the developing regions. This is apparent from the fact that today 80 percent of the world energy is consumed by only 30 percent of the population living in the industralized developed societies or regions. The U.S.A. with only 6 percent of the world population, consumes 36 percent of the world energy; and while the average annual consumption of coal per capita of the rest of the world is 2 tons, that of the United States is 12 tons. If we were to raise the world average per capita consumption of energy to the same as that of the United States, it would mean a six-fold increase in the world average consumption. While the population of the world has been on the increase at the rate of 2 percent per year, the global annual mean of energy consumption also has been on the increase at the rate of 2 percent, but the per capita consumption figures for the less developed regions have been almost constant.[1] This means that the per capita energy consumption of the developing regions has decreased and, the annual increase of energy consumption is almost entirely a function of the growth demands of developed nations. The gross national product shows a rough correlation with energy consumption and the policy of continued economic growth. In fact, industrial economics would break down, if growth ceased or even slowed down. But if we allow the present rate of growth to persist, the total ecological demand will increase by 32 percent over the next sixty-six years, and we will be able to do nothing about energy conservation. One thing is certain—the total world consumption of energy will double in the next eighteen years, because there will be more people to use more energy and their requirements per capita will be greater. This will mean that the gap between the "HAVE NOTS" (developing regions) and the "HAVES" (developed regions) will grow wider. In order to close this gap, four principal global budgets must be balanced, says Ronald Alber.[2] These are energy, material resources, environmental pollution and population. The spiralling nature of the demands on energy increases, food increases and demand on resources derived from the original use of solar energy is evident.

The cities of the developed regions of the world, the urban metropolitan areas of the developing regions or countries have a much higher likelihood for encountering severe critical shortages of food, resources and energy. This crisis situation is mostly due to the unprecedented increase in the population of these urban regions from 3 percent to 10 percent per year. The coupling of the large increases in urban populations with the sharp increase in commercial energy use, principally for transportation, electric power, manufactured goods and services, means that the developing countries will expect 400 percent-to-600 percent increase in energy requirements by the end of this century, although per capita the rise is still modest. This situation is in sharp contrast to the energy needs of the developed regions where 50 percent increase in energy is considered reasonable.[3] Herein indeed lives an energy crisis with critical portent for the entire world.

By examining the current studies in North America, we should be able to identify the sectors that promise an important reduction in total consumption. In Harvard Business School's recent publication, the key conclusion was that conservation is "a source of energy, that produces no radioactive waste, nothing in the way of petrodollars and very little pollution, it is the quickest and cheapest way to reduce dependence on imported oil." According to Robert Stobaugh, the United States could cut energy use by 30 percent to 40 percent by the year 2000—with little slowing of economic growth.[4] Probably an equally large savings can be realized through minor changes without affecting amenities to any significant degree. A reduction to less than half of the current consumption of energy is available without total sacrifice and essentially no change in lifestyle.

Efficiency and conservation should be utilized aggressively. Energy planning policies for both current operations and short-and long-term planning should be coordinated by: urban and regional planning; designs and renovations that attain satisfactory living standards at reduced levels of energy consumption; priority for energy supply systems that offer opportunities to improve and sustain the quality of life in urban regions; and the recognition that the appropriate solution will vary according to individual situations.

A planner can substantially affect energy demand in many ways. Promotion of urban and regional planning that will achieve multiple objectives including energy conservation, as well as conservation of other scarce resources such as land, water and the natural environment, is very necessary.

Through planning in a chosen area, a planner can constrain residents to adopt energy conservation or energy wastefulness. It is very important for planners to expand their way of thinking on energy, so as to identify its potential impacts on every element of comprehensive planning. It is not sufficient to consider only direct requirements of energy consumption in the form of gas, oil or electricity, but we must also consider the indirect needs of a region or community so as to supply the inhabitants with all the goods and services they demand. Everything has an energy cost: energy is a necessary input to virtually every phase of production process for all goods and services. If we consider this energy as "embodied" in the goods and services, we enlarge our view of energy flow through a region.

Energy Perspectives

Non-industrialized societies are heavily dependent on the traditional energy sources such as local solar energy that is made available through the agencies of food, work, animal feed, nonmineral fuel (wood, dung and agricultural wastes), wind power and direct waterpower. Energy consumption per person is very small; only a few energy units are required to sustain life.

In contrast, industrialized societies use large quantities of fossil fuels (coal, oil, and natural gas) and electricity, and consumption of energy per person is as much as a hundred times the energy contained in food. About 75 percent of the energy used by man is used in urban regions covering only 0.1 percent of the earth's surace. Over half of the total fuel energy for transportation is consumed in urban regions. Job-related energy consumption has gone up as more factories and offices have been built. The fraction of the population employed in factories and offices amounted to only about 10 percent in 1850, but rose to about 30 percent a hundred years later. Since 1960 it has risen to about 36 percent, as more and more women have taken jobs outside the home.

ENERGY CONSUMPTION

Economic growth has been progressively dependent on the use of energy since the Industrial Revolution. The mechanization of agriculture and industry, as well as internal urban growth and transport have required increasing expenditures of energy. The relatively high economic growth rates achieved by the developed countries have been accompanied by a greatly increased utilization of energy. However, recent events have led governments to question whether or not the interrelationship of the two trends is in fact a necessary condition for economic growth. Barry Commoner explains how economic growth involves complex interactions among the three basic systems—the eco-system, the production system and the economic system, which, together with the social or political order, govern all human activities.[5]

We can see that, as affluence increases, partly, through more jobs per family, more energy tends to be consumed in the home for comfort heating and cooling, etc.[6] In 1971 roughly one-fourth of the energy budget was devoted to each of the four major economic sectors, namely, transportation, residential/commercial, industry and electric utilities.

TRANSPORT

The transport of materials is an important function in every industrial society. Not only must people reach their work place in order to keep the wheels of industry spinning; in addition, the products of industrialization must be transported to distant markets, and food and resources must be transported to the urban conglomerations where people and machines are concentrated. Fuels, wastes and water must be transported to and from cities. In short, an effective system of transportation is crucial to the continued operation of industrial society.

Today transportation accounts for about 8 percent-to-25 percent of the total energy consumption. But it also is responsible for the largest difference in per capita energy, accounting for 16 percent of the 37 percent per capita differential between the United States and Sweden (the level of transportation demand within a country or state includes land area and population density and the gasoline price fixed). Transportation includes all forms of transport—air, road, rail, inland waterways, etc.

HOUSEHOLD AND COMMERCIAL SECTOR

More than half the energy consumed by this sector was used for space heating and cooling. Most of the remaining energy used in the residential/commercial sector was employed for water heating, cooking and refrigeration. Suburban households use more energy than either urban or rural ones, and families living in single-family houses use more than those living in apartments.

The commercial sector includes such diverse institutions as stores, office buildings, hotels, service stations, schools, hospitals, theatres, restaurants and sports arenas. Commercial energy use has grown at the rate of 5.4 percent per year since 1960. Its rapid growth rate reflects the growth of commercial and service activities themselves which have outpaced industrial growth consistently over the last decade. Office buildings built in recent years use far more energy, on the average than do older buildings with an equivalent amount of space. The difference can be traced to higher lighting levels, sealed windows, (requiring twenty-four-hour mechanical ventilation), glass curtain walls (allowing high levels of heat loss and gain), and the proliferation of computers, elevators, escalators, electric typewriters and duplicating machines.

The energy consumption in the household and commercial sector turned out to be very well explained by population and weighted average price of fuels. It accounts for 20 percent and 30 percent of the total energy consumption in Sweden and in the United States respectively. The major technological differences in the United States and Sweden comparison are: district heating and hydropower.

INDUSTRY

The heart of an industrial society is its industries. In 1970, three industries accounted for more than half the energy used in this sector: the metal, chemical and oil-refining industries. About 77 percent of all energy used in manufacturing feeds six highly consumptive industrial groups—food processing; paper; chemicals; petroleum refining; stone, clay and glass products; and the primary metals (aluminum and steel). Nearly half the energy used to power industry was obtained from natural gas, and its use in the industrial sector was growing at the fastest rate. About half of that energy went for heating processes, either through direct burning of fuels or through the production of steam. Much of the rest was used for running machinery, electrolysis, lighting, and for "non-energy" purposes—that is, as raw materials, for feedstacks, for manufacturing processes. Shift to plastics also has contributed to the growth of industrial energy use.

The explanation of per capita differential for industry energy consumption turned out to be more complex than the other two sectors. The major factor involved in explaining industrial demand is the level of Gross National or State Product. The composition of the industrial sector within the state or country is measured on the percentage of the industrial activity of major energy consuming industrial sectors, and the weighted price of industrial energy. The industrial sector accounts for 35 percent-to-51 percent of the total energy consumed. Japan showed to be an exceptionally heavy consumer in this area.

ELECTRIC UTILITIES

The large utility companies that generate and sell electricity comprise the fastest growing sector of the economy in terms of energy use, reflecting the increasing dependence of industrialism on electricity. The conversion of primary fuels to electricity always involved a substantial energy loss. In the process of electrical generation and transmission, about 65 percent of the energy content of the fuel is lost. When there is a choice between burning fossil fuels directly or using electricity, as in home heating, direct fuel is generally more efficient and less expensive, but unfortunately not so clean. So people prefer to use electricity rather than fuel, even though fuel is far cheaper. By 1990 the United States government expects the utilities to consume 39 percent of the total energy budget. Electricity is also the only form of energy in which atomic energy can be used.

The multitude of factors ranging from climate to population, density and style of living turned out to be insignificant in explaining why United States used more energy than Sweden, on a state-by-state basis. The most significant factor in this study was the price of energy in determining the level of demand. In every case it was found that the higher the price of energy the less energy was consumed, considering other things were equal. In the United States, the energy waste appears to be in response to the energy price.

Barry Commoner's study of work task versus work value in relation to the principles of thermodynamics shows that the amount of work obtained from an energy source depends on how well the source is matched to a work-requiring task. His study of work efficiencies reveal that in North America most tasks are rather poorly coupled to their energy source. About 28 percent of the national energy budget is used for low-temperature tasks such as space heat, water heating, cooking, air-conditioning and refrigeration, on an overall national basis. The estimated efficiencies range between 3 percent and 6 percent. About 25 percent of the energy budget is used for various forms of transportation, chiefly automobiles and trucks. Their efficiencies are estimated at 10 percent. About 37 percent of the energy budget is devoted to industrial use: most of them requiring a high quality source. The estimated efficiencies are 25 percent to 30 percent. Small percentages of the energy are consumed for the conversion of petroleum and natural gas into petrochemical products such as synthetic fibers, plastics, pesticides, fertilizers, etc. His own estimate of their efficiencies is nearly zero. Energy efficiency in North America is probably no more than about 15 percent.[7] There is a very large gap between the minimum amount of work needed to produce the goods and the services that we now enjoy, i.e., a much larger amount of energy is used to accomplish tasks. In theory, Barry Commoner explains, North America could save 85 percent of the energy now used. With practical limitations, we could still save up to 55 percent-to-60 percent of the energy now consumed. It seems clear that there is room for a drastic reduction in energy consumption without reducing the standard of living.

In October 1977, former President Carter stated—

> Ours is the most wasteful nation on earth . . . we are simply wasting too much energy . . . With about the same standard of living, we use twice as much energy per person as do other countries like Germany, Japan, Sweden, other countries of that kind. So, we have got to cut down our waste through conservation measures, voluntary action, and a realization of the seriousness of this question.[8]

Energy Conservation

One way to meet the energy crisis is to slow the growth in consumption. A large amount of energy is employed wastefully. This waste must stop. It is necessary to declare an absolute per capita minimum and maximum for energy requirements. Actions to conserve energy were foreign to our thinking until recently.

The Energy Policy Project Report of the Ford Foundation is an integrated national policy study and constitutes a major contribution to the understanding that conservation is as important as supply. It states that North America has entered a new age of energy, but has failed to adjust to the habits, expectations and national policies of the new age. In the twentieth century, developed nations have added a fourth essential to the old necessities of life. Besides food, clothing and shelter, they must have have energy. It has become an integral part of the nation's life support system. The objective of the Energy Policy Project has been to explore the range of energy choices open to North Americans and to intensify policies that match the choices. The range of choices is broad, both because the national resource base is diverse, and because there is plenty of room for improving energy use efficiency. The trend of recent years has been towards rapid growth in energy use, rapid enough to double consumption every fifteen years. The project has given special attention to the possibilities for saving energy. This study demonstrates that slower energy growth will not undermine the standard of living, but will exert a powerful influence on the environment and other closely intertwined energy problems.[9]

The Science Council of Canada also interprets the energy issue as a complex one, merging environmental, political and economic concerns. It emphasizes that we are at a critical juncture in our history and the survival of the natural system (of which we are a part) is at stake. The Science Council argues that we can begin to overcome our pollution and waste disposal problems, avoid capital shortages and economic dislocation, and live within our biological, social and physical resources limitations by adopting a conserver approach. The conserver society, as opposed to the consumer society, is of the principle against waste and pollution and its touchstone is doing more, with actions, for ourselves and for society.[10]

ENERGY FUTURE ALTERNATIVES

To assist in the analysis of energy choices, the Ford Foundation Energy Policy Project has constructed three different versions of possible energy futures for the United States of America through to the year 2000. Three alternatives scenarios were developed:

1. The Historical Growth Scenario
2. The Technical Fix Scenario
3. Zero Energy Growth Scenario

The three alternate futures, or scenarios, are based upon differing assumptions of growth in energy use. They are illustrative and intended to help test and compare the consequences of different policy choices in order to clarify the implications of different rates of energy growth. Questions considered are: What are their effects on the economy, environment, foreign policy, social equity, and life styles? What policies would be likely to bring about changes? What resources are needed to make each of them work? How much energy must be provided?

Certain common characteristics have been considered in all three scenarios. They all include enough energy to provide the population with warmth in winter and air-conditioning in summer, several basic appliances that would seem the height of luxury to most people in other parts of the world but are basics to North Americans, cars for most families, as well as other means of transportation. They are based on full employment and steady growth in gross national product and personal incomes. All three scenarios include enough energy for more material prosperity than the country now enjoys. The lower energy growth scenarios provide major savings in energy with small differences in the GNP from Historical Growth trends. Employment opportunities are better, and the real GNP is more than twice what it is today.

THE HISTORICAL GROWTH SCENARIO

The Historical Growth Scenario assumes that energy use in the United States would continue to increase till the end of the century at about 3.4 percent annually. It assumes that no deliberate effort would be made to alter our habitual patterns of energy use. Rather, this scenario assumes that a vigorous national effort would be directed towards enlarging energy supply to keep up with rising demand.

THE TECHNICAL FIX SCENARIO

The Technical Fix Scenario is an attempt to anticipate the results of long-term energy prices and government policies to encourage greater efficiency in energy consumption. The annual energy consumption increases at an average rate of 1.9 percent per year between now and the year 2000. This energy budget can provide essentially the same level of energy services, i.e., miles of travel, quality of housing and levels of heating and cooling, manufacturing output, etc. If the nation adopts specific energy saving technologies, such as better insulation and better auto fuel economy it would mean a savings as large as four-fifths of the current total consumption. It is a leaner and trimmer scenario, yet basically on the same track as Historical Growth Scenario.

ZERO GROWTH SCENARIO

The Zero Growth Scenario represents a modest departure from the accepted track. It would not require austerity, nor would it preclude economic growth. The real GNP in this scenario is approximately the same as in Technical Fix, and it actually provides more jobs. It includes all the energy-saving devices of Technical Fix, plus extra emphasis on efficiency. Zero Energy Growth does not mean zero economic growth. In a Zero Energy Growth future there would be a greater emphasis on service education, health care, day care, cultural activities, urban amenities such as parks—which generally require much less energy per dollar than heavy industrial activities or primary metal processing, whose growth would be deemphasized. It is directed towards the structural changes of post-industrial society.

Its main difference lies in small but distinct redirection of economic growth, away from energy intensive industries toward economic activities that require less energy. As energy excise tax, making energy more expensive, would encourage the shift. Compared with the other energy future, a Zero Energy Growth future would have less emphasis on making things and more on offering services—better systems, more parks, better health care. About 2 percent of GNP would be diverted through the higher energy taxes to these public services designed to enhance the quality of life.

WHY ZERO GROWTH?

The slower growth in energy consumption permits more flexibility and a more relaxed pace of development. The nation could halt growth in at least one of the major controversial domestic sources of energy—nuclear power, etc. Zero Energy Growth would then be able to allow more choices in the supply from conventional sources. This scenario also could permit use of cleaner, renewable, and smaller-scale energy sources such as wind, power, roof-top solar power, and recycled wastes to meet a larger share of the total energy demand. Zero Energy is actually part of achieving a sustainable economy. It can stem from a general desire to save what is left before it is too late. Reduced energy consumption offers a way to reduce total pollutant output. If energy consumption is stabilized, we can achieve a desired level of air quality, and we can do so at far less expense. Zero Energy Growth future provides the option of minimizing and even avoiding catastrophic accidents in energy supply systems. If the United States were to adopt a Zero Energy Growth policy, it would be easier to reduce imports, thereby easing the pressure on world energy supplies and making it easier for developing nations to grow.

ECONOMIC IMPACTS OF ZERO ENERGY GROWTH

The Zero Energy Growth scenario is an attempt to avoid shortages by a gradual tapering of energy growth. Because growth would decline over a period of decades, energy producers and consumers would be able to plan accordingly. It provides curtailing energy growth at a level higher than at present, perhaps 10 percent higher per capita. Technical progress and changes in a society can lower dramatically the level of energy required.

POLICIES FOR ZERO ENERGY GROWTH

The policies needed to bring about Zero Energy Growth include all the major policies of the Technical Fix Scenario plus specific economic policies needed to bring about a shift in the mix of the GNP. The major policy actions needed are as follows:

1. An energy sales tax. This tax would be imposed gradually, on a predetermined schedule, so that purchasers of energy-consuming equipment could plan accordingly. The tax would begin in 1985 at 3 percent of the retail price of energy, increasing at every fraction of 1 percent of the retail price of energy, and increase a fraction of 1 percent each year to about 15 percent in the year 2000. The tax would raise the price of energy-intensive goods and services relative to non-energy-intensive activities and thus would use traditional market mechanisms to reduce energy consumption.

2. Other policies that would have to be adopted to offset these effects. Some of the funds bought into the treasury could be directed toward public services that would facilitate and enhance zero energy growth:

a) public transportation
b) health care
c) housing
d) urban amenities, including clean streets and parks
e) education
f) cultural activities
g) day care centers
h) nursing homes, and old age benefits
i) law enforcement.

3. Decreasing automobile gas mileage. This is essential to the achievement of energy savings in the transportation sector. A legal performance standard or a heavy tax imposed on inefficient automobiles as part of the purchase price on automobiles would be required to achieve an average vehicle efficiency of 20 mpg. by 1985 and 33 mpg. by 2000.

4. Tightening building codes, lending requirements, and improving capital availability to ensure optimum building design. These policies are the same as in the Technical Fix Scenario.

5. Expansion of urban mass transit systems and development of a system of bicycle ways.

6. Implementation of airlines energy conservation requirements to raise load factors and slightly reduce cruising speed, which to the airlines should be a financial advantage in light of current fuel prices. Higher energy sales tax for shorter flights should be introduced.

7. Upgrading of rail services, including fast passenger service for short-

haul intercity runs. Granting of increased rate flexibility to the railroads.

8. Recycling. This also would be encouraged by requirements to mine urban wastes for scrap materials, and by providing funds for demonstrations of recycling technology.

9. Implementation of an aggressive government program to ensure research and development of technological improvements in energy consumption, which would become economically attractive through the energy tax.

10. An encouraging government attitude toward investments in foreign countries or states, which would have the effect of shifting some of the growth in energy-intensive industries to economically favorable areas.

The study concluded that substantial savings in energy are possible in the U. S. economy without significant changes in the present structure of economy, and without sacrificing the continued growth of real income. This review indicates the short-run impacts may be negative, but over the long run GNP may increase. Total productivity should not suffer so that economic impacts, over the long term should be acceptable.

Today, development in technology has become very energy intensive. Only 1 percent of the driving force in modern manufacturing industries is human energy, while 99 percent is geochemically stored and transformed energy. Why do we need so much energy? Basically because market forces fix the relative prices of human and natural energy. Natural energy until recently was freely available at a cheap cost and so was substituted for human labor. A single barrel of oil contains the energy equivalent of 6,000 hours of human labor. It is calculated at $15.00 per barrel, the "energy slaves" bound up in oil work for less than a penny per hour. Thus today, in all industrialized regions, fossil fuels have displaced human labor.[11]

Urban decisions have a great impact on the future demand for energy and the way in which different forms of energy will be used. Energy is used in the built environment in building materials and construction industries, for the transportation of people and goods, and especially, in the operation of residential and other buildings (heating, cooling and lighting of premises, hot water supply, electrical appliances, etc.). Policies and strategies related to the planning, construction and use of urban regions therefore influence directly or indirectly the consumption of energy in all the main sectors of industry, transport and other uses. In total, around 50 percent of all energy is utilized within, or for urban regions in the developed areas. The developed area as a whole, accounts for approximately 75 percent of the world consumption of purchased energy, and urban regions use up nearly 40 percent of world energy supplies.

Relationships Between Land Use And Energy

Land use regulation represents a novel approach to energy conservation. Land use patterns are the basic patterns of life around us—patterns of structures and activities they support. But is energy use sensitive to land use management variables? Emphatically this is so as the following two-step relationship illustrates.

LAND USE $\longrightarrow$ BASIC ENERGY DEMANDS $\longrightarrow$ FUEL

First, land use patterns influence basic energy demands, i.e., the materials and energy are both directly needed to support activites and structures. In turn, basic demands affect the types and amounts of fuel consumed. In addition, land use patterns directly influence the interrelationships between basic energy demands and fuel use. Land use patterns can affect the amounts and types of energy used for all basic energy demands in the residential, commercial, industrial and transportation sectors. It can reduce direct energy use for all purposes in the transportation sector and for the major purpose, in the building sector, such as space heating and cooling. Land use controls also induce indirect impacts in the industrial sector through market demands for goods and materials, but the magnitude of this impact is unknown. In addition, land use modifications can conserve critical fuels by increasing the efficiency of their utilization or allowing for substitution. They are linked by a complex set of energy production, delivery and utilization systems that interact reciprocally with land use patterns. For example, solar collectors require unobstructed air space and proper orientation.[12]

Urban Form and Energy Consumption

The components that determine the efficiency of urban forms are:
1. Geometry
2. Density
3. Land Use Patterns
4. Orientation
5. Transportation and Communication

GEOMETRY

Geometry and urban form have a significant bearing on effective land use and the environment of cities. The geometry of streets and lots often comprises the framework of urbanization. In fact the grid of streets and lots which comprise the city is like a net placed or thrown upon the ground. If the pressure for floor space increases, the logical solution is to increase the scale of the grid so that the balance is retained.

Many old cities now considered natural cities, i.e., new towns of the Middle Ages—were in fact based upon a systematic grid-iron plan and this so-called artificial grid is, in fact, a method, by which new towns are organized in any developing or colonizing situation. The planned grid town is not a prisoner of an architectural past but the best orderly pattern. The existing checker-board arrangement of buildings and streets is only one conceivable pattern. But, if the geometry of the pattern is changed, precisely the same amount of floor space can be accommodated with the same general height of buildings, but with a considerable increase in open space. It seems that in most towns which appear to be overcrowded all this land is available only if the right organizational principles are used to find it.

The order imposed by the appropriately scaled grid is not restrictive nor does it exclude the possibility of organic development with a wide range of choices and opportunities. An inappropriately-scaled grid, or no framework at all, on the other hand, is restrictive. Organic growth without a structured framework leads to chaos, resulting in high energy consumption and inefficiency of urbanization. We can easily illustrate inefficiency created by an inappropriate scaled grid of an existing urban form, e.g., the street grid of Manhattan island. New York is far too small, thus causing a preponderance of built forms. The resulting skyline of tall buildings presents an extremely large surface area to unit volume which affects the climatic conditions, causing unnecessary summer solar heat gain and winter heat loss. Most urban forms found in western cities are analogous to the design of radiators which attempt to maximize surface area. If we are truly concerned with low energy requirements, this situation should be avoided. Such disorganized patterns of high-rise buildings in the inner core of a city not only play havoc with winds, light and climate, but are also inefficient in land use, when applied to the city core as a whole. It is as possible to have a high density, dispersed pattern as a low density, concentrated pattern. The planned linear development of appropriately-sized grid allows urban development of high density, without incurring the usual penalties, and allowing agricultural and open spaces to be conserved.[13] The energy implications of such a linear medium density network of development affords possibilities for democratic, farsighted local planning policies.

DENSITY

The density of a settlement depends upon many factors: internal and external space standards, the standards of internal environment, the way the buildings are laid out and the nature of the site. The most basic function of a settlement is to provide shelter. Each type of facility needs space and space needs land. The forms of a settlement depend on the way these different land uses are related as well as on the shape, size and density, and on the way communications are arranged in that settlement.

Studies have indicated that the dwellings in high-rise blocks were substantially more expensive to build than those of a two-story block, and the range of densities possible with two-story housing is much wider than multistoried housing. More land and building resources could be saved by building one-two-and three-story houses of greater densities.[14]

Furthermore, space-conditioning economies and efficiencies are frequently achieved through locating the units in large planned structures which are served by a central, common heating and cooling system. Energy requirements for transportation are reduced under high-density conditions because employment and shopping areas are located close to residential areas and travel requirements are therefore reduced.[15] Also, high-density conditions make it possible to employ more energy-efficient modes of transportation. Utility system requirements in higher densities tend to use less energy per capita because economies can be attained through shorter transmission lines and larger conversion or production plants which serve the larger number of consumers present under higher densities. In most buildings, a saving of about 35 percent-to-50 percent is possible if there is a more selective use of the installation in its unaltered form. Energy-saving depends on the base with which you start.[16] The report of the American Institute of Architects' Energy Steering Committee has developed a national program for energy conservation in buildings, aimed at attaining the greatest potential savings of energy offered by the built environment.[17]

A recent publication[18] suggests that because there is a less number of exposed walls and windows, a single-family townhouse required half the amount of energy for heating or cooling than an equally-sized single-family, detached bungalow and semi-detached houses reduce the heat loss by 30 percent because of less exposed walls. While the concept of density can have a number of stipulative definitions, there is a general concensus that density does have an effect upon energy consumption. The general view is that as gross density increases, per capita energy consumption decreases. In one case study it was found that as density increased from the urban fringe towards the urban center, energy consumption per capita decreased until it reached the central area where density was the highest and energy consumption again increased.

It is also possible to increase efficiency with forms of housing that result from higher density. However, above a certain density, housing forms require higher energy input in the construction process. It is the task of the urban planner to find those forms that generate high densities with low energy renewable resource materials. For example, in Canada, this would mean timber construction which would limit construction heights to four stories and create a rebirth in Georgian planning. We must not, however, ignore the fact that low energy/high efficiency urban forms should be a pleasure to live, work and play in and around.

THE HIGH-DENSITY/LOW-RISE ALTERNATIVE

The high-density/low-rise option resulted in minimizing motorized transportation within the urbanized area, the distances never being greater than about 1 km, (single-level transportation networks minimize infrastructure investment). This is the basic prerequisite for encouraging non-motorized relations. The low-rise option, i .e., average building of three-stories, also induces limited use of elevators. This option emphasizes small groups of "townhouses." The over-all density of buildings and population achieved by "townhouse" apartments, even with small individual gardens, is the same as the high-rise buildings separated by collective empty spaces.

The high-density/low-rise concept has an immediate effect in the case of improving the microclimate, which is in itself a source of energy conservation. The concept entails groups of small building units which provide improved natural heating, cooling and lighting systems. This also allows for greater flexibility for future use of appropriate technologies as they become available on the market, e.g., heat-pump, heat storage, etc.

Obviously, the high-density/low-rise type of layout would have desirable economic advantages. Saving in energy per capita and a lower cost per square meter of used space would make this option especially worthy of consideration for new developments in any country.[19] It is well-suited to urban extensions as well as to new settlements. Its very flexibility also allows for the optimum use of local materials, local craftsmanship and local techniques. The savings in investments, together with the linear pattern of development, constitute a "generative" type of growth which can, according to economic circumstances, accelerate or stop altogether without undue debts for capital repayments.

Such a concept obviously will not determine life styles or immediately change consumption attitudes or behavior but it does create the physical framework for changes, such as contacts between people and the grounds. The low-rise/high-density alternative is, in short, the basis for a humanly built environment.

The findings in the *Costs of Sprawl*[20] deal more directly with capital and operating costs and land use. It supports the notion that higher energy efficiency is found in higher density development. The study also considered environmental factors such as air and water pollution. In order to make the comparisons, three types of communities were analyzed: the "low-density sprawl," the "combination mix," and the "high-density planned" communities. They are defined as follows:

1. Low-Density Sprawl—The entire community is made up of single-family homes, 75 percent sited in a conventional pattern and the rest clustered.

2. Combination Mix—This community consists of a housing mix of 20 percent of each of the five types of dwellings, half located in planned unit developments, half in conventional subdivisions.

3. High-Density Planned—In this community, housing is composed of 40 percent six-story apartments, 30 percent walk-up, two story apartments, 20 percent townhouses, and 10 percent clustered single-family homes. All of the dwelling units are clustered together into continuous neighborhoods, much in the pattern of a high-density, like-new town, self-contained community.

CONCLUSION TO THE *COSTS OF SPRAWL* STUDY

1. Land Use - Though the communities cover the same area, over 50 percent of the land in the high-density community remains undeveloped whereas it is totally developed in low-density sprawl. The amount of land for public buildings and schools remains the same for all communities but high-density uses half as much land for transportation compared with low-density sprawl.

2. Economic Cost - Total investment cost of high-density was 21 percent below combination mix and 44 percent below low-density. The savings resulted directly from differences in development gross density. The largest cost in roads and utilities were 55 percent lower in high-density than in low-density.

3. Environmental Costs - High-density community produced 45 percent less air pollution than low-density, because less fuel is burned for heating and automobile transportation. Low-density sprawl has more paved and roof area creating more problems of storm water pollution and sediment and downstream flooding. High-density and combination mix give greater freedom through design and planning to preserve natural vegetation and wildlife. High-density concentrates noise problems and puts more demand on urban designers to create pleasing development.

4. Energy Consumption And Water Use - High-density planned, can show an energy savings up to 44 percent per housing unit over low-density sprawl because of reduced heating, air-conditioning and changed transportation mode. Planning alone could save 14 percent by adjustments to low-density sprawl. Water for lawn use is reduced by 35 percent with high-density planned.

5. Personal Costs - Less travel time, lower cleaning and maintenance time, fewer traffic accidents. However higher crime is alluded to higher density, but this is not quantified.

The study concluded that higher-density results in lower environmental costs, lower personal costs and reduced natural resource consumption (land and fuel) for a given number of dwellings. Higher-density/low-rise can provide conditions favorable to encourage affordable technology for improving conversion efficiency. Options include: the utilization of energy corridors, total energy systems, district heating, heat recovery from incineration of refuse, wood waste, wood harvest (biomass), waste heat from industrial production or electrical generation, and the mini-electrical generation sales to existing utility companies. More compact patterns provide economic control district heating.

ORIENTATION

A building properly designed and oriented can greatly reduce the demands on heating and cooling systems, which, in turn, reduce the needed area of expensive solar collectors. Reducing the initial costs of solar systems, will speed acceptance and implementation of solar energy utilization.

The adaptation of building to environment has been a continuous prob-
blem throughout the centuries.

> The symphony of climate . . . has not been understood. . . The sun differs
> along the curvature of the meridian, its intensity varies on the crust of the
> earth according to its incidence. . . In this play many conditions are created
> which await adequate solutions. It is at this point that an authentic regionalism
> has its rightful place.[21]

The major elements of climatic environment which affect human comfort
can be categorized as: air temperature, radiation, air movement, and
humidity. The regional climate analyses contain: thermal, solar, wind,
precipitation, humidity, and topography analysis.

A building design that ignores the impact of the natural environment will
almost always have to use energy in the form of mechanical, structural, or
material interventions to compensate for the resulting discomforts and inconveniences of adverse natural conditions. Clearly, then, an urban and
regional planning project should start with a thorough analysis of the
assigned region or potential site alternatives. Planners should understand
and anticipate the effects of a particular site or climate on the energy flow
of a building, if a design is to use the environment to its advantage. Where
energy conservation is a major goal of a building design, sun is perhaps the
single most important natural element to consider.

SPATIAL ARRANGEMENT

There are a range of micro-climatic conditions that most people regard as
comfortable. It would seem logical that a structure should try inherently to
insure these conditions.[22] The comfort zone can be achieved through using
natural means inherent in good design or through mechanical means which
require a certain amount of energy; the amount of energy needed by
mechanical devices. Energy conservation standards meeting the physical
demands of the environment allow maximum flexibility in the choice of
materials, construction techniques, and design to encourage a reduction in
energy consumption of up to 20 percent for heating, cooling and ventilation.[23] In addition, urban growth, consciously based on principles that correlate building form with natural variation for purposes of energy conservation, will exhibit a diversity of building forms. When transferred into functional terms, these forms usually will produce the close-contact diversity
essential for a rich and humane community life, without overreliance on
mobility.

The Knowles study suggests energy conservation should be the basic criteria of individual buildings and building groups as they might be sequentially built in the future.[24] The study concentrates on reducing the susceptibility of individual buildings and groups of buildings to environmental pertubation as a way to conserve energy. The object is to use the shape and structure, the geometry and scale of buildings to help maintain equilibrium within a building under the stress of multiple cyclic forces in nature. An important step would be to shape buildings so as to reduce seasonal variations in sun energy and to equalize insolation from summer to winter. The energy-resource condition today requires that we consider alternative modes of growth with more initial control and higher growth costs but also with the promise of longer-range stability, lower maintenance costs, and a lessening of control over the life of the system. Such alternatives would exhibit higher diversity than we see now. This will provide an opportunity to build urban form arrangements on a large scale which are more responsive to variations and to the natural environment.

However, the orientation of a building is often dictated by streets, lots and set-back requirements. The orientation of the building should be as close to the east-west axis as is practically possible in order to minimize the effects of the sun. The optimum exterior dimensions for a structure vary in each type of climate.[25] Many studies provide information on the advantages of designing houses with regard to climate and solar impact. The creation of a three-dimensional geometry which provides for passive solar energy conversion and easy adoption to forthcoming active solar equipment will be a key consideration in urban planning of the future and an important addition to the conventional land use and density approach. Laws respecting the right to sunlight will have to be enacted to prevent the overshadowing of one building by another. An ecological awareness of land forms which receive maximum energy will have to be sought, such as south slopes of valleys and natural wind protected micro-climates. Moreover, urban patterns which generate low-rise, high-density buildings in simple linear or grid patterns will most probably be seen to contribute to the efficiency of energy conversion in built form. It is hoped that these new attitudes and planning frameworks will be carried over to local governments and builders.

The size and arrangement of buildings has a considerable impact on their energy requirements. Optimal arrangements from an energy point of view certainly can be determined, but it is precisely the mix of style, appearance and arrangement of buildings that contribute considerably to the liveliness and charm of a city. Urban planning and urban design increasingly should take energy aspects into consideration in the design, layout and spatial arrangement of buildings. In the spatial distribution of essential urban functions, such as living, working and recreation, consideration should be given to the amount of energy needed in maintaining and operating the urban forms. In deciding on the size and density of a city or on the question of changes in existing cities, its future energy consumption for heating and services should be taken duly into account and viewed together with the planned method of supplying energy.

For both, housing and other building sectors, the potential benefits in terms of energy efficiency are enormous. In the United States the profile of energy consumption in buildings has been summarized as "vital statistics"— 34-75-50, i.e., 34 percent of all energy use occurs in residential and commercial buildings.[26] 75 percent of this energy is used for space heating or cooling and for water heating, and 50 percent of the energy consumed in new buildings could be saved by better design of buildings and equipment.

SITE SELECTION

The initial selection of a suitable site is of fundamental importance to energy conservation and to the creation of favorable micro-climatic conditions. The main climatic factors for site selection are: sunshine, and radiation in relation to slopes and exposure, wind frequency distributions, as well as combined effects of wind and temperature.

The topographic vegetation and orientation of a site are all effective factors in altering the balance of the general climate. Slopes having different orientation and inclination receive different amounts of solar radiation. In regions where the elevation of the sun is low (such as the sub-arctic regions in Canada) the ground is cooled by radiation from the sky, because since cold air is denser, it flows downwards. Valleys and depressions collect the cold air at night. So neither valleys nor hilltops are favorable community sites: valleys are cold and hilltops are windy. A sloping site with a southerly aspect is the most favorable choice if maximum benefit from solar radiation is the objective.

Ideally, there should be as much protection as possible from the prevailing cold winds of winter, without undue loss of winter sunshine. Greater comfort and energy conservation will result if buildings are built fairly close together and are interspersed with belts of trees or other windbreakers in close formation. It is more rational for buildings to provide the maximum degree of shelter from the cold winds than to be widely scattered with little or no shelter. Moreover, it is advisable to refrain from designing long straight roads, especially in the direction of adverse winds. In winter, when the maximum benefit from solar radiation is desirable, a southerly orientation is most favorable. This orientation is not only important in reducing fuel costs (by catching maximum daylight hours in winter time and reducing heat costs), but also for psychological reasons because it lightens the home and creates a cozy atmosphere.

A departure from conventional planning and design practices is necessary to demonstrate the interaction between man-made and natural environments. Any building activity, in fact, every new house, creates a number of separate climates, replacing the ground climate on which it sits, in relation to the direction. Thus, the south wall's micro-climate will be more favorable than that of the north wall, and those of the east and west sides will be different again. If the creation of micro-climates is inevitable in man-made environments, then these micro-climates ought to be positive ones. All that is necessary to bring about favorable conditions is an understanding of the main climatic factors, namely, sunshine and radiation in relation to slopes and exposure, and the frequency distribution of wind and the combined effects of wind and temperature. The realization of these facts is indeed important in the different micro-climatic regions.[27]

Energy conservation and maximum use of passive solar energy are only the initial steps in the evolving process of planning and designing a community that will lead to active solar energy utilization. Planning based on micro-climate concepts will foster energy conservation.

SITE PLANNING FOR CONSERVATION

Site planning is basically the planning and organization of a site for optimum human use and enjoyment. It deals with the distribution of specific activities in a general way, so that the interaction, the best relationship and the optimum location in relation to other activities on a given site are best utilized in the existing and surrounding areas. Proper planning of orientation and location could contribute to full utilization of solar radiation and energy conservation.[28] The utilization of solar collection devices calls for a rethinking of the cost of land development so as to avoid the energy-intensive architectural and mechanical modifications necessary to correct errors in site selection. For purposes of solar radiation utilization, buildings should be on the south or west slopes and not on the north. To utilize solar radiation and to conserve energy in the cool temperature, hot-arid and hot-humid regions involves two major issues:

1. access to the sun and
2. location of the building on the site to reduce its energy requirement.

LAND USE PATTERNS

If we are to affect the long-term growth rate of energy demand, we must study the structural and developmental components of our society and their effect on growth. The physical component most closely associated with energy use in society, can best be analyzed through an understanding of land use. If we want to affect the structural changes that determine energy growth, eventually we must begin to affect the evolution of land use patterns.[29]

There is no doubt that major reduction in future regional energy expenditures can be achieved through the proportionate use and configuring of land use activities.[30] Adopting zoning policies which encourage clustered residential arrangements will result in significant savings in energy expenditures per household. Permitting larger areas of land to be zoned for multifunctional purposes will result in lower transportation requirements: institutional incentive structures which will attract industries requiring lower energy input per dollar will result in a decrease of industrial energy demands. The concept of separation of urban functions by zoning controls, green belts, etc., in fact the whole principle of segregation of land uses and urban activities, has been questioned increasingly. Changes in the planned over-orderliness has drained the vitality out of our cities and has made them inhuman and dangerous parts which do not function as an integrated whole. Studies of energy consumption in Sweden, for example, where the per capita energy consumption is 60 percent less than that of the United States with roughly equivalent levels of per capita GNP, lend further support to the concept that energy consumption patterns are intrinsically tied to higher density and to more intimate residential, commercial and industrial activities. Today in many instances the original 18th century city has become the central area of today, with vast sprawls of suburbs engulfing it. This central area has developed as the main dominant center containing most of the employment, shopping and entertainment facilities, leaving the suburbs as largely dormitory areas from which people commute to work in the central area. This has resulted in serious congestion, inconvenience, and waste. The displacement of residents from the central area has aggravated commuting problems and has turned the center into a dead crime area in the evening and weekends. This process has gone too far, particularly in many North American cities.

Another important factor influencing energy use characteristics in an urban region is land use arrangement. In recent years, there has been a reaction against the functional segregation of city activities such as housing, work, transportation and recreation. As oil and gas have replaced coal, and as offices employ more people than factories, the need to separate home and work has diminished. Meanwhile, the problems involved in moving people from one function to another have assumed a dominant place as far as planners are concerned. The old pattern of walk to work, shopping, education, etc., seems attractive again. Consequently, besides energy conservation, a better living solution accounts for seeking new and more effective land use arrangements. Due to this anti-urban segregation, major efforts are being made to revitalize the city center and bring residents back from the suburbs and create a residential neighborhood with comprehensive social and cultural facilities. It is, however, even more important to restructure our cities in a way so that the dominance of the single center is replaced by a polynucleated structure. We could initiate such an effort by identifying the small towns and communities which have merged together over time to become unities but which once again could be developed into important local centers. We should constrain the further growth of the existing center and develop the many existing local centers, building up a series of readily accessible, strong, new centers with local employment, shopping and social facilities.

In the process, a number of important objectives can be achieved. First, the need to travel long distances to work can be reduced, at least for a proportion of the population, by bringing homes and jobs closer together. Second, the anonymous vast scale of the city can be broken down into a number of places with their own identity but at the same time the ability to take advantage, through effective transport systems, of being part of the larger city. Finally, as a result, mass transport, such as bus, rail and underground services, should be improved considerably as the daily morning overload of inward commuters to the central area would be balanced by counterflows of passengers to these new centers. A recent metro-Toronto development study proposes decentralization and predicts that in 1990 the urban mileage will drop from an estimated 71 million miles daily to 55 million miles, and electric transit vehicle-miles will increase threefold. This concept of internal decentralization, or at least urban restructuring along these lines, has already occurred in recent years in many cities all over the world.[31]

Some recent investigations have attempted to quantify the effect on energy use of different patterns of urban land use. For example, in Long Island, New York, the current population of about 2.5 million is expected to reach 4.0 million by the year 2000. Two patterns of land use were compared: a continuation of "urban sprawl" and a pattern of corridors, clusters and centers.[32] The energy demand by the additional population would be 19.0 percent less than the expansion through "corridors, clusters, and centers." The largest potential savings were calculated to be in transportation, where energy consumption would be less than half the urban sprawl figure. If these savings were combined with the larger savings, possibly using more efficient automobiles, the total energy consumption by transportation in the region (including that of the 2.5 million who already live in urban sprawl conditions) would, in the year 2000, be only 65 percent of the consumption in 1972, although the population would have risen by 65 percent.

TRANSPORTATION

Since ready accessibility to jobs, shopping, educational, health and leisure facilities is the key to a good urban life, transportation must be of paramount importance in the planning of these facilities in relation to homes. Traffic is responsible for approximately 30 percent of the energy consumption in an urban area. The energy consumption pattern is highly concentrated at or near the center, falling away sharply towards the perimeter; such a high concentration of energy consumption is a major contributor to urban heat and pollution. Breaking down consumption further, it is apparent that 50 percent of transportation energy is used for private passenger transportation, mostly automotive. Energy accounting shows that about 20 percent is spent on personal transportation, as the urban resident drives an average of 20 miles per day. The automobile is by far the largest consumer of energy in Northern American cities.

Petroleum is the only source of fuel used by the present-day automobile, which uses 75 percent to 80 percent of the energy service. The reasons for the high percentage of energy use by the automobile-dominated transportation sector can be divided into two categories. The first is the inefficiency of the automotive transportation system, and the second is the growing role of the automobile in North American society. The auto is so closely woven into the life style of the society that any sudden interruption in its use would be catastrophic to many people. The auto has altered habits and perceptions. It has given a spatial independence to the arrangement of cities and altered the relationships between home, work, shopping, social interaction and recreation.

Solutions to the transportation dilemma can be found in a variety of approaches. One basic solution is to make the existing system as efficient as possible. This includes such tactics as making vehicles smaller, lighter, and more economical, reducing congestion and speed limits, and encouraging greater occupancy and use. Further along the lines is the recent research in the development of engines not dependent on petroleum for fuel. On an average, fuel consumption would decrease by 40 percent if these engine improvements were made available to the consumer.[33]

The next category of solutions requires reduction of these geographical distances. Unfortunately, our metropolitan areas are becoming more dispersed. Most mass transit systems are incapable of meeting the basic needs of a spatially dispersed society. Longer trips are needed to get to work, stores and other activities and so dispersed residential patterns are extremely costly to serve with convenient forms of public transportation.[34] Today's housing and activity patterns are far from the optimum as far as transport energy requirements are concerned.[35]

It would therefore appear that long-range methods for reducing dependence on an energy-intensive transportation network is to encourage life styles that are not integrally tied to such systems. It would involve recreating communities on a more human basis, where many of the day-to-day activities could be carried out within walking or bicycling range, and where the relationship between home, work and shopping would be closer. Thus, less transport would be required, and human energy would begin to replace fuel, constituting a savings of 50 percent.

At present, electricity is the most efficient energy source for transport by rail, tramways, subways, while road transport is more economical when based on gasoline or diesel fuel. This may be, however, only a temporary situation, and electricity appears to be the energy source of the future for transportation. Therefore, it may be justifiable to develop transport systems based on petroleum. At the present time, this should be done in such a way as to facilitate a transfer to electric vehicles in the future without large conversion costs.

In view of the economic and environmental advantages of public transit as compared to individual vehicles, urban regions should be planned so as to facilitate the efficient operation of public transit systems. This includes such matters as route planning, easy adjustment of service frequencies, and incentives to use the facilities, including the gradual introduction of free travel within the urban area.

However, while public transit is more energy efficient than the private automobile and, as such, offers opportunities for energy conserving changes, it must be recognized that these prospects may not be as great as those promised by greater efficiencies in the automobile; for example, in the United States, it has been estimated that a greater return in terms of energy efficiency can be obtained by improving car mileage (fuel consumption ratios) than by feasible shifts from private to public transit. If the shift is pursued, it should be done primarily as a means of improving urban living conditions (better air quality, reduced congestion and more mobility for those without cars). The Ford Foundation study estimated that, by switching to more efficient automobiles, energy consumption by automobiles could be cut by 20 percent in the United States between 1970 and 2000, although the number of automobile miles travelled per capita would decline by only 20 percent.[36]

THE CREATION OF THE POLYCENTRIC CITY—URBAN FORM AS A CONVERTER OF ENERGY

For the past twenty years or more, a polycentric urban structure has been emerging that consists of several clusters of activities located some distance from the downtown or central business district of the region. The CBD (central business district) becomes only one of the several destinations of importance in the region and in most cases, its relative size is not expected to increase in the future. The basic idea is that a city that consists of several relatively high-density destinations can utilize service better than one which contains any one dominant destination. Such a transit service would consist of:

1. good local service to a few centers of activity (including CBD)
2. good express service between centers and
3. good internal circulation service in a few high-density centers.

The rationale behind the major diversified centers is multifaceted. The polycentric urban form would reduce the travel requirements of the region, conserve energy and reduce air pollution. It would make an area-wide transit system more economical. Some low-income persons could find housing in such centers, and greater levels of urbanity and self-sufficiency would come to the outer city. Survey results show that the polycentric city concept is widely used by metropolitan planning agencies all over the world. The polycentric city is the city of the future. If transit is to play an important role today, it must be reoriented to serve this emerging urban form.

In general, the Polynucleated City required substantially less travel than the other forms. The residents of the Polynucleated City were found to consume 57 percent less gasoline for passenger travel than those of the Concentric City and 44 percent less than those of the One-Sided City. Their average trip lengths were 30 percent shorter than those in the Concentric City and 23 percent shorter than those in the One-Sided City.[37]

Today's road transport consumes about 36 percent of the total oil-derived energy supplies, and 71 percent of this is consumed by passenger carrying vehicles. Alteration of patterns of land use and transportation in cities is one approach to reducing oil and other energy consumption. In order that alternative patterns may be studied, it is necessary to model the urban system under various development scenarios. A study[38] was done in Melbourne to compare energy consumption patterns in 1976 and 2000 under scenarios such as:

1. present trends with extreme limits of zero population growth and 4 million at 2000;
2. increased fuel prices;
3. free public transport;
4. higher density development;
5. increased vehicle occupancy rates;
6. increased vehicle fuel efficiency;
7. satellite development.

The key variables are the density of development, the modes of transportation (road, rail, bus, pedestrian and cycling), and the relative locations of residential, employment, education, shopping and other activities.

This study showed that distance of travel plays an important role in determining transportation energy consumption. First, the consumption of energy by a particular transport mode is roughly proportional to the distance travelled, except at very low speeds for motor transport. Second, as distances are shortened, the traffic volume increases, allowing substitution of more energy-efficient modes, e.g., bicycle, pedestrian, bus or rail transport, as an alternative to private road transport. Third, the more efficient modes consume less space in terms of networks, servicing and parking, thus allowing more efficient use of land and hence higher gross densities of development.

Technique for the Optimum Placement of Activities into Zones (TOPAZ) is a planning model developed to evaluate alternative forms of urban systems in terms of infrastructure and transportation costs and benefits. A city is first divided into a set of zones and the activities which occur in the city, e.g., employment, residential, shopping and education, are formed into another set. The unit costs and benefits of establishing the operation of each of the activities in each of the zones, such as those of water, sewerage, drainage, electricity, gas, telephone, roads and land amenity, are calculated. Then the unit costs and benefits of trips between each activity in each zone, i.e., journeys to work, school, shops, etc., are estimated on the basis of existing behavior. This unit cost and benefit information provides an economic evaluation of an existing city. The report shows the zones adopted in the study with the inner zones (A,B,C, 1-4) being high and medium density industrial and residential areas, and progressively changing to lower densities with distance from zone A. A number of radiating corridors have been selected as future growth areas with the zones in the outer parts of these corridors being empty or sparsely populated at this stage. The study illustrates that:

1. Continued population growth will increase both total and per capita energy consumption.
2. Increased spread of the city also will increase per capita energy consumption.

In view of the impending exhaustion of petroleum energy resources it would appear wise to promote:

1. increased densities for new development;
2. redevelopment of high quality to a higher density;
3. increased use of public transport;
4. increased fuel prices;
5. increased dispersion of new development in a regional level of new growth centers or growth shifts to small cities;
6. possibility of adopting self-contained neighborhood amenity patterns to reduce trip lengths. A planned community would generate 30 percent vehicle miles of travel compared to one characterized by traditional sprawl. The interaction between land use and transport and determining urban forms and transport networks would provide maximum benefits to the community at a level of energy consumption that it can afford. Micro-planning is the art of relating the private dwellings to other community-used facilities. Micro-planning brings together all that offers major energy savings.

COMMUNICATION

Recent advances in telecommunications technology has eased the plight of the cities by upgrading life in rural communities. It could make possible a voluntary decentralization of people, business and government, creating the necessary conditions to bring the urban and energy crises under control. "The New Rural Society" project has conducted a broad study of the role of telecommunications as a tool to aid in solving social problems. The development of viable rural communities is absolutely vital and communication technology could make a significant contribution to the following social points:

1. employment
2. health care
3. continuing education and vocational training
4. community interaction and planning.

Telecommunication systems could be used as an alternative way of performing functions that presently necessitate travel. It is recognized that many forms of technology serve city needs and, at the same time, other forms contribute, perhaps inadvertently, to city problems.[39]

Advances in teleconferencing and office automation may bring us closer to an era in which white collar work can be increasingly decentralized and jobs can be brought to workers rather than workers commuting to jobs. In essence, we would be substituting telecommunications for transportation as a network technology business interaction. Richard Harkness explained a hypothetical city with a radial transportation network and computed the savings in miles travelled and in energy consumption for a number of alternative configurations. His findings are that persons working in typical central business districts travel twice as far and spend 2.25 times as long getting to work as do persons in the suburbs.

Harkness' calculations indicate that capital costs computed for the alternatives represent the transport investments that society could avoid if new office growth did occur and regional transportation plans could be scaled down accordingly. However, with growth, society can choose between alternatives and a potential for large savings. For example, he estimated that placing a growth increment of 50,000 new office jobs downtown and serving them with new freeways, and feeding a new rapid rail system would cost the society $551 million in 1969 dollar value. This is not the total cost of these systems but the fraction attributable to the 50,000 new office employees. One alternative was placing them in four satellite centers and using freeways and personalized rapid transit. This would cost $137 million. Another alternative is commuter sheds split among 50 office centers. This would shorten work trips sufficiently so that existing streets could handle them.

Advances in telecommunications pose both a threat and an opportunity to urban planners. The threat is that their impact will not be recognized until the people change work styles and land use patterns enough to upset regional plans and undermine major investments in high-rise office buildings and rapid rail systems which serve the central business districts.

The opportunities, however, are many. There is a potential for reducing the practical difficulties of transporting citizens to work, bringing the home-work environment back to a neighborhood scale, and opening new options of life style.

As an urban subsystem, telecommunications has been the invisible and relatively silent system on the road, while transit, building, sewer and other subsystems have dominated urban planning. Perhaps it is time for telecommunications to emerge and become a recognized part of comprehensive planning. Pehaps we should start thinking of a mode split between walking, bicycling, auto, and transit, and make telecommunications tools for creating a good urban form.

TELECOMMUNICATIONS FOR URBAN IMPROVEMENT

Telecommunications should be considered a potential tool for shaping urban development, and through it and other factors it may be possible to divert expected CBD office growth to suburban locations, and if this is so, society could save several hundred million dollars in transport investments in large cities. Advances in telecommunications propose opportunities to urban planners. New patterns of employment may reduce commuting, save energy, reduce transport costs, and simultaneously create a good urban environment. The potential effects of telecommunication advances on urban travel demand costs will be:[40]

1. basic technical advances plus application engineering that
   will increase effectiveness and reduce costs of two-
   way audio, video and facsimile systems, thus making
   teleconferencing an attractive substitute for travel
   and face-to-face meetings;
2. office organization, that, no longer tied to the central
   business district or CBD for face-to-face contact,
   will gain locational freedom and could decentralize
   to suburban locations;
3. suburban office jobs that will impose different transportation demands than CBD jobs, thus requiring different transport techniques and investments;

4. the better application of telecommunication technology to current city functions that can improve city living while stimulating favorable patterns of regional development.

CITY FUNCTION AND IMPROVED CITY LIVING

Present city communications systems—telephone, radio, television and broad band cable networks—all have unique characteristics and functions to support important parts of the total city environment. These systems can serve as building blocks for an increased spectrum of both basic and innovative city services. In planning for the future role of telecommunications, the city would require us to "organize" our thinking into viewing the city as a large information processing system in which much of the work going on is in the access, processing and exchange of information either for direct use or for indirect service to the physical functioning of the city.

Overlaying this conceptual information framework, there are many shared physical networks that now provide a means for management, services, coordination and feedback to take place. For example, there is the network of streets and public transportation by which people carrying or seeking information move to points of access. There is the telephone network, available as an alternative when the informational content is low. There are the private communication networks of two-way radio used so effectively by police, fire, street maintenance, emergency services, etc. And there is the general one-way informational network of TV, AM, FM broadcast, providing information on items of special interest and importance to the city public. These networks are vital to the city, just as a nerve system is vital to the body. Communication technology could play an important role in creating a new and better network pattern and concurrently in alleviating some of the problems of existing cities. Imaginative applications of communication technology to business, education, government, health care and cultural pursuits might stimulate a full utilization of the land resources for living.

Inventions could be solicited to shape the design of special communication systems which will allow these activities to be conducted more effectively in small communities clustered throughout the urban regions. It is important to stress that the intent is not to de-urbanize, but to give the next 100 million people the option to live and work in an improved urban, or a new rural, environment. First, regional planning must pinpoint small communities according to the availability of space, utilities and other social local services where business and population growth can be planned carefully to insure the desired quality of life. The vast majority of people in most of the western countries live in major urban regions. Smaller communities should be encouraged to grow more rapidly than larger ones, and attention should be given first to the communities which are in attractive rural environments and sufficiently distant from existing population concentrations so as to prevent possible fusion with these concentrations.

Second, the operation of business, education and government, as well as the means to provide for health care and culture, could benefit. The objective is to apply the proper communication systems to these functions and to permit business and government to move to attractive rural areas while maintaining effective interaction between their widely separated offices. At the same time, provision must be made for the development of cultural activities which often are considered absent in rural community life.

The chief effect of telecommunication developments will be to increase choices: the employee will be able to select the environment in which to live; the company or government operation will have a choice of areas in which to locate.

The technology to build these networks is well in hand. One of the ways in which 150 million people could be distributed over the next 30 years among existing communities of 100,000 population or less, is available. Under closer scrutiny, we see the problem basically as a city with its multitude of individuals somehow making the difficult decision between a desire for identity and the absolute necessity to accommodate to the organization and discipline required for coexistence in close living conditions.

Over the past several years there has been evidence of increasing interest in the needs of evolving urban spatial patterns upon resource utilization and specifically energy resource utilization. Achieving the benefits of any energy resource will necessarily involve some risks. Today, our storehouse of conventional fuels is rapidly dwindling and a continued dependence upon them would result in exhaustion of these fuel supplies within a relatively short period. Alternative sources are either environmentally risky (such as nuclear energy) or technologically unavailable in economic quantities for use (such as solar energy). These constraints make necessary the development of a radical energy conservation policy that may encompass most, if not all, of the following elements:

1. lower levels of individual energy consumption;
2. lower over-all growth of the economy;

3. lower, and eventually zero, population growth;
4. the restructuring of existing land use and transportation patterns to minimize resource, and, especially, energy use.

Today we have reached a situation in which our future transportation pattern and even our life-style shall be guided by two major planning objectives:

1. the establishment of an acceptable level of energy use for conservation purposes and
2. the creation of a pollution-free environment around us.

Three such issues are:

1. regional coordination of development;
2. central city revitalization; and
3. political conflict between central city and suburban constituencies.

Each of these issues are interrelated. Regional coordination of development within the framework of more intensive land use and transportation utilization patterns will rebound to the advantage of declining central cities. The effects upon urban spatial organization of three of these energy utilization levels or options are:

1. given continued current energy utilization levels, existing urban spatial organization patterns will remain dominant probably resulting in a spatial pattern similar to that suggested by Doxiadis in his "Ecumenopolis" model.[41]
2. given a major change in energy utilization levels, tending towards greater use of collective rather than individualized transportation modes, existing spatial organization patterns will be altered in the direction of greater land use and population density to facilitate the collection process.
3. given a moderate change in energy utilization levels, tending toward greater use of collective, rather than

individual, transportation modes, spatial organization
patterns will approximate a middle position between
the extremes of dispersion and concentration.

New energy conservation imperatives will result in a new land use and
transportation pattern for urban areas emphasizing both higher densities of
population and land use and the use of more cooperative means of urban
transport. Consequently, a limit to the growth of urban areas will be imposed upon by the causal factor of stationary or declining per capita energy
consumption. This new era of limited or zero growth in new land development will necessitate the introduction of several new policies with respect to
the administration and planning function of government. Paramount
among these will be:

1. a new coordinative goal of limited or zero land use
   expansion in urban areas;
2. the reorganization of governmental units within urban
   areas to achieve the objective of regional federalism
   necessary to effectuate the two-level decisional process;
3. an emphasis within the planning function upon the
   cooperative use of transportation and other facilities such as recreation, shopping, etc. to facilitate
   the lowering of per capita energy consumption levels.

Time to Choose a Path

FUTURE SCENARIOS

Today communities stand on the fringes of society as bold social experiments, but in an energy-scarce future they could comprise the
mainstream. History may have thrust upon nations the challenge and the
opportunity of building a new order. The greatest challenge of transition is
the redistribution of wealth within and between nations.

The following five scenarios encompass the basic policy assumptions for
the different world regions:

1. Limited-Growth Scenario—low population growth, low
   economic growth.

2. Muddling-Through Scenario—high population growth, low economic growth.
3. Resource-Allocation Scenario—low population growth, moderate economic growth.
4. Individual-Affluence Scenario—low population growth, high economic growth.
5. Expansive-Growth Scenario—high population growth, high economic growth.

The entire process should be examined with different development regions of the world until internal consistency is achieved. This will allow the global cooperation project to develop a framework that will make it possible to recognize and interrelate diverse and seemingly incompatible categories of information without resorting to oversimplification or suppression of variety. A critical turning point for solving the world's problems must be established for a global society of economic justice and equity. Change comes from areas of the world, which have not participated fully in past successes. We must work in those areas which participated less in the industrial era and are therefore able to see its flaws more clearly. We can change our ideas and our benefits and thus change the world in which we actually live.

The future interrelationship between energy and urban regions is not simply one involving the estimation of future costs, but of energy supply to urban regions and the people who live in them. The following eight points show the strong relationship between the supply of energy and the urban regions:

1. Availability: what forms of energy are available in different regions and to consumers in different types of building or transportation systems;
2. Location: of energy resources within the country of use, and abroad;
3. Prices: the future trend of prices for different fuels, measured both in terms of time-series and in terms of comparisons of equivalent heat content;
4. Convenience: simplicity in the end-use.
5. Reliability: vulnerability to interruptions in supply;
6. Flexibility: the ability to substitute one energy source for another, e.g., at the macro-level, electricity supplies from coal, oil, nuclear or hydro

resources or, at the micro-level, the ability of a
dwelling unit to meet heating requirements by alternative means if the normal method becomes unavailable
for any reason;
7. Environmental Impact: the pollution characteristics
of different fuels, in central and point-of-end-use
consumption systems;
8. Permanence: depletion of nonrenewable resources, etc.

Despite the uncertainties, despite the great differences between countries,
and despite the early stage at which the developed regions stand at present in
this adjustment, it is possible to identify a limited number of general
scenarios that are guiding energy planning in most developed countries.
Some countries may appear at present to be closely identified with a particular scenario, but more often present policies and actions in individual
countries may be compatible with more than one scenario. The scenarios
presented here do not represent the actual nature of energy policymaking,
but demonstrate that different energy policies are possible, and examine the
implications for urban and regional planning of these different policies. The
four scenarios are:[42]

1. Minimum disturbance and good housekeeping
2. The nuclear-electric option
3. The "soft technology," non-nuclear option
4. The two-stage transition.

MINIMUM DISTURBANCE AND GOOD HOUSEKEEPING

At present this also might be described as the 'wait and see' option, since
the energy supply situation is so confused and uncertain that major depar-
tures from past patterns cannot be justified at present. For some years to
come, countries will be engaged in long-neglected data collection, research
and exploration for new energy resources, and will experiment with new
techniques.

The basic assumption in this scenario is that the heavy dependence of
most developed countries on oil and gas will change only slowly over a time
period. It is recognized, of course, that this dependence on oil and gas
ultimately must be substantially reduced, since depleting reserves will be
high priced and may need to be allocated to uses where substitution is not
easy, e.g., the petrochemical industry. But this is seen as a gradual process,

reaching a climax in the twenty-first century. We hope that before then, the opportunities for new energy supplies and more efficient energy use will be much clearer. For example, current problems in nuclear fission technology, including nuclear waste disposal, may have been solved by then, or the prospects for nuclear fusion will be better understood, and similarly the viability of "soft" technologies using renewable resources, e.g., solar energy, wind, biomass, wastes, waves, etc., should be much clearer than it is now.

In this scenario, the energy-related activities of governments would have therefore probably four principal objectives during the next ten to fifteen years:

1. To develop known resources of conventional energy and ensure that exploration of new resources is maintained at a high level

2. To adjust energy-pricing policies to reflect—
   a) general world price levels for different fuels;
   b) calorific content rather than cost of supply; and
   c) relative scarcity of different fuels
   The overall impact of such considerations is likely to be a rise in energy prices, but the relative importance attached to these and other factors in determining price is likely to vary from country to country.

3. To provide substantial encouragement for research, development and demonstration of new energy supply technologies, and for more efficient techniques of energy use. This involves support across a broad range of activities including both "hard and soft" technology. In particular this would encourage the evaluation of alternative energy supply-and-demand strategies so that when, later in the century, major policy choices have to be made, governments are well equipped to make them.

4. To recognize that present energy consumption patterns involve considerable waste of resources. It has now become necessary to take firm steps to ensure conservation. Since estimates of present resources are based on high consumption rates in energy, the life of present resources can be extended considerably by effective measures to reduce that growth.

As far as the impact on urban and regional planning is concerned, the last of these objectives will be of the greatest importance in the present century. Changes will take place in energy pricing, in fuel substitution, and in conservation measures. Over a decade or so, new construction standards and retrofitting programs will bring recognizable benefits to individual energy users, and make a substantial contribution to reducing the rate of growth in energy demand. The process, aided by programs of education and information about energy issues, will be self-reinforcing, and efficiency in energy use will be demanded as a matter of course in buildings, transport and other elements of urban regions.

Price changes and other actions by governments will encourage the use of more efficient transportation systems, probably including the widespread use of electric vehicles for special purposes. Although much higher standards will be required in new buildings, economic factors will still play a major role in determining these standards. Performance standards gradually will replace component standards. Energy issues will influence the form and type of new building development, and great savings will be possible in all types of development. Choice of high-rise, or low-rise buildings, will be made primarily in relation to other urban planning criteria. Similarly, district heating schemes may overcome economic and institutional barriers and may force new developments into particular patterns.

The Nuclear-Electric Option

This scenario rejects the "wait and see" approach, or indeed any approach based on gradual change of energy supply patterns spread over several decades. The prospect of continuing rises in energy prices and the eventual exhaustion of nonrenewable resources is both clear and imminent, and the time to make the necessary adjustments is *now*. Indeed, the faster the conventional, nonrenewable energy resources such as oil and gas can be replaced by other forms of energy, the longer we can continue to use oil and gas for those purposes for which substitution is difficult.

This option rests on the belief that nuclear power generation has proved its feasibility and reliability, and that uranium supplies are adequate for the foreseeable future. Adjustment by the consumer will be negligible. In broader economic terms, the adoption of a nuclear-electric policy may be justified on the grounds that electricity is a much more appropriate form of energy for a post-industrial economy and society, than oil and gas.

The implications of this option for urban and regional planning are very different from those noted in the previous scenario. On the energy supply side, planners are likely to be concerned closely with such matters as the location of nuclear power stations, fuel reprocessing and nuclear waste disposal facilities, electricity transmission and distribution facilities, and sometimes with uranium extraction.

With regard to energy consumption, the nuclear-electric scenario offers a number of advantages. In the planning of urban communities, for example, recent attempts to achieve a closer integration of housing, places of employment, services and recreation, are assisted by a greater emphasis on electricity as the main form of energy. The exception to this is in the transportation systems where public transit may gradually return to tramways and trolleybus systems, at the expense of diesel buses; and mini-trams and mini-trolley buses are available in closely integrated communities. Battery-powered vehicles will at the same time become widespread, especially for local delivery and similar purposes.

A fundamental requirement in the nuclear-electric scenario is high standards of energy performance in both new buildings and the existing stock. These standards will ensure the most efficient use of electricity, which will become the main form of space heating in buildings. Since it is much more difficult to retrofit the existing building stock to high standards, the nuclear-electric scenario may accelerate the replacement rate by increasing obsolescence. It also may encourage high-density row housing and apartment complexes, which have a lower residual demand for energy rather than one-or-two family housing.

THE SOFT TECHNOLOGY, NON-NUCLEAR OPTION

This option is a reverse of the trend established in recent decades of large and expensive systems of energy generation. Instead, it returns to simpler, localized systems based mainly on renewable resources. These systems would be cheaper because they could avoid the expensive technology of centralized systems.

In practical terms this scenario encounters—or indeed challenges—major institutional barriers, since the trend towards high-technology, centralized systems of energy supply is well-established and highly institutionalized in both market and centrally-planned economies. The scenario would require energy supply utilities to be transformed from agencies designed to meet rising consumer demands for energy to organizations seeking to reduce (substantially) the dependence of the economy, society and individuals on centralized supplies. The keynotes of the "soft" technology decentralized approach are based on variety and flexibility, so that appropriate technologies may vary with location. Solar systems may be widespread, wind power may be feasible in a number of locations, refuse-burning district heating schemes may serve larger communities, and so on.

Soft technologies do not at present appear capable of providing a significant contribution to transportation energy needs. This may change in the future but at the present soft technology is mainly directed towards meeting the energy requirements of buildings, especially for space and water heating. This would enable oil and other nonrenewable resources to be conserved for transportation, etc. In such a setting, careful land use planning and other measures designed to reduce the need for transport, will play a significant role in making this approach successful. In terms of urban form, the planning and development of community energy needs will become an integral part of overall community planning and development.

THE TWO-STAGE TRANSITION

For most countries of the developed world, the 1970s has been a period of growing dependence on imported energy resources, or of growing concern for the price and reliability of supplies. For these countries, therefore, the main problem is that of finding ways to utilize the relatively brief period of energy abundance in order to prepare for the necessary transition to other sources early in the twenty-first century—or a two-stage transition. This transition entails a shift from dependence on solid fuel to greater use of oil and natural gas, and then to renewable resources. Therefore, nations must devise policies which will enable them to capture the benefits of the immediate future without causing urban regions and other consuming sectors to become overdependent on energy sources of only a limited life span. This scenario emphasizes that the rising standards of comfort and convenience in urban regions be maintained, through rising energy prices and other actions to encourage the steady growth in the efficiency and economy of energy use. As in the first scenario, long-term decisions about future energy supply and the consequences for urban regions may not be taken or felt for a decade or more.

The four scenarios discussed above have different impacts on urban regions; consequently they generate different problems and priorities for urban and regional planning and development. Whatever the scenario adopted, energy conservation and the elimination of energy waste in urban regions are of great importance. Most studies indicate that investments to achieve energy savings will have a higher return on investment and thus a more positive effect on Gross Domestic Product (GDP) growth, and employment than may be so with the supply expansion alternatives.[43] Similarly, if expansion of energy supply systems can be reduced or avoided by energy conservation programs, there are obvious environmental benefits.

The importance of energy conservation in urban regions is particularly strong in the "nuclear-electric" and "soft technology" approaches. The elimination of waste will make only very small demands on purchased energy supplies and on urban regions.

In all these options it is evident that significant changes in energy supply and consumption patterns will take place during the lifetime of communities and buildings which already exist or are now being created. Since the urban planner has a responsibility to the future as well as to the present, it is desirable that he should recognize the certainties of changes and plan for them. By undertaking the technically feasible option, it should be possible to achieve an ultimate saving of 15 percent-to-20 percent in the annual consumption of primary energy in building services, which would not impair environmental standards. This is a very large proportion of total national consumption. It is obvious that urban planning can have a significant influence on energy policies, and vice versa. As a result, the two need much closer integration and coordination than is currently practiced in most countries.

INVENTED FUTURE

One of our great dilemmas is that all knowledge is about the past, but all decisions are to be made about the future. So what do we do—what can we do—to prepare our school children and our university young people for the world of tomorrow? I believe in constructing two kinds of models about the world future. The first represents the kind of future we seem to be moving toward on the basis of on-going trends, and therefore the kind of future we are likely to drift into in the natural course of events, if we are passive and make no special effort to guide these events according to some creative and constructive plan.

The second model is an invented future. It illustrates three possible futures for the present high-energy society, the energy sources on which each might depend, and the levels at which those energy sources would support a stable population. In Future I, energy consumption does not reach the stability level because of social and environmental disruptions. In Future II, energy consumption and stability level ultimately coincide. In

Future III, energy consumption oscillates above and below the stability level because of the influence of climatic changes on energy supply, mainly food.

We live in an unbalanced world situation. The imbalance is between the natural world which we have inherited, and the artificial, man-made world we have created through human ingenuity. It is between the richer countries and the poorer ones. It is between man's onrushing capabilities in the physical sciences, his application to industrial technology, and his lagging capabilities in social and political management. We know how to dominate machines, how to indulge ourselves in every possible comfort and convenience. What we don't know yet is how to manage all these capabilities in a wise and practical manner. We have a long way to go to learn how to plan the future of our planet and our species, how to bring the world into a cooperative relationship, and how to persuade each nation to work fruitfully with the other so that we can make progress toward a stable world civilization, and so that we can live in a state of equilibrium with our planetary environment.

We have no idea where we are going but we are going there fast. It seems we are entering the future backwards. Anxiety about the future, anxiety about the unknown, are the root of the present crisis of our society. But anxiety can be a creative force, provided we have the courage to analyze lucidly its causes and to transform it into tangible proposals for action. To act is to think ahead, to think of the future here and now.

Notes

1. Thring, M. W. *Energy and Humanity*. Southgate House, Stevenage: Peregrinus Ltd., 1974, p. v.

2. Alber, Ronald. *Human Geography In A Shrinking World*. Belmont, CA: Duxbury Press, 1975, p. 49.

3. Meier, R. L. *Urbanism and Energy In Developing Regions*. Berkeley, CA: University of California, Center for Planning and Development Research, 1978.

4. Stobaugh, Robert. *Energy Future*. New York: Random House, 1979.

5. Commoner, Barry. *The Poverty of Power—Energy and The Economic Crisis*. New York: Bantam Book, 1977, p. 1.

6. Lapedes, W. D. *Encyclopedia of Energy*. New York: McGraw-Hill Book Company, 1976, p. 7, 9.

7. Commoner, Barry. *The Poverty of Power: Energy and The Economic Crisis*. New York: Bantam Book, 1977, p. 145.

8. Doernberg, A. *Comparative Analysis Of Energy Use In Sweden And The United States*. New York: U. S. Energy Research and Development Administration, 1975, p. 4.

9. Ford Foundation. *A Time To Choose: America's Energy Future*. Cambridge, MA: Energy Policy Project, Ballinger Publications Co., 1974.

10. Science Council of Canada. *Canada As A Conserver Society: Resource Uncertainties And The Need For New Technologies, Report No. 27, Sept. 1977.*

11. Davis, Jackson. *The Seventh Year : Industrial Civilization in Transition.* New York: W. W. Norton & Co. 1979, p. 174.

12. Priest and Kenneth. *An Overview and Critical Evaluation Of The Relationships Between Land Use and Energy.* Washington, DC: Federal Energy Administration, Office of Energy Conservation and Environment, 1976.

13. Stone, P. A. *The Structure, Size and Costs Of Urban Settlements.* Cambridge: The University Press, 1973, p. 33.

14. Stone, P. A. *Housing Town Development And Land Cost.* London: The Estates Gazette Ltd., 1962.

15. Housing and Urban Development Association of Canada. *A Building Guide To Energy Conservation.* Toronto: HUDAC, 1975.

16. Stein, R. C., *Architecture and Energy,* New York: Anchor Press, 1977, p. 77.

17. American Institute of Architects. *A Nation of Energy Efficient Buildings By 1990,* Washington, DC: 1735 New York Ave., 1974.

18. Regional Plan Association, and Resources for the Future Inc., *Regional Energy Consumption.* New York: Regional Plan Association, 1974.

19. Epstein, G. *Planning Forms for 20th. Century Cities.* Vol. 1 The Environment of Human Settlements. Toronto: Pergamon, 1976, pp. 157-166.

20. Real Estate Research Corporation. *The Costs Of Sprawl.* Washington, DC: U. S. Government Printing Office, 1974.

21. Corbusier, Le. "Building An Entire New City: Chandigarh, India," *Architectural Forum* (Sept., 1953). pp. 142-149.

22. Olgyay, Victor. *Design With Climate: Bioclimate Approach To Architectural Regionalism.* Princeton, NJ: Princeton University Press, 1963, p. 22.

23. Cerritos City Planning Commission. *Ordinace, No. 475.* Cerritos City Planning Commission, Cerritos, CA: 90701, 1974.

24. Knowles, Ralph, L. *Energy and Form: An Ecological Approach to Urban Growth.* p. 1.

25. Egan, David, M. *Concepts in Termal Comfort.* Englewood Cliffs, NJ: Prentice-Hill Inc., 1975, p. 25.

26. Achenbach, P. P. *Effective Energy Utilization In Buildings: Energy Utilization And The Design Professional.* Washington, DC: American Society of Heating, Refrigeration and Air Conditioning Engineers, 1974.

27. Shellard, H. C. "Microclimate And Housing Topographical Effects," *The Architectural Journal* (Dec. 1965).

28. American Society of Landscape Architects Foundation. *Landscape Planning For Energy Conservation.* Reston, VA: Environmental Design Press, 1977, p. 108.

29. Carroll, T. Owen, *Land Use And Energy Utilization.* Springfield, VA: National Technical Information Service, U. S. Dept. of Commerce, 1975, p. 1.

30. Doernberg, A. *Comparative Analysis of Energy Use In Sweden And The United States.* Upton, NY: Brookhaven National Laboratory, 1975.

31. Urban Transportation Development Corporation Ltd. *Energy Efficiency In City Design.* Toronto: 1976.

32. Brookhaven National Laboratory, and the State University of New York. *Land Use And Energy Utilization.* Stoneybrook, NY: 1975, p. 99.

33. Shonka, D. B. *Transportation Energy Conservation Data Book.* Superintendent of Documents, Washington, DC: U. S. Govt. Printing Office, 1977.

34. Keln, Dough (Chairman, Metropolitan Transit Commission), *Urban Transportation And Energy: The Potential Savings Of Different Modes.* The Congressional Budget Office Report, Washington, DC: U. S. Govt. Printing Office, 1977, p. 74.

35. Sharpe, R. *The Effect Of Urban Form On Transport Energy Patterns.* Urban Meterology Conference, Macguire University, Sydney, Australia: 1977.

36. Ford Foundation. *A Time To Choose*. Cambridge, MA: Ballinger Publishing Co., 1974, p. 443.

37. Peskin, Schofer. *The Impact Of Urban Transportation And Land Use Policies On Transportation Energy Consumption*. Washington, DC: U. S. Dept. of Transportation, Office of University Research, April 1977.

38. Sharpe, Ron. *Energy Conservation In Land Transportation*.

39. Dept. of Housing and Urban Development. *Communications Technology For Urban Improvement*. Washington, DC: National Academy of Engineering, 1971.

40. Harkness, R. C. *Telecommunication Substitutes For Travel*.Office of Telecommunications, Washington, DC: U. S. Dept. of Commerce, 1973.

41. Doxiadis, C. A. *Anthropolis City For Human Development*.New York: W. W. Norton, 1974.

42. U. N. Economic Commission for Europe. *Habitat And Energy*. New York: 1977.

43. *Energy Conservation In The International Energy Agency*. 1976, Review, OECD, 1978, p. 8.

Pro-Development Policies for the Eighties

LAWRENCE O. HOUSTON, JR.

Introduction

FIVE shifting economic and demographic forces will shape America's new housing and communities for the remainder of this century. They are:
1. modest increases in per capita real incomes,
   accompanied by periods of declines;
2. inflation, fed by low rates of productivity,
   soaring costs of housing and high interest rates;
3. chronic oil shortages, competing claims for
   energy and rising prices;
4. unprecedented rates of new household formation
   during the 1980-85 period and soaring demands
   for housing;
5. continued reduction in the size of the average
   American household.

At the same time, the development industry sees a mounting tide of government regulation among the major uncertainties that limit its capacity to produce housing and associated development at reasonable costs to consumers. These constraints range from predictable land use proscriptions such as zoning to ones which often require developers to hold land for unpredictable periods while agencies and courts resolve environmental disputes. A probable battle over the use of land for food versus its use for shelter is one of several issues that will find its expression in higher costs of development.

Much of the regulatory expansion of the past two decades affecting development carries with it the assumption that development in growth areas is not of significant national concern as compared, for example, with environmental protection. It also assumes that growth that is blocked in one place will be accommodated elsewhere without significant additional costs. Those assumptions are no longer valid, if they ever were. Nevertheless, absent the imposition of new policies, the regulations that were spawned by the priorities and assumptions of the past three decades—an era of larger households seemingly without resource constraints—will shape America's development for the next two decades. This would be a prescription for economic and political disaster.

In the long run, the new economic and demographic forces will prove the unworkability of hand-me-down policies and regulations. If, however, we attempt during the next five to ten years to act as if those forces were not important, we will add unacceptable burdens on the economy, overall energy resources, household budgets and the governments which must service new development. Families, for example, will necessarily have to sacrifice such discretionary purchases as recreation and education in order to meet the added and inescapable costs of housing and transportation built into future extensions of low-density, scattered, single-purpose development patterns.

There are, however, development choices that, if widely applied, will mitigate the effects of some of these forces and convert others into positive opportunities for family betterment, government savings and profitable development. They will, moreover, contribute to the nation's energy independence, increased productivity, and the fight against inflation.

As in so many other areas of human endeavor, we know far more about the resolution of these difficulties than we apply in practice. Specifically, we know how to design and market large-scale, mixed-use development projects that conserve energy, materials, transportation and land, while providing added amenities at affordable prices. We also know a good deal about why we aren't producing more such development today. What we haven't yet faced is the importance of supporting socially beneficial and profitable development through positive public policies. Having missed that point, we are not yet ready to apply the new values and priorities of the coming decade, changing obsolete rules and practices that otherwise will produce socially less desirable and less profitable shelter and communities.

The momentum of building practice, for example, is worrisome, but it can be changed. The AIA Research Corporation analyzed a sampling of buildings constructed in 1975 and 1976 and found that if the buildings had been designed with already proven and available technology, their energy efficiency could have been increased 30 to 50 percent.[1]

This article discusses the changing circumstances that require new development policies and briefly describes examples of efficient development that have been constructed *de novo* or incorporated into an existing town or city. It also addresses some regulatory problems that work against efficient development and suggests different approaches that might be taken.

Today, there is a new recognition of the limits of resources, including limits on government's ability to create, operate and maintain over extended distances both community infrastructure such as roads, water and sewer systems and essential services including police and fire protection. Also new is a recognition—slowly and painfully accepted—that there may not be in every family's future a newly built, detached home surrounded by a quarter acre of grass—although ownership of some type of home will probably be available to an increasing share of American households.

Behind these modifications of the American dream lie two stern and related economic considerations. First, the economic growth rates that characterized earlier post-war decades are unlikely to resume. Gross national product and real disposable income will increase at much more modest rates. The Congressional Joint Economic Committee's 1979 Mid-Year Report stated flatly: "The average American is likely to see his standard of living drastically decline in the 1980's unless the United States accelerates its rate of productivity growth. . . America's standard of living is threatened by slow growth policies. . . A stagnating economy will mean that fewer Americans will be able to afford the necessities of life such as a decent home—America can do better, but it must (reduce) reliance on foreign energy and address the tax and regulatory barriers to production. . . "[2] The moods of the eighties most likely to respond to these trends are personal thrift, public belt tightening and regulatory reform. We may expect smaller houses on smaller lots, for example. Some believe that trend has already started.

Second, the crisis of limited energy resources is beginning to be recognized as chronic and of long duration. Scholars debate as to whether or not future technological improvements and more modest proportions will contain the operating expense of cars or whether development patterns will change solely as the result of changes in the cost of gasoline.[3] These arguments miss the point. In fact, the significant crisis concerns the availability of energy supplies to support the nation's economy. When compared with the need for reasonably priced energy for the production of food, goods and services, for the expansion of jobs, for home heating and the movement of goods and people by rail, bus and truck, many current uses of the private auto appear trivial and unnecessary. When significant amounts of any source of energy are wasted, this loss adversely affects overall productivity and efficiency.

There is little doubt that we ultimately face auto use limitations, whether by regulation, taxation, or increased prices, if only to assure the availability of petroleum for more essential purposes. European countries have taxed the auto within some bounds, primarily because they do not choose to pay the social and direct governmental costs of accommodating the auto en masse. Any form of rationing, however, must prove politically difficult to sustain in the face of existing development patterns that make auto use unavoidable for most Americans as well as increasingly expensive for all. While some Americans live in places as convenient as Manhattan, a great many more reside in places that are hopelessly inefficient, requiring massive amounts of energy simply to earn a living, secure an education and feed one's household.

There are no quick or easy answers to these difficulties. The nation may temporarily escape facing part of the problem through higher automobile fuel standards, thereby enabling people to drive approximately according to their accustomed habits. Considerable conservation has already been achieved by businesses and more can be accomplished through insulating existing homes. These measures, however, fail to deal satisfactorily with the issue of the proper allocation of energy resources to various priority purposes. Nor do they take advantage of an opportunity in the context of the popularly recognized energy problem to deal positively with such related problems as city revitalization; environmental protection; holding down the cost of housing, local taxes and household transportation; and protection of farmland and open space.

Improving the efficiency of development patterns in the United States will not result from policies that simply load more restrictions on the use of land. The choices confronting us are not resolvable through polarized "growth" or "nongrowth" positions. On the contrary, all levels of government will need to accept that growth and change are as necessary as they are inevitable. What is at issue is the quality and location of that growth and its true cost.

Three design features, if widely applied, would make important contributions to the efficiency of the housing and communities that will be constructed in cities, suburbs and nonmetropolitan areas in the next twenty years:

1. Mixing employment, commercial and residential
   uses to facilitate pedestrian movement and the
   use of low cost, internal transit systems;
2. Siting buildings so as to make maximum use of
   solar heating and cooling opportunities, and
   applying the best state-of-the-art energy
   conservation techniques within buildings; and
3. Locating these mixed-use, energy-efficient
   projects so as to facilitate public transportation
   and shared private transportation among centers.

Taken together, these three elements could become the cornerstone of a new development policy, a companion to the emerging energy policies. Although these elements are rarely applied together by private developers or public agencies, there is sufficient practical experience with each of them to warrant widespread public support.

Current Settlement Patterns

In central cities, thinning populations have contributed to the inability of mass transportation systems to function without rising subsidies. Redevelopment projects emphasize construction of office buildings and downtown retail centers dependent largely upon suburban workers whose spending ends at the close of the workday, thereby curtailing restaurant, theater and other evening activities. While expanding private housing rehabilitation in most central cities testifies to strengthening housing markets for the middle class, few cities have used their powers and substantial federal subsidies to produce new, middle-income, in-town housing with recreation and parking features competitive with suburban communities.

Sixty years of suburban development has so segregated activities that shopping requires driving even to so-called "convenience stores." Almost no work sites, victims of the outdated assumption that such are universally "incompatible" with residential living, are allowed near residences. Building and zoning codes frequently require more interior and exterior space than is warranted by new markets. Some of this regulation was successfully designed to stop all or most growth in order to hold down taxes or for exclusionary purposes.

In the 1970s, the new growth regions of the country are outside metropolitan areas. The nation's extensive highway network, together with substantial increases in the number of motor vehicles per household, today offer nonmetropolitan workers individual labor market and shopping areas extending 45 and even 60 miles radially in all directions from their homes. Most new housing is located outside older small towns either in small tract developments or individual homesites of one, two or even five acres purchased from farmers. Auto dependence is now growing faster in the same areas where America's growth rates are highest, an ominous economic and political trend for the next two decades.

Auto Dependency

The states with the fastest rates of growth in motor vehicles are also characterized by low-density development patterns. Alaska, Colorado, Idaho, Nevada and Tennessee top the list. The highest percentage of driving age population holding licenses are persons living in unincorporated areas, also among the lowest-density places.[4]

Gasoline consumption has risen much faster than can be accounted for by population growth. In the first four months of 1979, gasoline consumption jumped 2.6 percent. The number of cars on the road (one for every two Americans) increases about 10 percent yearly. An Amoco poll found that the average household drove 20,400 miles in 1978, up 11 percent since 1974.[5]

Automobile trips are primarily for four purposes: earning a living, family business, social and recreational, and civic, educational and religious (Exhibit 1). Home-to-work commutation per household averaged 444 trips per year or 4,183 vehicle miles annually in 1969-70. Such trips averaged 9.4 miles in length. In these figures one can see the growing importance of dual wage-earning households. Substantial economies can be effected when the workplace of at least one member of the household is close to home, as opposed to the situation common today of having no work sites near residential areas.

Pedestrian-based, mixed-use developments, especially in nonmetropolitan areas, would serve the objective of reducing the number of miles driven per household in several categories of trip purposes. About 10 percent of home-to-work commutation trips might be eliminated for households where job and residential sites are sufficiently close to permit walking to work or shuttle or para-transit service. The second largest subcategory of trips, shopping, might be cut 80 percent where residential sites and shopping center sites are combined or placed in close proximity. Where shopping malls include medical and dental services, banks, movies and some government offices, additional reductions in trips or in trip length could be expected. The inclusion of public recreation and educational facilities in higher-density residential projects would accomplish reductions in these categories as well. More efficient location of settlements—near highway interchanges, for example—would facilitate car pooling and reduce the number of separate destinations. Some developers believe that "social and recreation" trips also can be reduced by increased investments in common recreation facilities. Mixed uses could substantially lessen the need for second or third cars and reduce the number of vehicle miles traveled per household.

Density Increases

The term urban density has been popularly associated with the wretched building forms and poverty of 19th century slums. One hundred years of suburbanization in turn has come to mean flight from density itself. In fact, it is more accurately seen as flight from badly managed density, physical decay or unfavorable social conditions.

In part because it appears to be changing, it is not clear today what the popular perception of appropriate density levels may be. Persons surveyed who lived in modern, well designed, medium-density residential projects find such projects attractive for reasons of economy, public safety, avoidance of home maintenance and added recreational opportunities.[7] On the specific question of density acceptance, projects with the highest marks for overall owner satisfaction included, first, one with 10 units per acre and, second, one with 3.9 units per acre. On the other hand, those receiving the lowest ratings tended to range from 10-to-15 units per acre. While the design differences make any conclusions risky, there does appear to be more acceptance of considerably higher density levels than are typical of most suburban development.

In 1975, the Real Estate Research Corporation (RERC), pointed to "drastic shifts in the types of dwelling units being built in response to economic factors." RERC suggested that, in the Washington metropolitan area, "the major form of dwelling units in the future may be the townhouse."[8]

Radical changes in densities are not necessary, however, in order to produce greater convenience and energy savings. Mixes of detached, semidetached, row houses, garden apartments and mid-rise apartments may be suitable in suburban communities. Projects in central cities would have few

Exhibit 1

Distribution of Automobile Trips, Vehicle Miles of Travel and Trip Length (miles) by Trip Purpose.[6]

Trip Purpose	Vehicle Miles per Household (annual)	% of Auto		Trip Length (miles)
		Trips	Travel	
a. Earning a living				
Home-to-work		31.9	33.7	9.4
Related business		4.3	7.9	16.1
Subtotal	5,166	36.2	41.6	10.2
b. Family business				
Shopping		15.2	7.5	4.4
Medical and dental		1.8	1.6	8.4
Other		14.0	10.2	6.5
Subtotal	2,401	31.0	19.3	5.6
c. Civic, educational and religious	612	9.3	4.9	4.7

Exhibit 1 (Continued)

Distribution of Automobile Trips, Vehicle Miles of Travel and Trip Length (miles) by Trip Purpose.[6]

Trip Purpose	Vehicle Miles per Household (annual)	% of Auto		Trip Length (miles)
		Trips	Travel	
d. Social and recreational				
Visiting friends and relatives		8.9	21.1	12.0
Pleasure driving		1.4	3.1	20.0
Vacations		0.1	2.5	160.0
Other		12.0	15.3	11.4
Subtotal	4,094	22.4	33.0	13.1
e. Other and unknown	150	1.1	1.2	9.4
Total	12,423	100.0	100.1	8.9

detached units and those in fringe areas few mid-rise apartments, depending on land costs and market opportunities.

The scale of projects, of course, will be similarly affected. Neighborhood units could relate to the population locally considered necessary to support an elementary school and/or a convenience food store.

As recently as five years ago, many believed that American families would not purchase row houses in new developments. Today, in response to the desire and ability of families of modest means to purchase new homes, row houses in medium-density tracts are being marketed not only in central locations, but in scattered, metropolitan fringe counties as well. What American families "want" in terms of location, housing style and amentities is conditioned by what is available at prices they can afford. Today, some relatively remote new suburban developments are offering impressive on-site recreation packages. Tomorrow, others may offer pedestrian access to schools, childcare, shopping and public or cooperative transportation. Demand shapes supply, yes; but supply also shapes demand. We may be entering a new era in which large scale projects offering higher densities, mixed uses, pedestrianization and low energy and maintenance costs will prove increasingly popular and profitable.

Location Opportunities

While acceptance of higher-density living appears to be growing, the tendency toward sprawl, skip or scattershot development also continues, making development location policies a major challenge in the United States. As the basis for an analysis of how energy consumption relates to land use patterns, RERC compared areawide energy consumption estimates under each of six scenarios in the Washington area. Significant differences in the probable rates of energy consumption were found among the different land use patterns. "Automobile energy consumption increases by 60 percent between the base and forecast years with 'Sprawl' development, while with 'Transit–Oriented' development, consumption is increased by only about 28 percent." The difference is attributable to three factors: shorter trip lengths, fewer trips, and a higher level of ridership on transit.

"The least consuming option ('Dense Center') annually provides enough fuel savings compared to the most consuming option ('Sprawl') to supply 109,500 additional single-family detached dwelling units. . . . "

Energy Efficiency in Contemporary Communities

The nation has a limited but important stock of communities that, because of location, mixes of use, and pedestrian opportunities, are in varying degrees energy efficient. Columbia, Maryland, the planned new town created by the Rouse Corporation, and Davis, California are two that contribute important experience for planners, developers and government policymakers.

Columbia is a mixed-use community that incorporates many energy conservation features that warrant study and application elsewhere. On the other hand, Columbia's residential density (2.2 units per acre overall) is low and is allocated as follows:

1. High-density areas (includes townhouses) —
   11 percent of total land — 12-to-15 units to the acre.
2. Medium-density — 20 percent of total land — four
   units to the acre.
3. Low-density — 12.7 percent of total land — two units
   to the acre.[9]

Columbia's smallest unit is the neighborhood. These house 2,000 to 5,000 persons and offer a swimming pool-recreation complex, a school and a convenience food store, all linked by off-street walkways. Columbia's management claims that 40 percent of those using neighborhood convenience stores are pedestrians.[10] Groups of neighborhoods are served by a shopping center and specialized facilities serving villages of 7,000-to-11,000 persons. Six loop bus routes connect to serve the neighborhoods, villages and shopping centers to the downtown office/hotel/entertainment area.

Columbia is partially circled by clusters of light and heavy industrial parks and other work centers. These locations are sufficiently close to facilitate short drives to work and car or van pooling or bus service. An "early bird" commuter system enables workers to connect with buses to major work centers outside Columbia. There is, however, no bus service from neighborhoods to Columbia's fringe employment complexes.

Davis, California, is a growing town of more than 30,000 which has successfully used the considerable powers of local government to shape development to meet its policy objectives. Energy conservation, environmental improvement and cost reduction are among those purposes.[11] The town increased residential densities and emphasized passive solar heating and cooling in building design and siting. Davis encourages new mixed-use projects, multiple uses in old neighborhoods and even home-based employment and business. After only a few years, Davis already can measure success in terms of substantially reduced use of gas and electricity and substitution of walking and biking for auto use.[12]

The city's Development Director points to "high density residential use" (e.g., 12 units to the acre) and contiguous development as central to the Davis plan. Minimum lot sizes are now as low as 3,000 square feet and development that requires excessive auto use is not permitted. Davis makes extensive use of the planned unit development (PUD) technique, allowing maximum flexibility "with absolute control" by the city over private development. Both the community and developers "give in order to gain."

In Davis' PUDs, "streets are of widths only to serve the necessary functions, neighborhood commercial and medical/dental facilities are within one quarter mile or walking distance of the residences, lots are oriented to facilitate use of solar devices (as many north-south as possible and the rest wider to allow homes to be oriented north-south), and fence and house setbacks allow the largest yards on the south side of the houses. Connecting pathways for bicycles and pedestrians encourage these forms of transportation over automobiles. . . ."

"Industrial sites can be situated close to residences so that more persons are encouraged to walk or ride bicycles to work (with proper landscaping, berming and other blending devices, many industrial sites are not incompatible with residences). The mix of multiples, attached and detached housing encourages the densities necessary to support transit systems and car pooling, as well as neighborhood variety and small commercial centers."

Builders' reactions have been mixed, although experience has tended to increase their support. One former skeptic claims that the city's rules are good for business. They sell more houses, he says "because people are concerned about their energy bills."[13]

Large-Scale Development Forms

The Urban Land Institute defines a mixed-use development project (MXD) as a "relatively large-scale real estate project characterized by:
1. three or more significant, revenue-producing uses
   (such as retail, office, residential, hotel/motel,
   and recreation);
2. significant functional and physical integration
   of project components (and thus a highly intensive
   use of land), including uninterrupted pedestrian
   connections; and

3. development in conformance with a coherent plan
   which frequently stipulates the type and scale of
   uses, permitted densities, and related items."[14]

The outlook in the 1980s for MXDs is generally favorable, although some developers feel that projects will become smaller unless more government assistance is available to lessen risks and front-end costs and to increase mixes of use. Needed subsidies most often take the form of parking garages and water and sewer lines.

While residential elements have received only secondary attention in MXD projects thus far, one developer clearly stated that he had "little faith in MXDs revitalizing downtowns without a residential component." However, the relative emphasis on retail outlets, reflecting the shopping center boom of the past twenty years, may decline as the result of a "flattening out of demand" for shopping centers as predicted by George Sternlieb and others. Including residential segments may become more attractive to shopping center developers in the next five years as they attempt to provide close-at-hand consumers for retail outlets.

Other experts point to new nonmetropolitan trade areas of only 10,000-to-50,000 persons as the "real new frontier" for centers. Some urge recycling old centers and more mixed-use development projects created by a "new breed" of developer, one who is urban oriented and able to cope with regulation and to exploit public-private arrangements.[15]

Woodlands, in the Houston, Texas metropolitan area, is a large-scale (20,000 acres) project, with plans for industrial, office, residential, retail and recreation uses, lessening auto dependency somewhat through the relative proximity of nevertheless separated uses. Large projects of this sort, while they do not tend to reduce the number of trips per household, can shorten those trips substantially, and, even without a strong pedestrian emphasis, may serve as relatively efficient communities.[16]

Similarly, Reston, Virginia, although located at considerable distance from the metropolitan center, contains many features that make it relatively energy efficient. Reston's management claims, for example, that if one lives "in a typical home in Reston," one would need to walk only an eighth mile to recreation, a quarter mile to school and work and a half mile to shopping.[17]

To make Reston more efficient for its 32,000 residents, the management has been attracting businesses. "More than 10,000 people have an opportunity to work in the same neighborhood where they live." Forty percent of the jobs in Reston are held by Reston residents. In addition, the Reston commuter bus service carries 3,000 riders a day to Washington and other employment centers.

"Yet recent studies indicate that it is not the task of driving back and forth that puts the most strain on the family car," Reston management claims. "According to surveys, more than 75 percent of all automobile traffic involves ordinary trips that are part of everyday living." Reston points to residential clustering, mixed uses and pedestrian access to a rich recreational mix as the factors that keep residents out of their cars.

While most development today is inherently energy inefficient and costly to residents and governments alike,[18] there are examples where the principles of efficiency are being successfully marketed. More widespread and accelerated application of these principles, however, will require various public incentives if they are to be applied in time to shape the new housing needed in the next five years.

Progress in altering development patterns may not prove to be so slow as many believe. A remarkable amount of change occurs in our built environment each year. Loss and replacement of housing occurs constantly. Each year, for example, the nation adds between 1 percent and 2 percent to its housing stock. About 600,000 units are lost annually for various reasons as approximately 1.5 to 2 million are added. Thirty-seven percent of the housing stock in the year 2,000 will be built between now and then.[19]

Moreover, household formations are expected to jump from an annual rate of increase of 1.5 to 1.7 million (1975-80) to as much as 1.9 million in the 1980-85 period.[20] The next five years should be an extremely favorable time to shape development and to design and place buildings so as to emphasize efficiency and conservation. States, counties and municipalities can act now to build energy savings into codes, plans and laws to benefit consumers and to reduce future costs of government through more compact, more efficiently located projects. Above all, government and industry leaders will need to cooperate to assure that planned development is profitable as well as attractive.

The "Most Favored Development" Strategy

A recent study provides an indication of the extent to which the cost of development is increased because of delays attendant on government regulation.[21]

The Rice Center points out that Federal Housing Adminstration (FHA) implementation of Section 102 (2)(b) of the National Environmental Policy Act (NEPA) primarily affects large-scale development (500 or more housing units). Developments under the 500 unit threshold require only a "special" or "normal" environmental clearance, the application for which is estimated by FHA to require thirty minutes of preparation. Large-scale development, however, requires preparation of a full Environmental Impact Statement (EIS). Preparation time is said to range from 30 to 125 days; approval time ranges from 10 to 12 months. The study, which took into account all regulation save zoning and building permits, reported an average regulatory-based delay of 5.5 months. The estimated cost of delay for all projects surveyed ranged from $390 to $590 per lot. The estimated 1976–77 "development duration increase" of five and one half months cost the average homebuyer a minimum of $560 to $840 per lot.

The evidence thus far suggests that the nation is likely to secure more energy-efficient development from relatively large-scale projects than from individual construction or most small-scale projects. The same also may be said for such policy objectives as open space and cropland protection, housing opportunities for low-and moderate-income families and holding down government and household costs. While small home builders can participate in large-scale projects, such scale is often required in order to apply the skills of engineers, landscape architects, economists, planners and others needed to produce quality as well as efficiency. Therefore, rather than increasing the costs of large-scale development to developers, home purchasers, renters, local governments and even regulatory agencies, targeted regulatory reform should provide subsidies in the form of reduced regulatory costs to projects of any size so long as they contribute to these objectives. Streamlined permitting and early decisions on applications would constitute financially attractive inducements similar to the zoning exceptions granted by New York and other cities in return for the provision of public amenities by developers.

States should be encouraged to experiment with regulatory reform. Where it would not conflict with constitutional home rule powers, states might change the legislative basis for separation of uses in local zoning. Rather than the present presumption of incompatibility of different activities in most zoning, state law might assume that mixed uses are desirable unless a specific finding of incompatibility is made by a local government.

The nation has long tilted its tariff rules to favor imports from countries from which we expect other benefits. Projects that meet the previously listed conservation objectives should be granted "most favored development" status, thereby favoring socially desirable projects through accelerated permitting.

Government in growing areas must no longer stand apart from private developers in the decisionmaking process. The public-private cooperation that has been applied successfully in redeveloping central city business districts is unknown in most metropolitan and nonmetropolitan growth areas because it has been assumed that there are no needed public benefits that cannot be supplied under standard zoning and building codes. In areas where growth requires public subsidies (including acquisition and regulatory adjustments) to secure for the public good benefits that otherwise would not flow from the private market, collaboration should be accepted under safeguards contained in state law.

Pro-development Policies

In summary, development policies for the 1980s should seek to:

a. Reduce per capita energy consumption for residential and personal transportation purposes.

b. Reduce household and individual transportation costs; household heating and utility costs; and taxes related to the cost of providing services and facilities for new development.

c. Reduce per-unit government service and facilities costs in new development.

d. Reduce cropland and open space loss caused by low-density, scattered development.

New development should facilitate public and pooled transportation among planned development and other centers. It also should reduce household trips and trip lengths by increasing the proportion of development in cities, suburbs and nonmetropolitan areas in close proximity to access points on major transportation systems.

Individual projects should:

(1) Improve pedestrian access to activities and facilities in daily use by:

(a) substantially increasing density levels through infill and new projects over those typical in each case of city, suburban and nonmetropolitan development;

(b) increasing the proportion of development of appropriate scale, economic viability and mixed uses—i.e., composed of residences, shopping, recreational and educational facilities and work sites—placing those uses within walking distance of, or transit access to, each other;

(c) creating efficient and safe walkway systems within communities, separating auto and pedestrian modes to the extent feasible and reducing the proportion of land devoted to auto uses and storage.

(2) Improve nonauto access to activities and facilities within and among settlements by integrating public or cooperative vehicular transportation systems into project designs; provide such transportation as an on-going community service.

(3) Minimize household energy consumption by applying market accepted, efficient, state-of-the-art insulation, heating and cooling equipment; passive and active solar techniques—north-south siting, deciduous and coniferous plantings, etc.

Such new development would:

a. Eliminate the need for more than one auto per household;
b. Enable households to hold the average annual number of trips and annual mileage to half the national average;
c. Reduce the percentage of commuters using private automobiles to half the state average;
d. Reduce the number the number of private autos with a single commuter to half the state average;
e. Reduce the number of auto trips for shopping to half the state average;
f. Assure that the distances from individual residences to schools, recreation, shopping, transit and employment will not exceed 10-to-15 minutes walking time;
g. Reduce average household energy consumption for heating and cooling by 20 percent.

The Federal Government could:

(1) Legislate "fast track" permitting procedures for environmentally sound, high-density, pedestrian based, energy-efficient, multiple-use projects to reduce lead time and associated costs of development.
(2) Establish priorities favoring such projects in awarding federal subsidies for residential, employment, recreation, transportation and commercial development projects, including water and sewer facilities.
(3) Devote a significant proportion of existing research and demonstration funding to produce practical experiments in multiple-use, higher-density, profitable developments of various sizes, mixes and locations.

Involve private developers, state and local governments in planning and implementing these projects.

State Governments could:

(1) Establish "fast track" permitting systems to lessen costs of meeting state requirements for approved settlement projects.

(2) Assure adequate land acquisition powers and capabilities among states, counties and other local government units most likely to grow so as to facilitate large-scale, multiple-use developments at optimal locations.

(3) Identify optimal locations for approved projects.

(4) Apply state transportation, facility siting approval and investment policies to reinforce efficient settlement patterns.

(5) Encourage public transportation systems to serve approved developments.

(6) Clarify trade-offs in state cropland and open space protection programs so as to favor high-density, multiple-use development (even at the loss of some land) when necessary to avert greater losses resulting from low-density, scattered development.

(7) Assure flexibility of state and local land use control legislation to facilitate negotiated decisions between local governments and private developers.

(8) Legislate tax incentives for energy-efficient communities.

(9) Establish statewide development goals and substate planning objectives.

Municipal and County Governments could:

(1) Establish standards for approved development and apply "fast track" permitting systems for approved projects.

(2) Conduct competitive procedures for selecting developers, favoring those that best meet local standards for energy-efficient communities.

(3) Adopt a "negotiated decision" process to shape new growth.

(4) Offer tax incentives for projects that hold down government service costs.

Sooner or later, the nation must come to acknowledge that negative restraints on growth by themselves are unworkable and unhealthy. Such constraints imitate the illusory effects of a girdle; they change the location of growth (at some cost), but they don't positively affect the underlying problem. American governments must soon come to terms with development and developers, formulating positive policies to assist desirable development in growth areas as well as in declining cities.

Pro-development policies are needed as companions to the nation's energy policy. Most of the elements exist in current development practice; the principle missing element is positive government support leading to widespread integration of residential, retail and work sites. Because community development requires an effective combination of skills and because of the inherently higher risks associated with large-scale, complex development, priority use of existing government assistance is required in growth areas as well as declining cities.

Such policies would have in the development industry a considerable national constituency which identifies its interests with clearer, more consistent government standards; more favorable regulatory treatment and more clearly defined assistance in growth areas.

The economic, demographic, resource and energy challenges of the next twenty years can be converted into favorable opportunities for public-private cooperation to serve the interests of householders, local governments and profit-making businesses, as well as the nation's energy security and economic well-being. If we wisely capitalize on the most favorable and conserving trends, we can build better communities for the next generation and help existing communities face the visitudes of resource shortages and limited revenues.

Certainly, we will need to face some important environmental trade-offs that have been avoided in the past. This may not prove so difficult as some believe. In the foreword to a recently published book on high-density development in Europe, for example, William K. Reilly, president of the Conservation Foundation, has written:

"Our (conservation) opportunities may go unfulfilled . . . unless we learn to reconcile environmental concerns with the need for more homes . . . The need in the 1980s is to go beyond stopping growth and to encourage and accommodate acceptable growth at acceptable places . . . Simply saying no to growth is . . . an insufficient response to a dynamic society's need for development. . . . Conservation and creation must be integrally linked if we wish to see the United States consciously develop more desirable land use patterns."[22]

While it may stretch some imaginations to envisage government, developers and environmentalists cooperating on housing projects for the 1980s, it is nevertheless true that changing circumstances and more mature views of our common destinies will reduce some of the issues that have previously divided them. More widespread recognition of the stake we have as a nation in such collaboration will help pave the way for adoption of balanced, constructive pro-development policies at all levels of government.

Notes

1. Bremer, Duncan. AIA Research Corp. *The Washington Post.* November 11, 1979.

2. *Mid-Year Report and Staff Study.* Congressional Joint Economic Committee, August, 1979.

3. Kain, John F. Paper delivered at the Urban Land Institute Mid-Year Conference, Dallas, Texas: 1979.

4. *Nationwide Personal Transportation Survey.* Bureau of the Census, 1969–70.

5. *Time.* May 28, 1979.

6. Bureau of the Census. Nationwide Transportation Survey, 1969–70.

7. Engstron, Robert, and Marc Putnam. *Planning and Design of Townhouses and Condominiums.* Washington, D. C.: Urban Land Institute, 1979.

8. *Energy, Land Use and Growth.* Council of Governments, August, 1975.

9. Columbia, Maryland Information Center.

10. Interview with Elliott Jacobsen, Columbia, Md. Real Estate Department, August 3, 1979.

11. McGregor, Gloria Sheppard. "Energy Conservation in Urban and Regional Planning: The Davis Experience." City of Davis, California: undated.

12. Interviews with Gloria McGregor, Community Development Director, Davis, California, 1979.

13. *The Wall Street Journal.* May 17, 1978.

14. Witherspoon, Robert E. *Mixed Use Developments: New Ways of Life.* ULI, 1976.

15. "Shopping Centers in 1988." ICSC. September, 1978.

16. "The Woodlands Metro Center Energy Study" (Preliminary), U. S. Department of Commerce, 1979.

17. Advertisement, *The Washington Post.* September 8, 1979.

18. Real Estate Research Corporation. *The Costs of Sprawl.* April, 1974.

19. Downs, Anthony. The Brookings Institution, ULI Conference, October, 1979.

20. Bureau of Census. *Current Population Estimates.* 1979.

21. *The Cost of Delay Due to Government Regulation in the Houston Housing Market.* The Rice Center. ULI, 1979.

22. Lefcoe, George. *Land Development in Crowded Places.* Conservation Foundation, 1979.

Energy Constraints and Development Patterns in the 1980s

GEORGE STERNLIEB
JAMES W. HUGHES

Introduction

Do the new energy constraints—destined to exert their full force in the decade of the 1980s—involve a wholesale revision of the spatial patterns of land use that have dominated the post-World War II era? Have history and market experience lost their value as guideposts for future development, i.e., have the geographic and economic parameters of the recent past been rendered obsolete? With building cycles now typically taking upwards of four years from initial land purchase to finished product, and with the results fixed in place and exerting their influence for many decades, these questions are not philosophical abstractions but rather the essence of required present day decisionmaking.

In broad terms, land use in the United States has moved in a single mature lifetime through three stages. The first period was one in which there was a concatenation in space of industry, commerce, recreation and residence. Its historic manifestation was the central city. The second stage of the life cycle was an initial spatial fragmentation of these functions, with inexpensive transportation providing the linkage. The development of classic suburban bedroom communities was the most visible symbol of this evolution, but the central city still preserved the major role of sheltering most nonresidential urban activities. The third stage of the life cycle, which is presently with us, is founded on the development of a critical mass of population and buying power in selected suburban ring areas, such as to either replicate or provide appropriate surrogates for activities which at one time were dominated by the core. The latter area, in turn, in all too many cases, has found it very difficult to maintain its "pulling power" against the frictions that have arisen in it—not least of them racial—and in the face of competing alternatives.

It is striking to note in this context that nonmetropolitan territories of America have recently shown the greatest level of activity. The nation's population—and by definition, housing—has moved from city to suburb and increasingly, out of both into a whole new dimension of dispersal. Much of this redistribution is the consequence of shifting employment concentrations to outer suburban locations, particularly along metropolitan circumferential freeways. Land which had once been remote from the central city—both inside and outside the formally designated metropolitan boundaries—gained increased residential attractiveness as the 1970s employment dispersal gained momentum.

What can we anticipate from the fourth stage of development? In our estimation there is an enormous inertia of events, of historic capital investment patterns, of job and population shifts, which once set in motion as powerfully as has been the case in the United States require not a decade, but literally a generation, to be shifted in a meaningful fashion.

Changes of a much more radical and rapid nature could very easily bring into question the entire political foundation of our government and its institutions. The United States has invested the bulk of its capital development since World War II in an increasingly centrifugal fashion. We cannot declare this obsolete without bankrupting the country. Within these limitations, however, we envision a relatively modest but significant alteration of the basic trendlines which are currently in existence. Presently, the concentration of employment at the metropolitan rim—along circumferential freeways—has reached such thresholds as to foster rapid residential growth in more dispersed nonmetropolitan rimlands. Energy realities, with appropriate public policy actions, present substantial opportunities for significant land use readjustments. Essentially these will be concentrated on intensifying land use in and along the great circumferential bands of development that now ring our central cities. The latter, in turn, will not be the scene of great revitalization, certainly in comparison with their past levels of activity and population holding functions. Instead we see in selected areas a much more limited, but nevertheless most meaningful, recycling potential.

Most central cities will be but one of a constellation of activity nodes of the post-industrial metropolis; a return to the central city dominance, to a land use pattern of 50 years past, is unlikely. There is a host of lesser adaptive mechanisms which can buffer all but the most acute of energy strictures: carpooling, for example; the rise of much more efficient vehicles; and a multitude of others are well within our grasp. We do not have to invent the future. Its first growth is presently with us.

An Age of Scarcity Versus a Latent Reservoir

First and foremost in any forecast is appropriate awareness of the new age of scarcity that suddenly is evident, and most particularly, of the housing component of land use development. It is the unparalleled levels of affluence of the American housing consumer that has been one of the great facilitators of dispersed land use arrangements. Will diminished standards of living abort the assumptions of the past?

To put it bluntly, Americans are no longer as rich as we once thought we were. The consumer's shrinking dollar now has to stretch in an increasingly painful fashion to encompass the supermarket and the energy costs of work-related transit. As a result, the resources available for housing will be subjected to increasing constraints. Given appropriate expansion of employment opportunities, more and more the backbone of the market must be the two-worker household. While current statistics indicate soaring labor force participation rates for women in general, and wives in particular (see Exhibit 1), this phenomenon has yet to be exhausted. As of 1977, only 17 percent of all husband/wife households had both members working full time, year round.[1] The higher figures conventionally cited in this regard are misleading in that they incorporate part-time workers and those who work for a limited part of the year.

This is an extremely important phenomenon, since the household at the top of the income ladder—as shown in the data of Exhibit 2—is the husband/wife family with the wife in the paid labor force. In a post-shelter era, then, with housing as investment an established convention, the potential for residential mobility will be substantial for select household types.

So, our prognosis starts off with the reality of this still very substantial latent reservoir of housing demand and an assumption that the economic world will be hard—but not come to an end—and that there will be jobs for the able and willing.

EXHIBIT 1

CIVILIAN LABOR FORCE BY SEX AND MARITAL STATUS,
MARCH 1970 AND 1978

| | Civilian Labor Force | | | |
| | Number (in thousands) | | Labor force participation rate | |
Sex and marital status	March 1970	March 1978	March 1970	March 1978
Both sexes, total	81,693	98,437	59.1	62.2
Men, total	50,460	57,466	77.6	76.8
Never married	9,421	13,978	60.4	69.2
Married, wife present	38,123	38,507	86.6	81.6
Married, wife absent	1,053	1,703	61.3	77.4
Widowed	672	667	31.9	30.5
Divorced	1,191	2,711	76.0	80.7
Women, total	31,233	40,971	42.6	49.1
Never married	6,965	10,222	53.0	60.5
Married, husband present	18,377	22,789	40.8	47.6
Married, husband absent	1,422	1,802	52.1	56.8
Widowed	2,542	2,269	26.4	22.4
Divorced	1,927	3,888	71.5	74.0

Note: Because of rounding, sums of individual items may not equal totals.

Source: Beverly L. Johnson. "Changes in Marital and Family Characteristics of Workers, 1970–78," *Monthly Labor Review*. (April 1979), pp. 49–52.

EXHIBIT 2

INCOME SHIFTS BY FAMILY TYPE, U.S. TOTAL 1970 TO 1977

Family Type	1970 Income	1977 Income	Change 1970 to 1977 Number	Change 1970 to 1977 Percent
Total Families	$ 9,867	$16,009	$6,142	62.2%
Male Head Total	10,480	17,517	7,037	67.1
Married, Wife Present	10,516	17,616	7,100	67.5
Wife in Paid Labor Force	12,276	20,268	7,992	65.1
Wife Not in Paid Labor Force	9,304	15,063	5,759	61.9
Other Marital Status	9,012	14,518	5,506	61.1
Female Head	5,093	7,765	2,672	52.5

Source: U.S. Bureau of the Census, *Current Population Reports*, Series P-60, No. 118, "Money Income in 1977 of Families and Persons in the United States," U.S. Government Printing Office, Washington, D.C.: 1979. U.S. Bureau of the Census, *Current Population Reports*, Series P-60, No. 80, "Income in 1970 of Families and Persons in the United States, U.S. Government Printing Office, Washington, D.C.: 1971.

The Secular Market Pattern

Within this overall context, what are the areal dimensions of growth going to be? Certainly, the events of recent history provide both the constraints and foundations for future alterations. The long-term patterns of population dispersal and growth are summarized by the data of Exhibit 3, which indicate the tremendous rapidity with which central cities have declined as population loci in the United States. The 31.5 percent share of the nation's population captured by central cities in 1970 diminished sharply to the 28.2 percent level by 1977. This was not simply a consequence of much more rapid growth in suburban and nonmetropolitan territories, but also for the first time, the absolute decline in the population of central cities. Over the seven-year period, the aggregate losses accruing to all central cities approached 2.9 million people, or 4.6 percent of the 1970 base. The bulk of

EXHIBIT 3
POPULATION BY PLACE OF RESIDENCE: 1970 AND 1977
(Numbers in Thousands: 1970 Metropolitan Definition)

| | 1970 | 1977 | Change: 1970 to 1977 | | Percent Distribution | |
			Number	Percent	1970	1977
U.S. Total	199,819	212,566	12,747	6.4	100.0	100.0
Metropolitan Areas	137,058	143,107	6,049	4.4	68.6	67.3
Central Cities	62,876	59,993	−2,883	−4.6	31.5	28.2
Suburban Areas	74,182	83,144	8,932	12.0	37.1	39.1
Nonmetropolitan Areas	62,761	69,459	6,698	10.7	31.4	32.7
Central Cities in Metropolitan Areas of 1 Million or More	34,322	31,898	−2,424	−7.1	17.2	15.0
Central Cities in Metropolitan Areas of Less than 1 million	28,554	28,095	− 459	−1.6	14.3	13.2

Note: The metropolitan and nonmetropolitan totals differ from Exhibit 1 due to different date of metropolitan delineation (Exhibit 1, 1977; Exhibit 2, 1970).

Source: Center for Urban Policy Research Analysis of Data Presented in: U.S. Department of Commerce, Bureau of the Census, *Current Population Reports*, Special Studies P-23, No. 55, "Social and Economic Characteristics of the Metropolitan and Nonmetropolitan Population: 1977 and 1970," November 1978.

the decline was suffered by the central cities located in metropolitan areas having one million people or more, presumably those with the population thresholds justifying investment in public transit facilities.

In contrast to the performance of the central cities stands the vigorous growth of metropolitan-suburban and nonmetropolitan territories. By 1977, the two together accounted for 71.8 percent of the nation's populace. Almost 153 million of the nation's 213 million people in 1977 lived outside of central cities.

That this trend has persisted despite the energy shocks of 1974 is evidenced by the data of Exhibit 4, which tabulates migration to and from central cities by families between 1975 and 1978. A net out-migration of 751,000 husband-wife families was experienced over this three-year period, almost 10 percent of the 1978 total. And for the more affluent subset—with incomes $15,000 or above—approximately three central city families migrated to the suburbs for every family that moved from the suburbs to the central city.[2] Clearly, a back-to-the-city movement is difficult to discern, and the inertia of established trendlines is suggested.

A BROADER PERSPECTIVE

But central city population trends are enveloped within a matrix comprising a number of causal dimensions—demographic, regional and metropolitan-nonmetropolitan.[3] The demographic sector is marked by the increasing importance of migration, compared to net natural increase, as a determinant of local growth. The regional axis is evidenced by the high growth rates in the South and West, and a relative stagnation in the aging Northeast and North Central states. The metropolitan partition shows a marked resurgence of nonmetropolitan territories, aborting a trendline which had been in effect for over half a century. The data in Exhibit 5 encompass the broader attributes of these three phenomena.

For the United States as a whole, the deceleration of population growth is gauged by an average annual compound growth rate in the 1970-to-1977 period (0.9 percent) significantly less than that (1.3 percent) of the preceding ten years—1960 to 1970. The major determinant of population change in this context is the decline in net natural increase (births minus deaths) over the two periods indicated, while net migration held relatively constant. And it is the increasing relative importance of the latter which translates into special shifts, one dimension of which centers on metropolitan clusters.

EXHIBIT 4
CENTRAL CITY MIGRATION BY FAMILIES: 1975 TO 1978

Family Income	Number of Families[1]	Movers From:			Central City Net Change
		Central City to Suburbs	Suburbs to Central City	Abroad to Central City	
Under $5,000	295	46	32	20	+6
$5,000 to $9,999	905	175	81	70	-24
$10,000 to $14,999	1,407	244	104	36	-104
$15,000 to $24,999	2,983	692	245	48	-399
$25,000 and over	2,049	423	168	25	-230
Total	7,639	1,580	630	199	-751

Note: [1] Number of husband-wife families with head 14 to 54 years of age, residing in central cities in 1978. (Numbers in thousands)

Source: U.S. Bureau of the Census *Current Population Reports*, Series P-20, No. 331, "Geographic Mobility: March 1975 to March 1978," U.S. Government Printing Office, Washington, D.C.: 1978.

The decade of the 1960s was one in which the relative concentration of America's population in metropolitan areas probably reached its pinnacle. While metropolitan areas, for example, experienced average annual increments of 1.6 percent, nonmetropolitan counties lagged considerably (0.4 percent). Within the former, there was only minor variation as a function of size, with the growth rates of the largest metropolitan areas (over 3 million people) generally comparable to metropolitan areas in total. Within nonmetropolitan counties, those increasing in population at the most rapid rate typically were closest to metropolitan centers, as evidenced by the commutation profile indicated in the exhibit.

When the focus shifts to the 1970-to-1977 period, an abrupt change in trendline appears. The annual growth rate of nonmetropolitan counties (1.2 percent) experienced a remarkable upswing to almost twice the level characterizing metropolitan areas (0.7 percent). Within the latter category, it was the largest metropolitan areas (over 3 million people) which were suddenly transformed into virtual no-growth contexts, with a barely discernable growth of 0.1 percent per year. While counties which have the heaviest incidence of commuting to more concentrated areas were at the forefront of the nonmetropolitan resurgence, their growth rates were nearly matched by noncontiguous areas. Even nonmetropolitan counties in which less than 3 percent of workers commuted to metropolitan areas secured gains at ten times the rate of the previous decade.

Even more salient are the net migration data of Exhibit 5. Migration is a telling criterion of location shift by choice, indexing the locational preferences of Americans. Population gains secured via migration are, in effect, immediate gains of economic markets; population growth through net natural increase does not directly produce job holders, homebuyers, or immediate transit users, at least within short-run contexts. Hence migrational patterns can be viewed as signals of market shifts, pointing to locations of eventual economic vacation (net outmigration) or of expanding support thresholds (net inmigration).

The shifting patterns in this regard are reasonably clearcut. From 1960 to 1970, nonmetropolitan counties lost considerable population through outmigration to areas of greater concentration—the metropolitan nodes. The pattern was most accentuated in those areas in which commuting to the latter was the least significant. In contrast, all the major metropolitan areas benefited from net inmigration. This period may well have represented the terminal point in the shift of population from the land as a function of the final stages of the agricultural revolution.

The reality documented by the 1970–1977 data has rendered obsolete long held spatial conventions. The migration ledgers of metropolitan areas as a

EXHIBIT 5

POPULATION AND COMPONENTS OF CHANGE FOR SELECTED GROUPS OF METROPOLITAN AND NONMETROPOLITAN COUNTIES: 1960, 1970, AND 1977

(Numbers in the thousands)

Metropolitan areas, nonmetropolitan counties, and regions	POPULATION			AVERAGE ANNUAL PERCENT CHANGE[1]					
				Population		Natural Increase		Net Migration	
	July 1, 1977 (provisional estimate)	April 1, 1970 (Census)[2]	April 1, 1960 (Census)[3]	1970 to 1977	1960 to 1970	1970 to 1977	1960 to 1970	1970 to 1977	1960 to 1970
United States	216,351	203,305	179,311	0.9	1.3	0.7	1.1	0.2	0.2
Metropolitan[4]	158,550	150,291	128,328	0.7	1.6	0.7	1.1	0.1	0.5
Over 3,000,000	53,260	52,864	45,766	0.1	1.4	0.6	1.0	-0.5	0.4
1,000,000 to 3,000,000	42,035	39,341	32,403	0.9	1.9	0.7	1.2	0.3	0.8
500,000 to 1,000,000	24,088	22,548	19,386	0.9	1.5	0.7	1.2	0.2	0.4
250,000 to 500,000	20,051	18,262	15,838	1.3	1.4	0.8	1.2	0.5	0.2
Less than 250,000	19,116	27,276	14,935	1.4	1.5	0.8	1.2	0.6	0.3
Nonmetropolitan counties by commuting to metropolitan areas[5]	57,802	53,014	50,982	1.2	0.4	0.6	0.9	0.6	-0.6
20 percent or more	4,549	4,013	3,663	1.7	0.9	0.5	0.8	1.2	0.1
10 to 19 percent	10,039	9,209	8,607	1.2	0.7	0.5	0.8	0.7	-0.2
3 to 9 percent	14,796	13,644	12,944	1.1	0.5	0.6	0.9	0.5	-0.4
Less than 3 percent	28,418	26,148	25,768	1.1	0.1	0.6	1.0	0.5	-0.9
Northeast	49,299	49,061	44,678	0.1	0.9	0.4	0.9	-0.3	0.1
Metropolitan	42,140	42,481	38,609	-0.1	1.0	0.4	0.9	-0.5	0.1
Nonmetropolitan	7,159	6,580	6,069	1.2	0.8	0.4	0.8	0.8	-

EXHIBIT 5 (Continued)

EXHIBIT 5

POPULATION AND COMPONENTS OF CHANGE FOR SELECTED GROUPS OF
METROPOLITAN AND NONMETROPOLITAN COUNTIES:
1960, 1970, AND 1977
(Numbers in the thousands)

North Central	47,941	56,593	51,619	0.3	0.9	0.6	1.0	-0.3	-0.1
Metropolitan	40,221	39,661	35,073	0.2	1.2	0.7	1.2	-0.5	0.1
Nonmetropolitan	17,719	16,932	16,546	0.6	0.2	0.4	0.8	0.2	-0.6
South	69,849	62,813	54,961	1.5	1.3	0.8	1.2	0.7	0.2
Metropolitan	44,907	40,032	32,755	1.6	2.0	0.8	1.3	0.8	0.8
Nonmetropolitan	24,942	22,782	22,206	1.2	0.3	0.6	1.0	0.6	-0.8
West	39,263	34,838	28,053	1.6	2.2	0.9	1.3	0.8	1.0
Metropolitan	31,281	28,118	21,891	1.5	2.5	0.8	1.3	0.7	1.3
Nonmetropolitan	7,981	6,720	6,162	2.4	0.9	1.0	1.2	1.5	-0.4

Notes:
1. Based on the method of exponential change.
2. Includes officially recognized corrections to 1970 Census counts through 1976.
3. Adjusted to exclude 12,520 persons erroneously reported in Fairfax County, Va. (Washington, D.C. -Md. - Va. SMSA).
4. Standard Metropolitan Statistical Areas (SMSAs) or, where defined, Standard Consolidated Statistical Areas (SCSAs) and county equivalents of SMSAs in New England (NECMAs); as defined by the Office of Federal Statistical Policy and Standards, Dec. 31, 1977.
5. Classification based on 1970 Census data on percent of workers reporting place of work who commuted to metropolitan territory as defined in 1977 (see footnote 4). Of the total 2,455 nonmetropolitan counties, the four groups specified included 178, 331, 479, and 1,467 counties, respectively.

Source: U.S. Bureau of the Census, *Current Population Reports*, Series P-20, No. 336, "Population Profile of the United States: 1978," Washington, D.C.: U.S. Government Printing Office, 1979.

whole were in virtual balance. However, the aggregate totals mask the growing variation as a function of size. While the larger metropolitan formations experienced either net outmigration or diminished levels of inmigration, the smaller metropoli (under 500,000 people) actually experienced positive migration gains. Concurrently, nonmetropolitan areas were transformed from origins to destinations in the overall migration process, with the scale of the transition largely correlated with the proportion of workers in the nonmetropolitan region who commute into metropolitan areas (but, as will be noted later, *not* necessarily to the central city).

It is clear that the nation had shifted into a new /phase of growth as the decade of the 1970s reached its end. The emerging growth poles are suburban and nonmetropolitan areas, as well as smaller metropolitan places. Certainly, part of the nonmetropolitan phenomena may be subsumed under the label of exurbanization. One can view this process as merely a continuance of the dispersion from the core city, first to suburbia and subsequently to more peripheral patterns of settlement. From an energy and transportation perspective, however, given the scale of the processes at work, there are a number of issues which are raised. Evident among a wide spectrum of implications is the impact on journey-to-work patterns and trips to the central core.

As shown in the data of Exhibit 6, only 10.5 million workers, out of a total 51.8 million residing in suburban (outside central cities) and nonmetropolitan areas, were employed in the central city. Indeed, 19.3 million out of 28.9 million suburban resident workers were employed in the suburbs.

The central city's diminished pulling power mirrored by this data can be interpreted as a function of the frictions inherent in transportation versus the unique lures and employment opportunities which are available in the city center. To the degree that the data on spatial diffusion suggest longer trips, in the absence of new transportation facilitators, clearly frictions are increased. Secondly, as population expands farther from the historic population concentrations, it reaches critical mass, providing the threshold for the development of alternatives competitive to central city attractions. Indeed this may well intensify the development which has already occurred in terms of regional suburban shopping centers, multicinema units, largescale suburban hospitals, and the like. The data on journey-to-work reflect the growth and impact of the circumferential highways and the thickening of population concentrations in outer metropolitan areas and adjacent territories.

It is also evident from the data shown in Exhibit 5 that the largest metropolitan areas—those which have the greatest absolute potential for in-

EXHIBIT 6
PLACE OF RESIDENCE BY PLACE OF WORK, UNITED STATES: 1975

(Workers 14 years old and over; number of workers in thousands; SMSAs as of the 1970 Census)

Place of residence	All workers	Reported a fixed place of work					No fixed place of work	Place of work not reported
		Total	Inside SMSAs					
			Total	Inside central cities	Outside central cities	In nonmetro-politan areas		
All workers	80,125	72,733	51,507	27,116	24,391	21,226	6,724	668
Inside SMSAs	55,418	50,425	49,429	26,119	23,301	1,005	4,512	481
Inside central cities	22,760	20,846	20,568	16,528	4,040	278	1,700	214
Outside central cities	32,658	29,579	28,852	9,592	19,261	727	2,811	267
In nonmetropolitan area	24,707	22,308	2,087	997	1,090	20,221	2,212	187

Source: U.S. Department of Commerce, Bureau of the Census. *Current Population Reports*, Special Studies P-23, No. 99, "The Journey to Work in the United States: 1975," June, 1979.

tensive means of mass transportation—also have shown the least vigor of growth. At least through 1977, the smaller and more dispersed the metropolitan concentration, the greater the trendline of growth.

The changing regional population parameters intersect the above phenomena and generate an additional axis of variation (see Exhibit 5). The Northeast and North Central Regions shifted to a net outmigration position in the 1970s, with declining net natural increase just sufficient in magnitude to maintain overall population stability. The South, in contrast, secured sharp gains both inmigration and total population, while the West retained its position as regional growth leader.

Within this context, the metropolitan areas of the Northeast in the 1970s have lost population while those of the North Central Region are rapidly approaching stability in size. Only in the South and West are there substantial positive annual metropolitan growth rates—in part perhaps as a function of annexation which may obscure the basic centrifugal forces which are at work even in those regions.

Alternative Futures

It is against the backdrop of these established patterns that alternative models of future development must be evaluated. Conceptually there are four major options to be considered.

1) *A Revitalization of the Central City.* The thesis has been advanced that the central city represents an infrastructure which in an age of high energy costs cannot be replicated as casually as it has been done in the past. And further it provides many forms of energy efficiency in terms of housing density, employment centralization and journey-to-work logistics. We would suggest that, despite these potentials, mass central city revitalization is wishful thinking.

There is, nonetheless a significant residual of potential for *selected* central cities. The key element here is the employment base. With few exceptions, the city's role as a dominant industrial workplace is over. The old smokestack cities will probably not rise again. At the same time there is the development of a very few national cities which pull people with particular attributes and generate unique economies. Particularly striking is the new dominance of consumption functions over those of production in some of the most successful settings of core "rebirth." But these new focal points are rare indeed.

The relatively monolithic center-of-the-market housing demand of the last thirty years, characterized by the mass developments of Levittown and the like; have given way to *consumer segmentation*. This in turn generates specific strata of population with quite differentiated tastes. Loft conversion is a meaningful and significant business in New York City, with more than 30,000 units already completed. It does not mean, however, that loft conversion in Youngstown, Ohio would make equivalent sense. Chicago's North Shore development is a thing of beauty, and assuming that the economy doesn't default, it will continue. It does not mean, however, that one can envision a similar scale of rebirth in Cleveland.

Race still plays a very major issue, and the new central city market as a mass phenomenon has yet to meet the test of resales. *The key element that planners must remember is that the housing element of land use must not only provide shelter, but also must provide for resale. Values have to be present not only to extant consumers, but also to those that they expect to sell to -- and at a profit.*

Until and unless government policy literally forces employment back into the central city, significant questions must be raised concerning major widespread central city revitalization, subject to the specific cases and their equivalent that we have cited. Few energy savings can be realized by residing in the central city if one is employed in a suburban/exurban location.

2) *The Inner Suburbs.* In the midst of central city stagnation and decline, inadequate attention has been focused on the aging of the older suburbs in America. Some are beginning to replicate the traumas of the central city —declining school systems, rapid white outmigration, tax base erosion and the like. Nonetheless, there is still excellent housing available within them. Small vacant tracts are available and their logistics make reasonably good sense. But many of them are locked into the central city commuting pattern —too distant from new employment bands—and therefore subject to the vagaries of employment in the core. Yet, most strikingly, there is no way the housing values that exist within them could be replicated in terms of new construction. Selective rehabilitation in this context may represent a far larger potential than is presently appreciated. As the scene of major new construction, however, we would have some hesitation. Suburban infill may provide select opportunities for development for those geared to the limitations and possibilities of that demanding field. But it is difficult to see the latter as a mass phenomenon, subject to the employment qualifier presented earlier.

3) *The New Ring City.* These are the areas typically straddling the circumferential freeways. They have evolved into dominant growth bands for jobs and, where zoning is available, housing in our society. We envision these as the real keystones to future residential development. An expansive, modern employment infrastructure has been recently set in place, and will

remain whatever energy limitations ensue. Encompassed are the nation's newest and most competitive (in an international sense) industrial activities. The journey-to-work is far from insurmountable—if housing can be provided reasonably close by (and if Detroit would finally provide the kinds of cars that are going to be required).

There is, however, a significant problem in such regions in that residential land availability is restricted because of zoning and subdivision controls. *Workplace and residence must be granted tighter linkages—not the jobs of the past (those of the central city) but the jobs of the here, now and future on the ring highway.* The only way to accomplish this is to institute much more sensible land use planning, tolerating a greater variety of forms and much more intensive housing development than presently holds true in a good many areas. The consequence of the failure to recognize this linkage is indicated by our next area.

4) The Exurban Spread. Observers who view this type of spread as a *perversion* of land use (we use strong language here because they do!) fail to comprehend the new job-related facilities of America. *The commute to work to the ring highway for most of these locations is relatively easy. Indeed the journey-to-work to urban locations for central city residents is as long in time as that of suburb-to-suburb commuters.* The extreme level of dispersion to the nonmetropolitan fringes, however, is a tribute to restrictive zoning which has inhibited even greater efficiency of housing allocation to job sites.

Given the level of investment that has been made in the outer regions, it is difficult to believe that their development will abort suddenly. Americans have been sensitized, however, not merely to energy costs, but, through repeated energy shortfalls, to *availability.* The market here, therefore, may be much more subject to irregularities, of boom and bust, of subdivisions that simply do not secure traffic, of holding patterns while the energy crises which will characterize the next decade are lived through. Buying the truck farm and waiting for development to come may require more time—and cost—than used to be the case. Land speculation is always hazardous—but doing so in the farther reaches now may impose even more rigorous strain, and pain.

But this inefficient (in energy terms) market pattern may persist if appropriate policy responses are not forthcoming. Indeed, present casually considered efforts to restrict growth in the burgeoning ring zones does not direct development activity back to the central city, but forces a "spillover" effect into adjacent nonmetropolitan areas, and actually reinforces a land use pattern of greater dispersal.

Conclusion

Thus the decade of the 1980s has far more uncertainties than its predecessors. While the last wave of the baby boom will still generate housing demand through the period, its wherewithal will be much more subject to question.

Greater cognizance will be needed for variations in *housing types*—of party wall construction and of multifamily units—and of *juridical status*—of condominiums and co-ops, either alone or in combination with fee simple holdings. In turn, these will create even greater pressures on conventional control of land use.

From an areal perspective, however, we see the zones of growth which became more and more rigorously defined in the 1970s largely continuing into the decade to come. Energy *costs* will not alter America in our period; the lack of energy *availability* may. There is a time for heroes and a time for caution. The latter is at hand.

The thickening of the new suburban/exurban ring city in the United States will take place with even greater rapidity. Along this bank we envision crucial nodes—now in part nascent, in part becoming obvious—arising as even more significant entities. Their particular centers will vary depending upon opportunity and past levels of commitment. We can envision, for example, older trading centers taking on new vitality, such as White Plains in the New York Region, which has parallels elsewhere. Other forms of aggregative magnets may very well be provided by the regional shopping center, taking advantage of prime locations, significant acreage, the conversion of hitherto sprawling parking lots into more intensive forms of development, while their past function is secured by multistoried or multiuse facilities. There is no rationale for a hundred acre parking lot given the competing values of alternate utilization. Again these formats do not have to be invented; they are presently in the ground.

We can envision with no great stretch of the imagination adjacent intensive office and residential development. The condominium format will play an essential role in our era of post-shelter housing—and afford the occupants of both kinds of accommodation, inflation-proof forms of investment as well as shelter.

The key issue of the next ten years will be the control of land use, moving it away from local parochialism and paranoia about the future and toward the levels of flexibility which we have projected here. As yet we do not see any great wisdom on the part of higher levels of government in terms of land use recommendations which would give us confidence in their participation. The struggle may therefore be much more the continuance and intensification of private entrepreneur versus local zoning community, with the courts increasingly called on to act as adjudicators in the face of legislative incompetence. Standing in the way of a true coming to grips with national land use policy is the spector, and the memory, of the central city. As yet, purposeful, higher level land use policy implementation has been much more dedicated to the cause of things past than of things present and much less concerned with the future.

Squeezing Spread City—A Second Look

ANTHONY DOWNS

Background—The Forces Affecting
Future Levels of American Growth

THE long-range trend in American cities towards decentralization in the form of suburban sprawl and non-metropolitan fringe growth has its origin in multiple factors. It is primarily, though not exclusively, a result of lower transportation costs. As such, the impetus towards decentralization will not be fully removed by increased transportation levies.

Another cause of metropolitan decentralization is the widespread preference among all types of households for low density living. This is heightened by a historic rise in real incomes. Americans have a preference for single-family dwellings on large lots and as their real incomes increase they move farther out in metropolitan areas where land is available to secure this additional space. A third factor encouraging sprawl is the relative obsolescence of central city residential and nonresidential structures and public service systems and capital facilities. This is at a time of rapid increase in the number and types of firms that desire horizontal space for product assemblage and distribution. Each of these factors underly a sustained tendency towards decentralization.

A final cause of decentralization, rarely discussed in America, is (1) a powerful desire among middle class Americans to segregate themselves spatially from the poor, and (2) another such desire among whites to segregate themselves from blacks and other minorities. These desires have profound effects on establishing in each metropolitan area a socio-economic hierarchy of neighborhoods. They also have a tendency to keep government powers fragmented so that upper income suburbanites can more easily indulge in exclusionary behavior. Such desires further make redevelopment of abandoned inner city areas extremely difficult because of the reluctance of those who can afford new housing to live in revitalized neighborhoods near the poor and minority groups from whom they wish to remain separated.

All of these factors that have encouraged decentralized urban settlement are currently operating today with the possible exception of rising real in-comes—real incomes may begin rising again in the future although more slowly than they have risen in the past. *Therefore, there will still be strong pressures for sustained decentralization of urban settlements in the 1980s despite higher energy costs.*

A second background factor concerning urban growth involves future population increases in the United States. Using the Census Bureau Series II projections, a population increase of 21.4 million is expected in the 1980s with an additional 16.8 million during the 1990s. (These estimates exclude any allowance for illegal immigration. In 1977 alone, over one million per-sons were caught trying to enter the United States illegally). In the 1960s, approximately 91 percent of all population growth occurred in metropolitan areas. During the first half of the 1970s that fraction dropped to 64 percent. Using an intermediate estimate of approximately 71 percent of all future growth occurring in metropolitan areas over the next two decades, produces an estimate of 27.2 million persons being added to metropolitan areas dur-ing the period 1980 to 2000. This is the same numerical gain as took place during the decade of the 1950s — the period of the greatest suburbanization of population in American history. Furthermore, most of the metropolitan population growth will not occur in the Northeast and Midwest but rather will take place in the South and West where cities, both existing and future, are likely to be laid out in very low density patterns.

In the early 1980s, many households without children will be forming as was the case in the latter 1970s. As time passes, household formation will re-main at this level but an increasing number of households will be producing children who soon will be of school age. This will occasion a housing search oriented towards suburban, low-density housing areas where better schools traditionally have been found.

A third factor is that in spite of significant future urban growth, a major-ity of all the urban structures that will exist in the year 2000 are already in place. Even if there could be built an average of two million new housing units per year for the next two decades, some going towards replacement and the remainder for net additions to the existing stock, 63 percent of all housing in the year 2000 would consist of units that are in place today. How we manage the housing that we already have is as, or more important than how we manage additions to the stock; although the latter also will be significant insofar as influencing energy use is concerned.

Energy and Land Use

There are three major scenarios concerning future urban growth vis-a-vis energy availability. One assumes higher fuel prices yet continuous fuel availability. The second also assumes higher prices but, in addition, recognizes that there may be intermittent periods of acute fuel shortages. The third also assumes higher fuel prices, yet couples this with prolonged absolute shortages requiring cutbacks of 25 percent or more in gasoline consumption. The first scenario can be met by changing transportation patterns and some conservation of home heating/cooling without noticeable resultant shift in urban settlement patterns. Absolute shortages will impact settlement patterns especially if there are prolonged dry spells requiring severe energy cutbacks. Under those circumstances, people will be unwilling to locate in places where they may be isolated in the absence of gasoline. However, we will not begin to rearrange our settlement patterns until after such shortages have not only taken place but have remained in existence long enough to convince us that they are going to be there for the foreseeable future.

This is related to a second energy factor. The impact of settlement forms and travel patterns for energy use is potentially large. Twenty percent of all energy consumed in the United States is for the heating or cooling of residential and commercial structures; 26 percent is used in transportation. But the impact of energy savings upon the costs of building new urban settlements or of even relocating others are not nearly as significant. So cost considerations alone will not cause developers to radically alter new development locations, even if energy costs rise very sharply.

A third energy factor is that significant savings of energy can be accomplished in ways other than behavioral changes involving urban settlement. One of these is alterations in building construction or site orientation. This can save significant amounts of energy connected with heating and cooling structures without any basic shifts in density or urban settlement patterns. New houses are being built experimentally without furnaces in climates as far north as Providence, Rhode Island, with savings in energy use of up to 80 percent of normal operational consumption. Similarly, major savings in gasoline can be achieved by shifting travel behavior without changing future settlement patterns. These involves less trips, better trip planning, more car pooling, and shifts to smaller cars. Most Americans appear more willing to change their travel behavior than to change their settlement locations.

Another avenue for saving energy through urban development is by shifting to higher density settlement patterns. This is the one that is most often talked about and yet least likely to happen. In theory, settlement patterns at high density involve less energy consumption. This is true because people are closer to services, shopping and their jobs and additionally, require less heating because they occupy smaller housing units with common walls. Planned in-fill development instead of exurban sprawl also would reduce overall travel. However, if we opt for more compact urban development and if this is done by closely matching up in each year the amount of land absorbed into housing with the amount made available through supplying the land with utilities, then there would be reduced competition among landowners to capture development. This might increase the monopolistic position of landowners and drive up the price of land faster than would occur under sprawl, as the latter involves multiple, competing supplies of land available simultaneously to developers. This tends to limit price increases. Thus, if we try to cure our energy problem by forcing compact development, we may be driving up the price of land and resultantly, the price of housing which occupies that land.

Energy Availability and Settlement Patterns

Given only higher fuel prices and no absolute shortages, there may be some squeeze effect upon far-out fringe growth because of increased commuting and movement costs. However, that depends, in part, on what happens to other costs, especially those related to housing. Housing costs have gone up quite rapidly in the past few years. The median price of existing single-family homes has increased 10 percent compounded annually since 1970, and 17 percent over the past year. If housing prices continue to rise at the latter rate, this may cause households to move farther out to get cheaper housing despite escalating commuting costs. This is true because housing is so much more expensive relative to the costs of commuting.

Commuting costs and housing preferences in the Cleveland Metropolitan Area are being studied at The Brookings Institute. In the Brookings study, the impact on residence/workplace relationships of immediately doubling the price of gasoline from $1.00 to $2.00 per gallon is examined. For commuters traveling the longest distances in the Cleveland Metropolitan Area, if their gasoline prices doubled to $2.00 a gallon, they would add approximately $600 per year to travel costs. A reduction of that amount would cut the price of housing they could afford by about $6,400. Housing prices have been rising, on average $5,700 per year for existing homes since 1970 and are up $7,000 per unit for last year alone. If housing is cheaper far out than close in for the same size and type of unit, people may still move to peripheral areas to get lower cost housing despite price increases in gasoline. Similar evidence of peripheral price savings exceeding increased costs of commutation also could be cited for the Los Angeles Metropolitan Area.

A second conclusion regarding energy availability and land use patterns is that there will undoubtedly be much more use of public transit in the future — the bulk in the form of highway-oriented public transit. There will be sufficient interruptions of supply to cause more people to turn to back-up transportation so as not to be isolated in the absence of gasoline. Since 1977, according to the *Annual Housing Survey*, 71 percent of all household heads drove to work alone; another 16 percent were part of carpools. Only 5.4 percent used mass transit. Therefore, if only 6 percent, one out of sixteen, of all persons now using an automobile for the work trip shifted to mass transit, this would double the current patronage of mass transit.

It does not take a very large number of people shifting away from the automobile as the primary work trip mode to cause severe demands on existing mass transit systems. On a two-dimensional grid, most of this added patronage will involve circumferential or horizontal trip destinations, not vertical trips to the downtown core. The central business district employs a very small percentage of the workers in metropolitan areas. Thus, new mass transit demand will not take place for fixed-rail transit serving existing corridors but rather for on-road vehicle systems, busses, vanpools, etc. serving diverse locations. There will be great pressure to extend public transit to exurban areas not now served well by this form of transportation. Since public transit is not self-financing, a shift in transportation resources from gasoline funds to mass transit financing is also very likely. We still have to pay a great deal of attention to roads and highways because that's where both new mass transit additions and an increased number of more efficient automobiles will be running.

A third conclusion is that travel patterns (how and how often one goes to work) will be altered in response to higher energy prices before settlement patterns (the place from which one goes to work) are altered. The main adjustment will continue to be in travel patterns unless prolonged shortages of gasoline take place.

A fourth conclusion is that select older neighborhoods, in now declining cities, will be revived by rising demand for housing but this will not necessarily lead to a general revitalization of the cities in which these neighborhoods are located. It is misleading to look at booming downtown office space markets and the regentrification of a few older neighborhoods in large cities and to conclude that an overall revival of cities is now underway. Places like New York, Philadelphia, New Orleans, Chicago will lose population and tax base despite the revival of some of their more exclusive neighborhoods. A few cities where decay was minimal in the first place, like Portland and Seattle, are having enough revival to effect positively the whole city's tax base and economic balance. Still others—Newark, Detroit, East St. Louis—will experience little revival in any of their neighborhoods. The tremendous diversity of American cities makes any generalization on how much revitalization is going to occur, very difficult. Nevertheless, select revitalization of individual neighborhoods will take place in several of the classic older cities.

A final conclusion about future urban settlement is that any major increase in density will be more likely to take place in the middle band of suburban locations rather than in second order cities or abandoned portions of large cities. More new jobs are being created in suburban areas than ever before. Outer areas have more vacant sites with fewer unfavorable adjacent environments, and the development of such areas perpetuates the socio-economic segregation to which many American households have become accustomed.

Metropolitan Form and Increasing Fuel Prices

JOHN F. KAIN

Introduction

A survey of the available evidence about higher energy prices indicates that there is considerable uncertainty about what the future world price of petroleum is likely to be. Concurrently, it turns out that this range of uncertainty is much narrower than one might surmise by looking at the popular discussions of the issue. It's very difficult to find responsible analysts who would predict more than a doubling in real terms of petroleum prices over the next twenty years. The outer bounds are somewhat on the order of a three-fold increase and there are respectable arguments that suggest that world petroleum prices in twenty years could actually be lower than they currently are today.

The Real Price Increase of Gasoline

The first point to be discussed is the doubling of the real price of petroleum. What is meant by real prices must be emphasized. The point that is being made here is that the price—the world price of petroleum—corrected for inflation, is likely to be no more than double what it is today. Now that's very important because in periods of rapid inflation such as has been the case for the last decade or so, it's extremely easy to lose perspective on price increases. For example, many of us fail to recognize that the real price of gasoline in 1979, before the most recent upsurge, was 5 percent *lower* than it was in 1960 and 12 percent *lower* than in 1940. When one talks about one dollar (per gallon) gasoline, it must be recognized that this is really 50 cent gasoline relative to a decade ago.

Further, when we talk about prices being twice as high twenty years from now, we're talking about a doubling in real terms rather than in nominal or money terms. Even so, doubling is, at first blush, a big increase. It turns out that the more you examine this doubling, the more you find out that it is really a much smaller increase than it actually appears. The implications of this doubling of fuel price are really much less than one might expect and further, translate into much less than a doubling of the cost of car ownership. For instance, a doubling of the price of crude oil translates into a 30-percent increase in gasoline prices at the pump. This is a reflection of the fact that there are a number of items that go into gasoline pricing in addition to crude oil prices. When you pare this 30-percent increase by the fact that gasoline and oil consumption account for only 30 percent of the total annual cost of owning and operating an automobile, a doubling of crude oil prices translates into, at most, less than a 10-percent increase in the real cost of automobile ownership. This assumes no increase in automobile efficiency, i.e., that the fuel consumption of automobiles will be similar to what we have right now. Of course, this is not going to be the case. Significant increases in the fuel efficiency of automobiles is technologically feasible. Indeed, federal efficiency standards are such that automobile operating costs in ten-to-twenty years could well be lower than they are today even with a doubling or tripling of crude oil prices. In fact, an unintentional consequence of the federal fuel standards is that automobile ownership and the per mile costs of automobile operation may be lower in ten years than they are today.

Increasing Fuel Prices and Metropolitan Form

A general conclusion is that anticipated increases in crude oil prices are not likely to have much of an impact on auto ownership/use or on patterns of metropolitan development. This conclusion is further strengthened when it is realized that the out-of-pocket costs of automobile ownership represent only a portion of the transport costs associated with urban travel. The value of the time spent in commuting, which has increased steadily over the last century, is equal to, or more important than, the daily transportation cost outlays. For example, since the end of World War II, the hourly time costs of the journey-to-work have increased by close to 75 percent. Recognition of this is largely responsible for the fairly steady shift which has taken place from transit to automobile use within metropolitan areas. This is so because the automobile has been able to reduce journey-to-work trip time, in some cases while increasing allowable commuting distances.

In projecting American adjustments, we are fortunate that we can study the several foreign countries that have had much higher gasoline prices than the United States for a considerable period of time. If their historical adjustments to fuel price increases are examined, the previously-stated general conclusion regarding the probable adjustment by American households appears even more valid. To give some indication of the present magnitude of existing differences in fuel pricing at the international scale, in February 1979, the price per gallon of gasoline in the United Kingdom was twice as high as the price in the United States, while gas prices in Belgium, Denmark, France, Germany, Italy and the Netherlands were between 2.7 and 3.2 times the price in the United States. Thus, from an international perspective, there are locations where gasoline prices are currently much higher than they are in the United States and this has been the case historically. Both casual empirical analysis and careful econometric studies reveal that these higher gasoline prices have had a major impact on the size and fuel efficiency of private cars in these countries but they have had virtually no effect on the levels of automobile ownership per capita and only modest impact on the numbers of vehicle miles traveled.

Another reason why you might not expect much in the way of settlement pattern impact from higher petroleum prices is the fact that transport costs and accessibility are only two of the factors that are responsible for the trends in urban development which have taken place over the last century. In fact, increases in both per capita income and household income are far more fundamental determinants of metropolitan growth patterns than are the basic costs of transportation. The rapidly-rising increases of auto ownership and low-density rather than high-density living are both consequences of higher per capita and household incomes.

Since per capita incomes are expected to continue to increase over the next twenty years, albeit at a slower rate, the past trends in auto ownership, journeys to work and metropolitan development are also likely to continue unabated. For example, recent projections of automobile ownership, which assume a doubling or more of world petroleum prices by the year 2000, suggest an increase in auto registrations per capita from .50 to .60 by the year 2000. These projections and most other projections of this kind indicate sustained steady increase in automobile ownership.

Other analyses which employ econometric models of auto ownership and use, also indicate that rising per capita incomes produce very rapid increases in both auto ownership per capita and vehicle miles traveled. Sensitivity analyses, examining the effects of higher gasoline prices vis-à-vis higher levels of income, indicate that future income growth is a much more important factor affecting automobile growth and use than is the future level of gasoline prices. For example, a comparison of automobile ownership and use projections for two alternative cases, one assuming high future gasoline prices and flat income increases, the other flat gasoline prices and increasing incomes, reveals that a 40-percent reduction in the growth of real per capita disposable income would have much more of an impact on future auto ownership rates than a 300-percent rise in real gasoline prices. The impact of petroleum price increases on automobile ownership and use, for all of the aforementioned reasons, is very, very modest at best.

The type of analysis discussed above concentrates on the relationship between accessibility and metropolitan development yet neglects another line of possible influence i.e., the effects of higher heating, lighting and other power requirements on the geographic placement of both residential and nonresidential structures. While this particular problem has not been examined to the same extent as the accessibility issue, it is very probable that the impact of higher heating costs on metropolitan development trends is also likely to be small. Higher prices will encourage fuel source switching and encourage both firms and households to conserve energy. Builders will be more conscious of the siting of their properties, use more insulation and other forms of structure sealants, de-emphasize glass surfaces or when glass is used, employ double glazing or reflective/tinted surfaces, etc. There are numerous technological improvements that can be made in the design, construction and arrangement of individual structures that will produce significant reductions in fuel costs.

Conclusion

It is more likely that these are the types of changes that will take place and less likely that there will be major shifts in the spatial arr ngement of metropolitan areas as a consequence of higher fuel and other energy costs. Similarly, the design of home heating systems, of various types of energy-consuming appliances and lighting will emphasize energy efficiency. As with the automobile, increased energy efficiency will greatly dilute the impacts of higher energy costs. Modest decreases in the size of residential and nonresidential structures will more than compensate for any of the remaining potential energy increases.

In the long-run, however, higher incomes will enable Americans to buy more of everything including the peripheral, lower density residences that the bulk of them seem to prefer.

Solar Energy, Land Use, and Urban Form

T. OWEN CARROLL AND ELISSA B. UDELL

Introduction

HISTORICALLY, neither land use planners, developers and regulatory bodies who affected the form of land use development, nor utilities, oil companies and manufacturers of automobiles and appliances who determined the structure of the energy system have considered the land use implications of producing and consuming energy. This is not to say that the two areas of decisionmaking did not interact. The current patterns of land use in and around American cities could only have developed in the presence of a long and extended period during which various forms of energy were available at comparatively low prices. The existence of "spread city"[1] also depended on the plentiful availability of highly flexible energy carriers such as oil, natural gas and electricity. At the same time, a centralized energy production-distribution system was developed in response to demands imposed on the energy system by the structural elements in land use. These elements require the use of the automobile and life styles compatible with large amounts of personal space both inside and outside of the home. In the sense that these patterns of land use development were not constrained by energy prices and the forms of energy delivered, there was little need to focus attention on the land use implications of alternative energy systems.

156

Past conditions in which energy and land use were not substantially tied to one another have been replaced by a new set of factors which suggest important interactions between the energy system and land use planning. Following the "energy crisis" in 1973-74, we began as a nation to adapt to a world in which the supply and price of energy was subject to change.

In order to determine the influences that the energy situation will impose upon urban form, it is important to recognize two distinct characteristics of this "energy crisis" and the continuing energy situation in the nation now. First, while there may be temporary problems with the availability of gasoline for autos, natural gas for certain industries, or heating oil, these shortages tend to appear sporadically and often result from special circumstances involving the movement of particular fuels. Viewed in the overall context of the energy situation, there is *not* a shortage of energy but rather the development of a permanent condition in which energy in the forms we use it will be priced well above current costs. There is not a lack of energy, but the energy which we use will be expensive even by today's standards. Second, energy technology is undergoing rapid change. At the same time, the costs of utilizing traditional fuels are rising rapidly, there is an equally rapid transition toward more energy efficient devices and changing energy forms. Home heating systems are being improved, while solar technology for home heating is becoming cost competitive.

Over even one decade the energy system is expected to undergo substantial change both in terms of lowered consumption of conventional fuels and a shift toward solar and other technologies. Cities, however, evolve over longer periods of time. Consequently, in asking questions about the influence of today's energy situation upon the future of cities, we must be careful to address questions not of how the city will respond to today's energy situation, but rather how will the city respond to the changing character of the energy system itself. Today's transitory state of the energy system may have little effect upon the long-term future of urban form, whereas that configuration of the energy system which evolves ten years from now may have a far more vigorous and lasting impact upon urban form. It is this last view of the interaction between energy and urban form which we adopt in this paper.

Cities in Transition

A city is an area of relatively dense population, with some form of interdependence and specialization of functions. The earliest known cities consisted of clusters of dwellings which were situated near river deltas. These cities were dependent on agricultural systems which were sustained by human and animal labor. City sizes were constrained by land and labor limitations, reflecting a lack of fossil fuel energy use, which later prompted industrialization.

The first industrial cities formed around the year 1800, in England. They are characterized by widespread communications, mass education, and shifts of some residents to the city outskirts.[2] Mechanized factories were built for mass production of goods, and an occupational structure based on highly specialized knowledge emerged. Mass industrialization was initially dependent on the use of coal and coke, since Britain had very little wood. Later, dependency switched to oil.

At the same time, in America, small urban centers were established along the Atlantic coastline. By the year 1700, the town of Boston had a population of 7,000, the largest in the United States, while Philadelphia and New York each had populations of approximately 4,000. By the end of the eighteenth century, 5 percent of all Americans (about 250,000 people) lived in twenty-four urban places, of populations of 2,500 or more.[3] These early Americans met their energy needs primarily by burning wood which was readily available.

During the eighteen hundreds, significantly more urbanization occurred. The populations of Boston, New York, Philadelphia and Baltimore expanded to more than one-half million. At the same time, coal became the main source of power on the commercial level. Railroads, particularly, consumed large quantities, allowing people and goods to move easily between cities.

Between the years 1890 and 1930, a population shift began to occur. People who previously lived within cities began moving out into suburban rings. Oil, which was more easily distributed and clean-burning, began to replace coal as the predominant industrial, commercial and residential fuel. Between 1920 and 1930, the increasing use of gasoline-powered automobiles and trucks stimulated growth rates of the suburban belt to first equal, then exceed those of the core cities. Since then, the gap in growth rates has progressively widened in favor of the suburbs.

From 1940 to 1950, suburbs grew at twice the annual rate of central cities.[4] Between the years 1950 and 1960, the population of the nation's suburbs grew at the rate of 5.5 percent while the central core cities grew at only 2.8 percent.[5] By the end of the year 1966, more people lived in the suburban belt than in the city core.[6] Because suburban land costs are less than in cities, mortgages have been more readily available, and the tax burden generally lower, suburbs almost totally dominated metropolitan area employment and population growth in the recent past, while the central cities suffered serious economic decline.[7]

Most suburban industries depend on trucks for transport of goods.

Lower transportation costs are realized at outlying sites, especially for locations near major interchanges of metropolitan freeways. Because many industries and businesses have relocated in the suburbs, suburban commuting to central cities is declining as a proportion of all work trips.[8] Instead, many suburbanites choose jobs close to their homes and use the automobile for transport. The automobile itself played a major role in affording individuals the option of leaving the cities and moving to suburbs. It is preferred to other forms of transportation, such as the railroad and bus, because it permits privacy and point-to-point convenience. In addition, it allows access to the many business districts and employment opportunities which are located off the route of main public transportation networks in suburban areas. These recent patterns of business and residential relocation to the suburbs were encouraged also by declining real prices for most fuels over a longer period.[9]

The trend toward the suburbs, and even exurban migration to nonmetropolitan areas, is expected to continue. Various reasons for this are: the transportation and communications advantages of central cities have declined severely; computers, data links by phone and remote copy machines speed the flow of information which once passed by hand; and cable television and other audio and video systems promise homeowner use for shopping as well as business conference applications.[10] Experiments in moving clerical jobs to suburban satellite offices have proven successful.[11] Recent energy shortages and the evolution of the energy system itself will create new pressures influencing urban-suburban migration.

The Energy Situation

Perhaps the most difficult problem we face today is that the energy system is changing. What will be its form in the near future? Will fossil fuels be replaced by nuclear electricity? Will solar power become widely utilized? Will oil continue to be available for transportation needs or will we shift extensively to alcohol or to electricity? These are questions which are difficult to answer with precision. However, it is possible to identify the broad outline of elements in the energy system of the future.

Conventional energy technology continues to increase in cost. Only a few years ago coal-fired plants cost about $300 per kW and delivered electricity to customers at about 3 cents per kWh. Environmental emissions controls for coal-fired plants have added substantially to these costs.[12] In addition, the cost of coal has increased from 10 dollars per ton in 1973 to about 30 dollars per ton in 1978.[13] The nuclear power industry is experiencing the same type of cost escalation, especially in the area of captital equipment. From just a few years ago the cost of nuclear construction has almost doubled to about $1,500 per kW.[14] Recent safety additions which will be considered for nuclear reactors are likely to add to these costs. Construction costs for both labor and materials are continuously increasing. Site delays and environmental impact statements for power plants are subject to lengthy approval procedures which increase costs substantially. The same types of cost-escalating factors occur in the energy resource development process as well. Oil prices are approaching $30 per barrel and are expected to double in the 1980s. Oil exploration and field development costs have risen rapidly as well as those for coal field development, well drilling and geothermal plants. These all reflect the fact that most opportunities for the easy exploitation of the earth's energy resource base have been utilized and that further expansion of conventional energy resources will become increasingly costly. While some advancements in technology are expected to lower the cost of exploration, development and technology for utilizing conventional resources, most of the conventional fuels' energy system represents technology which has matured over a long period of time and is unlikely to undergo any substantial cost reduction.

Solar technologies have recently entered the energy picture in substantial ways. Solar thermal home and commercial heating systems are readily available from a number of manufacturers.[15] At current prices, the solar thermal systems are already cost competitive with electric resistence heating in most areas of the country, though they remain somewhat more expensive than oil and/or natural gas-fired systems.[16] Solar photovoltaic systems hold promise for electricity generation. As more efficient methods are found to handle the semiconductor materials, the costs are expected to continue to decline. Also, semi-conductor technologies lend themselves to potential "breakthroughs" which could provide substantial cost reductions. Texas Instruments recently announced such a breakthrough in the construction of photovoltaic devices, and it anticipates marketing home electricity systems by the early 1980s.[17] Though some of the materials and labor costs for solar thermal systems and photovoltaic energy devices are increasing, there are substantial economies of scale and technological innovation yet to be realized within the solar industry. Solar technologies are early-on in their "learning curve" and it seems most reasonable to expect continuing declines in the cost of solar systems. The recent success of major firms including Grumman, Reynolds, and others, as well as the large number of smaller firms operating in the solar market adds to this expectation. Solar thermal and photovoltaics can be expected to play a major role in the energy system in the future.

Large-scale central solar systems, the heliostat for example, receive continuing attention from utility companies.[18] These systems use large arrays of collectors which focus on a single source to produce high temperature steam to drive turbines. There are no commercial systems available today, though there are several small demonstration projects. This type of electricity generation requires tracking of the collector system, high temperature boilers, and other advanced technology which is subject to an increasing cost. For the most part, these systems require large amounts of land and are expected to be placed well outside urban areas. If these large-scale solar systems evolve they should have relatively little impact upon the form of the city.

There is a variety of other solar technologies which may play a part in the energy system. Biomass, that is agricultural residue and human and animal waste, can be used under some conditions for direct burning, methane production, or alcohol production. The cost of producing alcohol is currently well in excess of market prices for oil under even the best conditions and is expected to remain so.[19] Its only real advantage is in the utilization of waste which otherwise has no economic value. Wind energy is quite suitable and appropriate in remote locations. However, even in high wind coastal areas the potential generation is quite small even for household needs.[20] Geothermal energy resources are utilized in many countries, wherever economical sources have been found.[21] Ocean thermal energy conversion utilizing the temperature gradient in the oceans has potential, but requires the development of a wide range of new engineering and construction skills to carry out practical electricity generation.[22] All of these solar-related technologies will be important but probably small contributions to a diversified energy system based upon renewable technologies.

In asking where the energy system is going in its evolution, the inescapable conclusion is that the future energy system will be much more substantially based upon solar technologies, rather than today's reliance upon fossil fuels. Technologies for utilization of conventional fuels have matured and seem unlikely to undergo cost reduction. However, solar technology even today is becoming competitive with conventional energy and offers the expectation of both economies of scale and technological innovation to substantially reduce its costs. Like all new technologies, it becomes important to ask questions about solar energy's eventual impact upon the form of the cities. And, given the surprises of the past involving technological innovation—the way the world changed in response to the introduction of rail travel, the automobile, television—there is a suspicion that solar energy technology may have far more substantial and significant impacts in the future than now imagined. Consequently, in this paper, we will speculate about some of the more far-reaching potential impacts of solar upon the evolution of cities.

The Equivalence of Land and Energy

Solar energy is a distributed resource. Northern areas in the United States receive 60 percent as much insolation as more solar intense southern regions.[23] As a result, the potential for effective utilization of solar energy is far more dependent upon local climate and physical environment than upon general location in the country. Because of its low-energy density, large collector surfaces are often required in practical application of solar energy. The built environment itself acts as both receptor and barrier to insolation. It may provide physical structures upon which to mount collector surfaces or, alternatively, structures which eliminate the solar potential of the surrounding environment.

Many influences shape the form of the built environment. Building type, whether residential, commercial, institutional, or industrial, has much to do with size, height and volume enclosed per unit roof area. This volume-roof area relationship is a critical determinant of the potential to satisfy building end use energy demand with solar. Roof mounting of large collector arrays is easiest for buildings with large roof areas relative to their enclosed volume. Zoning laws and land use regulations can be used to facilitate the collection of sunlight with respect to building height, side yard restrictions, structure orientation, population density, etc. These regulations may be used to encourage solar energy development by requiring communities to look ahead and plan, so that building and land use relationships promote solar energy use. Variations in the layout of streets also affect building orientation and spacing. Land devoted to the automobile, including streets, parking lots, driveways, represents a large fraction of urban and suburban land use. Transportation policy will, therefore, have an effect on potential land areas for solar collection. These aspects of the built environment suggest a wide range of interaction between land and solar energy systems.

To obtain a concrete portrayal of the effect of widespread utilization of solar energy on the mix and spatial arrangements of land use activities at the community level, we analyze the land use requirements of a "self-contained" community designed to rely exclusively on the use of solar energy for meeting its stationary energy demands. The introduction of on-site energy technologies such as solar and wind energy, is often assumed to occur within the context of purely residential and/or commercial development. Industrial development is, by and large, ignored. While such examples are certainly representative of existing land use patterns and are useful for estimating the potential for the introduction of such technologies, they are of limited use in considering integrated communities. It is often the industrial sector, for example, which provides basic employment for the community residents. To ignore the character and magnitude of the industrial demands produces, in our view, somewhat misleading results. Therefore, we define our community as an integrated residential-commercial-industrial area. This self-contained community is not intended to represent in any sense, a prototypical community. It is, rather, a construct in which we include an agglomeration of residential dwelling types, commercial and public services, and industrial activities sufficient to house and employ the population, to provide it with public services, and to satisfy its shopping requirements. Consequently, the results of the analysis of utilization of distributed energy technologies in the community is not meant to predict how, or if, these particular technologies will be employed. The analysis is fashioned exclusively to focus attention on the dynamics of the interactions between the utilization of solar energy and land use, particularly the transition processes involved in moving to regional land uses consistent with distributed energy, urban redevelopment and the role of central energy systems, and increased requirements for planning on a regional level.

The community land use design chosen for analysis is based on an average population density of 6,300 persons per square mile, which is typical of that found in suburban communities in the San Francisco Bay and Los Angeles Metropolitan areas. The population of the community is 10,000 persons. This particular figure was chosen primarily for convenience. Family size is typical of suburban development, ranging from four per household in single family to two per household in high-rise structures. The community itself occupies some 925 acres of which about 2/3 are in residential development and the rest is split between commercial and industrial activity. We assume a housing mix which is 60 percent single-family, somewhat below that typical of present suburban development. The commercial and industrial employment and floorspace requirements are established under existing planning design criteria, as well as overall requirements for the infrastructure of supporting roads, utility rights of way, schools, parks, and other public facilities and services.[24] The overall land use requirements have been derived assuming fully saturated use of land. Saturation typically runs from 20 percent at suburban fringe areas to above 90 percent for inner-city areas. Were the population of the community to be scaled up to 50,000 persons and above, the overall proportion of land use would remain much the same, though some shifts toward additional high-rise housing and the formation of downtown shopping areas might be expected.

Residential heating and cooling loads in Exhibit 1 are based on new construction practices and 1,500 square feet of living space on 1/4 acre-zoned parcels. The values are typical of those used for design and evaluation of solar installations in areas of California.[25] Commercial energy demands are somewhat similar to those for residential structures, with the exception that construction practices normally lead to heating demands double those in the residential sector and cooling demands considerably larger than for residential structures. Industrial energy demands are somewhat more difficult to estimate because they are dependent on the type of manufacturing industry, its specific plant site, industrial process used, and equipment employed in the manufacturing process. However, energy input/output analysis can be used to establish an approximate breakdown of process heat, space heat, air conditioning, electricity, and other industrial needs.[26]

For solar thermal applications, the collector and thermal storage and its exchange systems are sufficiently simplified that they can be characterized quite easily.[27] However, in the consideration of integrated solar heat/cool/electricity for which turbine cycles may be chosen at a wide variety of temperatures and associated efficiencies, the thermal/electric load balance becomes a determinant. Mixed thermal/photovoltaic systems are conceptually simpler and yet have similar collector area requirements. While such considerations are critical to the design of the energy system for a specific community, they are less important in a first estimate of the overall land use requirements to support the energy system. Consequently, we adopt a straightforward design example for solar thermal and photovoltaic electricity and use the appropriate efficiencies for devices associated with this system.[28]

In this analysis, we assume short-term thermal storage. Should seasonal storage become available, the daily-average design becomes one utilizing average-annual energy demands. Since energy collection and storage then extend throughout periods of modest heating and/or cooling demand, the collector areas for residential and commercial sectors would be reduced by 40 percent. On the other hand, large-scale seasonal storage to support industry needs is unlikely, and as a result, total land use requirements for the community energy system would be reduced by less than 20 percent through seasonal storage. Consequently, land use implications of solar energy, while not insensitive to the particular technical features of elements of the community energy system, are determined for the most part by the need for large land areas to harness substantial quantities of solar energy.

The relationship between the community land use allocation and land re-

Exhibit 1

SOLAR COLLECTOR REQUIREMENTS

	Load (24 hour average demand per unit floorspace)			Collector/Floorspace Ratio	
	Cold Climate (5000 DD heat)	Warm Climate (3000 DD heat)	Electricity	Cold Climate	Warm Climate
RESIDENTIAL					
Single family	.91 kWh/m²	1.2 kWh/m²	.11 kWh/m²	.7	.8
Multi family	.66	1.4	.11	.4	.9
COMMERCIAL					
Retail	1.5	5.0	.19	1.1	2.6
Office	1.6	4.4	.19	1.2	2.4
Other	1.4	3.6	.19	1.1	2.0
INDUSTRY					
Light Manu-facturing 350°F		.4	.27	1.0	.8
Heavy Manu-facturing		.5	.34	2.3	1.0

Source: Based on data from T. Owen Carroll, et. al. "The Planners Energy Workbook." BNL/SUNY Report No. BNL 50633. Upton, NY: Brookhaven National Laboratory, June 1977. Solar insolation (24 hour average per unit collector area) taken as 6 kWh/m² with thermal efficiency .4 and photovoltaic efficiency .07.

quirements for solar energy usage based upon solar thermal/electric technology supplying all homes, businesses and industrial plants on an individual basis, is illustrated in Exhibit 2. The community bounded by solid lines is divided vertically into land use sectors and horizontally into the categories of land use activity within each sector. The built space is denoted by shading, and shaded plus unshaded areas represent gross acres allocated to specified land use. Collector areas, shown by heavy dotted lines, exceed the built space in most of the land use sectors. Indeed, total collector requirements approach or exceed the gross acreage in use in all areas, with the exception of single-family residential areas.

Several immediate conclusions can be drawn from Exhibit 1 and Exhibit 2: (i) solar energy land use for collector space is modest but conflicts with the built environment in many instances. For example, integrated solar system collector areas for shopping malls exceed built floorspace by a factor of 2 and would cover virtually the entire gross acreage dedicated or zoned for mall development, (ii) on a community basis, large areas must be dedicated to collector space. Community central energy systems would require, for example, a minimum of 400 acres dedicated to collector area, or 40 percent of the land area of the entire community, (iii) satisfying industrial energy needs, particularly for process heat applications, constitutes a large component of the land use demands within a complete self-contained community. Solar energy systems place demands on land use that clearly do not easily fit into present land use planning design criteria.

At the community scale, this integration of solar energy systems into the built and natural environment creates problems in terms of available land area which has been the subject of several studies. Comparison between studies is somewhat difficult due to varying assumptions about the mix of residential, commercial and industrial land uses, densities and the type of solar system used. However, Exhibit 3 displays various community characteristics and collector areas required for six modeling efforts.[29] A large range in results is evident. Collector area requirements at higher densities represent a greater proportion of total land area, but are similar in terms of amount of collector area required per person at lower densities. The inescapable point is that providing a community's energy needs with solar energy will require significant amounts of land. The Office of Technology Assessment found that 40 percent of the total land area would be covered with collectors to supply thermal energy and electricity to a residential-commercial community; most of this area could be found on roofs, parking lots and streets.[30] These results confirm the community analysis discussed above and suggest that land use density is an important determinant of the viability of solar energy.

EXHIBIT 2

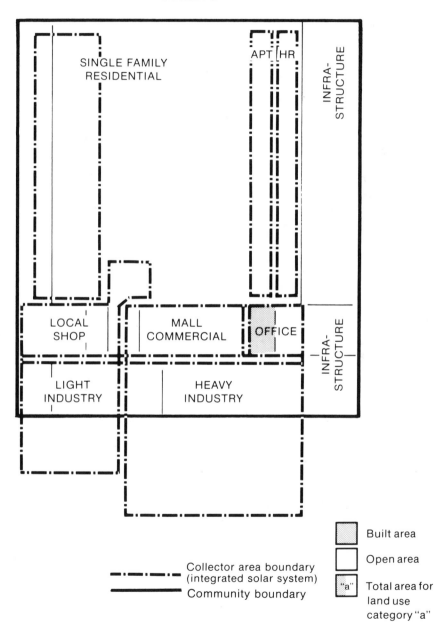

SINGLE FAMILY RESIDENTIAL

APT HR

INFRA-STRUCTURE

LOCAL SHOP

MALL COMMERCIAL

OFFICE

INFRA-STRUCTURE

LIGHT INDUSTRY

HEAVY INDUSTRY

Built area

Open area

"a" — Total area for land use category "a"

Collector area boundary (integrated solar system)

Community boundary

Exhibit 3

LAND USE CHARACTERISTICS OF HYPOTHESIZED SOLAR COMMUNITY ENERGY SYSTEMS

Researcher	Solar System	Community Characteristics	Population	Land Area (Acres)	People*/Acre	Dwelling Units/Acre	Collector Area	Collector Area/Person	Collector Area/Total Land Area
Carroll/Nathans (10)	STES**	Residential 640 acres Commercial 120 acres Industrial 165 acres (California)	10,000	925	15	5 (4 to 20)	1,491,000m²	149.3m²/p.	40%
Marshall/Sandia (8)	STES	1,000 sfd (New Mexico)	2,800	352	8	3	28,000m²	12m²/p.	2%
Harrigan/Sandia (9)	STES	2,000 sfd (New Mexico)	6,580	226	29	10 (6 to 54)	81,300m²	12m²/p.	9%

Exhibit 3 (Continued)

LAND USE CHARACTERISTICS OF HYPOTHESIZED SOLAR COMMUNITY ENERGY SYSTEMS

Researcher	Solar System	Community Characteristics	Popu-lation	Land Area (Acres)	People*/Acre	Dwelling Units/Acre	Collector Area	Collector Area/Person	Collector Area/Total Land Area
OTA (7)	STES	Single-family dwellings(sfd) 1864 bldgs; 8-unit townhouse, 232; 36-unit low-rise, 72; 196-unit high-rise, 20, shop. ctr, 1 (Albuquerque, New Mexico)	30,000	640	46	16	900,000m²	33m²/p.	35%
Kahn/ LBL (11)	District Space & Water Heat	Residential & Commerc. bldgs (California)	1,000	40	25	9	6,000m²	7.4m²/p.	4%
Lovins (12)	District Space & Water Heat	1,000 sfd (California)	2,800				4,000m²	1.4m²/p.	

* When necessary, a conversion of 2.8 people per household was used (1978 U.S. Bureau of Consensus)

**Solar Total Energy Systems

Spreading "Spread City"

Since the use of solar energy to meet stationary community needs creates some competition between the energy system land requirements and traditional land use activities, the density of land development becomes a critical issue. In Exhibit 4, for example, we show per capita household consumption for changing density in the New York area.[31] At higher densities in the urban core, per capita consumption declines because floorspace per capita decreases and attached, multistory buildings require less energy input per unit of floorspace. This reduction of space conditioning requirements is common to all urban areas. The solar energy system, however, is limited by collector areas and can only supply decreasing portions of urban energy needs at higher densities, as shown in Exhibit 4. Above about 6,000 persons per square mile, the site space occupied by the household becomes insufficient to meet collector needs. The urban core therefore must continue its reliance upon the import of appropriate energy forms. Consequently, whereas distributed solar power lowers energy costs in suburban areas, the urban center continues to draw upon older conventional technology, or, possibly, begins to draw from remote, centralized, large-scale solar facilities. These are both expensive options. Solar energy, therefore, becomes most cost effective and enhances household ownership economics in the suburbs.

Up to this point, we have ignored the transportation requirement, also associated with land use density. The location decision of individual households is assumed to consider both housing and transportation costs, and these reflect to some extent the levels of energy consumption for each of these activities. In Exhibit 4 we show energy consumption per household for transportation.[32] Again, while these are figures for the New York area, they can be expected to apply more generally to other urban areas. Because trip lengths are shorter and the modal choice involves heavier use of mass transit, per-capita transportation energy consumption decreases in the urban center. The net residential energy need minus the solar contribution plus transportation energy consumption, represents the total commercial energy requirement per household. We notice a minimum in the range of 5,000-8,000 persons per square mile. At higher densities the solar energy contribution declines and, consequently, must be replaced by conventional energy forms. At lower densities, though residential needs are met by solar energy, the transportation requirements associated with dispersed patterns of land use increase conventional energy requirements.

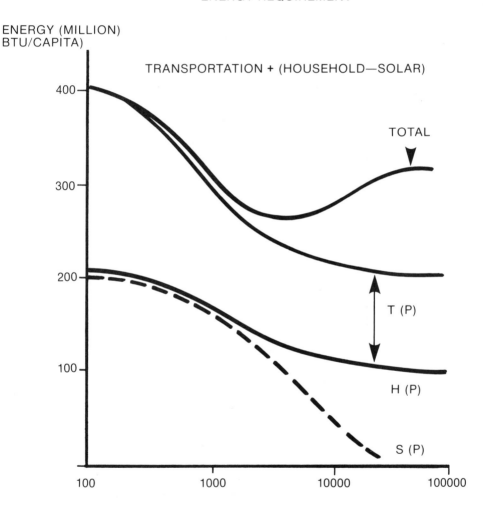

EXHIBIT 4
ENERGY REQUIREMENT

ENERGY (MILLION)
BTU/CAPITA)

TRANSPORTATION + (HOUSEHOLD—SOLAR)

TOTAL

T (P)

H (P)

S (P)

DENSITY (PERSONS/SQ. MI.)

Transportation considerations may not provide any restraint on the way in which solar energy adds to pressures for suburban and exurban movement. Increasing energy prices would most likely encourage the trend of business and industry relocation from the central city into the suburbs because transportation energy savings would accrue to the suburban labor pool under such movement. Moreover, even a quick analysis of energy and time spent travelling, shown in Exhibit 5, indicates the important role that travel time plays within the choice of mode.[33] The ranking of transportation vehicles by passenger-miles per gallon is often used to suggest the desirability of mass transit modes of travel, but does not explain current use. However, if the estimated vehicle speeds are multiplied in, we obtain a measure of rapidity of travel (passenger-miles per hour) *and* energy efficiency (miles per gallon) which correlates well with present patterns of transit mode utilization. This "time-energy efficiency" emphasizes the implicit economic value of time which must be counted in decisions on migration to the suburbs. Since travel times are considerably shorter for intra-suburban auto commuters, we might expect continuing dominance of the private auto as the mode of movement. As a nation, the largest time-energy savings in the transportation sector may then derive from the movement of both jobs and people into suburban areas.

The brief look at solar energy within higher density environments suggests ties between the energy system and land use. Also in the community analyses, we find that collector requirements are sufficiently large that the placement of collectors and their interaction with the traditional pattern of land use in a community becomes important. In this sense, it might be said that *through associated land use requirements in a renewable-based energy system, energy demands have been converted to land use demands. As such, the energy system must then compete with all present and projected land use activity and becomes subject to the broad range of economic, political, and social and institutional constraints traditional in governing the use of land.* There is also a range of specific impacts in terms of the need to alter the building codes, subdivision standards and zoning to reflect reduced energy loads for structures themselves as well as consideration of sunrights and other actions which might encourage patterns of land use more compatible with the utilization of solar energy.

Planning Issues

There is presently a lack of interaction between decisions affecting land use and those affecting energy production and delivery. The situation en-

Exhibit 5

TRANSPORTATION "EFFICIENCY"

Energy Efficiency		Time-Energy Efficiency		
Mode	Pass- Mi/Gal	Mode	Mi/Hr	$\left(\dfrac{\text{Pass-Mi}}{\text{Hr}} \cdot \dfrac{\text{Miles}}{\text{Gallon}}\right)$
Commuter Train	120	Jumbo Jet	400	11200
Motorcycle	110	Jet	400	8000
Volkswagen	58	Corp Jet	300	4500
Bus (Urban)	40	Commuter Train	30	3600
Auto (General)	35	Motorcycle	30	3300
Taxi	32	Volkswagen	30	1740
Jumbo Jet	28	Auto (General)	30	1050
Jet	20	Train (Intercity)	50	850
Train (Intercity)	17	Helicopter	60	480
Liner	15	Taxi	15	480
Corp Jet	15	Bus (Urban)	10	400
Auto (Urban)	10	Auto (Urban)	20	200
Helicopter	8	Liner	10	150
Yacht	6	Yacht	15	90

countered in utilizing solar energy represents almost an opposite extreme, where the interaction between mixes of land use activities and energy systems needed to meet their demands becomes critical. These are minimized where energy support systems are exclusively located in or around individual buildings. As long as zoning regulations are such that this requirement will be met, there is considerable flexibility in adapting the mix of land use activities to community needs and desires. However, as soon as we deal with the matching of resources to demand imposed by more centralized energy support systems located on one or several sites separated spatially from where the energy produced is consumed, then the sensitivity of the community production-distribution system to the land use mix increases. Not only must such systems take into consideration the energy demands of existing land use activities, but that which will exist in the future. These differences between the modes of technology and utilization will have a number of implications for regional land use development, and in particular for the transition to a distributed energy technology future.

Local physical conditions will affect the amount of solar energy input both on a local and seasonal basis. It also will determine the extent to which building settings and architectural designs can be utilized to take advantage of passive solar systems. If we assume that a mix of land use activities may come to be dictated by energy considerations, locations having accessibility to the higher inputs of solar energy may come to be reserved for the more energy-intensive commercial and industrial activities in the region. Locations having lesser resource inputs would have to be restricted to activities with correspondingly lower energy requirements such as single-family residences. These considerations may result in land areas having high aesthetic and convenience value being restricted to energy production activities. The results may mean locating industries in areas now reserved for residential and recreational uses. In addition, for a region as a whole, there may not be sufficient numbers of high resource input locations to accommodate the energy demands of the land use activities that normally would be attracted to the region as a result of non-energy related factors. For example, if we calculate the acres of open land required for a number of energy-intensive industries in order for solar systems to provide their process energy requirements, they would need land areas two-to-five times their current requirements. These kinds of industries would be forced to fill the entire open space around the plants with solar collectors. Finally, if there is a scattering of existing structures already present in the region, they may preclude obtaining sufficient acreage to dedicate to community energy facilities.

Focussing attention on the transition from our current system of satisfying regional energy demands to one dependent on distributed solar energy, rather than the outcomes, suggests a different set of land use planning considerations raised by redevelopment of existing communities. It is evident that any transition of the magnitude we are discussing here will result in a renewal of much of the current building stock in existing metropolitan areas. Although this renewal may take place over a period of fifty or more years, it will involve not merely construction in sparsely settled areas, but redevelopment of areas already densely occupied.

Redevelopment seldom has resulted in accessibility to very large vacant areas. It usually involves replacement on a structure-by-structure basis, or on a block-by-block basis. Moreover, where larger areas are opened up, they have been in older downtown areas. A continuation of these renewal practices would restrict the manner in which distributed solar energy could be used to displace centralized energy systems. For example, the land area of the self-contained community of 10,000 people used in the analysis above requires approximately one thousand acres. Such an area is not likely to be available under current redevelopment policies. Nor is there any assurances that the vacant land which does become available will be located on sites of high resource input, due to the local physical terrain or nearby building obstructions. Moreover, unless zoning laws are revised, there may be no possibility of including a sufficient diversity of activities needed to optimize the utility of community solar support systems. Finally, in many "downtown" areas, the existing density of land use is such that insufficient acreage can be set aside for energy collectors. Such concerns can lead to conflicts between community-level and regional-scale planning.

Existing central energy systems and the newer distributive energy systems will have to coexist over an extensive time period. Because of the substantial need to coordinate the activities of the two systems during the transition, the jurisdiction of land use planning boards will almost certainly have to be extended to the central as well as the distributed energy systems. Neighborhoods served by the central or distributive systems may have to be chosen not by their competitive economic advantages, but simply as a matter of public benefit. For example, in the case of certain older neighborhoods containing declining building stock, there may not be merit in trying to meet their energy demands by solar until the building stock reaches a point where it can be replaced. The difficulty with adopting such a measure as general policy is that it may produce a situation in which older neighborhoods in the region end up being served by an unreliable central system which is in the process of being phased out. Consequently, the issue of how communities are to be identified as the first in line, and last in line for conversion to distributed solar energy has considerable potential for producing conflicts between local community groups and the regional planning authorities.

Present strategies for effecting the transition to solar energy rely upon fiscal incentives, to purchase units. In central cities, however, renters predominate, particularly among lower-income groups. Financial incentives favor movement to solar energy among more affluent suburban dwellers who have both available capital and sufficient site space to supply their energy needs. To the extent that solar technology reduces homeowner costs, it acts to encourage suburban migration. Since more affluent groups predominate in this movement, the tax base in the cities declines steeply, including the ability to raise and apply capital for the transition to solar. At the same time, the utilities experience declining load factors when providing only backup to solar users. As their generation picture becomes more difficult and rates increase, the central city, which remains more heavily dependent upon centralized energy technology, feels additional financial burdens. Both the direct energy costs of solar and its secondary impacts upon the utility industry will encourage escape to the suburbs.

A Final Comment

A pattern or scenario of increasing use of solar energy seems to have emerged. The economics of using solar continues to improve, boosted in part by the entry of major corporations into the market. However, collection requires large areas. In high-density areas, considerable control and planning is required to assure solar access; even so, solar has limited potential in the central city. The most rapid expansion is likely to occur in suburban and exurban areas where individual lot access is most easily obtained and the spread of commercial and/or industrial solar collector areas can be most readily accommodated. Housing costs and business energy requirements can be reduced. Suburbs become more attractive. Left behind is the city core, served by expensive centralized energy systems which rely upon traditional fuels. Replacement of equipment or plants will substitute newer centralized systems for the older systems, but with few cost reductions possible. As we move to a solar future and migration from the central cities increases, the economic gap between the suburbs and the urban core can only widen.

There are few factors concerning the energy picture of today and tomorrow which show promise for the continued growth and health of the urban core. Instead, there are indications that the evolving energy system will add to already existing pressures for out-migration from urban areas and expansion of suburbs in America.

Notes

1. Downs, Anthony. *Squeezing Spread City*. Proceedings of the Energy-Sensitive Land Development Conference. Washington, D.C.: 1979.
2. Palen, J. John. *The Urban World*. New York: McGraw Hill, Inc., 1975, pp. 19, 20.
3. Mohl, Raymond A., and Neil Betten. *Urban America in Historical Perspective*. New York: Weybright and Talley Co., 1970, p. 29.
4. Wirt, Frederick M., et. al. *On the City's Rim*. Lexington MA: D.C. Heath and Co., 1972, p. 18.
5. Pell, Clairborne, (Senator). *Megalopolis Unbound*. New York: Frederick A. Praeger Publishers, 1966, p. 23.

6. Wirt, et. al., *Op. Cit*, p. 18.

7. Downs, Anthony. *Opening Up the Suburbs*. New Haven: Yale Press, 1973, p. 17.

8. Downs, *Op. Cit.*, p. 19.

9. Pindyck, R. *The Structure of World Energy Demand*, MIT Press, Cambridge, Mass.: 1978.

10. Golany, Godeon. *Innovations for Future Cities*. New York: Praeger Publishers, Inc., 1976, p. 23.

11. Ibid, p. 29.

12. Landsberg, H. H. *Energy—The Next Twenty Years*. Cambridge, MA: Ballinger Publishing Co., 1979.

13. Munson, J.S., and R. Stern. "Regional Energy-Environment Data Book—Northeast" BNL Report 24867. Upton, NY: Brookhaven National Laboratory, October, 1978.

14. Stephenson, L., and G. R. Zachar (ed.) *Accidents Will Happen*. New York: Harper and Row, (Perennial Library), 1979, p. 207.

15. See, for example, *Solar Age Magazine* (monthly).

16. Bright, R., and H. Davitian. "The Marginal Cost of Electricity Used as Backing for Solar Hot Water System," *Energy*, 4 (1979) p. 645.

17. Soft Energy Notes. International Project for Soft Energy Paths. San Francisco: Fall, 1979.

18. Redfield, D. "Solar Energy and Conservation." Industry Applications Society Conference Record. IEEE-IAS Annual Meeting, 1978.

19. Chambers, R.S., et al. "Gasohol: Does It or Doesn't It Produce Positive Net Energy," *Science* 206 (November 16, 1979), p. 789.

20. Reddock, T.W., and J.W. Klein. "No Ill Winds for New Mexico Utility," IEEE *Spectruym* (March, 1979). p. 57.

21. Howe, J., et. al. "U.S. Energy Policy in the Non-OPEC Third World." Washington, D.C.: Overseas Development Council, July, 1979.

22. Isaacs, J. D., and W. R. Schmitt. "Ocean Energy: Forms and Prospects," *Science* 207 (January 18, 1980). p. 265.

23. "Solar Energy Utilization for Heating and Cooling." ASHRAE Handbook of Applications. 1974.

24. DeChiara, J., and L. E. Koppelman. *Planning Design Criteria*. New York: Van Nostrand Co., 1969.

25. Christensen, M., et. al. "Distributed Technologies in California's Energy Future." LBL Report 6831. California: Lawrence Berkeley Laboratories, University of California, September, 1977.

26. Carroll, T. O., et. al. "The Planners Energy Workbook." BNL/SUNY Report 50633. Upton, New York: Brookhaven National Laboratory, June, 1977.

27. Hewett, R., and P. Spewak. "Systems Descriptions and Engineering Costs for Solar-Related Technologies." Volume II-*Solar Heating and Cooling of Buildings*. MITRE Report MTR 7485. McLean VA: MITRE Corporation, June, 1977.

28. *Op. Cit.*, p. 23.

29. Pollock, P. L. "Planning for the Solar Transition: Solar Utilization and Land Use Intensity." SERI Report JT (2)8-28. Golden, CO: Solar Energy Research Institute, September, 1978.

30. Daddarro, E., et. al. (ed.). "Application of Solar Technology to Today's Energy Needs." OTA Report OTA-E-66. Office of Technology Assessment, U.S. Congress, June, 1978.

31. "Regional Energy Consumption—Second Interim Report." New York: Regional Plan Association, August, 1975.

32. Ibid.

33. Rice, R. A. "System Energy and Future Transportation" in M. G. Morgan (ed.) *Energy and Man*. New York: IEEE Press, 1975.

Direct Use of Solar Energy in the Compact City

PETER LEWIS POLLOCK

Introduction

The long-term consequences of our failure to make the transition to renewable energy resources in our cities may prove unbearable in terms of economic dislocations, unemployment and adverse environmental consequences. (Hayes 1979)

LAND-USE planners and policymakers have a significant role in the commercialization of solar energy technologies. If solar energy is to be used in a decentralized fashion; that is, collected near the point of end use, careful integration of solar technologies and the buildings within which the energy is used must be achieved. Direct conversion of solar energy into useful space and water heat and electricity requires that technologies have adequate area and insolation available to meet end-use needs. Since planners have some control over the development of the built environment, they also have some control over the viability of solar technologies.

Many urban planners and policymakers are skeptical about the practicality of solar energy technologies in urban areas (Basile 1978). They believe that the direct use of solar energy could be more easily accommodated in relatively dispersed cities characterized by urban sprawl. As land-use densities increase, less roof area is available for solar technologies as a function of building volume, and compacting buildings means that solar access is harder to ensure. Denser, more compact cities consume less energy than sprawling ones due to lowered building space conditioning requirements (smaller units, shared walls) and transportation requirements (closer match between trip origin and destination, improved viability of transit). Will meeting the energy conservation goals with denser cities make solar energy impractical in an urban setting?

This paper seeks to demonstrate that compact cities and the use of solar technologies are compatible goals. No argument will be made concerning the benefits of compact cities, nor will the economic and technical viability of solar systems be discussed. Rather, it will be assumed that compaction is a worthwhile land-use goal and that solar technologies will find widespread acceptance as other energy sources prove too expensive, unsafe, or unavailable. Only those solar technologies that use the direct energy of the sun will be considered. These include active and passive space and water conditioning and photovoltaics, the direct conversion of solar energy into electricity. Wind, biomass, and hydroelectric technologies will not be directly addressed.

Solar Access and the Density Question

What is solar energy's value? To what extent should solar access be protected? What densities can be achieved if solar access is preserved? Unfortunately, no firm answers exist for any of these questions.

As alternative fuels become increasingly expensive and scarce, the value of solar energy as a fuel source should increase. The present concentration on the protection of solar access for energy conversion, however, tends to neglect sunlight's value in enhancing our quality of life. A sunny spot to grow vegetables or flowers, sunbathe, hold picnics, and dry clothes can be just as important to our well-being as a solar hot water system. Clearly solar energy's value will increase and the demand for protecting access to the sun will grow.

Essentially three levels of solar access can be protected: rooftops, building faces, and yards. With each succeeding level of protection, building development on adjacent parcels is increasingly constrained. The solar technology being considered determines the required extent of solar access protection. Rooftops are likely sites for hot water systems, usually requiring two or three 4 ' x 8 ' panels for residential applications. Similarly, active space heating and photovoltaic panels may be roof-mounted. A number of passive space heating systems, including direct gain systems and attached sunspaces, require solar access to the south face of the building. If rooftops are shaded (perhaps intentionally for cooling purposes) or if the roof area is insufficient to meet a building's energy demands, solar systems may need to be sited away from buildings. In addition to yard-mounted collectors for individual buildings, community systems involve common collectors, storage, and district distribution networks for a number of end users. Groundmounted solar collectors generate the most constraints on adjacent development to protect solar access.

No easy generalizations can be made about the location of solar energy systems. Active space heating systems may be mounted vertically on building faces and passive systems can utilize rooftop storage. Use of insolation for interior lighting may present significant energy savings in office and commercial buildings, necessitating access to all building faces.

Solar access protection is compatible with urban densities. Using a concept called a solar envelope, the largest volume in which a building will not shade adjacent parcels, Professors Khowles and Berry at the University of Southern California demonstrated that quality moderate-density development is achievable while protecting solar access (Knowles and Berry 1980). In various case study designs for commercial office buildings in Los Angeles, floor area ratios (FAR) of 2.25 to 4.73 were achieved. Averaging results from six different sites indicated that for moderate-density housing, a density of 52 dwelling units per acre was achieved. These findings indicate that urban densities and solar access are possible if the form of solar development responds to sun angles and surrounding development.

Lower densities may result in cities with a more northern latitude (in our hemisphere), although the exact densities have not been determined. Higher or lower densities can be achieved depending on the "rules" used to generate the solar envelopes themselves by the amount of shading permitted. A trade-off exists between higher building densities and solar access protection, but a balance encouraging both can be achieved.

New Cities and Towns

Requiring that a new development provide sites and access for solar energy systems is easier than to retrofit an existing development. Buildings can be properly oriented, sized, and spaced to use passive solar design. Several legal tools exist to protect solar access after proper site planning, such as covenants and permitting systems (Erley and Jaffe 1979). California has amended its subdivision map act to encourage "maximum utilization of natural heating and cooling opportunities" (Calif. Energy Commission 1979). Lincoln, Nebraska, now offers a density bonus for developers who design solar subdivisions (Stangl 1979).

Since few towns are being built, except those associated with western U. S. energy development and new military bases, how would new solar towns compare with today's cities? A federally-funded research project, the Technology Assessment of Solar Energy (TASE), asked that question (Ritschard 1979). The Urban Innovations Group at the University of California at Los Angeles compared three solar urban futures for the year 2000: one with 6 percent of demand met by solar technologies, the second with 25 percent of its energy supplied by solar technologies, and the third a hypothetical city built *de novo* to maximize solar energy collected on site. These three "cities" were identical in terms of population (100,000), land use, goods and services produced, and energy demand. Their differences were compared in terms of physical layout, environmental quality, socioeconomics, and quality of life.

In the first two cases, the residential, commercial, and industrial sectors all met the on-site energy collection requirements of the given supply scenario. For the first case, 3.3 percent of total residential roof area, 3.9 percent of total commercial area, and 12.3 percent of industrial land area are required. The second option requires 20.2 percent of total residential roof area, 16.4 percent of total commercial area, and 83.7 percent of industrial land area.

The third case places larger demands on land use. The residential sector can be totally self-sufficient if 80.7 percent of available residential roof area is used. The commercial sector can collect 67 percent of its energy demand by using about 50 percent of available parking area and 100 percent of available rooftops. The industrial sector can collect only 18 percent of its energy needs on site. However, if land area in the hypothetical city is increased 34.5 percent (from 10,000 to 13,450 acres or from a gross density of 10 persons/acre to approximately 7.4 persons/acre) all three sectors can be energy self-sufficient. The authors note that the resultant area of the hypothetical new city is still less than the median area (14,780 acres) of 23 existing U. S. cities with the same population.

Retrofit of Existing Cities

Integrating solar technologies into an already developed area presents many more problems than are encountered in a new development. Building and roof orientation may not be optimal, although variations in solar collector orientation of up to 30° reduces efficiency minimally (Total Environment Action 1980). Variations in building height and building compaction present problems with solar access. Energy demands in high density areas may be too great for any significant contribution from solar technologies.

Could a city with a typical land-use mix meet the Domestic Policy Review of Solar Energy's maximum practical goal of 25 percent solar contribution by the year 2000 using on-site solar technologies? A team from the University of California at Berkeley researched the questions as a part of the TASE project (Ritschard 1979). Maximum, unshaded on-site collector area (including passive) was determined from aerial photographs for six representative land-use types. This included an evaluation of the potential for shared collector area among several buildings.

This study produced these results:

> Assuming a typical mix of the land-use types studied, a community can achieve the 25 percent goal for the year 2000 using on-site technologies with current performance.

> Only the commercial central business district cannot achieve the goal on site. This deficit can be more than offset by surplus solar collector area in other land-use types.

> Central-city industries require the use of energy sources other than direct solar technologies to meet the goal. Other renewable energy sources, such as cogenerating sytems using biomass, could make up this deficit.

> Environmental impacts of this scenario include disposal of hazardous working fluids, visual intrusion (especially in high density areas). and removal of 15-35 percent of the tree canopy in the low-density residential area.

Another recent study assessed the retrofit potential for photovoltaic systems in the west San Fernando Valley of California (Bryant 1979). Nevin Bryant of the Jet Propulsion Laboratory designed a procedure "to produce maps and tabulations revealing the amount of rooftop area available for establishing solar collectors and the proportion of energy demand that could be potentially supplied by solar photovoltaics." His conclusion? "For the sixty five square mile study area, the results showed that *with half the available flat and south facing roofs used* [emphasis added] and assuming the availability of energy storage, 52.7 percent of actual energy demand (kWh) could have been met in 1978 using photovoltaic collectors."

The TASE research team from the University of California at Berkeley produced a useful concept in analyzing the potential for on-site solar energy use: surplus and deficit solar collection areas. Their study notes that an urban central business district had a deficit that could be made up with surpluses from surrounding land uses. In technological terms, these systems would be similar to district heating systems but would use direct and indirect solar energy as their fuel source. Since higher density land uses exhibit more problems with on-site solar collection, what is the availability of undeveloped land to site neighborhood solar systems? This questions is being addressed in another study at the University of California at Berkeley. The legitimate concern is that the areas that need undeveloped land for community systems have the least available due to intensive development pressure and high land costs.

Factors that May Mitigate Land Use Impacts Associated with Solar Energy Use

Use of on-site solar technologies affects land use in various ways. However, factors that may mitigate these impacts are:

Increased end-use efficiencies. If a water heater is insulated and flow restrictors are placed in faucets, less hot water and solar collector and storage area will be required. If more efficient appliances and daylighting are used, less electricity and photovoltaic collector area is needed. Increased end-use efficiencies may tend to lengthen pay-back times, but will lessen solar collection area requirements.

Increased efficiencies in solar technologies. The state-of-art in solar building design and technologies is rapidly changing. Improved efficiencies can be expected in all solar technologies (including storage), thereby lowering area requirements. In addition, lessons will likely be learned from applications in urban areas, especially with large-scale passive design and district heating.

Energy supply from outside the city. Many studies push the use of on-site solar systems to the upper limit, but society will continue to rely to some degree on conventional energy systems and increasingly on solar technologies sited outside cities. Solar technologies possibly located outside cities include wind energy conversion systems, large and small scale hydroelectric facilities, and bioconversion systems that produce ethanol and methane.

Potential sites for solar energy systems other than on buildings. The potentials and problems associated with the use of undeveloped land for neighborhood solar systems was mentioned. A variety of common land uses, most of them associated with automobile use, could also serve as sites for neighborhood solar systems. These include drive-in theaters and restaurants, gas stations, parking lots, and cemeteries. The potential for multiple use of these land areas should be explored.

Changing land-use patterns. Land-use patterns are dynamic. When change occurs, as in building redevelopment or retrofit, the opportunity to plan for solar energy is created. By designing and positioning buildings with solar energy in mind, the long-term viability of the investments being made in our cities can be ensured.

Conclusion

Arguments may persist over the near term feasibility of solar energy systems, but subsequent generations must deal with the effects of current decisions. Buildings last at least fifty years. If research and development goals are met, residential use of photovoltaics will be cost effective in five years (U.S. DOE 1979). It is not too early to plan for widespread use of solar energy.

Jon Van Til has pointed out that as our energy prospects diminish, options for the spatial organization of cities decrease as well. Renewable energy sources can *expand* these options once again (Van Til 1979).

Consumers have shown they will use solar energy systems even during the nascent stage of commercialization. Local governments have shown they understand their citizen's desire for energy self-reliance. Land-use planners and policy makers must respond with creativity and understanding in order to achieve the exciting prospect of the solar age.

References

Basile, Ralph J. 1978 (May). "A View From Here: Solar Energy and Land Use." *Environmental Comment.*

Bryant, N. A. 1979 (Dec.). *Urban Solar Photovoltaics Potential: An Inventory and Modelling Study Applied to the San Fernando Valley Region of Los Angeles.* Pasadena, CA: Jet Propulsion Laboratory; Abstract.

California Energy Commission, Solar Office. 1979 (June). *Solar Access: A Local Responsibility; A Supplement to 1978 Legislation.* Sacramento, CA: California Energy Commission.

Erley, Duncan; Jaffe, Martin. 1979 (Sept.). *Site Planning for Solar Access: A Guidebook for Residential Developers and Site Planners.* HUD-PDR 481. Chicago, IL: American Planning Association. Available from: Superintendent of Documents, U.S. Government Printing Office.

Hayes, Denis, Executive Director, Solar Energy Research Institute. 1979 (17 Oct.). Testimony before the Subcommittee on the City, Committee on Banking, Finance and Urban Affairs, U. S. House of Representatives, Hearing on "Renewable Energy and the City."

Knowles, Ralph; Berry, Richard. 1980 (Feb.). *Solar Envelope Concepts: Moderate Density Applications.* Golden, CO: Solar Energy Research Institute. Available from: NTIS.

Ritschard, Ronald L. 1979 (Oct.). *Assessment of Solar Energy Within a Community: Summary of Three Community-Level Studies.* Washington, DC: U. S. Department of Energy.

Stangl, Debra. 1979 (Approved 8 Oct.). *Design Standards for Density Bonus for Energy-Efficient Housing, Summary.* Lincoln, NE: City of Lincoln, NE.

Total Environment Action, Inc. 1980 (Mar.). *Passive Solar Design Handbook, Volume I, Passive Solar Design Concepts.* Washington, DC: U. S. Department of Energy. Available from: NTIS.

U. S. Department of Energy, Office of Conservation and Solar Applications. 1979 (1 Oct.). *Federal Photovoltaic Utilization Program Plan.* Washington, DC: U. S. Department of Energy.

Van Til, Jon. 1979 (July). "Spatial Form and Structure in a Possible Future: Some Implications of Energy Shortfall for Urban Planning." *Journal of the American Planning Association.* Vol. 45 (No. 3): pp. 318-329.

SECTION II

ALTERNATIVE LAND USE MEASURES TO LIMIT ENERGY CONSUMPTION

Energy and Land Use

BRUCE HANNON

Introduction

ENERGY from the earth's mineral resources is available at a rate limited only by our technology, but is in finite supply. Energy from the sun is available at a finite rate because of the fixed surface area of the earth, but the supply of the sun's energy is unlimited. The United States may now be at the point of turning from an economy based on stored energy toward one based on energy flow. If we do proceed, land will become increasingly more important as an energy collector, as will the lakes and oceans, and land will return toward its former level of relative importance as a factor of production.

The emphasis in this paper is on the connection between energy conservation and land-use patterns. Certain land-use changes greatly facilitate the conversion to a lower energy using society. These effects of changes are based on the results of recent research on energy and employment demands coupled with a series of economic forecasts regarding the extremely complex effects on the economy wrought by energy scarcity [8]. Energy scarcity also should produce lower real income in the long run and, therefore, implementing energy conserving land-use patterns should enable the society to live with less future economic depravation. Our choice is to anticipate the effects of energy scarcity and plan for it or adopt a laissez-faire attitude and let inflation and periodic unexpected shortages mal-distribute the onerous economic effects.

World Energy Use and Our Need for Conservation

The data in Exhibit 1 show the relative distribution of after-tax income and energy use in the U. S. in 1960 and 1961. A comparison is made with world energy distribution in 1971. In such figures, the closer a line of distribution lies to the straight line, the more nearly perfect is the equity of distribution. United States energy use was distributed better than income in 1960-61, probably because it was so inexpensive. Since income distribution has improved in the ten years after 1960–61 [1], energy distribution has likely also improved.

As seen in Exhibit 1, the U. S. system clearly distributed energy far more equitably among its own members than energy was distributed (ten years later) among the nations of the world. In the world energy use curve the U.S. is represented by the last 6 percent (94 to 100 percent) of the world consuming population and about 38 percent of world energy consumption.

In the 1950s Russia lost the last of its great charismatic leaders. Khruschev vowed in 1956 to unite the nation through economic and industrial growth [2]. The aftermath of the death of Mao in China has seen the identical thrust [3]. Thus Russia and China, nearly half of the world's population, now have major plans for increased economic growth. While the connection between economic growth and the rate of energy use may be weak in the mature industrial countries, the connection is certainly stronger in the developing countries. Therefore, we can expect major increases in the world demand for energy, particularly oil.

And yet the known world supply does not seem to be increasing very rapidly. In light of the connections, the data in Exhibit 1 takes on new meaning. The world competition for fossil and uranium reserves should continue and will probably increase. The U. S. can turn to its own supplies of coal, oil, gas and nuclear power. But these fuels are either decreasing in proven reserves or faced with great uncertainty regarding environmental, safety and economic regulation [6]. We can turn to solar energy as a supply but this energy source is a function of our land area.

Land use in the U. S. is currently about 21 percent crops, 27 percent rangeland, 32 percent forests, 8 percent urban, industrial and recreation land, and 12 percent undeveloped [7]. Of the total 2.3 billion acres, 40 percent is publicly owned. To capture the sun's energy for use in homes and industry could produce substantive conflicts in land use with the current production of food and fiber.

For example, I have calculated that to provide the U. S. residential, commercial and industrial energy used in 1970 by growing especially suited vegetation would require nearly all of the U. S. crop and forest land.[1] In

EXHIBIT 1. DISTRIBUTION OF U.S. CONSUMER INCOME AFTER TAXES
AND ENERGY USE AND WORLD ENERGY USE.

Source: R. Herendeen, "Affluence and Energy Demand," *Mechanical Engineering*
October, 1974), pp. 19–22.

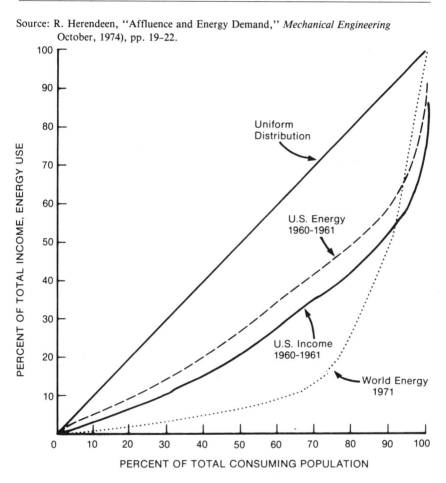

contrast, if the entire 1970 energy demand were to be supplied by strip-mined coal, about a million acres would be required [8]. In about fifty years[2] of strip mining at this rate, we would have cleared an area roughly equal to the size of Pennsylvania and West Virginia, and would have exhausted our presently known reserves of strip-mineable coal. There would also be a decline in soil fertility in the strip-mined areas.

The conflicts in land use for both solar and coal energy are obvious. A more promising strategy seems to lie with energy conservation and in the connection between land use and energy demand. But conservation has its own special set of problems.

The Dilemmas of Energy Conservation in the U.S.

The patterns which will be adopted in the United States are certainly not clear. However, there are three dilemmas which confound a rapid and easy transition to a lifestyle based on lower total energy uses than at present [9].

First, labor, capital and land have been replaced historically by energy.[3] To save energy, the cost of labor, capital and land must be reduced relative to the cost of energy.[4] This substitution means real wages and real returns to capital and landholders must decline on the average, relative to energy cost. Only energy-resource owners would derive economic benefits from such a change. Thus, labor and capital and landholders plus all of the associated institutions would be likely to resist a rise in the relative value of energy. Effective resistance would block a more equitable income distribution.

Second, total energy demand and income appear to be almost linearly connected (Exhibit 2); therefore, if energy use is to be reduced, wage and income reduction cannot be compensated for by longer work weeks, improved returns on investments or increased land productivity. But the principle of rising expectations, so ingrained in each of us, provides a basis for considerable social unrest if real income must be reduced.

Third, most energy-reducing plans would free dollar flows which, in turn, could be used for other purchases, thereby also demanding energy. The net reduction in energy demand may be small or nonexistent. Thus rearranging the pattern of expenditures is marginally effective but can even lead to a net increase in energy consumption.

Therefore, barring any major introduction of labor, capital and energy-saving new technology, we seem to be faced with decreasing real per capita income as the principal long-range means of reducing energy use.

Thus, a need develops for rational land use, not only to allow us to reduce our energy use and cut the volatile foreign energy ties, but also to allow us to cope with a possible future of declining income. We can intelligently plan for such a future. The planning strategies are basically two: a detailed set of specific and consistent governmental intervention and regulations, or a single act by government to artificially raise the price of energy in anticipation of eventual increases.

EXHIBIT 2

DIRECT, INDIRECT, AND TOTAL CONSUMER ENERGY IMPACT
PLOTTED AGAINST CONSUMER INCOME, 1960 AND 1961.

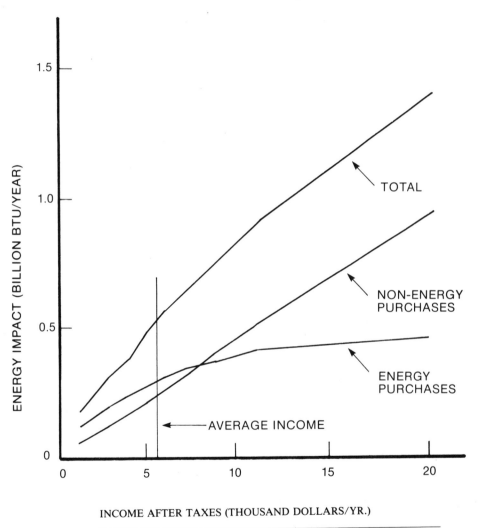

INCOME AFTER TAXES (THOUSAND DOLLARS/YR.)

Source: R. Herendeen, "Affluence and Energy Demand," *Mechanical Engineering*
October, 1974), pp. 19–22.

Means to Conserve Energy

In general, there are three basic categories in which energy conservation practices can be placed: the efficiency of direct energy use, product use (indirect energy) efficiency, and the limitation of total economic activity. The three procedures can be viewed respectively as having an increasing impact on the lifestyles of the three main energy-consuming groups: individuals, industries and government. Individual consumers are the main subject of this paper, since their decisions presently control about two-thirds of all U. S. energy use [4].

Direct energy-use efficiency can be improved, for example, by using more efficient systems for space heating and cooling, designing automobiles with greater fuel efficiency, and employing better insulation. The focus in this part of the paper is on the connection between urban living space, population density and direct energy use.

Product-use efficiency requires a more sophisticated view of energy use. Since all things that are made and delivered require energy, the judicious substitution of one product for another can reduce energy use. For example, traveling by buses or trains instead of by cars and planes will reduce energy use; so will the substitution of labor for mechanical manufacturing processes. Refillable bottle beverage systems use a third of the energy and demand far more labor than the throwaway container systems. Substituting vegetable for animal protein will save energy and release a considerable amount of prime farm land.

If these efficiency changes cannot meet the desired societal need for the planned reduction in energy use, control of the total economic activity represents the last resort. In general, total spending would need to be reduced and measures that equitably permitted lower-cost living would need to be sought. One of the most sensible choices would appear to be a carefully planned fragmentation of the large urban areas, discussed below.

DIRECT ENERGY USE EFFICIENCY

In 1970, all residences directly consumed about 20 percent of all the energy used in the United States for space temperature control, cooking, lighting, water heating, and appliance and machine operations [10]. In addition, the fuel used by personal autos in urban areas accounted for 2.9 percent of all U. S. energy use, excluding trips to work [11]. Thus, the noncommercial urban-area activity directly accounted for about 23 percent of the total energy used.

Two basic changes with a direct impact on energy and land use would be to decrease urban sprawl and to provide more efficient transportation systems. A study by the Real Estate Research Corporation [12] showed that the direct use of energy could be reduced by 44 percent if the urban residential fringe area were a planned high-density one, as opposed to the present low-density sprawl condition.[5] This figure represents the maximum energy savings rate. To estimate the average potential savings, I assumed that all present residential use is made up of "mixed sprawl" [12], a combination of single-family homes, standard apartments, and high-rise buildings. Accordingly, direct energy use could be reduced by about 31 percent if present conditions were converted to high-density planned communities [12]. This would reduce total annual U. S. energy use by about 13 percent, excluding the energy investment required to rebuild the urban area. These savings are produced by consolidating the external surfaces in the design of homes and commercial businesses, thereby reducing the area exposed to large temperature differences,[6] and reducing the average trip length for autos when not used for travel to work.

The noncommercial urban area comprises about 4.5 percent of all U. S. land [7]. Assuming, as we do here, that the average present development can be represented by a "mixed sprawl" condition, then the switch from the current urban land use to one of a high-density planned nature would free about 50 million acres of urban land. This land could be used to slowly decentralize and relocate some of the nation's industrial facilities, thus allowing for shorter trips to work. Such land could also be used to develop community recreation areas, thereby reducing recreational travel needs while focusing individual concern on local environmental quality.[7]

Although the high cost of capital and historically low cost of energy have fostered centralized industrial production [9], certain changes in the cost structure could help realize a general decentralization. First, raise the cost of energy relative to capital through an energy tax or energy rationing, which would effectively limit the return to capital holders. Second, reduce the wage differentials within urban areas by setting minimum and maximum wage levels. Third, have the cost of employee transportation borne explicitly by the employing industry, causing workers to cluster around their workplace or giving the industry an additional reason for decentralization [13].

If high-density planned development displaced the current patterns of urban residential and commercial development, individual dollar costs for home and community services could be reduced by about 32 percent[8] [14].

These dollar savings along with those achieved by denser housing would likely be respent on the consumption of other things, consequently requiring energy, land, capital and labor to be consumed in other sectors of the economy. This important respending phenomenon produces a net impact on the factors of production, as is discussed below. The direct effects of reduced sprawl and reduced auto use on energy and land use are so great that any conceivable respending scenario is unlikely to reverse them. The net impact on labor is not clear; but aside from the reconstruction labor, the demand for general labor would be largely unaffected.

Suggesting a change to higher-density communities also calls for an analysis of why sprawl has occurred. Certainly greater per capita affluence, the low cost of land on the fringes of urban areas relative to the cost of land near the center, the desire for neighborhood socioeconomic uniformity, and relatively cheap gasoline are parts of the sprawl mechanism. But the driving forces appear to have been a growing population and, in particular, the phenomenon of family unbundling.

Apparently, there is a desire to reduce the population density inside the home as well as in the neighborhood. In earlier times, members of two or three generations often lived under one roof. But gradually, grandparents, parents and older children sought and found separate housing. Further increases in the demand for separate housing probably stem from changes in rates of divorce and separations.

The unbundling pattern is depicted in Exhibit 3. The data show a decline of 24 percent in the average number of persons per room from 1940 to 1970. Since then, the decline in persons per room on the average, seems to have leveled off. The number of housing units ooccupied at densities of over 1.5 persons per room has steadily decreased, while the number of those at the lower density has risen dramatically.

The number of persons per housing unit continues to decline, indicating a trend toward housing units with fewer rooms. This phenomenon is somewhat verified by the rate of multiple-unit construction, compared to the construction of single-family homes. In 1964, 38 percent of all housing units were for multiple-family occupancy; by 1972, the figure was 45 percent [16].

Regardless of the causes of family unbundling, however, the trend is probably reversible by high energy costs alone. In fact, rebundling in a variety of forms could occur very quickly in response to a severe energy crises. The conflict of the desire for unbundling with land use could also be controlled by well-enforced zoning laws, better coordination between urban and rural governments, and a tax on present nonurban land which becomes severe if the land were to be converted to urban use. The transfer of fringe area development rights into the denser part of the urban area is another well known, but untested, solution.

EXHIBIT 3: THE TREND TOWARD LOWER DENSITY HOUSING

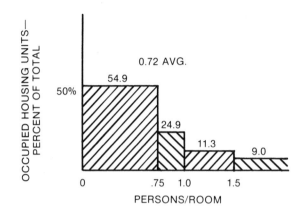

1940
(34.85 MILLION OCCUPIED HOUSING
UNITS—3.79 PERSONS/UNIT)

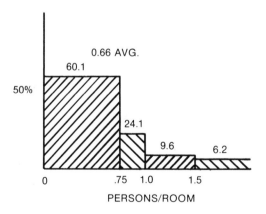

1950
(42.97 MILLION OCCUPIED HOUSING
UNITS—3.51 PERSONS/UNIT)

EXHIBIT 3: THE TREND TOWARD LOWER DENSITY HOUSING (CONTINUED)

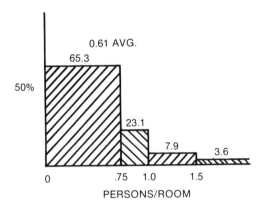

1960
(53.02 MILLION OCCUPIED HOUSING
UNITS—3.36 PERSONS/UNIT)

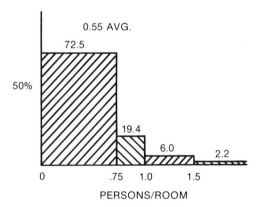

1970
(63.45 MILLION OCCUPIED HOUSING
UNITS—3.18 PERSONS/UNIT)

EXHIBIT 3: THE TREND TOWARD LOWER DENSITY HOUSING (CONTINUED)

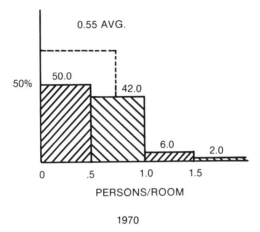

1970

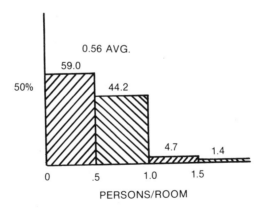

1973
(69.34 MILLION OCCUPIED HOUSING
UNITS—3.00 PERSONS/UNIT)

Family unbundling obviously increases material consumption per capita. However, the psychological drive behind unbundling may be competition.

PRODUCT USE EFFICIENCY

Many examples of alternate product uses that would save energy and increase employment have been examined [9]. The alternatives include the various items of personal consumption, a variety of personal and freight transportation systems, the use of home appliances, and the effects of a variety of federal programs concerning construction and service activities. Changing from one of these products to another would, however, have a nearly neutral effect on land use, since they all derive from roughly similar industrial processes.

As an example of an energy-saving production substitution, consider the aluminum beverage can. This container is considerably more energy-intensive and less labor-intensive than the refillable glass container [17]. Another example is the use of energy intensive herbicides in row-crop agriculture to displace farm labor. Probably the most important product substitution is the increased use of the urban bus instead of the car [11]. The car is 3.5 times as capital-intensive, 1.7 times as energy-intensive and only half as labor-intensive as the urban bus on a passenger-mile basis. Total passenger-mile cost for the bus, including subsidies, however is 1.5 times that of the car. This dollar cost difference stems from the fact that bus drivers are paid labor while auto drivers are not. Nevertheless, companies which operate largely in dense urban areas are more profitable than those which do not and their passenger-mile costs are lower than that of average urban cars [11]. It is also possible to substitute lower-energy-using video communication systems for a portion of urban travel [19].

The product choices that would affect energy and land use are those involving foods and fibers. For example, the sustained yield of the average forest acre in the U. S. is about 860 pounds of paper products per year, at an energy cost of about 19,000 BTU[9] per pound [17]. Thus, every 1,000 pounds of paper products not used spares about 1.2 forest acres from sustained cutting, and saves the energy equivalent of 152 gallons of gasoline. As an additional example, in reducing urban sprawl, high-rise apartment buildings, requiring masonry, cement and steel, are substituted for single-family houses, often made of wood. So reducing sprawl would have an indirect impact on forest land use.

The most significant single product choice affecting both energy and land use is the method of protein production and consumption. The chief sources of U. S. protein are animals, principally beef. For most of the rest of the world, the main protein sources are vegetables and fish. The Energy Research Group has determined the total dollar, energy and employment costs and the agricultural land consumed by three protein delivery systems [18]. The results are shown in Exhibit 4.[10]

The three basic systems are beef protein production (in three major areas of the U. S.), processed soybean protein production (as a meat extender and as a complete meat substitute), and the direct home consumption of unprocessed soybeans. The costs are compared on the basis of a unit of net utilizable protein, an empirical standard devised by nutrition experts to allow an equitable comparison of different protein-use efficiencies.The initial results are not surprising: switching from beef protein to a soybean meat substitute saves the consumer money while also reducing energy use, employment and the demand for agricultural land. In terms of beef raised in the cornbelt, a complete soybean meat substitute is one-sixth as energy-intensive when compared on a unit protein basis; direct soybean consumption is one-eighth as energy-intensive.

But a switch from beef to vegetable protein also would reduce the employment required and the total consumer dollar costs. If total U. S. spending is to be maintained, as likely in the short run, the consumer will spend these dollar savings on something else, and that consumption will require energy and employment increases which would tend to offset the energy savings and employment losses obtained within the food industry.

If we assume that the consumer spends his dollar savings on general personal consumption [21], then the *net* impact on the economy of a voluntary shift from beef to vegetable protein is an *increase* in energy consumption and a *decrease* in total employment: approximately 53 million BTU and 1/3 of a job per 1,000 pounds of net utilizable protein. The reduced demand for agricultural land of 23 acres of cropland per 1,000 pounds of protein would probably remain unchanged as consumers spend their savings on other forms of personal consumption.

We do not know the direct and indirect demands for land through the various forms of personal consumption. However, under the average responding scenario, it is conceivable that the demand for forest and cotton land would increase slightly as more lumber, paper products and clothing would be required.

Since the average person in the U. S. directly consumes about 8.75 pounds of net utilizable beef protein each year [18], the total effect of a voluntary switch from beef to the soybean meat substitute would be a decrease of about 597 thousand jobs, an increase of some 16 million barrels of oil (energy equivalent) each year, and a decrease in the use of more than 40 million acres of cropland (considerably more grazing land).

We also could assume that the average consumer would focus the dollars

Exhibit 4

The Dollar, Energy, Employment, and Land Costs of Protein Production in the U.S. (1973), Per Pound of Net Utilizable Protein.[a]

(costs cover the entire system from seed planting to the table; corrected for caloric content differences using impacts of soybean oil.)

SYSTEM DESCRIPTION		DOLLAR	ENERGY (100,000 BTU)	EMPLOYMENT (Thousandths of a Job)	LAND Hundredths of an Acre (% Grazing Land)	PERCENT PROTEIN (on the table)
BEEF (Cow-calf and Feedlot)	Inter-Mountain	10.01	6.65	0.95	431.00 (98.9)	17.7
	Texas	10.52	5.81	0.81	53.80 (96.3)	17.7
	Cornbelt	10.47	6.01	1.03	5.70 (61.4)	17.7
PROCESSED SOY BEAN	T.S.P. (Textured Soybean Protein— an additive)	1.48	0.84	0.080	0.010 (0)	52.0
	Unitex (a complete meat analogue)	1.57	0.93	0.089	0.010 (0)	52.0
UNPROCESSED BEAN		1.04	0.55	0.079	.019(0)	38.0

[a]Net utilizable protein is an empirically determined value based on growth of test animals per unit of protein consumed

Source: B. Hannon et al. "The Dollar Energy and Employment Costs of Protein Consumption," *Energy Systems and*

saved in the switch to vegetable protein on the nonenergy items of average personal consumption. In this case, about 60 million barrels of oil (energy equivalent) would be *saved* each year and about 564 thousand jobs would be lost. The net result on the demand for land would be essentially unchanged from the previous scenario. Capital investment increases would be needed only to supply the increase in general personal consumption.

Of course, the above calculations are based on the average and not on the marginal costs of protein production. Even if average and marginal costs are equal under current production, they would not remain so as beef production declined and vegetable protein production increased. The difference in energy costs, however, would probably increase as the shift occurs, since the former will become less efficient and the latter more efficient. The dollar cost difference is not likely to change appreciably as long as some market competition prevails.

The ratio of the energy cost differences to the differences in dollar costs will increase relative to the energy intensity of personal consumption. Therefore, the probable lower bounds of the effects of the voluntary shift were the ones calculated. By similar reasoning,[11] an energy cost increase should decrease the net energy and increase the net labor required per pound of substituted protein.

This responding effect is a difficult dilemma for a nation bent on reducing energy use, and it appears in all examples of our research. What are the possible solutions? The government could ban meat production and tax vegetable protein, then spend the tax on an activity such as postal services which is sufficiently labor-intensive to offset those jobs that would be lost and uses small enough energy amounts so that a net energy savings and an employment increase would result. The average wage would have been lowered under such a change, especially in relation to the cost of a unit of energy. The tax could be used to subsidize the construction of new energy supplies, such as electric power plants. But here the money would create fewer jobs and use more energy than it would in personal consumption, thus exacerbating the above dilemma.

The government could absorb the tax as a reduction in the money supply, but this would reduce total economic activity and, in the short run, further reduce employment via the multiplier effect. Under this scenario, wages would slowly decrease until full employment is reached. Again wages would be reduced relative to the cost of energy.

Anything that would raise the cost of energy, such as an energy tax or an energy-rationing program, would speed the process of substituting labor for fossil energy. The resulting increase in the labor intensity of the economy could possibly be structured to completely offset the loss of jobs resulting from the switch in protein sources. The revenue produced by an energy tax should be returned to the consumer as a reduction in the income tax [9]. Presumably, the consumer would lower his overall energy demand by redirecting purchases to less energy-intensive goods and services. In the short term, this behavior would be equivalent to an increase in energy efficiency. But in the long term, I suspect that the effect of the tax (and energy rationing) would lower real income because it would make the economy more labor intensive.

It is also conceivable that the urban space which would be freed under a sprawl-reduction plan and not used by relocated decentralized industry and commerce could be used to produce food for the surrounding population, where climate and soil conditions permit.[12] Apparently, vegetable protein production would be more practical than meat production on this limited space.

In the high-density urban areas, high-efficiency energy systems are feasible. Small electric generators (servicing 5,000 to 10,000 people) whose waste heat is available for space heating and cooling could be made with thermal efficiencies nearly twice those of the present electric generating systems [22]. However, without high-priced energy relative to capital, without high-density urban areas, and without the appropriate mix of housing and industrial and commercial buildings in these areas, high-efficiency total energy systems are probably not feasible.

CONTROLLING ECONOMIC ACTIVITY

One principle is now clear. Energy can be substituted to an extent for capital, labor and land. Fossil and nuclear energy is the only nonrecycleable primary factor of production and, therefore, it is in finite supply.

As long as energy is plentiful and inexpensive relative to land, labor and capital, it will be substituted for them. As the supply of energy becomes short, its dollar cost will rise and the factors just listed will be substituted for energy. When labor is substituted for energy, we, as workers, will of necessity share less in the fruits of production per unit of our effort. So, too, will the land and capital holders share less per unit of investment. To the worker and to the shareholder, this means a decline in real wages—getting poorer in a material sense. The core problem becomes: How do we become poorer without sacrificing present equity in material well-being?

Convincing people that they need to reduce energy use in order to save money does not guarantee the preservation of equity. Individuals with higher incomes are not attracted by the argument as much as those with lower incomes.

On the other hand, it is conceivable that until the advent of higher-priced energy, energy was used by the poor as a surrogate for income in order to establish status. If so, this pattern will not be given up easily. Guarantees of equity, if such exist, must come from the government. Therefore, solutions to the question of reducing net energy consumption must at the very least possess the capacity for equity.

Under the suggestions for improvements in the efficiency with which energy is used, I have pointed to the examples of reducing urban sprawl and meat consumption. Both of these scenarios reduce demands for energy and land directly; but together with the income-preserving effects, could reduce the net energy saved and possibly net decline in the demand for labor, under the current technology of production. A declining demand for labor implies a lowering of real wages if full employment is maintained. It is possible to reduce wages sufficiently to induce full employment through a technological change, to an economy that is more labor-intensive and less energy-intensive. But, theoretically, individual income could be preserved by longer work weeks.

Exhibit 2 shows the connection between energy and income during the 1960-1961 period, the latest available complete data set on personal consumption in the U. S. [4]. Direct energy use tends to saturate with rising income, but on the average, total energy use is nearly linear. From this, I infer that if income were lowered, total energy demand would be lowered almost proportionately. The total energy curve in Exhibit 2 is nearly linear because consumption patterns change as incomes rise. For example, as income goes up, so does the consumption of beef, and, consequently, energy and land.

To blunt the hardships of accepting reductions in income, we should seek general ways of reducing the cost of living as in the examples given earlier of substituting a high-density pattern for urban sprawl and vegetable protein for red meat. Exhibit 5 shows an intriguing connection between the cost of living and the population size of the consuming community [20]. The cost of living rises as the urban population goes up. Some of the cost of living differential noted for the Southern region reflects differences in heating and clothing costs.[13]

Income per person is also higher in larger cities [17]. In fact, wages appear to increase faster than the cost of living (except in the small- and moderate-sized Northeastern cities and the small Southern cities) as city population increases [20]. This phenomenon may have provided an economic basis for urbanization. But smaller population centers somehow, either through greater economic efficiency or because of a less frenetic

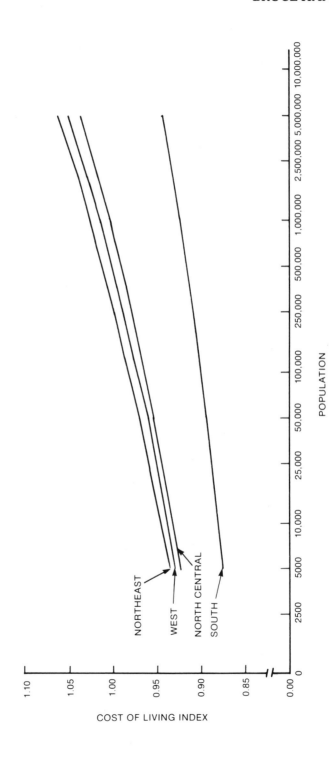

EXHIBIT 5: COST OF LIVING INDEX, BY POPULATION SIZE AND REGION (1966)

lifestyle, allow for people with lower average incomes, compared to larger population centers.

Exactly what population level would be desired for a minimum cost of living is open to question. There has been considerable debate about optimal city size and the basis for determining it [20, 23, 24, 25]. Nearly three-quarters of the U. S. population was urbanized in 1970, 10 percent more than in 1950 [23]. In both 1960 and 1970, the "average" urban area had about 20,000 people [23]. Compared to 1950, the most rapidly growing urban areas in 1960 were those with 10,000 to 25,000 population. But in 1970, the most rapidly growing incorporated urban areas were cities of 5,000 to 10,000 population [23]. These smaller communities may well be subsidiaries of larger cities, and the residents may not be paying for all the costs which they incur (for example, commuter-generated air pollution). Residents of smaller communities working (and possibly incurring some unpaid costs) in larger communities are likely to at least pay for their own community's cost, plus the extra cost of long-distance transportation. Therefore, their total costs could be reduced, along with the intrusion costs to the larger city, by working near their residence.

Historically, large cities are thought of as possessing an organizational structure that provides more benefits than costs to the average participant. The question is: Who should and who actually does financially support this structure? If all beneficiaries—rather than simply the residents—can be made to support it, then the structure can be viable. The viability of large cities would be assured if these beneficiaries were also making the necessary long-term commitment to continue their appropriate support. It is unclear whether current users of the large cities in the U. S. possess either the willingness or the commitment to pay their share of the costs [26].

Local governmental expenditures per capita seem to be positively correlated with urban size [24, 25]. Municipal expenditures per person more than quadrupled for a three-order-of-magnitude change in population level in 1967. Municipal plus county per capita expenditures were the lowest at a population of about 10,000. For county expenditures alone, the lowest per capita figure occurred when the population was about 100,000 people [25].

Local expenditures per capita for police and recreation rose with city size, but highway expenditures and the costs of sewage treatment facilities per capita declined with rising city size [24]. The optimal city size based on public costs seems to be about 100,000 people [24, 25]; however, the data in Exhibit 5 shows no such optimum because the curves are only indicative of the cost of living [27] and do not include all taxes.

With rising rates of urban growth per capita, local governmental expenditures went up but increases in personal income declined [25], indicating that zero urban population growth is also required to hold down the cost of living.

Federal expenditures at the city level, like revenue-sharing, would appear to be a subsidy for smaller cities, since the funds are generated from income taxes and distributed on the basis of population. Unfortunately, the spending of revenue-sharing funds is not constrained to activities which would aid in controlling city size.

Reducing the dollar cost of living also implies the need to reduce intercity transportation, particularly by people. The need for the interstate and urban freeway system would diminish, and it is doubtful if such a system could survive on the rather highly subsidized large truck traffic alone.[14]

A decline in land used for intercity transportation is probably the largest change in land use that would result from reducing the size of the typical urban place. The entire 43,000-mile interstate highway system alone occupies slightly more than a million acres [29]. This land is more economically productive in a transport mode than as farm or timber land, as long as energy and food remain inexpensive.

If large urban areas could governmentally and productively fragment to mimic the function of smaller cities, the total cost of living (private plus public) for their inhabitants should decline, as would individual income. The cost could be further reduced if each subcity would also retract its sprawl, thereby traveling less and by more efficient means. In fact, the fragmentation of political control in large urban areas is a necessary condition for changing to a high-density pattern. The high-density pattern is, in turn, required for a more efficient transportation system. As cost is reduced, then income and, consequently, total energy and land use also can be reduced. Also, production and service technologies must become more labor-intensive in order to create full employment.

But there is surely a limit to increasing urban population density. This limit must be based on the economic and social condition of the residents and such things as the quality of the surrounding environment, the amount and quality of the nearby open space and the general sense of community.

A major planning tool for accomplishing these goals is a rebated primary energy tax; a tax on energy as it moves from the ground or is imported [9]. By rebating the tax on a per capita or on an income basis, perceived equity can be preserved or even enhanced as an energy use is decreased. A recent study [30] concluded that a 20 percent increase in energy prices due to a national primary energy tax would reduce energy use by 5.7 percent, increase employment by 0.7 percent and create a rebate of $40 billion (1967). Such a tax anticipates a future rise in energy price perhaps caused by an external political force and perhaps exacted without concern for the social trauma which it causes. Thus the tax can condition our economy rather predictably and further, it blunts demand, reducing the anticipated energy price increase.

Summary

As the cost of energy increases relative to the cost of labor, capital and land, work trips will become shorter as will family business and recreation trips, and they will occur more by public mass-transit modes than by the use of personal cars. Per capita housing space and sprawl also should decrease. Vegetable proteins will be substituted for animal proteins. If an accompanying high level of unemployment is to be avoided, average wages along with average incomes and average total energy would decrease. Land use would probably also decrease as a result. To allow the smallest decline in the quality of life under a condition of decreasing income, the absolute size of all urban places should drop to about 100,000 people, and a near zero rate of growth would be required.

Reducing urban sprawl improves the direct energy use efficiency, frees urban land to allow employment locations (including food production) to be near residences, and reduces the dollar cost of living. Changing protein consumption from beef to a processed soybean protein is an example of improving process energy efficiency. This change in protein source reduces the demand for energy, land and employment. The change would also free consumer dollars, which if spent in an average way would increase energy use but would not restore the lost employment, unless the technology of food production were to become more labor intensive.

Switching from a sprawled urban area to a denser one and changing from beef to vegetable protein would free relatively large quantities of urban and crop land, about 50 million acres each. This change could mean a reduction in the dollar value of urban and certain rural lands.

If total dollar income is maintained, direct and process energy efficiency improvements are not likely to lower overall U. S. energy use significantly. Then the absolute level of personal spending and taxes must decline. Because the cost of living and general level of personal income are lower for small cities versus larger ones, the political and productive reorganization of large cities into smaller ones (perhaps 100,000 people or less) is recommended.

The relative independence of these smaller communities can theoretically be established by a judicious organization of the community boundary to maximally include the services and jobs demanded by those who presently live there. Accepting a job in another community, for example, would then require a move to that place. Only under such requirements can a large urban area be fragmented so as to accurately mimic the lower per capita costs of the small community. Sprawl reduction could then be slowly accomplished within each "new" community, further reducing the cost of living. Physical capital increases would be needed to decentralize present industry but capital demand would decrease in the areas of personal transportation and housing.

I believe this scenario is needed to protect and insure equity as energy becomes more and more scarce. Planning for an energy shortage or even setting the controls on city growth and development makes the transition to a lower-energy lifestyle less of a traumatic one. It seems to me an experiment is in order. It may be possible to subsidize a change in a small urban area to generally conform with the above requirements; sprawl reduction, mass transit use and all jobs held within the community. Careful measurements of the change in costs for the residents and their products should be made to check the argument.

On the other hand, a federal energy tax, geared to the unit costs of labor, capital and land, or energy rationing alone should produce the desired effects, as a surrogate for a host of coercive and more direct plans. The higher the price of energy, the more a labor-intensive economy would be induced, meaning fuller employment and probably the best that can actually be done to preserve equity. The tax or outright energy rationing [9] should be considered now at an introductory level, in anticipation of a future for the United States with a declining availability of energy that becomes more and more expensive. Capital tax credits should be allowed only if the change produced would be more labor- and less energy-intensive per unit output.

The new lifestyle in the smaller, denser, more nearly independent community would also tend to simplify what many seem to view now as an already overly complex and perhaps unmanageable society. Such a lifestyle could restore to each person a greater sense of place and perhaps, to labor, a greater chance for dignity. Personal talents could be turned to solving local problems, rather than being diluted in attempts at overall national societal control.

Intelligent energy and land use [31] could become the focus for a convergence to the plethora of conflicting values now so common in this society. The benefits of extensive urbanization have not been without their associated costs. The separation of these benefits and costs by the energy-intensive process of a geographical discounting (as opposed to temporal discounting) has produced a perception of net benefits in favor of large and sprawled cities. What has been lost is each individual's awareness of his connection to the land and to the frail finiteness of energy.

Notes

1. Throughout this paper, I calculate the bounding condition (upper or lower) in order to illustrate the effects of the logical extremes.

2. Long before this time, the costs of strip-mined coal would be likely to rise so high that solar energy would have become a greater reality and overall energy demand would have dropped.

3. Our research clearly shows that the products of the U. S. economy in the 1960s are either energy-intensive or labor-intensive but not both [9]. A redistribution of demands for these products could therefore produce an overall increase in the demand for labor and a decrease in the demand for energy.

4. If e is the unit cost of energy, L is the unit cost of labor, C the unit cost of capital and l the unit cost of land, and we order the factors, energy, land capital and labor in order of decreasing scarcity, then the inequality: $\frac{e}{L} \frac{e}{C} \frac{e}{e}$ 1 sets a constraining condition for proper factor substitution. Thus governmentally-induced increases in the unit cost of energy must be coupled to rise with the unit costs of labor, capital and land.

5. Conversely, an increase in the dollar cost of energy would tend to reduce sprawl.

6. Such density also reduces the potential for employing solar heating and cooling.

7. It is also possible that the population density and socioeconomic condition of urban areas is responsible in part for a declining birth rate. If so, then reducing sprawl has an especial value over dispersing the population.

8. About $3,800 per average housing unit per year in 1973. This value includes differences in community and housing capital costs and community but not housing operating and maintenance costs.

9. One British thermal unit equals 1,055 joules.

10. To insure that the dollar, energy, labor, and land costs represent only those associated with deriving protein, the effects related to the least-energy-cost method of producing calories (soybean oil) were removed from the costs of each of the types of protein supplies, in proportion to their associated calories.

11. Here I assumed that the most energy- and labor-intensive of the two processes are increased and decreased in intensity the most, respectively, by an energy price increase.

12. Environmentalists also would argue that human wastes could be more easily applied to these local farmlands rather than released to the nearest river or shipped great distances to sparsely populated areas.

13. The significantly lower cost of living in the Southern region suggests the strategy of geographic relocation, in addition to city size reduction, as a means of lowering energy demands equitably.

14. Intercity truck transport is far more energy-intensive than railroad carriers [28], but rail use requires greater warehousing and increases inventory costs.

References

[1.] Radner, D. and J. Hinrichs. "Size Distribution of Income in 1964, 1970 and 1971," *Survey of Current Business* (Washington, D.C.: U. S. Department of Commerce, Bureau of Economic Analysis, October 1974), pp. 19-31.

[2.] Crankshaw, E., *Khruschev* (New York: Viking Press, 1966). See also: A. Brumberg, *Russia Under Khruschev* (New York: Praeger, 1962).

[3.] See for example, articles in *Far Eastern Economic Review*, Vol. 105, 27 and 33 (July 6 and Aug. 17, 1979), pp. 10-40 and pp. 51-57.

[4.] Herendeen, R. "Affluence and Energy Demand," *Mechanical Engineering* (October 1974), pp. 19-22.

[5.] Schipper, L. "Explaining Energy," Document UCID-3663, (Berkeley, CA: Energy Resources Group, University of California, June 1975).

[6.] Stobaugh, R. and D. Yergin, eds. *Energy Future: Report of the Energy Project* at the Harvard Business School (Random House, 1979). See also, their paper: "After the Second Shock: Pragmatic Energy Strategies," *Foreign Affairs* (Spring 1979).

[7.] Frey, H. "Major Uses of Land in the United States, Summary for 1969" (Washington, D.C.: U.S. Department of Agriculture, Economic Research Service, AER 247, December 1973).

[8.] For a breakdown of 1970 U. S. energy use see, Cook, E. "The Flow of Energy in an Industrial Society," *Scientific American*, Vol. 225, No. 3 (September 1971), pp. 138-139. I assumed that the solar capture efficiency would be 0.4 percent (Kemp, C. and G. Szego, "The Energy Plantation," Warrenton, VA: Intertechnology Corp., March 1975), that transport fuels would be provided in the form of industrial alcohol and that one half of all residential and commercial space and water heating could be provided by the sun directly (about five percent of the total U. S. energy use). The efficiency of land use might be improved by providing the alcohol by fermentation, by short-term storage of the sun's energy, in molten salts for example, and by direct generation of electricity. The latter is useful for peak demands primarily. In the latter two cases, the energy costs of distribution may be substantial and the net energy produced may not exceed the system's capital, operating and maintenance energies. Energy-intensive manufacturing facilities also could be relocated to desert areas where solar energy is more intensive and less interrupted than elsewhere.

[9.] Hannon, B., "Energy Conservation and the Consumer," *Science*, V. 189, No. 4197, pp. 95-102. See also, Hannon, B. "Energy, Growth and Altruism," Mitchell First Prize Paper, Limits to Growth '75, reprinted in Dennis L. Meadows, ed., *Alternatives to Growth-I: A Search for Sustainable Futures* (Cambridge, MA: Ballinger Publishing Co.), Chap. 4, pp. 79-100. See also, Note 28.

[10.] Hise, E. "Seasonal Fuel Utilization Efficiency of Residential Heating Systems," EP-83, (Oak Ridge, TN: Oak Ridge National Lab., April 1975), p. 2.

[11.] Hannon, B., R. Herendeen, F. Puleo, A. Sebald. "Energy, Employment and Dollar Impacts of Alternative Transport Options," in R. Williams, ed., Energy Conservation Paper (Cambridge, MA: Ballinger Publishing Co., 1975), pp. 105-130.

[12.] Real Estate Research Corporation. *The Costs of Sprawl* (Detailed Cost Analysis) (Washington, D.C.: U. S. Government Printing Office, April 1974), pp. 147, 194.

[13.] See the successful experience of the 3M Corp., Minneapolis, Minn., with "van pooling," where the company leases vans to certain workers who are responsible for the work trips of about eight to ten fellow employees. The next largest use for the urban auto is recreation, which, in most instances, could be handled by mass transit [11]. The use of the car for grocery shopping cannot be replaced by mass transit, however. In this important instance store-home delivery would appear to be a lower-cost (energy and dollars) alternative, providing that deliveries could be made along a precalculated route and the demand for this service was great.

[14.] See note 12, pp. 215 and 218. Includes capital operating and maintenance. Present value at 10 percent interest charge.

[15.] U. S. Department of Commerce, "1960, 1970 Census of Housing, U. S. Summary," Part XXIX, Table J, and; Part 1, Table 4, respectively; and "1973 Annual Housing Survey, U. S. and Regions," Part A, Table A-1 (Washington, D.C.: U. S. Government Printing Office).

[16.] U. S. Department of Housing and Urban Development," 1972 HUD Statistical Yearbook," Table 334 (Washington, D. C.: U. S. Government Printing Office, 1973).

[17.] Hannon, B., "System Energy and Recycling: A Study of the Beverage Industry," Document No. 23 (Urbana, IL: Energy Research Group, Office of Vice Chancellor for Research, University of Illinois, January 1973).

[18.] Hannon, B., C. Harrington, R. W. Howell, K. Kirkpatrick. "The Dollar Energy and Employment Costs of Protein Consumption," *Energy Systems and Policy*, Vol. 3, No. 3 (1979), pp. 227-241.

[19.] Goldsmith, A., "Telecommunications: An Alternative to Travel" (Washington, D.C.: U. S. Department of Transportation, Office of Telecommunications, June 1974).

[20.] Hoch, I. "Urban Scale and Environmental Quality," Reprint No. 110, Table 2, Resources for the Future (Washington, D. C.: August 1973).

[21.] Energy Intensity = 63,000 Btu/1973 dollar; (35,500 without energy sectors) Job Intensity = 6.9 Jobs/100,000 1973 dollars (7.1 without energy sectors) Estimated from the inflation and productivity changes calculated from "Handbook of Labor Statistics," 1974, 1972, 1973 and note 6, Table 5 (Washington, D.C.: U.S. Department of Commerce, Bureau of Labor Statistics). Table 83 (total private) 1972 and 1973 and note 6. Energy and Labor Intensity of Personal Consumption Without Direct Energy Purchases; Herendeen, R., B. Segal and D. Amado, "Energy and Labor Impact of Final Demand Expenditures 1963 and 1967," Tech. Memo. 62 (Urbana, IL: University of Illinois, Energy Research Group, Office of Vice Chancellor for Research, October, 1975), pp. 8, 9.

[22.] Widmer, T. and E. Gyftopoulas. "Energy Conservation and a Healthy Economy," *Technology Review* (June 1977), pp. 31–40.

[23.] Richardson, H., "The Economics of Urban Size," (Lexington, MA: D. C. Heath Co., 1973), pp. 49, 114.

[24.] Karvel, G., G. Petry. "Optimal City Size," Real Estate and Land Use Report Series, Monagraph No. 72-5, (Boulder, CO: University of Colorado, Center for Real Estate and Land Use Studies, Business Research Division, November, 1972).

[25.] Bradley, R., "The Costs of Urban Growth, Observations and Judgments" (Colorado Springs, CO: Pikes Peak Council of Governments, 1973), p. 16.

[26.] Bureau of Census. "Population Estimates and Projections: Components of Change Since 1970," Series P-25, No. 618, (Washington, D. C.: U. S. Department of Commerce, January 1976).

[27.] Sherwood, M., "Family Budgets and Geographic Differences in Price Levels," Monthly Labor Review (Washington, D.C.: Department of Labor, Bureau of Labor Statistics, April 1975), pp. 8-15.

[28.] Bezdek, R. and B. Hannon., "Energy, Manpower and the Highway Trust Fund," *Science*, V. 185 (August 23, 1974), pp. 669-675.

[29.] "1975 Interstate Cost Report Estimate (E), Rural, Urban and Total Miles by Right-of-Way Widths and Acreage Prevailing," Interstate Reports Branch, (Washington, D.C.: U. S. Department of Transportation, Federal Highway Administration), Program R 49.

[30.] Hannon, B., Robert A. Herendeen, Peter Penner, "An Energy Conservation Tax: Impacts and Policy Implications," Doc. 267 (Urbana, IL.: University of Illinois, Energy Research Group, Office of Vice Chancellor for Research, Jan. 1979 (Rev. July 1979), and Hudson E. and D. Jorgenson, "The Economic Impact of Policies to Reduce U. S. Energy Growth," *Resources and Energy*, V. 1 (1978), p. 208.

[31.] Leopold, A., "A Sand County Almanac" (New York: Oxford, 1949), p. 201.

Reducing Travel and Fuel Use Through Urban Planning

DALE L. KEYES

Introduction

Over one-fourth of all energy consumed in the U.S. is used to transport goods and people, and of this total, roughly one-third is devoted to travel within urban areas. The equivalent of almost three million barrels of oil is consumed each day to convey urban residents to work, to school, to the shopping center, to the tennis court, and of course, home again. We travel frequently, cover long distances, and choose the automobile as our primary mode. Moreover, our trips are overwhelmingly of the single-passenger variety. Surely, the fat can be trimmed from this largess.

Twice within a span of five years Americans have been reminded of their dependence on the thin gasoline supply line. Crowded buses, hour-long waits at service stations, and exhausted supplies of compact cars all attest to the importance we place on mobility and the adversity we tolerate in pursuit of it. They also attest to the limitations we face in transport options, at least in the short run. Since the need for mobility within metropolitan areas and, to some extent, the transportation alternatives available are determined by the spatial configuration of work, shopping and housing opportunities, it seems reasonable to assert that trimming the fat could be accomplished by redesigning urban America. But before we leap to unwarranted or at least unrealistic conclusions, a review of evidence linking urban development characteristics and metropolitan travel is in order.

The Determinants and Characteristics of Urban Travel

The pattern of travel in any metropolitan area is undoubtedly the product of complex interrelationships among land use patterns, transportation system characteristics, and the types of households which reside there. More specifically, the type and amount of travel is a function of (a) the spatial arrangement of desired destinations, (b) the pecuniary and opportunity costs of travel, (c) the income level of the traveler, and (d) the "supply" of travel opportunities in terms of the availability, speed and comfort afforded by alternative modes (sidewalks, bicycles, autos, buses, taxis, trains). The supply function, in turn, is specified by private and public investments in travel vehicles and transport facilities. The location of residential, commercial and industrial areas [(a) in the list above] and the nature of the transport systems [(d) above] are joint functions of current public policies, corporate decisions, household preferences and the set of policies/decisions/ preferences made in the past. One influences and in turn is influenced by the others, as illustrated in Exhibit 1. Our understanding of these interrelationships falls far short of being complete, but various studies conducted over the last two decades have brought the picture into clearer focus.

From an energy prospective four aspects of urban travel patterns are important: trip length, trip frequency, travel speed and travel mode. Aggregate distance traveled (the product of trip length and frequency) obviously bears a direct relationship to energy consumption. Mean speed per trip and variability about the mean affect the efficiency of aggregate travel; continuous travel at slow speeds is the most efficient, stop-and-go movement, characteristic of city travel, is most energy consumptive per mile. With respect to the mode of travel, energy efficiency varies widely among modes on a per passenger/pedestrian mile, full occupancy basis, with an approximate ranking from most to the least efficient as follows: bicycle, walking, buses, electric trains and autos. Discrepancies broaden when typical occupancy factors are considered, as illustrated in Exhibit 2. Each of these key determinants of the energy consumed for urban travel will be discussed separately.

TRIP LENGTH

As distances between typical urban origins and destinations increase, a

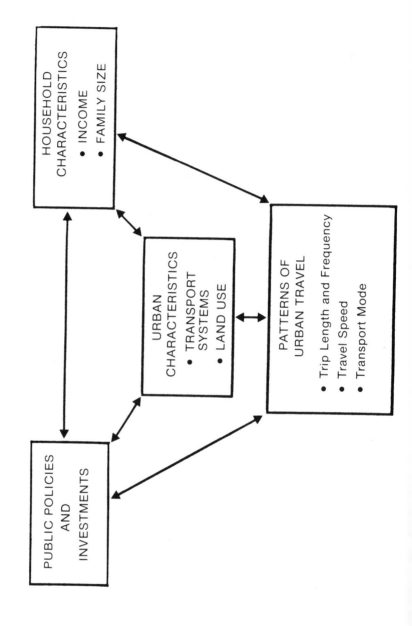

EXHIBIT 1
INTERRELATIONSHIPS AMONG DETERMINANTS OF URBAN TRAVEL

HOUSEHOLD CHARACTERISTICS
• INCOME
• FAMILY SIZE

URBAN CHARACTERISTICS
• TRANSPORT SYSTEMS
• LAND USE

PATTERNS OF URBAN TRAVEL
• Trip Length and Frequency
• Travel Speed
• Transport Mode

PUBLIC POLICIES AND INVESTMENTS

EXHIBIT 2
ENERGY EFFICIENCY OF VARIOUS U.S.
TRANSPORTATION MODES

Transport Mode	Energy Consumption (BTUs per passenger mile)
Automobile	
All Driving	4,800-5,400
Urban Driving	6,550-8,100
Rail	
All Travel	2,400-4,100
Urban Travel	2,150-3,700
Bus	
All Travel	1,360-1,600
Urban Travel	1,700-3,700

Sources: Eric E. Hirst (1972)
Stanford Research Institute (1977)
L. Schipper and A. J. Lichtenberg (1976)

corresponding increase in average trip length (in distance) is expected. Empirical evidence abounds to confirm this expectation. Using population size as a rough proxy for the separation of homes and travel destinations, several studies in the late 1960s and early 1970s reveal strong correlations of size with average trip length (Voorhees, 1968; Bellomo, et al., 1970). Trip length for individual households also has been shown to be strongly dependent on distance of residences from the Central Business District (CBD), (Voorhees, 1968). Since population decentralization frequently precedes employment redistribution in metropolitan areas, a peripheral residential location implies a long commute to work for many residents.

Zahavi (1976) has employed a direct measure of job-home separation in his analysis of automobile trips. Zahavi computes the change in the relative number of jobs as one moves away from the CBD and compares this statistic with the relative number of people at their place of residence. The difference between the two values measures the average disparity between jobs and home. Zahavi finds a strong correlation between his disparity and average auto work trip length.

However, trip length is also dependent on trip speed. Many have argued that trips are longer at the urban periphery because expressways and relatively uncongested highways allow for high speed travel. Perhaps people measure trip duration in time as well as distance. In this case, increased trip length is caused by both the need and ability to travel further within a given time budget.

TRIP FREQUENCY

The number of separate auto trips taken by metropolitan residents appears to be a strong function of population density, both at the neighborhood level and across the entire urbanized area (Smith, 1968; Deutschman and Jaschik, 1968). Where densities are high, destination points are clustered allowing for a degree of substitution of walking, transit and other modes for auto trips. Moreover, high population densities typically produce congested streets and limited parking conditions, further reducing auto travel.

If mode substitution is the principal explanation for reduced auto trips, then frequency of all trips may not decrease with increasing density. Deutschman and Jaschik (1968) found just such an effect in New York. In fact, households without automobiles who lived in high density areas took more trips than similar families in lower density areas. This is depicted in Exhibit 3.

TRAVEL SPEED AND MODE

As suggested above, travel speed and mode are intertwined with the length and frequency of urban trips. Travel in higher density areas and nearer the CBD is made at slower speeds and to a greater degree by foot or mass transit. Examined in greater detail, these general findings are found to be sensitive to other features of urban development as well.

Mass transit usage continues to be dominated by the residents of older industrialized metropolitan areas. In New York City, for example, nearly 40 percent of work trips and 25 percent of all trips are made by transit (U.S. Department of Transportation, 1976; U.S. Bureau of the Census, 1972). Similar figures for Chicago, Philadelphia and Boston are about 20 and 10 percent, respectively. These are all areas with large historical investments in public transportation facilities.

EXHIBIT 3

TRIP PRODUCTION PER HOUSEHOLD vs. RESIDENTIAL
DENSITY STRATIFIED BY AUTO OWNERSHIP FOR
THE NEW YORK METROPOLITAN AREA

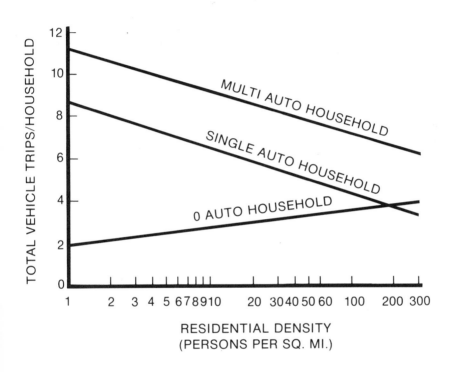

Source: Deutschman and Jaschik (1968)

Walking trips are even more sensitive to unique characteristics of individual urban areas. Although density is generally correlated with
pedestrian travel, the highest levels of nonvehicular travel occur where
military installations, universities, or other large employment facilities juxtapose workers' homes and jobs.

The conclusion must be that high density encourages non-auto tripmak-
ing but may be insufficient alone to significantly raise levels of pedestrian
and mass transit travel.

SUMMARY

Although not emphasized in the discussion above, the relationship between features of urban development and the determinants of urban travel have been established taking into account the economic status and demographic characteristics of the trip-makers. But it must be stressed that household income and car ownership characteristics play a profound role in determining travel behavior. Thus, when we hold these factors constant, the strength of associations between urban development characteristics and travel behavior may be quite low. One study ranked five household variables ahead of neighborhood density and distance from the CBD in ability to explain variations in urban travel (Stowers and Kanwit, 1965).

With these cautionary notes in mind, we can reiterate the findings sketched out above. Otherwise comparable households who live in low-density urban areas or on the periphery of large metropolitan areas tend to travel further, faster and much more frequently by auto than their counterparts in high-density cities. Only when local densities are substantially high are a significant fraction of all trips made by mass transit. At extreme levels, density serves to discourage auto use and even ownership, further reinforcing transit and nonvehicular travel. Finally, special situations which encourage or require people to live near their place of employment will stimulate non-auto travel beyond that associated with high population density and readily available mass transport.

These conclusions have been reinforced by a recent study conducted at the Urban Institute in Washington, D.C. (preliminary results in Neels, 1977). A sample of neighborhoods in eight metropolitan areas was used to develop quantitative relationships between travel behavior of households and characteristics of the neighborhoods they live in. These relationships will be used subsequently to estimate the size of potential energy savings realizable from altering land use patterns.

Direct Measures of Energy Used for Travel

Thus far we have examined the variation of individual energy-sensitive features of urban travel. To translate these variations into changes in energy use requires a second step—one which employs energy consumption coefficients by mode and speed of travel. Alternatively, gasoline sales in metropolitan areas can be measured directly and compared to characteristics of the urban setting. This approach shortcuts the linkage between development characteristics and fuel use by leaving implicit the separate links in the relationship.

SIMULATION STUDIES

Many of the studies which adopt the indirect approach to urban transportation fuel assessments employ hypothetical urban configurations and simple representations of urban travel behavior. In most cases, the transportation models assume that the average length of journey to work is sensitive to the separation of jobs and homes, but that the number of trips to other destinations is not influenced by their proximity. This automatically selects compact urban arrangements as energy efficient. The empirical evidence to justify the assumption regarding non-work trips is lacking—the results of empirical studies are mixed. But as long as the tentative nature of the model's underpinnings is kept firmly in mind, the results of these studies can serve a useful purpose in setting the bounds of the urban form—energy for travel linkage.

The results of one such study indicate that linear cities appear to be least, and the concentric ring cities most, energy consumptive, the difference in consumption being perhaps as great as a factor of two (Edwards and Schofer, 1975). Linear and polynucleated cities in the simulations are more densely settled than the concentric city, and the authors report that energy consumption, holding population constant, is negatively correlated with employment and population concentration. However, relatively large increases in concentration have to take place to achieve significant energy savings.

More recent experimentation with a slightly refined version of this model indicates that the degree of similarity between population and employment distributions is a key factor in determining work trip lengths. Furthermore, it appears that urban development characteristics work primarily through their intermediate effects on road congestion (travel speeds) and work trip length.

A second simulation examined the energy impacts of future growth options for one specific metropolitan area—Washington, D.C. (Roberts, 1975). The scenarios selected for examination are broadly representative of different development patterns. They include (a) "Sprawl"—low-density, noncontiguous residential growth at the fringe with new employment located primarily at the metropolitan center, (b) "Wedges and Corridors"—new housing located in a radial configuration along rapid rail transit routes and highway corridors, new employment located near the metropolitan center and around transit stations, (c) "Transit Oriented"—all new housing and commercial employment located along radial transit lines, (d) "Beltway Oriented"—all new residential and employment development located along a circumferential highway, (e) "Dense Center"—higher density, more concentrated residential and business development focused on the metropolitan center. The energy implications for future travel were estimated using a simple transportation model calibrated to explain current travel behavior in the area. No changes in energy utilization technology or in energy conservation efforts were assumed.

The results show that the increment of transportation energy used added between 1970 and 1992 by the most consumptive development pattern ("sprawl") would be about 1.7 times as large as that added by the "Dense center" and "transit oriented" patterns, due primarily to increased commuting distances and to decreased use of transit resulting from "sprawl."

It must be stressed that these figures are upper bounds of the feasible. What Roberts labels as "sprawl" and uses as his benchmark is a worst case scenario and not a continuation of current development trends. In addition, mass transit fuel use is considered totally insensitive to development pattern. And since alternatives which economize on auto travel do so, in part, by substituting transit for auto modes, this assumption overestimates the fuel savings of these scenarios. Contrast these findings with those of an evaluation of alternative growth options in Baltimore: maximum savings in transportation energy use between current trends and feasible alternative growth patterns is estimated at 6 percent (Baltimore Regional Planning Council, 1977).

EMPIRICAL STUDIES

The direct approach to transportation energy analysis is illustrated by two studies discussed in this section. The direct approach benefits from its reliance on actual patterns of energy use in realistic urban settings. However, it suffers from the confounding effects of fuel use by nonresidents of the areas investigated (tourists or residents of neighboring jurisdictions) and for non-travel purposes (by trucks and off-highway vehicles). Additionally, fuel use by transit vehicles is frequently ignored.

The data source for the studies discussed here is the Census of Retail Trade conducted every five years. Stewart and Bennett (1975) were first to use the data on retail gasoline sales in a study of urban development effects on transportation energy use in 134 metropolitan areas. Using regression analysis, the authors tried to explain variations in gasoline sales in terms of total SMSA population, rate of growth, degree of population centralization (percent of the population in the central city), and population density (average densities in the central city and in the rest of the metropolitan area). Various control variables also were included in the specification. Of the several development characteristics, only total SMSA population was found to be significantly (negatively) related to variations in per capita gasoline consumption. The failure to find a statistically significant relationship between urban development patterns, thus defined, and gasoline consumption may be interpreted in either of two ways: as evidence that there is no such relationship, or as further evidence of the inadequacy of these measures as descriptors of metropolitan form.

I lean toward the latter explanation since the density measures employed reflect the arbitrary delineation of jurisdiction boundaries as much as true variations in urban compactness.

The authors of this study did find that ports, and, to a lesser extent, areas with a rugged terrain, exhibited lower levels of gasoline consumption. This suggests that metropolitan areas which are spatially constrained in their development may economize on travel and/or make greater use of mass transportation.

To provide a better test of the effect of development characteristics on gasoline consumption, George Peterson and I specified a new regression incorporating variables more reflective of those urban features thought to be important for urban travel (Keyes and Peterson, 1977):

Total metropolitan area population
Proportion of the metropolitan area population living in high-density census tracts (10,000 persons or more per square mile); and
Proportion of jobs in the metropolitan area located in the Central Business District (CBD)

Alternative specifications contained a measure of the variation in census tract population density, reflecting one or more high-density nodes of people and perhaps commercial activity. Variables reflecting the extensiveness of the highway and transit systems were included to capture the independent effects of these features, as well as the way they are interrelated with and help explain the energy effects of development patterns. Measures of household income and gasoline prices completed the specifications.

Estimation of these equations using data from 49 metropolitan areas collected by the Bureau of the Census and other sources revealed statistically significant relationships between gasoline consumption and each of the development variables, (Keyes and Peterson, 1977). The results are summarized in Exhibit 4. Taken together, the estimated equations sketch out a picture of the energy-efficient city generally consistent with earlier views: small and compactly developed or multi-nucleated with a large proportion of its population living in high-density neighborhoods and a relatively

Exhibit 4
RESULTS OF THE GASOLINE CONSUMPTION ANALYSIS

Effect on Per Capita Gasoline Consumption

Variable	Direction*	Comments
Metropolitan Population	Positive	Significant**
Concentration of Population in High Density Census Tracts	Negative	Significant
Variation in Census Tract Density	Negative	Significant
Concentration of Jobs in the CBD	Positive	Significant alone; not significant in combination with transportation system variables
Household Income	Positive	Significant for real but not nominal income
Gasoline Price	Negative	Significant
Miles of Freeways	Positive	Significant
Line-miles of Transit	Negative	Significant

*A positive effect indicates that gasoline consumption increases with an increase in the explanatory variable
**"Significant" means statistically significant at least at the 0.10 level

Note: Details can be found in Keyes and Peterson, (1977)

uniform distribution of jobs and residences. Though an example of such a city is not available in the U.S., it represents the summation of those urban features which make possible less reliance on the auto or discourage its use outright.

Explanations for the energy-economizing effects of these features also can be fashioned from the interaction of development and transport network variables in the analysis. Since the effect of the high-density variables is weakened (but not eliminated) when the measure of transit coverage is added, it would appear that high density indeed acts partly by improving access to mass transit and partly by clustering trip ends. (The energy-reducing effect of the direct measure on population clustering further supports this explanation.) Thus, propinquity and congestion work together to discourage auto travel.

Job concentration at the urban core serves as a proxy in our sample for home-workplace separation, since homes are relatively dispersed in most areas. Where jobs are clustered in the CBD, commutes from the primary residential areas at the urban periphery tend to be lengthy and, since the strength of this variable is reduced when the highway coverage variable is added, this trip appears to be made primarily by auto.

Both of the empirical studies just described ignore transit fuel use. To the extent that travel economies are realized by substituting transit for auto use, the estimated savings are overstated. However, the results of both studies indicate that the reduction in VMT is accomplished more through decreases in auto trip frequency and length than through mode substitution.

The results of the transportation energy studies accord well with our understanding of travel behavior. Shorter trips and greater use of non-auto modes produce net energy savings. The extent to which trip length and modal choice can be altered through manipulations of land use patterns can now be addressed.

Potential Energy Impacts of New Urban Development

The findings described in the previous two sections, taken at face value, suggest that locating new residential and employment growth in urban areas so as to reduce average trip length and encourage use of non-auto modes would contribute to reducing transportation energy use. The specific nature of growth accommodation strategies designed to reduce energy and the extent of these potential reductions are the focus of this section.

Changes in urban development which tend to bring population and employment distribution into balance should be most desirable from an energy perspective. If jobs are currently concentrated in the CBD, then locating new residents at central locations should reduce the *average* commuting distance for the newcomers. If new employment is sited on the periphery of metropolitan areas which currently show a high degree of population decentralization, the effect on new work trips should be roughly the same. Where new development (either residential or employment) locates at sites with access to expressways or transit lines, commutes at more efficient speeds (assuming congestion will not be a problem) or via more efficient modes will be encouraged.

New towns and planned unit developments (PUDs) were once the favored children of national urban policy planners. Both, it was argued, offered the opportunity to juxtapose homes, jobs, shopping facilities and recreational opportunities. Within such an environment, people could ride jitney buses or bicycles or even walk to most of their destinations. I will attempt to evaluate both the incremental concept of energy-efficient growth accommodation and the "self-contained" approach embodied in new towns and PUDs.

In order to better understand the effectiveness of land use measures, let us review the quantitative findings of studies discussed in earlier sections. We will look first at possible changes in energy actually consumed for travel.

Exhibit 5 is drawn from the Urban Institute study of travel behavior in eight metropolitan areas (Neels, et al., 1977). Shown are estimates of typical auto trip lengths, household daily VMT, and percent of all trips made by transit for three metropolitan areas (small, average and large) and for three combinations of location and density within each area. These are hypothetical cases in the sense that the travel characteristics were not actually observed in each case but were estimated from regression equations representing conditions across all neighborhoods and metropolitan areas in the sample. Thus, a high-density fringe neighborhood (at least 10,000 people per square mile located approximately 25, 15 and 8 miles from the CBD for each representative area respectively) may not exist in either Los Angeles or Youngstown. In fact, a high-density fringe neighborhood is nothing more than an extrapolation from the general relationship developed in the study, and it lies outside the range of neighborhoods investigated. The other two neighborhood types are more realistic; examples of "inner high density" (three miles from the CBD) and "fringe low density" (less than 750 people per square mile) can be found in several of the areas sampled.

The values for trip length, VMT and transit use in Exhibit 5 were computed by assuming values typical of new suburban households for all variables other than land use-related variables in the regression equations: $12,000 income level (1970 dollars), 3.5 persons per household, and 10 percent black families. The metropolitan size, neighborhood density and neighborhood location variables were then specified and the travel values calculated.

With the limitations of this approach in mind, we can examine the results in Exhibit 5. In terms of daily VMT for auto use, travel could roughly double if new households located in low-density fringe neighborhoods as opposed to inner high-density ones. The absolute VMT differences are conditioned by the size of the metropolitan area, but the relative differences are roughly the same. In all cases, the fringe high-density alternative represents a midpoint on the VMT scale.

Exhibit 5
TRAVEL CHARACTERISTICS FOR
HYPOTHETICAL NEIGHBORHOODS

	Trip Length (Miles)		
	Inner High Density	*Fringe High Density*	*Fringe Low Density*
Los Angeles (pop = 8,350,000)*			
• Work Trips	16.4	19.3	26.3
• All Trips	12.7	14.6	17.2
Average Area (pop = 1,945,284)*			
• Work Trips	7.4	8.6	11.4
• All Trips	5.8	6.6	7.7
Youngstown (pop = 396,000)*			
• Work Trips	3.9	4.1	5.6
• All Trips	3.3	3.5	4.1
	Percent Transit Use		
Los Angeles	4.4	0.6	0.3
Average Area	2.3	0.7	0.3
Youngstown	1.6	1.1	0.3
	Daily Auto Vehicle Miles Traveled		
Los Angeles	49.7	73.5	101.9
Average Area	24.8	35.9	52.7
Youngstown	14.7	20.2	29.6

*Populations are 1970 values

Definitions
Inner High Density: A neighborhood of at least 10,000 people per square mile three miles from the CBD
Fringe High Density: A high-density neighborhood 25 miles from the CBD in Los Angeles, 15 miles in the "average area," and 8 miles in Youngstown
Fringe Low Density: A fringe neighborhood with a density of no more than 750 people per square mile

Source: Neels, et al., 1977 and other unpublished data from the Urban Institute

A review of the table entries for trip length and transit use reveals that variations in trip length are primarily responsible for variations in VMT. Transit use is minor for all development combinations except high-density inner city development in a large metropolis (Los Angeles). However, none of the metropolitan areas (i.e., older industrial cities) with high levels of transit patronage was included in the sample.

These results suggest that increases in total urban travel due to population growth will be sensitive to the location and density characteristics of that growth. Moreover, the magnitude of the auto VMT differences indicate that the impact on total metropolitan-wide travel may be significant.

The empirical analysis of gasoline consumption can be used in a similar way to approximate differences among alternative growth accommodation schemes. Using the coefficients of the development variables in the regression equations, the impact on metropolitan-wide gasoline use of changes in the key development variables can be estimated. For example, an increase in population of 500,000 (average size in the sample was about 1,140,000), a decrease of 0.05 in the proportion of metropolitan residents living in high-density areas (average value in the same was 0.15), or an increase of 0.03 in the proportion of jobs located in the CBD (sample average was 0.09) would each increase per capita gasoline consumption by ten gallons or about 3 percent from the mean value (330 gallons per person per year). A simultaneous shift of both HIDENS and CBDJOB in the direction of energy conservation by a full standard deviation would reduce per capita gasoline use by less than 40 gallons per person, per year, or approximately 12 percent. Changes in area-wide development patterns of this magnitude could only be brought about by drastic alterations in the location and density patterns typical of new development. For comparison, adding 100 miles to a transit system would reduce gasoline use an average of six gallons per capita.*

An example will illustrate the type of growth strategy required to achieve these region-wide changes. Given an urban area with characteristics equal to the average values in our sample, and assuming the annual growth rate to be 2 percent for population and 1 percent for jobs, a 10 percent reduction in energy use could be achieved by locating all new development for about ten years at the urban fringe, with new residential areas constructed at a density of at least 10,000 people per square mile. This is summarized in Exhibit 6. Since the total population increases with growth, the net impact on job decentralization and increased population density must be more than compensating.

---

*Inclusion of public transport fuel use would reduce total fuel savings somewhat.

Exhibit 6

ILLUSTRATIVE GROWTH STRATEGIES TO ACHIEVE A
TEN PERCENT METROPOLITAN-WIDE REDUCTION
IN GASOLINE CONSUMPTION

	Initial Metropolitan Value	New Development Value	Resultant Metropolitan Value
Population	1,140,000	250,000	1,390,000
Jobs	250,000	26,000	276,000
Fraction of Residents in High-Density Areas	0.15	1.00	0.34
Fraction of Jobs in the CBD	0.09	0.00	0.08

The two examples described above suggest what could be accomplished, not what is likely to occur. The trend for over thirty years has been toward low-density suburbanization. However, the potential for fuel economy appears to lie in locating new urban residents in high-density developments near employment opportunities. Given the current distribution of jobs in most metropolitan areas, this means a location near the urban fringe. New industrial development should follow suit. Certainly, new towns and PUDs may be the means for increasing densities at fringe locations, but "self containment" is a naive concept. Rarely do households live and work in the same neighborhood. In fact, travel behavior studies of residents in 15 new towns revealed no difference in total VMT per household for all but recreational trips when compared with 15 matched "bedroom" communities (Burby, et al., 1974). Instead industries in PUDs may save energy by virtue of their relative proximity to a large fraction of all families in the region.

Of course, these are generalizations; local conditions could dictate other opportunities for reducing fuel use. In-filling with high-density residential development may provide access to jobs and shopping via public transport, especially where employment remains centrally focused. Commercial, industrial, and residential development around transit stations is another example of energy-sensitive land use planning. The central theme which emerges from this review is that strategies to reduce urban travel and the attendant fuel consumption should be based on region-wide patterns of housing, employment and transport facilities.

Policy Implications

The extent to which new land use policy initiatives are called for will depend on the significance of possible savings in a national context of alternative measures to reduce fuel used for travel and the degree to which the desired land use changes may occur without new land use regulations. With respect to the significance of urban fuel savings, Hirst and Moyers have estimated that urban travel accounts for approximately 8 percent of national energy use (Hirst and Moyers, 1973). Assuming that a nationwide decrease of 10 to 15 percent in energy consumed per person for urban travel is feasible from changes in urban development patterns alone, national energy consumption would decrease by 0.8 to 1.2 percent per year once the changes had been realized. If this could be accomplished by the end of the century, by which time national consumption should approximate 110 to 120 quadrillion BTUs per year (U.S. Department of Energy, 1979), then the annual savings would equal one quad or more per year. This compares with an estimated one year's savings of 1.9 quads from mandatory 20-mile-per-gallon fuel efficiency standards for new cars (U.S. Federal Energy Administration, 1975). Doubling the real price of gasoline is likely to reduce auto gasoline consumption by 50 percent or more (a one year savings of about four quads by the end of the century), depending on the assumed long-term price elasticity of gasoline demand (-0.5 to -0.9) (Wildhorn, et al., 1974 and Chase Econometric Associates, 1974). Futhermore, the institution of strategies based on efficiency standards or gasoline price increases is likely to condition a land use approach. An increase in automobile fuel efficiency will shrink the difference between development patterns. On the other hand, price rises, while not affecting the magnitude of differences in transportation energy use among development patterns, diminish the importance of public control over land use in general; metropolitan areas will naturally grow toward more efficient forms as energy costs become a larger fraction of consumer expenditures. Dramatic increases in gasoline prices will be required to cause a perceptible movement in this direction if the past is a guide to future consumer behavior, but such increases are clearly possible in an environment of uncertain petroleum supplies. Perhaps too much attention has been paid to land use changes alone. Clearly, major public investments in transit facilities will lessen the reliance on the auto. And various measures to publicize and facilitate the use of existing facilities and various ride-sharing arrangements may reap additional benefits. Care should be exercised, however, to investigate the relative costs and fuel savings benefits of each effort.

Conclusions

Systematic differences in the level and type of travel by residents and thus energy use are clearly observable among metropolitan areas of different development patterns. Energy efficient SMSAs have high residential densities and similar spatial distributions of jobs and homes. Among SMSAs with these characteristics, the smaller ones are least energy consumptive.

However, the observed differentials in transportation energy use are modest when compared to the energy savings estimated for other conservation strategies. To achieve even the relatively small savings anticipated from these findings, large changes in historical urban development patterns would be necessary.

References

Baltimore Regional Planning Council. *The Effect of Urban Growth Alternatives on Travel, Air Quality and Fuel Consumption.* Baltimore, Md.: 1977.

Bellomo, S. J., R. G. Dial, and A. M. Voorhees. "Factors, Trends, and Guidelines Related to Trip Length," *National Cooperative Highway Research Program Report 89.* Washington, D.C.: Highway Research Board, 1970.

Burby, R., S. Weiss and R. Zehner. *A National Evaluation of Community Services and the Quality of Life in American New Towns.* Chapel Hill, N.C.: University of North Carolina, Center for Urban and Regional Studies, 1974.

Deutschman, H. D., and N. L. Jaschik. "Income and Related Transportation and Land Use Planning Implications," *Highway Research Record.* Washington, D.C.: Highway Research Board, 1968.

Edwards, J. L., and J. L. Schofer. *Relationships Between Transportation Energy Consumption and Urban Structure: Results of Simulation Studies.* Chicago, Il.: Northwestern University, Department of Civil Engineering, January, 1975.

Hirst, E. *Energy Consumption for Transportation in the United States.* Oak Ridge National Laboratory, March, 1972.

Keyes, D. L., and G. Peterson. "Urban Development and Energy Consumption." Working Paper No. 5049-1.5. Washington, D.C.: The Urban Institute, March, 1977.

Moyers, J. "Efficiency of Energy Use in the United States," *Science.* 179 (1973), 1299-1303.

Neels, K., M. D. Cheslow, R. F. Kirby, and G. F. Peterson. "An Empirical Investigation of the Effects of Land Use on Urban Travel." Working Paper 5049-17-1. Washington, D.C.: The Urban Institute, August, 1977.

Schipper, J. and A. J. Lichtenberg. *Efficient Energy Use and Well Being: The Swedish Example.* Report prepared for the U.S. Energy and Research and Development Administration (now Department of Energy), April 1976.

Smith, Wilbur, and Associates. *Patterns of Car Ownership, Trip Generation and Trip Sharing in Urbanized Areas.* Report prepared for the Bureau of Public Roads, U.S. Department of Transportation, 1968.

Stanford Research Institute. *Railroad Energy Study: Description of Rail Transportation in the U.S., Vol. II: Rail Passenger Transportation.* Report prepared for the U.S. Energy Research and Development Administration (now Department of Energy), January, 1977.

Stewart, C. T., Jr., and J. T. Bennett. "Urban Size and Structure and Private Expeneditures for Gasoline in Large Cities," *Land Economics 51.* (November 1975), 365–73.

Stowers, J. R. and E. L. Kanwit. "The Use of Behavioral Surveys in Forecasting Transportation Requirements," *Highway Research Record 106.* Washington, D.C.: Highway Research Board, 1965.

U.S. Department of Energy. *The National Energy Plan II.* Washington, D.C.: April 1979.

U.S. Federal Energy Administration. *Project Independence.* Washington, D.C.; 1978.

Voorhees, A. M. and Associates. "Factors and Trends in Trip Lengths," National Cooperative Highway Research Program Report 48. Washington, D.C.: 1968.

Zahavi, Y. "The Effects of Transportation Systems on the Spatial Distribution of Population and People" presented at Operations Research Society Meeting, November, 1976.

Energy Conservation Through Large-Scale Development: Prospects and Problems

DONALD E. PRIEST, LIBBY HOWLAND, AND ROBERT M. BYRNE

Introduction

THE issue to be discussed here is the role large-scale development can play in solving our energy problems. Land development options affect both the supply and the demand sides of the national energy equation. Energy-efficient buildings and site design-related reductions in the quantity of materials used increase energy supplies, in the sense that a barrel of oil or its equivalent saved is a barrel that is available for other uses. On the demand side, land development patterns are very important determinants of how much energy is needed for transportation. They affect both distances travelled and the mode of transportation used.

The points we will make in this paper are:

that large-scale development that is well located can conserve a significant amount of energy;

that the extent of large-scale development is limited by a variety of factors related to market uncertainties created by the current regulatory climate, the risk assessments of developers and lenders, and the financing situation;

that there are public policy actions that can provide adequate incentives to scale;

that the optimum incorporation of energy conserving features into large-scale projects requires market incentives and regulatory action; and,

that public policy should be concerned with large scale as an energy conservation technique.

Large scale is not defined in this paper in numerical (population or acreage) terms. The essential elements of a large-scale residential project in this context are (1) a site of sufficient size and project of sufficient density to permit flexibility in the arrangement of land uses and mix of dwelling units; and (2) a size sufficient to support some nonresidential uses—recreational, shopping, office, industrial. We are talking about planned but not necessarily self-sufficient neighborhoods or communities under the planning control of a single developer.

We begin this paper with an introduction to the present state of large-scale development, and to the opportunities and reasons for public policy actions to explore and encourage scale as an energy conservation strategy. Next, we outline the specific energy conservation opportunities that inhere in scale. The following section of the paper examines what the impediments are to developers' inclusion of optimum energy saving in large-scale projects. Finally, we consider policy implications for large scale and for the conservation in large-scale projects. The relationship of scale to energy consumption is a relatively unexamined aspect of energy research and policy. We hope that we can demonstrate the merit of further investigation of the subject and the utility of its inclusion in policy planning.

Large-Scale Development: Background and Opportunities

For many years the Urban Land Institute (ULI) has been concerned with the large-scale development form as a technique for improving land development patterns. We believe that the promotion of large-scale development as part of an urban growth strategy can help planners to establish a balance of land uses that maximizes public benefits and minimizes conflicts between development and the environment. During the 1960s and early 1970s, there was a pronounced trend toward increased scale in land development projects. Developers discovered, at least in theory, that larger projects with longer development cycles could result in increased profits, mainly through capture of land value increases. Public sector planners began to see that the planned arrangement of land uses in large-scale undertakings promoted such public goals as a mix of housing types and economic groups, the efficient and safe movement of people between land uses, a better quality physical environment, the efficient use of public facilities, and a salutary mix of tax ratables.

However, in the mid 1970s, economic forces and public policy demands on development projects combined to take the rosy glow from the emerging large-scale development form. Many large projects failed financially. Three factors predominated in these failures:

Large-scale developers had acquired or obtained options on very large tracts of land. Carrying costs were frequently tied by loan agreements to prime interest rates, and as these increased the costs of carrying this land became infeasible.

A growing trend toward shifting the burden of providing public improvements and services for future residents from public entities to private development entities had radically increased front end development costs. When the 1974 recession hit, creating a decline in housing and land markets, developers were caught in a serious cash squeeze.

New kinds of regulation—environmental and inventive land use controls—were being applied to development projects. The results of these were major delays in project approvals, major changes required in proposed projects, and an increasing possibility that projects could be stopped or significantly altered by the intervention of citizen groups and the courts. Increased uncertainties and delays had major cost implications. Sales could not be made to offset continuing debt service requirements, housing production could not be carefully timed with market conditions, and investors tended to increase interest rates to compensate for the increased risks associated with the uncertainty of approvals. The feasibility of large-scale was severely impaired.

Large-scale development as it was practiced in its brief heyday no longer constitutes a significant portion of development activity. Most knowledgeable development industry practitioners agree that the purchase of large tracts of land and massive investment in public facility improvements prior to the marketing of properties are not viable development techniques under conditions of national economic cyclical variations and extensive regulation at the local level. Privately funded development of projects on the scale of recently developed new towns will rarely be undertaken because of current and foreseeable difficulties involved in the financing of long-term commitments.[1]

Having taken a beating, most of the large corporations with internal capital sources that entered the land development field in the 1960s and 1970s have gotten out again. It is noteworthy that the firms most capable of undertaking large-scale projects today may be the Canadian development companies that are now penetrating the U.S. market in high growth areas. They are noted for their capital resources and their staying power, in contrast to some highly leveraged U.S. firms which have trouble maintaining long lead time projects. Canadian developers feel that American firms could benefit from observing the Canadian method of financing. Being balance-sheet oriented, they obtain bank loans based largely on their long-term earning records.

At least until very recently, it seemed that dramatic initiatives at the federal or state level in terms of financial incentives or tax incentives, or in terms of the imposition of planning requirements to favor large scale at the local level, were improbable. But we are aware of a new federal government interest in large scale, an interest that is concentrated on the potentials for energy conservation in scale. While renewed interest could result in policy initiatives to encourage scale, the lessons of past experience will dictate a cautious federal approach to large-scale policy.

At the local level, public policy actions can create very significant incentives to large-scale development, particularly through improvements in project approval procedures and through a resumption of historic local government roles in providing public facilities and services. The developer who wishes to pursue the large-scale option will have to rethink and adjust his operating procedures as well. To successfully carry through a large-scale project, developers will have to work more closely with local governments. Developers will have to devise means for balancing their need to maintain the project control required to ensure a return on investment with their ever more apparent need to operate in quasi-partnership with public agencies.

As we discuss in detail in the section on policy implications, effective means do exist for improving public-private cooperation in large-scale development. Local governments must first determine that the large-scale option is a technique suitable for implementing the growth plans and development goals that they have identified as being in the public interest. ULI has identified many of the public cost/benefit considerations associated with large-scale development.[2] On the plus side in the costs and benefits calculation is scale's considerable potential for saving energy, which brings us to the main point of this paper: should large-scale development be encouraged as a matter of public policy because of its potential energy efficiencies. Should energy planners devote time and other resources to large-scale development strategies? Are the energy savings associated with large-scale significant enough to warrant concerted public policy initiatives on its behalf? We believe that the answer to these questions is "yes." Besides the greater opportunities for energy efficiencies which large-scale physical planning and design provide and which are discussed in the next section, there are a number of compelling reasons for exploring and encouraging large scale as an energy conservation strategy, among them:

Large-scale developers, being relatively few in number and at the same time relatively sophisticated and receptive to current economic forces, are easier to reach and influence than is the universe of small homebuilders.

Through terms specified in their sales contracts with builders, through project construction guidelines, or through their own education programs aimed at both the builder and the consumer markets, large-scale land developers are in an advantageous position to encourage builders in the project to adopt energy conserving construction practices.

Innovations in large-scale development spin off to smaller developers. One of the easiest eays to get a new homebuilding idea across is to actually demonstrate its technical and economic/market feasibility. It has, for example, been suggested that the successful market reaction to townhouses in the new town of Reston, Virginia, directly stimulated townhouse development throughout the Washington, D.C. metropolitan area.

Large-scale developers are most likely to have the financial strength to underwrite the front end costs of energy conserving unit design and construction techniques. And economies of scale apply, lowering the per unit costs of energy conscious design and construction.

Large-scale projects can serve as a laboratory for testing new ideas[3] and for monitoring the magnitude of energy savings resulting from design sensitivities.

Many of the opportunities for energy conservation in large-scale undertakings, while not inherent in the form itself, are relatively simple to effectuate as a part of the development process and without increasing development costs. If these opportunities are generally taken advantage of, the result will be energy savings nationwide and cost savings for the affected households.

These aspects of large scale, its potential for accelerating the pace of energy innovation beyond the savings that it can achieve in and of itself, lead us to believe that the mechanism of scale is an appropriate and useful tool for energy planners and policymakers. There are, however, two important caveats to bear in mind:

Only if it is well located can large scale make a significant contribution to net energy savings. Transportation energy consumption by residents at far-out locations removed from such daily activities as jobs easily threatens to outweigh any energy savings achieved through project design.

Special efforts must be made to ensure that energy conservation is made an integral part of the project design. While large scale offers considerable opportunities for energy conservation, for the most part these are not inherent in the large-scale concept. Energy conservation considerations must be incorporated into project planning at the outset.

Energy Conservation Opportunities in Scale

It is generally conceded that, all other things being equal, large-scale development offers more opportunities for conserving energy than does small-scale development. "Generally, the larger the scope of the development, the greater the opportunities for energy conservation because of both the greater number of individuals involved and the greater possibility for interaction between systems."[4] The major energy conserving opportunities related to scale are:

site planning efficiencies,

activity mixes, and

housing unit mix and placement,

innovative technologies dependent on a critical mass.

The larger the scale, the easier it is for the planner to control land use relationships, to create energy conserving site plans. The larger the size of the project, the greater the variety of uses that can be included and placed in close proximity one to another. Large-scale projects can attain the critical mass needed to support centralized utility systems and other capital intensive energy efficient technologies.

SITE PLANNING EFFICIENCIES

Most of the techniques of site sensitive planning for energy conservation have been known for centuries, although attempts to quantify them in terms of cost effectiveness are relatively recent.[5] A comprehensive overview of climate maximizing or minimizing techniques is presented in *Landscape Planning for Energy Conservation*, which maintains that "optimum siting and site planning may be more important and, in most cases, less expensive than architectural or mechanical solutions in solar radiation utilization and in energy conservation."[6] A number of site-specific factors can be manipulated in the development plan to achieve significant heating and/or cooling load reductions for individual buildings. Solar incidence, wind velocities and directions, and variations in microclimates as a function of topography and vegetation are the important climatic variables. Energy-conscious site planning will allocate gross land uses to minimize or maximize natural forces and it will configure and orient the individual structures to take the best advantage of their specific sites and lots. Energy-conscious site planning also requires that the shading and wind pattern effects of the buildings to be constructed be taken into account.

Beyond the natural site conditions, the energy-conscious site planner must consider the relationship between project components. From an energy point of view, the most important elements are the Btus embodied in infrastructure construction and maintenance and the mode and efficiency of the circulation system. Street layout, lengths, and widths must be carefully planned.[7] A well thought out system of paths and bicycle routes can save considerable user energy. Efficiency in outdoor lighting is important. Well planned utility networks (water, sewer, telephone, electricity, gas, cable television) can reduce energy requirements and save development costs as well. Natural stormwater drainage systems can reduce energy needs, as well as enhance the environmental features of a site.[8]

ACTIVITY MIXES

Energy considerations can enter into the decision on what project components are provided. What nonresidential features are included in the project; how carefully they are planned to mesh with future residents' needs and preferences; how they are sized and integrated to allow the full capacity use of structures and facilities; how accessible they are to housing—these are project-planning elements that can have a major impact on the per capita energy use of future residents.

HOUSING TYPE MIXES

Operational energy consumption in housing is very much a function of housing type as well as of unit size.[9] In general, attached housing by virtue of common walls consumes less energy per square foot than detached housing. Clustering houses can reduce the heating and cooling loads of individual units. Low-rise multi-unit housing is potentially the most energy efficient form of housing that can be built. Decisions on the energy conservation features to be included in the housing and nonresidential units—insulation, materials, glazing, HVAC systems, weathersealing, etc.—play an obvious role in determining how much energy will be consumed over the lifetime of the project.

INNOVATIVE TECHNOLOGIES DEPENDENT ON CRITICAL MASS

Sufficiently large projects make possible, by virtue of their size and density, a range of energy-conserving options that are uneconomic or impractical for smaller-scale development. Community utility systems and co-generation plants that produce electricity and thermal energy are of considerable current interest. The technological and economic feasibility of such systems for large-scale projects appears to be rapidly improving. For large-scale commercial projects or very dense residential projects, sophisticated energy management systems using computers are proving increasingly cost effective. In the area of transportation, the coordination of large-scale development plans with community transit planning in an effort to improve service and ridership through such concepts as park and ride, shuttle service, or bus route modifications can have a significant energy payoff. Wastewater reclamation and reuse for residential lawns and nonresidential irrigation is energy conservative for projects located in regions that import pumped water. The economic feasibility of reclamation increases with scale and density.

IMPEDIMENTS TO ENERGY CONSERVATION IN DEVELOPMENT PRACTICES

The development industry's response to current and impending energy supply and cost constraints has not been as rapid as we might have hoped. Widespread commercialization of innovative practices in an industry as diffuse and diverse as the development industry is problematical in any case. There exist important impediments to the acceptance of energy-conservative techniques by real estate project developers. Once these impediments have been identified and evaluated, we can begin to devise programs and policies that will realistically and effectively address critical roadblocks to energy innovations in development.

In order to assess changes to date in development practices, we first discuss two studies that dealt with developers' perceptions of the impact of the energy situation on development practices. Then we identify what we believe are the most significant impediments to energy conservation in development: (1) the lack of usable and/or credible information on promising techniques; (2) the return on investment situation for developers; (3) market demand and market risks; and (4) public regulations. In a last section, we outline some of the impediments specifically relating to the provision of centralized utility systems in large-scale projects.

DEVELOPMENT INDUSTRY RESPONSES TO ENERGY CONSTRAINTS

Obviously the land development industry is sensitive to changes in market forces. In the case of the energy situation, its reaction has been perceptible although not overwhelming. In 1977, ULI conducted a survey of its members as to their perceptions of the effects of energy prices and supply prospects on development practices.[10] A more recent survey by the University of North Carolina's Center for Urban and Regional Studies measured consumer and builder perspectives on energy in housing development.[11] Both of these studies, as well as conventional wisdom, indicate that energy considerations are more and more important to developers, but that most of their efforts to promote energy conservation are focused on structures and on the operational HVAC components of structures. The macro elements of land development, such as project location, project size and site design have received far less emphasis.

The North Carolina survey asked forty-three land development firms in the state about the prospects of their adopting each of nine energy conservation ideas for subdivision design and development. For each idea, the respondents were asked to indicate whether it was very probable, somewhat probable, or not probable that they would use it during the next five years.[12] The results are summarized in Exhibit 1.

Eighty-four percent of the North Carolina land developers said that it was very probable or somewhat probable that their project locations would be influenced by energy considerations. Fifty-eight percent rated the energy conservation idea of including two or more land uses in their projects as very or somewhat probable in the next five years. In the ULI Survey, 22.9 percent of the respondents perceived that the energy situation has had a general effect on the location of development, while 23.7 percent noted a limited effect on location. Given the usually long time that it takes to effectuate and perceive changes in development patterns, "the perception of a general shift in location by almost a quarter of the ULI membership deserves attention."[13] Concerning the specific location effects of the energy situation, ULI members cited in decreasing order of importance more infill and redevelopment, the reduction of sprawl in fringe areas, a shift in favor of sites with public transportation, an increase in mixed-use projects and the decentralization of commercial and office projects.[14] See Exhibit 2.

Besides changes in project location and the mixing of residential and nonresidential uses, two other energy conservation ideas were selected as very or somewhat probable by a majority of the North Carolina developers: landscaping to shade streets in the summer and use of a cluster lot design. Close to 50 percent of the North Carolina developers saw some likelihood that they would include bicycle paths or footpaths in their projects. The

Exhibit 1

**PROBABLE USE OF ENERGY CONSERVATION IDEAS
FOR SUBDIVISION DESIGN**[12]

	Percent of Builder/Developers Who Rate Probability of Use During Next Five Years: (N=43)		
Energy Conservation Idea	*Very Probable*	*Somewhat Probable*	*Not Probable*
1. Location near existing community facilities, such as shopping, schools, employment	59	25	16
2. Landscaping with deciduous trees to shade streets in summer	49	26	25
3. Two or more land uses included in project, such as residential and retail, residential and recreation, residential and employment	30	28	42
4. Cluster design with community open space	19	32	49
5. Bicycle/walking paths	14	33	53
6. Orienting all or most lots in a north-south direction	16	19	65
7. Solar access covenants	9	14	77
8. Narrow street paving to reduce solar absorption	14	7	79
9. Installing one central heating and electric generating system for the entire development	0	2	98

Exhibit 2
Summary of Perceived Effects on Location of Development[14]

Effect	% of Respondents Who Perceived Location Effects	% of Entire Responding Population
Less sprawl in fringe area	39.0	16.4
More in-fill and redevelopment	65.9	28.0
Shift to sites with public transportation	27.9	11.7
Decentralization of commercial and office uses	10.0	4.0
More mixed-use projects	24.8	10.4

least probable innovations were more "radical" and farther removed from current trends in land development: orientation of most lots in a north/south direction, solar access covenants, narrower streets, and central co-generation facilities. ULI members perceived that among various development practices, site design elements were least influenced by energy constraints: only 18.2 percent of the respondents noted a general effect, and 27.2 percent a limited effect.[15] As shown in Exhibit 3 those site design elements that were perceived to be most responsive to energy considerations were foot and bicycle pathways, more attached housing, and more tree retention or planting. The energy situation was not seen as having as large an impact on street design.[16]

THE LACK OF USABLE AND CREDIBLE INFORMATION ON PROMISING TECHNIQUES

The body of knowledge on the magnitude of potential energy savings from individual land development techniques is growing. But it still lacks credibility in the development industry and is almost unknown by most consumers. We do not have convincing data on a project-by-project basis as to the dollar costs versus energy effectiveness of site design techniques.

Exhibit 3

Summary of Perceived Effects on the Site Design of Developments[16]

Effect	% of Respondents Who Perceived Site Design Effects	% of Entire Responding Population
Inclusion of foot and bicycle pathways	46.9	19.6
More attached and fewer detached single-family units	41.5	17.3
More extensive tree cover retention or tree planting	41.2	17.1
Greater mix of residential, commercial and industrial uses	33.6	14.0
Reduction in length and width of paved roadway	21.7	9.0
Fewer free-standing residential, commercial, and industrial structures	16.3	6.8

Evidence indicates that many site design techniques that are energy effective are also cost effective.[17] However, precious few examples exist of sizable projects which have been planned systematically for energy conservation, projects in which the design of the site, location of structures, and architectural detailing were grounded on climatic and topographic analyses for the purpose of saving energy.

Energy-conscious land planners are well aware of the need for a quantitative data base that would define the relationship between fuel consumption, unit type, unit size, location and occupancy characteristics.[18] The monitoring that would be required entails time and money. But quantification of the energy savings that result from design sensitivities would help public and private planners alike to include consideration of energy-conscious practices and policies in their decision-making processes.

A case study attempt to quantify the energy consumption characteristics of different project development options (known as SAND: Site and Neighborhood Design project) has recently been completed under U.S. Department of Energy auspices. In 1978, five ongoing large-scale projects were selected by DOE for a demonstration of ways by which developers could incorporate energy-conserving options into their planning and design processes. About $200,000 per project was allocated for preparation of an Energy Conservation Alternative Plan, which was to be compared to the conventionally developed plan. The five case studies showed up to 60 percent energy savings when sophisticated systems were used and up to 30 percent savings from site design alone. These savings were achieved without increases in developments costs.[19]

Using the SAND case studies as a starting point, ULI is working on the development of several monographs on energy conservation design techniques. These monographs will be tailored to the specific concerns of an audience of developers, lenders, appraisers, utility company representatives and others involved in land development. These monographs plus our plans to conduct conferences for development industry persons to disseminate the results of the SAND demonstration cases are aimed at overcoming the gap between what "energy analysts" know can be done and what operating real estate practitioners perceive can be done.

At least two of the SAND case studies pointed out that the developer's reaction to the energy conservation options presented in the alternative development plans was directly related to his knowledge of development experiences with energy conservation. Developers who had some previous knowledge were much more willing to cooperate with SAND recommendations. An education process based on real experience and the demonstration of "success stories" is needed to help overcome attitudinal constraints. As the North Carolina researchers concluded from their examination of land development practices: "In sum, it appears that while a number of possibly good ideas for saving energy through subdivision design are available, developers must be given more information about their usefulness in saving energy and their market acceptance. Otherwise, most will not make the difficult transition from concept to reality."[20]

RETURN ON INVESTMENT

Many techniques for achieving energy savings in large-scale development entail extra front end costs, which the developer has to pay. While all energy savings techniques that are feasible will, by definition, pay back in the sense that fuel savings will eventually exceed the amount of the initial investment (and maintenance) required, the nature of the development industry is such that it is not the developer or builder, normally, who realizes the technique's operating efficiencies. Where the builder's or developer's investment could be repaid through higher selling prices, he may not be willing to risk raising his prices in a competitive market. Without consumer demand for energy conservation, any money invested in it by the developer may become at best a pass-through investment rather than producing a return.

A technique that increases first costs without visibly improving the marketability or profitability prospects of a project involves a risk neither developer nor lender will find acceptable. It is difficult to reconcile paybacks over a long term with the opportunity costs involved in capital investment. This investment return problem occurs with the macro elements of development, i.e., site design, mixed-uses, or the comparative efficiencies of different housing unit types. For macro techniques, the most workable types of financial incentives are those that directly affect project profitability, incentives such as density bonuses or assurances that public approvals for future phases of a project will be forthcoming. These incentives are discussed later in the paper. Financial institutions can play a key role in this regard, by offering better loan terms to projects designed for energy efficiencies, but they will do so only when the risk of such products is sufficiently limited to warrant better terms.

MARKET DEMAND AND RISKS

Closely related to the problem of financial incentives, is the developer's need to match his product to the market and avoid undue risks. Land development is a highly competitive and risky business. The successful developer does not usually stray far from proven formulas. Innovations, particularly innovations that involve extra front end costs or that make visible changes in the appearance of the product are tried incrementally if at all. The fiduciary role of lenders and their lending guidelines typically put them in a position that reinforces the conservatism of developers and builders. In a context of complex processes and motivations, the problem of allocating the risks and benefits of investment in energy-conserving techniques is far from simple.

Market demand is the key in the development process; developers supply what consumers demand. Adoption of energy-conserving techniques will come only after consumers indicate that they want them and will pay for them. This market philosophy is evident in all stages of the development process, as noted in the Greenbrier (Chesapeake, Virginia) SAND study:

The land developer's role in relation to the residential market is, in many important aspects, similar to the builder's. The builder is the consumer of the product or improved lots, offered by the land developer. In this regard, the developer's success or failure depends on his or her ability to offer individual builders a product that their previous experience indicates will sell to the potential homebuyer, at a price which is competitive in relation to other lots within the market. If, and when, consumer demand places a premium on south facing lots for example, then builders will respond accordingly in their purchases, and land developers will make every effort to provide as many lots with southern orientation as the market can reasonably absorb.[21]

In the SAND studies, market risks were the most frequently cited reasons for the infeasibility of energy conservation options that were otherwise considered quite practical. The perception and assessment of market risk is largely a matter of subjective judgment on the part of developers and lenders, and much of it is based on experience, on what has worked in previous projects. To give some examples from the SAND studies:

For Radisson, a new town in central New York state, the proposed energy conservation plan involved clustering housing for solar orientation and wind protection. The developer was highly concerned about the marketability of an extensive cluster housing plan in an untested market area. For that reason, a less extensive clustering plan was devised. Compared to the more rigorous energy plan, this one resulted in a reduction in per-lot profits and an increase in annual operating costs for households.

The research team concluded that "due to the fact that cluster housing patterns are not currently used in the Central New York area, an educational effort must be made to inform consumers of the advantages of the benefits (costs, environmental, and energy) to be realized by cluster housing. Sampling of Radisson residents indicates that data as to actual energy savings of housing units built is required before cluster housing can be marketed in terms of its energy-conserving benefits."[22]

For the Texas new town, the Woodlands, a downtown energy conservation alternative plan called for the modification of land use locations—a mixed land use cluster concept "to achieve a smaller drain and the proper adjacent land uses, such that shared parking, walking trips between users, and off-peak utility loads can be utilized." However, the prime actors—developer, lender, and owner—were found to be wary of the mixed-use concept. The developer worried that his flexibility to respond to market changes and to manage the project's phasing would be impeded. The developer feared that by imposing development constraints which were incumbent in mixed use he would decrease his market competitiveness in securing commercial developers. Owners and lenders were found to anticipate security problems with mixed use and to react adversely to the possibility of nonuniformity in the level of amenity among adjacent uses.

A suggested reduction in parking ratios encountered lender and owner resistence. It was felt that the data on parking demand by land use was inadequate and that possible future requirements were not fully considered. Lenders insist on a safe margin of error. The idea of shared parking between uses and owners would be difficult to implement. Shared parking raises a number of difficult legal and administartive issues, and where these increased the complexity of sales transactions, they were viewed as impairments to the marketing program and pace of development. The additional site planning requirements and administrative structures that would be needed for specific measures to encourage transit were identified as significant constraints to transit options.[23]

At Burke Center, the developer objected on market grounds to such options as a change in building locations, shade trees that would block the visibility of merchants' signs, and a combination of office and retail uses. The study team pointed out that the type of options they had presented had actually been used in the market area and had proved marketable. However, this is the builder's business, and he is financially responsible for his decisions. It is one thing to say that we question these points, but we are still dealing with his business judgment . . . Although the builder will own and maintain the center over the anticipated 20-year payback period, he will not receive the benefit of reduced energy costs, except for possibly site lighting, parking maintenance, etc. The benefits will accrue to the tenants, and the builder believes he would take all the risks and still not be able to convince the tenants that they should pay a higher rent because they will be paying lower energy costs.[24]

PUBLIC REGULATIONS

For energy conservation in large-scale development, the regulatory problem may be more the lack of positive incentives to energy conserving practices than it is the existence of legal impediments.

In many areas, however, local land use and development controls do impede energy-conservative development practices. These controls involve rigid requirements for platting, the separation of uses, and, in some cases, excessive subdivision standards. Energy-conserving site planning requires flexibility so that structures can be related to climatic and topographic features. Conventional Euclidean zones prevent this. The typical provisions in many zoning ordinances related to setback requirements, side and rear yard requirements, lot coverage, the orientation of houses, minimum lot sizes, and the like seriously limit adoption of innovative site design practices. Well planned mixes of uses and of housing types conserve energy, but are not permitted in many areas covered by single use districting. The use of solar collectors can be impeded by zoning use restrictions on them, height or lot coverage restrictions, and the lack of public (or private covenant) mechanisms for assuring future access rights to the sun.

Jurisdictions which do allow site flexibility, clustering and mixed uses still may hinder energy conservation practices by virtue of their street standards, storm water management standards and parking standards. As many miles of residential streets may be built in the next thirty years as have been built during the entire history of this country. Construction of too wide streets and too thick pavements is a considerable energy waster, and is done often because of an attitude that is outmoded in conditions of resource constraints. As ULI stated:

> The process of converting from gravel or dirt streets to current practice has largely relied on experience and expertise developed in the highway engineering field. This process has largely been the filtering down and adaptation of standards developed for highways having high volume heavy vehicle use . . . The approach to setting standards is "when in doubt, pick the highest standard."[25]

The superior approach to residential storm water management in most cases is the use of natural overland flows and open channel and swale routings, rather than reliance on enclosed components.[26] Natural systems obviously use less energy by avoiding material usage and Btu's embodied in the materials. The reasons that many localities are wary of open channels is that they require periodic maintenance.

Many other physical standards that apply to development, e.g., parking standards, access standards, building code requirements, can mitigate against energy-efficient design solutions. Only when the requirements of local communities are assessed in a systematic way, will progress be possible on changing certain development practices to save energy.

COMMUNITY ENERGY SYSTEMS ENCOUNTER A MULTIPLICITY OF IMPEDIMENTS

The provision of energy supply and distribution systems to serve single large-scale developments could double the energy-saving potential of these projects. But such innovation involves significant changes in development practices, as well as the solution of many associated institutional, financing and marketing problems. This section discusses some of the issues surrounding community-wide energy systems.

Sufficiently large and sufficiently dense development can support its own on-site, centralized utility system. There are still a number of technical and economic problems to be worked out, but current research and experience, especially in Europe, indicates that on-site, small-scale district heating systems which can be conventionally fueled or which can use less common fuel sources such as solar energy, geothermal energy, or garbage, can be economically sound and highly conservative of energy. District heating plants that also produce electricity—co-generation facilities—can be even more energy efficient. Small plants can be built in one-to-three years and they can pay for themselves in terms of fuel bill savings in from 2-to-6 years.[27] There are, however, some major nontechnological impediments to the inclusion of district heating or co-generation facilities in large-scale development projects. These impediments are financial, institutional and market. Front-end developer financing of utility systems is out of the question in the great majority of cases. As it is now, developers feel themselves already overburdened by requirements that they supply such community facilities as roads, sewers, open space, schools or school properties, etc. The institutional roadblocks revolve around long-term ownership and operation of community systems, public utility regulation, and the relationship between community systems and public utilities. Utility rate structures and the ability or willingness of utilities to purchase excess electricity generated by local facilities are problems. Additionally, community energy systems might encounter market resistance. The SAND research team for the Greenbrier project in Chesapeake, Virginia, for example, felt that "the public was distrustful of utilities and that a thermal utility would be viewed by the public as another source of future utility problems, thereby affecting home sales."[28] Homeowners might feel that the centralized systems could constrain their independence as homeowners, because of arrangements to operate and maintain the systems over long periods of time. Reluctance would be similar to the initial difficulties condominiums faced in gaining acceptance. The Shenandoah SAND team, in studying the feasibility of a con-generation facility, noted that there were indeed very difficult institutional (ownership and operation) problems involved. Even if these could be solved, the researchers foresaw other impediments. The feasibility of the thermal distribution system would depend on builders' willingness to install the necessary heating/cooling connections. The developer did not feel it could compel them to do so, although it could use covenants or sales contracts provisions to encourage it. Bank valuation of the system could pose additional problems. The thermal distribution system would replace gas furnaces and electric cooling equipment, but electric heat pump back-up systems would be required for the individual units. Bank appraisers might question the sense of investing in two heating systems and therefore might appraise them at less than their total cost. "The economics of the system would have to be quite favorable for the system to be valued at true cost."[29]

Policy Implications

Currently practical site planning and passive solar design techniques offer considerable if largely unquantified opportunities for energy conservation in development projects of a certain size. The addition of "sophisticated" structural or mechanical systems to units in a project and/or the inclusion of community-wide energy supply and distribution facilities can double the energy savings realizable through site planning and design alone. This potential for significant per capita or household savings in energy consumption in large-scale projects merits public policy consideration of actions to (1) encourage project scale and (2) encourage energy-efficient planning in large-scale development.

TAKING ADVANTAGE OF LARGE SCALE: PUBLIC POLICY ACTIONS TO FAVOR SCALE

We have discussed the market sensitivities of the development industry, the financial constraints on developers or lenders for front-end investment in end-use operational energy conservation, the development industry's attitudinal constraints, and informational gaps regarding available techniques for energy conservation. Large scale, as a method of land development, in and of itself, addresses or can address some of these impediments to energy efficient development:

Land developers will not put "unmarketable" controls or conditions on the lots they intend to sell, but they are in a good position to encourage certain types of innovative practices, including siting for energy conservation as well as other energy-efficient construction techniques.

Developer programs to encourage energy efficiency through construction guidelines or covenants can be reinforced by strong marketing programs to stress the advantages to consumers, thus reducing the market risk of energy investments by developers. ULI has examined a number of real estate projects that were designed for energy conservation, and has determined that developer marketing of energy conservative features in speculative projects can be very successful.[30]

Homebuilding energy innovations stimulated by large-scale developers, to the extent they prove marketable, will encourage imitation in the homebuilding industry at large. This dissemination process can be compared to that for clustered and attached housing or to that for mixing housing types within a single project, which concepts were initially proven marketable in large-scale projects around the country and thereafter filtered down to smaller homebuilders.

Public policy actions that use large scale as a forum for technological innovation and demonstration should be repaid in public interest terms by increasing the rate of speed of adoption of energy-conserving techniques and technologies. Such actions would be based on the large-scale developer's superior financial strength and corresponding ability to underwrite front-end costs and on the lower per-unit costs involved in innovation on a large scale. Appropriate public actions would focus on improving the developer's rate of return on energy investments, through financing arrangements, guarantees, tax incentives, or joint development strategies.

Work done by ULI's research division has identified public policy techniques which have the greatest potential as incentives to scale.[31] The adoption of some or all of these techniques, in combination with specific actions to stimulate energy-conscious planning and design, can help realize the many opportunities for energy conservation that derive from scale. Public policy techniques that facilitate larger project scale are permit coordination, flexible development controls, public improvements publicly provided and jurisdictional adjustments.

Permitting coordination. At the beginning of this paper we note that environmental and land use controls as they were applied to land development projects in the 1960s and 1970s resulted in delays and uncertainties that severely impaired the feasibility of large-scale development. In order to deal with the risks implied by ever more stringent and unpredictable public controls and citizens' reactions to project proposals, developers limited the size and time frames of their undertakings. Jurisdictions which want to encourage scale as a matter of public policy will have to make sure that the regulatory climate is conducive to it. A coordinated and rationalized permitting process is a necessary step in this direction. Permitting reform should have these three goals:

to minimize costly delays

to provide guarantees that the future phases of a project will receive the necessary detailed approvals

to increase the public sector's ability to analyze all project impacts—physical, social, fiscal, environmental and energy—and to deal in a coordinated and early fashion with adverse impacts.

The extent to which permitting can be coordinated is limited by political and bureaucratic realities. Nonetheless, even modest progress in coordination can encourage scale by facilitating the project-planning process and by reducing the considerable costs of delay and uncertainty.

Flexible development controls. The advantages of large scale lie in its planned mix of land uses; and, planning for an optimal mix requires a great deal of flexibility. Planned unit development provisions or mixed use development provisions are local government development control devices that can institutionalize the planning flexibility that good large-scale development needs. Land use control devices can allow and even require the consideration of many development options for a particular site. Local governments can use flexible plan review procedures to encourage the incorporation of desirable elements into development projects. Among the elements they can encourage is energy efficiency, which is discussed shortly.

Public facilities. We also mentioned earlier that the trend toward shifting the burden of providing public improvements and services from public entities to private development entities adversely affects large-scale project feasibility. If large scale is to become a viable development form, the public sector must become more involved in the provision of services and facilities. "The shift of public service and facility costs from public agencies to the developer and then on to the consumer has resulted in a development process that is no longer viable. . . . Programs increasing public agency involvement in project planning or public service and facility provision must respond to the need to maximize the public benefits associated with scale and readjust the respective roles of the private developer and the public agency in the development process."[32] Large-scale development can be stimulated by a coordinated and positive use of public land use planning, capital improvements programming, public land acquisition, and zoning and subdivision incentives.

ULI's *Large-Scale Development* describes one public policy approach, used for Germantown, Maryland, which holds great promise as a technique for involving the public and private sectors in a joint large-scale effort. For the Germantown project, Montgomery County is using fairly conventional regulatory and spending authorities to encourage the development of a new community in an area with multiple landowners. A plan covering an 11,000 acre area has been developed specifying land uses, open space, sewer, water and transportation systems, and other public facilities; a program for financing and constructing public facilities has been devised; a program for the phasing of development in accordance with the project plan has been established; and, finally, the initiative for the development of nonpublic facilities (housing, etc.) within the framework established by the plan is left to the private landowner and developer. The success of this technique requires a strong public commitment to facility development.

Positive involvement by the public sector can help mitigate one major flaw in the large-scale approach as it has been carried out in the past, namely, that large-scale developers have been forced to locate at considerable distance from suburban fringes in order to assemble land at reasonable prices. Publicly provided incentives to large scale at more favorable locations would also help the marketability of the projects themselves.

Jurisdictional adjustments. Large-scale development often runs afoul of jurisdictional problems. Development at an appropriate scale may be discouraged or prevented by an overextended existing jurisdiction such as a large urbanizing county, by multiple competing jurisdictions, or by a combination of both these situations in a multi-tiered local government structure. Political boundaries rarely coincide with optimal, rational growth patterns. Solutions to these problems lie in jurisdictional adjustments that are specifically tailored to growth policy requirements, adjustments including incorporation, annexation and consolidation. The purpose of the adjustment in each case is to ensure that development areas are comprehensively planned, regulated in a coordinated manner, and financed in an equitable manner. Financing equity can best be achieved by financing public improvements through lower cost public borrowing and allocating the burden of improvements financing to new residents through taxation rather than through more costly mortgage payments.

PUBLIC POLICY ACTIONS FOR INCORPORATING ENERGY
EFFICIENCY INTO LARGE-SCALE DEVELOPMENTS

One approach to getting developers to include energy-conserving measures in their projects would be to *require* certain practices—the prescriptive building code approach. Its advantage would be to avoid the market acceptance problem by forcing all developers to adopt equally stringent practices. Davis, California has successfully initiated a community-wide energy conservation program that uses prescriptive and performance standard techinques, and the program has not seemed to have had any significantly adverse effects on the supply of housing. However, direct government manipulation of market forces in land development should be approached with great caution. Developers are unenthusiastic about regulatory approaches and the results in the past have often been unpredictable and contradictory to stated purposes. Public policies that combine incentives, constraints and education are more likely to succeed politically, less likely to add to inflationary burdens, and less likely to stifle innovation and creativity.

Local governments can program into their plan review procedures measures for requiring or encouraging energy-efficient development. The greatest incentive that the public sector can offer the developer is an increase in overall allowable density. Such density increases or bonuses could be negotiated in exchange for the inclusion of energy-conserving features in a development project. Density bonuses might be applied to, for example:

the provision of bikeways and pedestrianways

site plans that make maximum use of southern orientations, by locating development on south-facing slopes or by arranging streets so as to create correctly oriented building lots

site plans that maximize the wind buffering and/or breeze channelization capabilities of vegetation, topography and structure layouts

residential development plans that include nonresidential uses, such that projected daily vehicle miles travelled by future residents would be reduced

development plans that emphasize common wall-type housing

development plans that use nontraditional energy technologies for space conditioning and water heating, such as active solar systems, wind power, district heating, or co-generation energy facilities

efforts by the developer to coordinate project planning with community mass transit facilities and plans.

The PUD review process has not been systematically used by local government to promote the consideration of energy-conscious planning techniques by large-scale developers, through such devices as special bonuses or other incentives to energy-conserving practices. However, promotion of PUD itself, as an energy-efficient type of development, is something that some local governments consider in their planning for community-wide energy conservation. The San Jose California planning department, for example, is trying to encourage developers to use the PUD approach because it has the effects of reducing energy required for heating and for street lighting, and of reducing street lengths and cutting down on vehicle use.[33] At Shenandoah new town, site planning for energy conservation under the SAND project was facilitated by the area's designation under the zoning resolution as a "new community" district, permitting considerable flexibility in project design. "The major drawback to the resolution is that it provides no impetus for implementing any energy-conserving plan."[34]

It has been suggested that the consideration of energy conservation in large-scale project planning could be formalized through the imposition of an energy impact statement requirement, which would entail negotiation between the developer and the public sector to come up with techniques for mitigating adverse energy impacts. A better approach might be to compare forecasts of the energy that would be consumed in a standard development project of the same size according to some sort of "state of the art" guidelines. Developers should be required to design their projects so that "standard" energy consumption norms, such as the U.S. Department of Energy's Building Performance Standards, are not exceeded. Any reduction below the standard in total energy consumed would be rewarded by density bonuses according to a formula,—say, every 10 percent reduction in total energy consumption would result in a 5 percent allowable increase in developable space or gross density.

As alluded to previously, local governments can contribute to energy efficiency in large-scale development through a reexamination of certain development standards, most notably those for streets, storm water drainage, and parking. Without sacrificing the public health, welfare and safety that these standards were adopted to protect, local governments should be receptive to regulatory flexibility or review of standards in the interest of promoting energy-conserving site and building design.

Some of the more interesting and promising technologies of energy conservaion in large scale require front-end investments that are beyond the means of the developer. While concepts such as district heating can greatly reduce energy consumption and, more and more, can be demonstrated to be cost effective over the life-time of a project, they are not facilities that a project developer can or will provide under existing institutional and financial arrangements. We need some creative thinking in order to devise mechanisms for including the provision of innovative energy supply systems into the development process. When there is proven merit, these systems could be provided by the developer under special financing arrangements—loans, grants, bond issues. They could be underwritten by the government under a coordinated public-private development process. Or, governments and public utility regulatory bodies could endeavor to foster cooperative arrangements between developers and utilities for the provision of project-based systems.

Conclusion

Whatever public policy approaches are developed to encourage or require energy conservative practices in development, they must address the major impediments—the lack of a credible information base and of practical experience on techniques; return-on-investment considerations; market demand and risks; and regulations that hinder innovations in design. Government at all levels can play a role in overcoming these impediments. We believe that the large-scale development form offers a promising arena for effective public actions in energy conservation because of the opportunities for energy efficiencies that are inherent in it, because of its visibility, and because of the sophistication and development skills of its practitioners.

Notes

1. As contrasted with American firms that seek loans based on the value of individual projects.

2. Priest, Donald, et al. *Large-Scale Development: Benefits, Constraints, and State and Local Policy Initiatives.* Washington: Urban Land Institute, 1977, 135 pp. Management & Control of Growth Series.

3. For example, the Mission Viejo Company, developers of a 10,000 acre planned community 50 miles south of Los Angeles, is conducting research to determine the feasibility of solar-assisted street lights.

4. O'Donnell, Robert M., and James E. Parker. "Large-Scale Development: A Breeder for Energy Conservation," *Environmental Comment* (July 1977) pp.3-5.

5. Cheap fuel for heating and cooling, in fact, in the recent past lowered and perhaps obviated the cost effectiveness of most energy-conserving site design techniques.

6. American Society of Landscape Architects Foundation. *Landscape Planning for Energy Conservation.* Reston, Virginia: Environmental Design Press, 1977, p. 70.

7. For a guide to design concepts for residential streets see *Residential Streets—Objectives, Principles & Design Considerations.* Washington: Urban Land Institute, American Society of Civil Engineers and National Association of Home Builders, 1974, 48 pp.

8. For the advantages of natural storm water drainage systems see *Residential Storm Water Management—Objectives, Principles & Design Considerations.* Washington: Urban Land Institute, American Society of Civil Engineers and National Association of Home Builders, 1975, 64 pp.

9. Operating energy consumption also varies widely by user habits, an issue not under discussion here.

10. Byrne, Robert M. "The Impact of Energy Costs and Supply Prospects on Land Development Practices," *Urban Land* (September 1979). pp. 6–12.

11. Burby, Raymond J., et al. *Energy and Housing: Consumer and Builder Perspectives.* Chapel Hill, NC: University of North Carolina, Center for Urban and Regional Studies, 1979, 396 pp.

12. Ibid., p. 127.

13. Byrne, *op. cit.*, p. 7.

14. Ibid., p. 7.

15. The order of magnitude of positive responses to the question of whether the energy situation has had a general effect on various land development practices was as follows: (1) building design, 53.2 percent; (2) design and operation of transportation systems, 31.4 percent; (3) provision and use of alternative energy resources used in developments, 24.8 percent; (4) size or scale of development, 23.6 percent; (5) location of development, 22.9 percent; (6) site design of developments, 18.2 percent.

16. Byrne, *op. cit.*, p. 9.

17. See Land Design/Research, Inc. *Cost Effective Site Planning: Single Family Development.* Washington: National Association of Home Builders, 1976.
See also *Residential Streets—Objectives, Principles & Design Considerations. op. cit.*
See also *Residential Storm Water Management—Objectives, Principles, & Design Considerations. op. cit.*

18. Goldberg, Philip. *Planning With Energy.* Philadelphia: Rahenkamp Sachs Wells & Associates, 1975.

19. From a presentation by Gerald Leighton, Director, Community Systems Division, U.S. Department of Energy, to the Urban Land Institute's New Communities and Large-Scale Development Council and Development Systems and Services Council, Orlando, Florida, October, 1979.

20. Burby, *op. cit.,* p. 128.

21. Text missing.

22. *Site and Neighborhood Design Case Study: Radisson, New York.* prepared by The Reimann Buechner Partnership for the U.S. Department of Energy, 1979, p. 150.

23. *The Woodlands Metro Center Energy Study: Department of Energy Case Studies of Project Planning and Design for Energy Conservation.* The Woodlands, Texas: Woodlands Development Corporation, 1979.

24. Hrabak, *op. cit.*

25. *Residential Streets—Objectives, Principles & Design Considerations. op. cit.,* p. 47.

26. *Residential Storm Water Management—Objectives, Principles & Design Considerations. op. cit.*

27. Paparian, Michael. "Double Power for Your Energy Dollar," *Planning* (November 1979) pp. 18-20.

28. *Executive Summary "A Case Study of Project Planning and Design for Energy Conservation" Greenbrier, Chesapeake, Virginia.* draft, prepared by Greenbrier Associates for U.S. Department of Energy, 1979, p. 53.

29. *Case Study of Project Planning and Design for Energy Conservation. op. cit.,* pp. 6/13-6/16.

30. Seelig, Julie H. *Focus on Energy Conservation: A Project List.* Washington: Urban Land Institute, 1978; Howland, Libby, and Jane Silverman. *Focus on Energy Conservation: A Second Project List,* Washington: ULI, forthcoming, 1980.

31. *Large-Scale Development, op. cit.*

32. *Large-Scale Development, op. cit.*

33. Erley, Duncan, David Mosena, and Efraim Gil. *Energy-Efficient Land Use.* Chicago: American Planning Association, 1979, p. 18.

34. *Case Study of Project Planning and Design for Energy Conservation. op. cit.,* p. 6/3.

Maximizing Energy Conservation Through Site Planning and Design

DAVID B. CRANDALL

Introduction

WE have reached a point in land planning and design that if we collectively pool design and planning concepts and techniques tested and proven over time and combine them with technological advancements of the present, we can significantly impact long-term conservation of energy while upgrading the quality of our environment.

Since the advancement of technology in heating and cooling systems, and construction practices and products, we have adopted the fantasy that we can select sites, locate facilities, and design structures with little or no regard for natural elements or forces. We believed that technological advances could withstand the untiring forces of nature.

This belief, or attitude, has recently come under closer scrutiny as we begin to see the benefits of capitalizing on the forces of nature through passive and active solar systems, wind generation and earth sheltered structures.

This paper recognizes the tremendous savings in energy which can be affected through the development and implementation of new technologies in the generation, distribution and storage of energy. What this paper addresses is the need to:

1. Refine and implement changes in our land-use regulations which are counterproductive to conserving energy and a higher quality of environment.

2. Increase the awareness of land planning and design professionals to the positive benefits of incorporating land planning and design options into their commissions.

3. Assist review boards, public agencies, and the general public in seeing that these options are incorporated into their communitys' growth and expansion.

The options identified in this paper may not appear significant when viewed singularly. However when these energy savings are collectively assessed over time, and the benefits they automatically instill into our built environment are observed; their importance cannot be denied and they must be given greater support.

Energy conservation can be impacted in each phase of the land development process.

I.   Site Selection
II.   Design/Construction
III.   Operation
IV.   Maintenance

Each of these development phases can result in the conservation of either:

a. embodied energy: Energy consumed in the manufacture of materials and equipment used during the construction phase of a project, plus the energy consumed in the operation of equipment during that phase.

b. end use: The energy consumed during the operation of a completed project. (i.e., heating, cooling, hot water, and lights).

Embodied Energy

The conservation of embodied energies is worthy of greater attention, even though it may be less significant in terms of energies saved when compared to other passive and active systems. The options available to us not only conserve energy but they are cost effective, conserve our natural resources, and enhance the quality of our environment.

Since the options available to us are unlimited, I have selected four general areas for discussion:

1.   drainage
2.   grading

3. roadway or street construction
4. structure type

DRAINAGE

By allowing nature to handle as much of the runoff as possible through utilizing existing drainage courses and water bodies, creating or protecting water impoundment areas over permeable soils, and respecting flood plains; we can reduce the amount of energy consumed in the manufacture of pipe, grates, manholes, etc., the energy consumed to transport the materials to the site, and the energy consumed in the operation of equipment to excavate trenches and place and compact backfill.

GRADING

Through sensitive site design and grading, we can reduce the amount of cut and fill required, thereby minimizing the fuel consumed by earthmoving equipment. Through matching existing slope conditions with compatible land uses, we enhance the opportunities to sensitively design and grade a given site.

Well-executed grading plans, while preserving existing vegetation, can further impact energy conservation by minimizing clearing and grubbing operations. This minimizes equipment operation, hauling and replanting operations. The vegetation saved also can have a significant impact on the heating and cooling of structures which is discussed later in this paper.

Roadway or Street Construction

One of the more significant elements, in terms of construction costs and embodied energies, is roadway design and construction. Every foot of new roadway entails multiple construction operations which all require the operation of equipment in some form. The clustering of facilities (i.e., residential clusters) and designing to user-desire lines are options open to the designer of a particular development which will allow him to affect development costs and embodied energies by reduced road lengths and widths.

Roadway widths should receive close scrutiny, since two feet of width can mean a 7 to 10 percent increase in construction materials and costs.

Remember that, in most residential developments, reduced road lengths also mean reduced utility lengths.

STRUCTURE TYPE

A critical aspect of the planning and design process addresses the selection of building types and mix. For example: in a residential development, which type of unit should be used (single-family detached, single-family attached, garden apartments, high-rise) and to what ratio. Not only does the unit selection impact heavily on heating and cooling loads once occupied, but the unit type significantly affects the conservation of embodied energies through common walls, footings, utility runs, etc. Selection of the proper unit type and its physical form also allows the designer to minimize site disturbance (grading, clearing and grubbing) and its resultant impacts on energy and the environment.

If we are looking for built environments which have less asphalt and more open space, enhance the preservation of natural systems, and are cost effective; then these passive options, of which only a few have been mentioned, warrant further consideration.

End Use

We do not have to search very far for examples of how earlier civilizations designed and built their habitats to maximize the heating and cooling potentials of the wind and sun. (i.e., prairie house with sod roof, north African towns with narrow, shaded streets, the veranda hut for ventilation). However with the advancement of technology, we have become more and more reliant on energy-intensive technological solutions and less dependent upon natural systems. It is therefore essential that we reexamine the energy-conserving potential of natural energy systems and expand upon the opportunities they offer through passive options.

It has been suggested that the incorporation of passive options into the development process for a given project will cause financial hardships for the developer since specific data collection and analysis are necessary in the planning and design phases. Look at the type of data (vegetation, soils, slope, ground water, etc.) needed. I think you will agree these are not any different than what has been required by designers/planners over the years if they were providing a responsible service to their clients.

The options available to the designer/planner to impact the conservation

of energy consumed during the use and/or operation of a completed project are unlimited. (i.e., heating and cooling, transportation). For the purpose of this paper, we have isolated a few of the options as a starting point:
1. site selection
2. unit orientation
3. roadway layout

Before analyzing a given program and site to determine the optimum design/planning response, it is necessary to determine whether the principal problem of the region in which new development is proposed is heating or cooling. This is possible through a comparison of a "Heating Degree Day Index," which shows heating loads, with a "Discomfort Index," which shows cooling loads. Once it has been determined that it is necessary to minimize either heating or cooling requirements for a given development, we can begin to capitalize on the natural and/or built characteristics of a given site.

In order to be able to assess resource potentials during the site selection and unit orientation phases, the designer must be aware of fundamental characteristics of the sun and wind.

Sun 1. The angle of the sun varies from winter to summer with the sun approximately 45 degrees higher in the summer. This characteristic is important since it allows winter sun to reach farther into the interior of structures and makes the summer sun the easiest to block out. It further means flatter surfaces, such as roofs, receive the greatest radiation in the summer and vertical surfaces, such as walls, receive the greatest radiation in the winter.

2. The sun's path, from sunrise to sunset, in the summer is 240 degrees, allowing east and west facing surfaces to receive more sun. In the winter, the sun's path is only 120 degrees, allowing south facing surfaces to receive more solar radiation.

3. The sun is at its lowest angle on December 21st.

Wind 1. Wind velocities are greatest at the tops and windward sides of hills (i.e., wind speeds increase by 20 percent as they are forced up over hills or ridges).

2. Colder air settles to valley bottoms creating cold pockets.

3. Water bodies have a warming effect on surrounding lands in the winter and a cooling influence on adjacent land masses in the summer. This is caused by the cooling/warming effect the water body has on air movement (winds) passing over it. The benefits received seem to be in direct proportion to the size of the water body.

SITE SELECTION

Whether selecting a site for a proposed development (gross site selection) or selecting a site location (discreet site selection) for a given program item (i.e., single-family detached home), the process remains the same. The objective of the site selection process is to determine which of a number of sites offers the greatest potential for conserving energy.

The options which are available to the designer/planner to maximize energy conservation vary greatly as he moves from selecting a site for a new development to selecting a specific site and location for a single unit. For example: When selecting a site for a proposed development, opportunities will exist to capitalize on the benefits of sites near existing transportation systems; whereas when selecting a site for a specific unit, opportunities will come primarily from the design and construction of an efficient roadway (street) system which provides access to the optimum number of energy efficient sites.

Throughout the site selection process, a comprehensive approach must be maintained. All factors (natural and man-made elements) must be assessed for their environmental, economic and social impacts as well as their energy potential. In order that our efforts to implement energy-efficient developments be successful, they must be also financially feasible, marketable, environmentally sound, and respond to the social and cultural needs of the community. An energy-efficient community without a resident population can hardly be cited as a case study to encourage future energy-efficient development.

To be effective, we must first identify all the resources of a given site as well as any inhibiting factors.

a. Solar radiation gains derived from slope gradient and orientation—differences in slope gradient and orientation can affect site microclimates in the same way latitude affects the climatic regions of the United States.

b. Commuter trip distances to work and shopping, etc.

c. Vegetation—evergreen vegetation for buffering cold winds and deciduous vegetation for blocking out hot summer sun.

d. Elevation—valley bottoms are cold due to cold air settlement, and exposed ridges are cold due to wind exposure and an increase in velocity.

e. Subsurface conditions—such as soil classifications and water table, can have a direct impact upon the surface air temperatures and humidity, dependent upon the soil's moisture content and the proximty of ground water to the surface. Subsurface conditions are further meaningful when assessing a given site (group of sites) for opportunities to develop or expand upon an Integrated Community Energy System (ICES)—(i.e., ground water heat pumps, natural gas deposits, geothermal potentials).

f. Water bodies (natural and/or man-made)—impact air temperature and humidity (i.e., warmer winters/cooler summers, cooler days/warmer nights).

g. Identify both on-site and off-site resources which have potential for acting as a primary fuel (i.e., wood, wind, solar, industrial heat waste).

UNIT ORIENTATION

Once the optimum site within a given region has been selected for the proposed development based upon available resources, and the site assignments for specific elements of the program are made, consideration should then be given to building orientation (unit orientation).

It is essential that as a specific unit is designed and placed on a site, decisions are reached through careful consideration of both the wind's and sun's potential impact on the unit. It is also essential that we look at the opportunities offered by the site (vegetation, topography, existing structures, water bodies, etc.) to either maximize or minimize these impacts.

Even though a comprehensive approach to addressing the cooling benefits of the wind and warming benefits of the sun is necessary, for the purpose of this paper, the design responses are presented separately. Since the weighting of specific design responses varies from region to region, only basic principles are presented here:

Hot Humid Regions:
- Minimize east and west facing walls.
- Five degrees south, southeast building orientation preferred.
- Shade primary outdoor living areas.
- Maximize air movement to reduce effects of high humidity.
- Unit should be shaded whenever possible by vegetation.
- Minimize energy-intensive pavements and building materials whenever possible (i.e., asphalt streets and drives).
- Avoid areas with high water tables and the associated dampness and humidity.

Hot Arid Regions:
- Minimize east and west facing walls.
- Twenty-five degrees south, southeast building orientation preferred.
- Glazing on south facing walls preferred over east and west.

- Maximize the cooling effects of evaporation across water bodies.
- Orient active living spaces in the unit to the southeast for warming effects of early morning sun.
- Provide maximum shading.
- Block out hot dry winds.

Temperate Regions:
- Maximize southerly wall exposure for winter warmth.
- Seventeen degrees, 5 mins., south, southeast building orientation preferred.
- Optimize glazing on south facing wall and minimize glazing on north walls.
- Position inactive spaces along north building walls to buffer winter winds.
- Establish new or reinforce existing wind breaks along northerly exposures to buffer or deflect winter winds (i.e., vegetation, fencing, mounding).
- Minimize blocking of winter sun from reaching structure.
- Locate buildings in winter wind shadows.
- Minimize number of openings on northerly exposures and protect openings from prevailing winds.

Cool Regions:
- Maximize southerly wall exposure for winter warmth.
- Twelve degrees south, southeast building orientation preferred.
- Locate buildings in winter wind shadows.
- Minimize shading of structure in heating season.
- Establish new or reinforce existing wind breaks along northerly exposures to buffer winter winds (i.e., fencing, vegetation, mounding).
- Minimize number of openings on northerly exposures and provide protection of openings from prevailing winds.

From these principles, the optimum orientation of a given structure on the site is achieved. It is only after we have capitalized on the benefits by unit location, design and orientation should site-specific design responses be developed.

Keeping these basic principles in mind, the designer can then utilize the resources, both natural and man-made, available to him to further minimize adverse climatic factors and capitalize on favorable climatic factors. It is imperative that during the site selection process, these resources also are inventoried and the opportunities offered by them carefully weighed when determining the final site.

The most common of the resources available to the designer/planner at a site-specific design level are:

1. Vegetation (deciduous and evergreen).
2. Architectural elements (overhangs, fences, trellises, etc.).
3. Earth.
4. Structures.
5. Water.
6. Pavements.

Each of these resources can be utilized singularly or in combination to:

1. Maximize or minimize the cooling benefits of the wind.
2. Maximize or minimize the warming benefits of the sun.

For example:

1. Evergreen vegetation's primary value comes from its ability to effectively buffer wind, causing reduced air infiltration and transmission.

2. Deciduous vegetation's primary value comes from its ability to block out the hot summer sun and allow penetration of the winter sun.

3. Earth can effectively direct the wind over or around a structure through mounding, or responding to existing topographic features, and also can be effective in protecting structures by filling against the structure or lowering the structure into the earth.

4. Architectural elements such as fencing can be used to buffer/deflect undesirable winter winds and channel cooling summer breezes.

5. Trellises effectively block out summer sun and allow penetration of winter sun.

6. Existing structures are effective in creating wind shadows which the proposed unit can be placed in.

7. Vegetation and architectural elements can be effective, either singularly or in combination, in creating sun pockets which affect heat buildup and transfer the benefits to the structure.

8. Dark colors absorb radiation whereas light colors deflect radiation.

9. Water bodies, new or existing, can be effective in increasing humidity in hot arid regions and, if large enough, cool summer air movement and warm winter air movement.

10. Dark pavements or building materials can be used to absorb radiation when desirable (cool and temperate regions), whereas light-colored building materials can reflect undesirable radiation (hot humid and hot arid regions). Heat buildup from excessive street widths is also within the designer/planner's area of influence.

As these various principles and resources are applied in a comprehensive fashion and in balance with other develoment concerns (i.e., program, function, marketing, space requirements), the positive environmental benefits and cost-benefits will start to become evident.

For example: In an effort to capitalize on the buffering effects (wind and sun) of vegetation to conserve energy, the preservation of existing vegetation will increase and by creating sun pockets, or increasing shade, outdoor living spaces will get greater use. Also as we address embodied energies by eliminating excessive pavement widths, we will preserve more of our natural resources (gravel, stone) and increase the visual quality of the environment.

We may even find solar rights begin to give meaningful form to our urban skyline.

ROADWAY LAYOUT

Roadway layout has a place under this section because the design of our vehicular and pedestrian circulation systems is possibly the singularly most influential element in determing whether optimum energy efficient sites will be feasible. There is no sense in today's society to suggest sites which have the greatest potential for conserving energy if the roadway costs to reach those sites is excessive or if the sites are not accessible.

In addition to this basic premise, roadway design and construction can have significant bearing on the preservation of embodied energies as addressed earlier; and without proper layout, the energy conserving potential of many sites will be compromised.

A few basic principles of roadway(street) layout for conserving energy are:

1. Minimize roadway lengths in an effort to reduce vehicle trip distances, maintenance operations and utility lengths.

2. Maximize the amount of east-west residential streets in order to allow optimum unit orientation to the south. This principle is recommended due to the traditional homeowner's and builder's attitude that homes must face the street.

3. Maintain minimum roadway widths, particularly in the hot arid and hot humid regions, to minimize heat buildup around structures.

4. Street layouts which allow clustering of structures, in addition to the cost benefits and environmental benefits from increased open space, provide the greatest flexibility in locating structures in wind shadow areas, creating sun pockets through unit placement and unit orientation for optimum solar gain.

Increased efforts to quantify the energy savings of the passive options discussed earlier in this paper are being undertaken. Quantification to date suggests that even though singularly each option may not offer the potential savings that the more technical active systems or passive building systems offer, their cumulative effect warrants serious consideration.

Barriers to Energy-Efficient Land Planning and Design

REGULATORY CONSTRAINTS

Through both local and national efforts to preserve individual rights, enhance the quality of our built environment, and preserve our natural environment, we have unconsciously adopted policies, enacted legislation and developed habits that now impede the incorporation of energy-conscious design options in land planning and design into both public and private development efforts. We have land use controls that are so complex and rigid that opportunities for innovative solutions to today's problems are all but eliminated. This is not only true for energy conservation, but we begin to find that it is counter-productive to their original intent.

Since each community's land use controls, procedures and policies should be based on the community's characteristics (both natural and man-made), variations exist from community to community. The following constraints which we have witnessed in existing regulations, policies and procedures, and which are also counter-productive to a quality development are only a small sampling of what actually exists:

1. Setback requirements often prohibit the incorporation of passive/active solar options (i.e., solar greenhouses and detached collectors).

2. Under present controls, a significant percentage of structures will be built that never will be able to attain access to the sun's energy (setback, orientation).

3. Local roadway specifications often require excessive street widths and cross sections causing high development costs, unnecessary use of precious natural resources, and high amounts of embodied energy.

4. Restrictions affecting mixed land uses generate unnecessary trip distances.

5. Environmental review regulations discourage the tapping of new energy sources (i.e., gas, ground water).

6. Density regulations prohibit maximizing energy benefits obtainable from natural elements (i.e., sun, wind, soils, vegetation, ground water).

7. Regulations which limit residential building types often encourage the construction of energy-efficient structures.

8. Maintenance policies of state and local governments often create unnecessary expenditures of fossil fuels and embodied energies (i.e., mowing of rural interstate medians and shoulders). Consideration of conservation techniques such as reforestation in these areas for the development of wood as fuel is one alternative worthy of study.

9. Construction policies which encourage the use of certain materials, increase capital costs, lower the quality of development, and utilize valuable natural resources and embodied energy (i.e., paved highway medians).

10. Sufficient incentives do not exist to encourage state agencies or local communities to initiate energy efficient alternatives (i.e., inventory of natural or man-made resources available for resource subsystems, rezoning a community to energy efficient criteria).

11. Maximum conservation of energy can be achieved when various options are evaluated and incorporated into an integrated system. However with our system of specialized agencies (state and local levels)—(i.e., water sewer, planning), the opportunities to look at a comprehensive system are often overlooked.

12. Review agencies/boards lack the necessary appreciation for and understanding of energy-conservative options and techniques to reach sound conclusions. This unfortunate situation often leads to the adoption of land use policies and regulations which work against the best interests of the community.

13. The use of nonindigenous materials increases fuel consumption during transportation and impacts construction cost. Policies to incorporate indigenous materials into local construction practices should be encouraged.

If we are going to open our land use regulations to permit and encourage energy conservative developments, we need:

1. Regulations that direct, instead of inhibit development, and that are flexible and encourage innovative answers to today's problems.

2. Increased public appreciation for and awareness of the short-range and long-range impacts of a comprehensive planning process.

3. To educate members of review bodies to the positive and negative impacts of their decisions.

4. Quantification of comprehenvive energy-efficient planning approaches.

5. Expansion of the data base from which design decisions evolve.

Due to the subjective nature and complexity of the land planning and design process, each of these needs will require large expenditures of human energies and funds.

Conclusion

It is difficult to model and quantify the impacts of energy options once one goes outside the building shell, due to the scale of spaces, the variation in natural land forms from site to site, and the difficulty of dealing with the complexities of our country's diversified climates.

It is my concern that we will continue to be frustrated over land planning and design options and increasingly expend our efforts only towards the pursuit of energy options that are easier to quantify (i.e., highly technical active systems). We will leave land-use decisions to highly subjective judgments and again "pay the price" in terms of quality of environment, natural resources and potential energy savings.

We need to believe there is something to be gained from a comprehensive planning approach, to create an appreciation for the potential benefits by the general public; and then provide the designer/planner with the tools for implementation. If we do not; designers, planners, public agencies and review boards will continue to frequently reject positive development proposals and, even more frequently, accept development proposals which work against their own community's best interests.

Energy Efficiency in Existing Tract Housing: A New Challenge to Planners and Developers

ROBERT H. SOCOLOW

WE are getting used to the idea that a mature industrial society not only produces but also upgrades and maintains its goods. The road builders have the job of resurfacing existing highways, the chemical engineers have the job of recycling what used to be wastes, the librarians rebind their books, and educators have the job of educating at mid-career. What is the corresponding assignment of those of you who are planners and developers?

At least one important assignment is to go back to the tracts of housing you put in place over the last twenty years and cut their energy use in half. You are the custodians of what is potentially the nation's least expensive energy. You are challenged to produce it.

Drilling for oil and gas in our buildings[1] is a new national priority, motivated by economics, national security, and new discoveries in buildings research. American tract housing is a particularly attractive source of conservation energy, because its energy-costly features are relatively easy to fix, and these features are repeated hundreds, if not thousands of times, house after house. In an era of cheap energy, housing was built which used energy lavishly. In a market skewed to give noneconomic priorities to first costs, energy was slighted beyond what was socially optimum even at pre-OPEC prices. In an intellectual climate where rockets and reactors were glamorous

272

and engines and buildings weren't, American engineers did not notice that the models of buildings that underlay the few energy-related regulations in place failed to capture the essential roles of convective heat loss and of solar heat gain, thereby all but assuring that the design community interested in responsible vernacular architecture would lose its way. The result is housing that uses about twice as much energy for space heating as comparable housing in Sweden, normalized for climate.

Beginning almost immediately, new buildings will become sophisticated in a technical sense. Tracts will surely have a visible compass orientation, for example; the days are numbered where a single house design is casually rotated on a town plan, so that the same living room points north, south, east and west around a tract. The sunlight falling on south-facing surfaces and thus available to substitute for fuel in winter is simply too important to ignore. Changes in new construction, in planning within sites, and in siting the sites, all will have important effects on national energy use. But, in my opinion, the significance of the entire set of issues related to new housing is dwarfed by the significance of *retrofit*, the upgrading of the existing housing stock. For the remainder of this brief paper, I will try to forecast the character of the enterprise to save energy in existing homes.[2]

Quantitatively, the space heating of American homes accounts for roughly 12 percent of national energy use, approximately as much as is used in American automobiles. This use probably can be cut in half by sensible measures, a savings of roughly 2 million barrels of oil-equivalent per day. An oil province that produced 2 millon barrels per day would have a reserve of 7 billion barrels (at a reserve set equal to ten years production), so the delivery of retrofits to American homes is analogous to the development of a 7 billion barrel province. The province is larger (at least twice as large) if one includes the savings from commercial buildings, from residential water heating, from residential appliances, and from new construction.

This 7 billion barrel home-heating province is exceedingly dispersed. It is spread over 70 million homes, so that a single house is equivalent to a 100-barrel field and a 350-house complex is equivalent to a 10-barrel-per-day stripper well. Nonetheless, this conservation energy from house retrofits is potentially cheaper than most domestic fuels. The key challenge is marketing, packaging and introducing an innovation. The production of this energy will be job-intensive, requiring a national labor force of perhaps 300,000 workers if the retrofitting is spread over a ten year period. These jobs offset a much smaller number of jobs involved in capital-intensive energy production.

There is an important role for self-help, and a large leverage on costs (as conventionally calculated) if people can be taught to retrofit their own homes. There is a corresponding role for voluntary institutions and for organizations at a commmunity scale. Because of the similarities between what can sensibly be done in large buildings like schools and churches and what can sensibly be done in homes, such large buildings might become training grounds for a community-based labor force.

A nationally significant effort, however, will probably require institutional mechanisms that aggregate the market. William Rosenberg, the former chairman of the Michigan Public Service Commission, uses this analogy: "The model for blanketing our homes, if you will, is the gas companies' efforts in the 1950s and 1960s, when they went down the street, put in a pipeline, arranged for people to be home, and either hooked up a new gas furnace or converted an existing coal or oil furnace to gas. They went block by block, they put all of their costs into the cost of doing business as a gas company, and they made a profit on their return on that investment."[3] Financial mechanisms involving the gas and electric utilities would encourage an aggregation of the market; among these are rate-basing of utility investments in conservation and establishing joint ventures of banks and utilities. But there are other paths to aggregation, like franchises—modeled perhaps on H & R Block's technical assistance with income taxes, or Avon's door-to-door marketing of cosmetics, or Century 21's coordination of real estate agencies.

The rate of production of conservation energy may depend crucially on the availability of long-term financing at favorable interest rates. Utility rate-basing is attractive in this respect, as are various subsidies through the tax system, including—in the particularly knotty rental market—tax advantages like accelerated depreciation. There is apparently a very large source of long-term capital in the insurance industry, an industry increasingly interested in conservation.

The production of energy conservation in homes is critically handicapped by problems of credibility. Because physical units like therms and kilowatthours are forbidding, scorekeeping is almost exclusively in dollars, and current dollars at that. The result is that modest efforts at conservation that actually do achieve energy savings in the 10 to 20 percent range appear to the household in a particularly unsatisfying form, captured by the lament, "I went through this whole thing, and my energy bill is no less than it was a year ago." Compound this lack of positive feedback with the mystification created by a ramshackle market in unfamiliar devices of all kinds that will cut fuel bills by X percent, and you have an environment for the dissemination of innovation that could hardly be worse.

Four proposals address this set of problems: 1) keep score in physical units, 2) carry out and publicize well-disciplined demonstration programs,

3) train (and perhaps license) specialists in the diagnosis and treatment of houses, and 4) maintain a vigorous R & D effort that is alert to adverse side effects of conservation strategies.

Scorekeeping in physical units founders unless a single method is used nearly exclusively. The analogy is the scorekeeping in miles per gallon that facilitates the effort to improve the energy efficiency of automobiles. For household energy use, scorekeeping, in my opinion, ought to be in *gallons per year*, with gallons understood to mean gallons-of-fuel-oil-equivalent at 140,000 Btu (1.4 therm) per gallon. Conservation measures in a large house in an average climate or in a house of average size in a severe climate should reduce the space heating score from 2000 to below 1000; conservation measures in a house of average size in an average climate should improve the score from 1000 to below 500. By using *gallons*, one connects with the energy unit whose price is best known, the gallon of gasoline. Barring unusual tax preferences, the price of a gallon of liquid fuel is not likely to vary dramatically no matter what its form (gasoline, diesel fuel, gasohol, heating oil, etc.). Moreover, during at least the first part of the period when the scorekeeping is getting established, one has the added convenience that scores in gallons and in dollars are nearly the same. In choosing gallons *per year*, rather than, for example, per square foot or per degree day, one avoids a large number of problems of interpretation. (Automobile scorekeeping does not divide out the weight of the car for similar reasons.) The one potential embarrassment of gallons per year is its nationalism, but neither the liter per year nor the gigajoule (GJ) per year—two possible metric units of energy—is in the cards.

Several recent or imminent government programs are designed to facilitate the quantitative discussion of energy use in houses. The Residential Conservation Service is an already mandated program to audit all houses on demand. Special programs addressing rental housing (designed to give both landlord and tenant a common set of facts they might both trust) and addressing housing at the time of sale (designed to get a recent energy audit stapled to the closing papers) have been proposed by the Congress.

Demonstrations are the best method I can invent to persuade people that we really can save a lot of energy in buildings. I believe that demonstrations, or pilot projects, will be done in all representative climates and building types, trying out various organizational forms. A demonstration in an inner city will have a different set of actors and different financing from one done in a suburban community.

A demonstration program can be designed as a training experience for its participants. The house retrofit is challenging in an institutional sense because trained people and the necessary products are available to implement the second-most-effictive conservation strategies, such as installing insulation and storm windows, but not to implement what appear to be the most effective strategies of all, those which address, especially, the leakiness of houses. As Arthur Rosenfeld, Director of the Buildings Research Group at Lawrence Berkeley Laboratory, describes it: "The modern American house is built a little bit like a lobster trap, and instead of the water flowing through it, air flows into it, is heated, and then flows out of it."[4] The leaks were *designed* into the structure, at a time of nearly perfect unconsciousness about energy. They bring warm air from basement to attic through open passages in interior walls, through hollow masonry blocks, through open shafts around furnace flues. They bring cold air through electric sockets punched in exterior walls and through the oversized holes made to help the electricians and plumbers run their wires and pipes.

The leaks often bypass the insulation that is in place and degrade its value. Finding the leaks takes equipment that is not yet available commercially, and some on-the-job training; the ultimate cost/benefit ratio for a strategy of finding and fixing leaks and bypasses, however, is almost surely going to be one of the most impressive available, at least in American housing built in the last thirty years.

Those trained to diagnose houses with new kinds of equipment and to treat those routine problems that engender energy waste have sometimes been called house doctors. In other metaphors they are detectives or mechanics. Their identifying characteristic is a mission of treating the house and changing it. The current program in this country subordinates this activity to another (limited to describing a house and making recommendations for action) that requires people with different training, collars white instead of blue. Although neither activity precludes the other, a program for houses and households expressive of urgency, in my opinion, should have as one objective that the very first visit that takes so much time to arrange is a visit in which something important goes on that saves energy. The visitor is carrying a blower door that helps find leaks, and a clock thermostat, and a water heater blanket, and gaskets that go behind wall outlets, a low-flow showerhead, a caulking gun, and weatherstripping.

In addition to persuading onlookers and training participants, each demonstration will be part of an R & D effort, in the sense that it is designed to reduce the uncertainty about the feasibility of some retrofit strategy, or its acceptability, or its effect on energy consumption. One important result of some demonstrations will be the discovery that a particular strategy is unacceptably costly or unpleasant or ineffectual. We are still low on the learning curve for energy conservation in housing.

In that sense, it is fortunate that one of the salient characteristics of the buildings conservation program is that you couldn't and wouldn't want to do all of it at once. You can imagine making 70 million hoola hoops or beanies with propellers on the top and having every household try them for a year and then give them up. You wouldn't want to do that with buildings. If you go too fast, you run into every kind of bottleneck. You also lose the opportunity to go into houses a few years from now with a new set of ideas that weren't on the original lists.

Moreover, there are side effects of conservation strategies that aren't fully understood. It is worrisome that houses could become overtight, in the sense that air pollution from sources within the house reaches levels which have the potential for long-term adverse consequences for public health. Mechanical ventilation heat exchangers are becoming available that can provide fresh air at a much reduced energy cost. All the more reason to proceed in a way that allows us to keep making corrections.

To conclude, the near future will see a major effort to save energy in existing buildings, facilitated by new forms of financing, better information, and a labor force with new skills. What is still lacking is an effective delivery system for conservation. Delivery will be easiest by far, however, where many houses are alike in a single place. America's planners and developers are obvious experts in this task, for in order to bring tract housing into existence in the first place, you had to solve an altogether relevant set of tasks: to phase marketing and delivery of services, to coordinate a wide variety of building professionals, and to stay responsive to—even one step ahead of—customer demand. How are you now going to proceed?

Notes

1. This particular felicitous phrase is the title of Report PU/CEES 87 of Princeton University's Center for Energy and Environmental Studies, July 1979, by Marc H. Ross and Robert H. Williams. I have been very much stimulated by extensive discussions with both authors.

2. I am helped in this exercise by my notes from the Residential Workshop of the Alliance to Save Energy Conference Symposium at Dumbarton Oaks, Washington, D.C., October 17-18, 1979. I was the Technical Advisor at that Workshop. See also the book I edited based on our experimental research *Saving Energy in the Home: Princeton's Experiments at Twin Rivers* (Cambridge, Mass., Ballinger, 1978) and my more conceptual essays, "The Coming Age of Conservation," *Annual Reviews of Energy 2* (1978), pp. 239-289, and "Energy Conservation in Housing: Concepts and Options," in *Future Land Use: Energy, Environmental, and Legal Constraints* by Robert W. Burchell and David Listokin, eds. New Brunswick, N.J.: Rutgers University Press, 1975, pp. 311-324.

3. Private communication, October 1979.

4. Private communication, October 1979.

The Adoption of Energy Conservation Features in New Homes: Current Practices and Proposed Policies

EDWARD J. KAISER
MARY ELLEN MARSDEN
RAYMOND J. BURBY

Introduction

RESIDENTIAL environments account for a major portion—almost 20 percent—of total fuel energy consumption in the United States (Stanford Research Institute, 1972). Although residential energy use is expected to increase at an annual rate of 1.7 percent in the nation as a whole (Hirst and Carney, 1978), and at a much higher rate in states and localities where continued economic development is taking place, significant reductions in residential energy growth are possible if the adoption of energy conservation practices is widespread and rapid. In this paper, data and analyses on consumer and builder energy conservation behavior and intentions are presented. Based on this information, a variety of policies are proposed to make energy conservation potential a reality in the residential sector.

In recent years the base of knowledge about factors affecting energy conservation in the residential sector has been expanding rapidly. Major obstacles to energy saving in the residential sector appear to be "inertia and institutional obstacles, rather than technology or building economies" (Council of State Governments, 1976, p. 30). Most research, however, has focused on institutional obstacles rather than individual characteristics.

Characteristics of the housing industry produce a "self-reinforcing resistance to change" (Schon, 1967). Homeowners' and investors' use of energy saving features also has involved relatively little change in lifestyle or investment. More specifically, the trend toward energy efficiency in homes has been slowed by building codes which may have frozen outmoded practices into law (Thompson, 1978); the highly leveraged character of the industry, which leads builders and buyers to be very sensitive to the first costs of housing and long-term value, but not operating costs (Hirschberg and Schoen, 1974); the sensitivity of lending institutions to risk, which can lead to an aversion to financing innovative housing (Barrett, Epstein, and Harr, 1977); fragmentation of the building industry, which increases the number of decision points in the diffusion of an innovation (Council of State Governments, 1976); the orientation of building craft unions to traditional practices (see Beyer, 1965; National Commission on Urban Problems, 1968, Pt. III, Chapter 4); and legal uncertainties regarding "access to the sun," which may result in consumer hesitancy to invest in solar devices (Eisenstadt and Utton, 1976; Miller, Hayes, and Thompson, 1977; Myers, 1978). Perhaps most important is the complexity of the process of building for energy efficiency in homes. A variety of types of decision agents, performing a highly interrelated set of functions, each subject to influences from a vast array of sources, determine the energy efficiency to be realized in the residential sector (The George Washington University, 1978).

While information about economic and institutional constraints has been valuable in the initial development of policies to promote the adoption of energy conservation features in new housing and neighborhood design, information at a finer grain is needed if we are to move forward to a second generation of policy development. In particular, we need detailed information about the market (consumer demand) for various energy conservation features and about the builders who are most likely to respond to market signals. For example, to what extent has the use of energy conservation features in new housing been expanding? How much extra are consumers willing to invest to achieve greater energy efficiency in their next house? To what extent are builders responding to consumer interest in energy efficiency? Which energy conservation features are builders adopting most rapidly? What sources of information and adoption decision criteria do builders use in deciding to use various energy conservation features in new housing? Answers to these questions should provide a basis for more finely tuned and carefully targeted policies to increase the market penetration of energy conservation practices and new technology in the new housing segment of the residential sector.

The Data

The data reported were collected in the fall of 1978 through surveys of statewide random samples of households and home builders in North Carolina. The household survey involved a 30-minute telephone interview with the head or spouse in 604 households. The sample frame of telephone numbers was constructed so that cases were drawn from each county in the state in direct proportion to the county's proportion of the state population. Telephone numbers were then systematically selected from telephone books and the last two digits were randomized. The response rate for the household survey was 81 percent.

The sample frame for the survey of single-family home builders was constructed from two sources: (a) membership lists of the North Carolina Association of Home Builders; and (b) the "List of Licensed General Contractors" published by the North Carolina Licensing Board for Contractors. Personal interviews averaging 60 minutes each were obtained with a random sample of 100 home builders located in 62 communities across the state. The response rate for the builder survey was 73 percent.

Energy Efficiency Features in New Houses: Past, Present, Future

THE INCREASING LEVEL OF ENERGY CONSERVATION FEATURES OVER THE PAST TWO DECADES

The dramatic trend toward increasing energy efficiency in new homes is illustrated in Exhibit 1 which shows the extent of use of energy-saving features as original equipment in owner-occupied units over the past two decades, excluding consumer retrofitting by households, which has further contributed to energy efficiency of dwellings. The use of insulation, weatherstripping, caulking, storm windows or double paned glass, storm doors and heat pumps has increased substantially over the period. The decrease of one type of energy saving feature—attic fans—may be associated with the pervasive use of central air conditioning in new homes. By 1975, over three-fourths of homes had insulation in the ceiling, floors, or walls or had weatherstripping or caulking; over half had storm doors and storm windows or double-paned glass. None of the households in this study

EXHIBIT 1
AGE OF STRUCTURE AND USE OF ENERGY EFFICIENT EQUIPMENT[a]

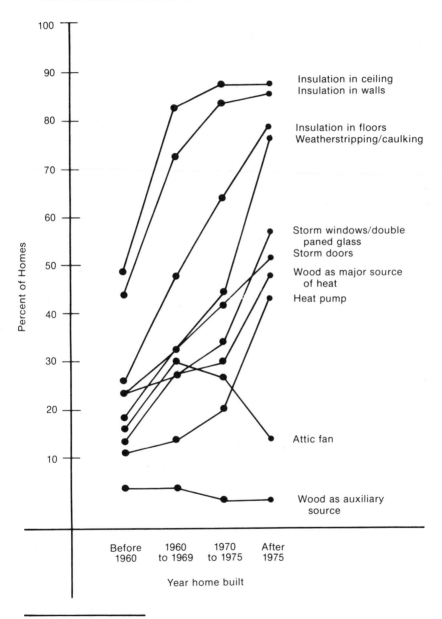

[a]Figures include only original equipment and exclude retrofitting by current owner.

used alternative energy sources such as solar heating. Although these figures exclude retrofitting of existing housing by homeowners and do not consider the rental segment of the residential sector, they do illustrate the dramatic trend toward energy efficiency in homes.

CURRENT STATUS: INCORPORATION OF ENERGY CONSERVATION FEATURES IN THE CONSTRUCTION OF NEW HOUSES IN 1978

Home builders' current use of energy conservation features in new homes in North Carolina is summarized in Exhibit 2. The twenty features summarized in the table are divided into the two groups: (1) heating and cooling equipment; and (2) construction features to improve thermal performance with respect to heat gain and loss. The left-hand column of numbers refers to the percent of builders in the sample who used the particular feature in their most recent speculatively built houses. The right-hand column weighs the use of the particular feature by the number of houses constructed by the builder in 1978. Thus, use by larger builders received more weight and the result is an estimate of the percent of new speculative houses that are incorporating a particular feature, instead of the percent of builders who are using the feature.

Examining equipment features first, it can be seen that both the sizing of heating/cooling systems to closely match but not exceed design loads (used by 88 percent of the builders and 96 percent of houses) and energy-efficient (compared to electric resistance heating) heat pumps (82 percent and 85 percent respectively) have been adopted widely. Among approaches to improve the thermal performance of the house, over 90 percent of the builders and houses were incorporating storm or double-glazed windows, over 80 percent were utilizing a square or rectangular shape, and about two-thirds had adopted insulated ceiling access panels, insulation exceeding the requirements (R-19 in ceiling, R-11 in walls and floors) of the North Carolina Uniform Residential Building Code, and had limited the glass area to 10 percent or less of the house's floor area. Thus, the evidence shows that builders are definitely taking steps to produce more energy-efficient homes. On the other hand, almost no builders were employing solar features or using reflective glass or insulated shutters.

FUTURE PRACTICE: FURTHER ENERGY CONSERVATION FEATURES MOST LIKELY TO BE EMPLOYED NEXT BY BUILDERS AND DEVELOPERS

To obtain a reading on emerging practice as well as current practice, builders were asked what feature they would be most likely to add if condi-

Exhibit 2

**ENERGY CONSERVATION FEATURES USED BY
THE HOME BUILDING INDUSTRY: 1978**

Energy Conserving Feature	Estimated Percent Builders[a]	Houses[b]
Heating/Cooling Equipment Features		
1. Heating/cooling system closely sized to match design loads	88	96
2. Heat pump heating/cooling system	82	85
3. Attic fan to draw in cool night air	25	11
4. Clock thermostat on heating/cooling system	14	29
5. Wood stove	5	2
6. Solar hot water heating	4	1
7. Active solar space heating	0	0
Construction Features Affecting the House's Heat Loss/Gain		
1. Storm windows/double glazed windows	93	90
2. Square- or rectangular-shaped house	82	81
3. Insulated ceiling access panel	67	66
4. Insulation exceeding building code standards	64	58
5. Glass area 10 percent or less of floor area	65	75
6. Landscaped lot with deciduous trees for summer shade	51	49
7. Fireplace which uses outside air for combustion	28	15
8. Insulated hot water pipes	23	34
9. Thirty-inch roof overhang to let in winter sun while providing summer shade	20	8
10. Passive solar heating using a maximum of south-facing glass	10	7
11. 2 × 6 framing for extra insulation	8	5
12. Reflective glass	4	0
13. Insulated shutters	1	0

[a]Percent of builders who used conservation feature on latest speculative house built by firm. N = 80

[b]This estimate is obtained by weighting the builder's use of the feature by the number of houses he built in 1978. Hence, larger builders received more weight in this "weighted average" approach.

tions warranted production of a more energy efficient dwelling. Exhibit 3 shows the results. The most likely next increment of energy-efficient practices will be, in absolute increases, even more insulation (17 percent) and an energy-efficient fireplace (15 percent). If one examines relative increases (the proportion of those not already using a feature, but who would go to that feature next), one finds that builders will continue to turn to storm windows and increased insulation to improve thermal efficiency first, then to heat pumps and working fireplaces as heating systems. In spite of these observations, however, the predominant impression is that builders will not turn to any one energy conservation feature. In addition, few are willing to adopt solar heating principles yet.

Although most attention to energy conservation in the residential sector has focused on individual homes, additional opportunities to conserve energy are present in the location and design of subdivisions. From a former official of the U.S. Department of Energy, energy savings through the energy-conscious design of new neighborhoods could result in a 5 percent reduction in national energy consumption by the year 2000 (Leighton, 1977). Realizing opportunities for energy savings through neighborhood design will require major changes in land developers' usual ways of subdividing land, as well as changes in local land development codes to allow clustering and energy-efficient streets and lot orientations. In order to explore developers' interest in new ideas for neighborhood design, we first asked those in the sample of home builders whether their firms engaged in residential land development. The 43 firms that were also land developers were then asked how probable it was that their firm would adopt each of a series of nine energy-saving ideas for subdivision design and development. Their responses are summarized in Exhibit 4.

A majority of the builder/developers indicated that it was either very probable or somewhat probable that their firms would use four of the ideas during the next five years. These four were: (1) location of the subdivision near existing community facilities; (2) landscaping with deciduous trees to shade streets in summer; (3) use of two or more land uses to reduce travel needs of the residents; and (4) use of a cluster lot design with community open space. These four practices, of course, will require the least amount of change in developers' current modes of operation. More radical ideas were much less likely to be viewed as probable candidates for adoption. Only 35 percent thought it was very or somewhat probable that they would attempt to orient lots in a north-south direction, while 23 percent indicated some likelihood of using solar access covenants, 21 percent thought they might use narrower streets than usual, and only 2 percent thought there was any chance of their using a central heating and generating system (ICES) for a

Exhibit 3

**ENERGY CONSERVATION FEATURES MOST LIKELY TO BE
ADDED IN THE NEXT HOUSE IF CONDITIONS WARRANTED A
MORE ENERGY EFFICIENT DWELLING**

Energy Conservation Feature	Percent of Builders Who Would Use the Feature Next (N varies)	
	As Percent of the Total Sample (N = 80)	As Percent of Those Not Yet Using the Feature
Insulation exceeding building code standards	17	47
Fireplace which uses outside air for combustion	15	21
Attic fan to draw in cool night air	10	13
2 × 6 framing for extra insulation	8	9
Solar hot water heating	8	8
Glass area 10 percent or less of floor area	7	20
Insulated hot water pipes	6	8
Storm windows/double glazed windows	4	57
Thirty-inch roof overhang to let in winter sun while providing summer shade	4	5
Heat pump heating/cooling system	3	16
Landscaped lot with deciduous trees for summer shade	3	6
Clock thermostat on heating/cooling system	3	3
Active solar space heating	3	3
Insulated ceiling access panel	1	3
Wood stove	1	1
Insulated shutters	1	1
Passive solar heating using a maximum of south-facing glass	0	0
Square- or rectangular-shaped house	0	0
Heating/cooling system closely sized to match design loads	0	0
Reflective glass	0	0

QUESTION: Now, thinking about the energy conservation features that you did not use in that house, which *one* would you most likely use if conditions prompted you to build a more energy-efficient house (referring to list of features on the card considered in the previous question)?

subdivision they might develop. In sum, it appears that while a number of possibly good ideas for saving energy through subdivision design are available, if they are to be adopted more rapidly, developers must be given more information about their usefulness in saving energy and their market acceptance. Otherwise, most will not make the difficult transition from concept to reality.

FUTURE CONSUMER DEMAND FOR ENERGY EFFICIENCY

Builders' increasing interest in energy efficiency in their housing products (if not their neighborhoods) is a reflection of growing consumer interest in energy-saving features of housing. To explore consumer demand further, households were asked whether they planned to move over the next two to three years and, if so, whether they expected to buy or rent. Households that were likely to be in the market for a new home were then asked about their interest in energy-related aspects of their next house.

Compared to the 45 percent of homeowners who considered energy efficiency in selecting their current homes (and the 22 percent who rated energy efficiency as a very important factor in their choice), 87 percent of the prospective home buyers we interviewed said that they would think about having more energy-saving features in their next house. See Exhibit 5. The energy conservation features prospective buyers said they would be looking for most were features to improve the thermal efficiency of the house, including insulation, storm windows and storm doors. Next most frequently mentioned were various energy-saving appliances. Finally, a few prospective buyers (less than 10 percent for each feature) mentioned heat pumps, wood stoves or fireplaces, solar heating or hot water, and the general design of the house. A remarkably high proportion—25 percent—when asked directly whether they had thought about having solar hot water heat in their next house said that they had considered having this energy-saving feature.

In addition to reporting that they would think about having various energy conservation features when they selected their next home, very high proportions of the prospective home buyers indicated that they would be willing to spend more on the house in order to achieve savings in their heating bills. Ninety percent said they would spend $200 more to save $50 per year; 85 percent would spend $600 more to save $100 per year; and fully 80 percent indicated that they would spend $1,200 more for a heat pump if they could save a third on their heating bills.

Exhibit 4

PROBABLE USE OF ENERGY CONSERVATION IDEAS FOR SUBDIVISION DESIGN[a]

Energy Conservation Idea	Percent of Builder/Developers Who Rate Probability of Use During Next Five Years: (N = 43)		
	Very Probable	Somewhat Probable	Not Probable
1. Location near existing community facilities, such as shopping, schools, employment	59	25	16
2. Landscaping with deciduous trees to shade streets in summer	49	26	25
3. Two or more land uses included in project, such as residential and retail, residential and recreation, residential and employment	30	28	42
4. Cluster design with community open space	19	32	49
5. Bicycle/walking paths	14	33	53
6. Orienting all or most lots in a north-south direction	16	19	65
7. Solar access covenants	9	14	77
8. Narrow street paving to reduce solar absorption	14	7	79
9. Installing one central heating and electric generating system for the entire development	0	2	98

[a]QUESTION: In addition to residential building, does your firm also subdivide and develop residential land? (If yes) Here is a list of energy conservation features that have been suggested for use in subdivision design and development. HAND CARD G For each, please tell me whether it is very probable, somewhat probable, or not probable that you will use it during the next five years. First, what about . . .

Exhibit 5

**PROSPECTIVE HOME BUYERS' INTEREST IN
ENERGY EFFICIENT HOUSING**

Indicator	Percent of Prospective Buyers (N = 79)
Thought About Energy Saving Features of Next Home[a]	
Yes	87
No	13
Total	100
Energy Saving Features Would Most Likely Look for in Next Home[b]	
Insulation, storm windows, storm doors	51
Energy saving appliances	19
Wood stove/fireplace	7
Heat pump	6
Solar space or hot water heating	6
Design of house	3
Don't know	8
Total	100
Thought About Adding Solar Hot Water Heating[c]	
Yes	25
No	75
Total	100
Willingness to Spend More to Save on Heating Bills Spend $200 more to save $50 per year[d]	
Yes	90
No	10
Total	100
Spend $600 More to Save $100 per Year[e]	
Yes	85
No	15
Total	100

Exhibit 5 (Continued)

**PROSPECTIVE HOME BUYERS' INTEREST IN
ENERGY EFFICIENT HOUSING**

Indicator	Percent of Prospective Buyers (N = 79)
Spend $1200 More for Heat Pump to Save One Third on Heating Bills[f]	
Yes	80
No	20
Total	100

[a] QUESTION: Has the cost of energy made you think about having more energy-saving features in your new home?

[b] QUESTION: (If Yes) What would you most likely be looking for? Open-ended responses classified and coded by research staff.

[c] QUESTION: Have you thought about adding a solar hot water heater?

[d] QUESTION: If you were able to save $50 per year on heating bills, would you be willing to spend $200 more for your new home?

[e] QUESTION: If you were able to save $100 per year on heating bills, would you be willing to spend $600 on additional construction costs for energy saving devices in your next home?

[f] QUESTION: Would you consider buying a new home with an electric heat pump which would cost $1,200 more to purchase, but would save you one-third on your heating bills?

Factors Influencing the Adoption of Energy Conservation Features in the Construction of New Houses

To explain the incorporation of energy features in new speculatively built homes by builders, three sets of variables related to the adoption/nonadoption decision process were measured and evaluated. These include: (1) primary stimuli or motivation to consider energy-efficiency features in the first place; (2) criteria used in making final adoption/nonadoption decisions; and (3) sources used in building awareness of initial stimuli and in final adoption decisions.

PRIMARY STIMULI

Builders' interest in using energy-conservation features in new houses produced by their firms is triggered by some stimulus. This may be market demand, the actions of their competitors, or an awareness of opportunities, perhaps stimulated by educational or incentive programs of professional associations, public interest groups or government. The relative roles of each of these factors is examined here.

Marketing and Cost Factors.— When asked open-ended questions about what caused them to consider using an energy feature in their most recent house and what factors affect the use of energy conservation features in general in their market area, builders emphasized market demand and cost factors above all others. See Exhibit 6. Two-thirds of the respondents chose either one or both of these answers as original stimuli in their most recent house and an even higher percentage, 84 percent, chose either one or both as the best explanations for builders' overall use of energy-conservation features in their market area.

While being market-oriented, most builders do not consider the adoption of energy-conservation features as innovations in the sense of creating a substantially different housing product to expand beyond their previous markets. Rather their motivation is "to maintain adequate sales and marketability of . . . houses" of the type they already build for whatever market they are already serving. Thus, the incorporation of energy-conserving features is part of the pattern of conservative, adaptive, economic behavior.

Cost and marketability appear to be closely linked. Cost considerations, both from the standpoint of initial costs (which the market must be willing to pay) and operating costs (which consumers hope to reduce by installing

Exhibit 6

**STIMULI TO CONSIDER PUTTING ENERGY FEATURES
INTO NEW HOUSES**

Stimulus Category	Percent of Builders Who Cited Stimulus	
	With Respect to Their Own Most Recent Speculative House[a] (N = 76)	With Respect to Builders in General in This Market Area[b] (N = 100)
Marketing ("sales," "customer demand," "consumer desires")	36	41
Cost factors (mostly cost of energy)	32	43
Awareness/knowledge of energy conservation features	21	25
Technical efficiency	17	NA
Government regulations	5	15
Builder ethics, pride	7	6
Other miscellaneous stimuli (none more than 4 percent)	26	16

[a] QUESTION: What was it that caused you originally to consider using (energy feature being probed about most recent speculative house)? (Feature is: extra insulation, 42 percent; heat pump, 39 percent; passive solar heating, 7 percent; 2 x 6 framing, 7 percent; solar hot water, 3 percent; fireplace with outside air, 3 percent.)

[b] QUESTION: What do you think are the major factors that affect the use of energy-conservation features by builders? (An open-ended question, coded later in the office; not limited to one factor if several mentioned.)

energy-conserving features) are key factors in determining whether energy-conservation features are marketable and therefore whether they are included in new homes.

Awareness and Knowledge.— Awareness and knowledge ranked third, behind cost and marketing factors, as a stimulus to consider installing energy conservation features in new houses (see Exhibit 6). Twenty-one percent of the builders cited it as a factor with respect to their own most recent speculatively built house and 25 percent cited it as a factor for builders in general in their market area. Although well below the percentages of builders who cited marketing and cost factors, it is well ahead of other explanations.

Exploring this explanation a bit further, we find that a high proportion of builders say they are aware and concerned about the energy situation. Sixty-one percent of the builders thought that the need to save energy is very serious and 68 percent thought that people do not have the right to use as much energy as they can pay for. However, builders get mixed grades on their knowledge about sources of unwanted heat gain and loss in a house. Ninety-one percent knew that the standards of the North Carolina Uniform Residential Building Code are required by law and not just recommended. Eight-nine percent answered correctly that windows are the biggest heat loss (compared to roof, walls or floor). However, only 63 percent answered correctly that cutting infiltration around windows and doors is more effective than adding more insulation over the ceiling. Still fewer (only 40 percent) answered correctly that a conventional fireplace does not reduce the winter heating requirements in a typical house in North Carolina (in fact it is a source of heat loss if operated while the heating system is on).

Based on statistical measures of association (not shown) and in spite of its ranking third among factors cited by builders, there is no evidence in these data that greater knowledge of the causes of unwanted heat gain and loss leads to increased use of energy-conservation features in new houses. Nor is there evidence that builders who hold stronger energy-conservation attitudes are more likely to use energy-conserving features in the houses they build. In fact, the greater the knowledge and the stronger the attitude toward energy, the less likely builders were to use energy-conservation features, although the associations are not statistically significant. The overall conclusion is that builder behavior is unrelated to general attitude and knowledge.

To summarize findings about initial stimuli and basic factors underlying adoption of energy-efficiency features, it is clear that straightforward conservative economic factors of marketing and cost considerations are the primary behavior motivations. Other factors, such as energy-conservation attitudes, technical knowledge about energy, governmental regulations and simple inclination to improve one's product probably play significant supporting roles, but they are clearly secondary to motivations related to sales considerations. The policy implications are straightforward: (1) encourage buyers to demand energy-efficient houses; (2) reduce costs of energy-conservation features; and (3) make it possible for buyers to pay for them more easily (say, through lenient interest rates) or; (4) raise the cost of energy that can be saved by employing conservation features.

FINAL DECISION CRITERIA

Size and Price of the House.—Clearly one of the most important factors in decisions about whether to use energy-conservation features in new construction is the size and price of the house to be marketed. Although only 50 percent of the builders stated that the price range of the house was important, Exhibit 7 shows clearly that price does make a difference. More expensive, large, generally two-story houses are much more likely to incorporate energy-conservation features. For example, over twice as many houses above $55,000 have eight or more energy-conservation features when compared with those priced below $55,000.

The principle behind the installation and marketing of energy-conservation features is that, although they add to the initial cost, they save money in the longer run. Unfortunately, they add more proportionately to the cost of lower-priced houses, where the customer's purchase decision is also much more vulnerable to this initial cost. Thus, it is those least able to afford higher operating costs later who are also least able to avoid them by investing in a more energy-efficient house to begin with.

Other Decision Criteria.— Builders were also asked explicity about criteria they used in deciding about energy features (as opposed to initial stimulus to consider a feature). Uppermost in their minds at the time of the decision was marketability, expressed as consumer demand, interest or at least acceptance. As shown in Exhibit 8, marketability was mentioned by almost 90 percent of the builders. Furthermore, almost half felt it was the single most important criterion in their final decision (see right-hand column in Exhibit 8). Performance reliability is also an important criterion when it comes to the final decision about incorporation of an energy-efficient feature.

To summarize the discussion about final decision criteria, marketability and costs are again the vital considerations for the builder. Uppermost in

Exhibit 7

THE INFLUENCE OF THE SIZE AND PRICE OF THE
HOUSE BEING BUILT ON THE NUMBER OF ENERGY
CONSERVATION FEATURES INCORPORATED

Independent Variables: Type of House	Dependent Variable: Number of Features Used[a] (N = 50)		
	Energy Efficient Equipment	Thermal Efficiency Features	Total Number of Features
Two-story vs. other (mostly one-story)	NS	.26	.23
Size (number of square feet)	.24	.20	.28
Sales prices	.23	.21	.28

[a]The statistics shown are Kendall's Tau$_c$.

NS = Not statistically significant at .05 level.

his mind is whether the added costs of the energy conservation features will be justifiable to the consumer. These considerations become especially critical, and more difficult to meet, as the price of the house goes down. That is, it is more difficult to market energy-conservation features on less expensive houses. Improving the availability, ease of installation and reliability of energy-conservation features would further reduce the financial risk to the builder.

SOURCES OF INFORMATION ABOUT ENERGY CONSERVATION FEATURES

Builders' sources of information are expected to influence the adoption of energy conservation features in two ways. The first is indirect and

Exhibit 8

DECISION CRITERIA USED IN MAKING FINAL DECISION ABOUT USE OF ENERGY CONSERVATION FEATURE

Decision Criteria	Percent of Builders (N = 75)	
	Used as a Criterion Perhaps Among Others[a]	Most Important Criterion[b]
Likely consumer demand, interest, acceptance	89	47
Performance reliability	69	28
Availability	40	1
Ease of installation	40	0
Cost in comparison with alternatives	36	4
Degree of change from houses company was building	28	1
Use by competition	27	0
Willingness/unwillingness of subcontractors to participate	24	0
Building regulations	20	8
Profit margin	19	1
Others	23	9

[a]QUESTION: Thinking back when you were deciding whether to use (NAME OF FEATURE), which of the following factors did you consider in making up your mind about it? (HAND CARD J)

[b]QUESTION: Again, thinking about the factors you considered in deciding to adopt (NAME OF FEATURE), which factor was the most important in your decision?

operates through the conditioning effect of information on the primary stimuli that induce builders to consider energy-related features of housing in the first place and, later, on decision criteria used in deciding to adopt particular features. The second is more direct. Particular sources of information, because they are particularly credible or because they provide more useful data, may themselves induce builders to construct more energy-efficient houses. If we can (1) identify the sources builders use to find out about energy and which are relied upon in making the final decisions about whether or not to use new features, and (2) determine which sources are most related statistically to the actual use of energy-efficient features in houses that are built, we will have a powerful tool for channeling additional energy information to builders and for further stimulating energy efficiency in new housing.

As shown by the first column of figures in Exhibit 9, builders use a wide range of written material and personal contacts. Every source offered as a possibility was used by at least 40 percent of the builders and 10 of the 16 sources were used by more than half of the respondents. The three sources marked by asterisks are the ones builders reported they relied on most in their final decisions about the use of energy-conservation features, regardless of how many other sources they used or how often they used them. These three key information sources were the electric power company, consumers, and the National Association of Home Builders' handbook on energy conservation.

The remaining three columns of the table show the statistical relationship (Tau_c) between the use of each source and the number of energy conserving features incorporated in the most recent speculatively built house. While the electric company, popular magazines, TV, newspaper, consumers and trade publications were all used by 60 percent or more of builders, their use does not seem to be related statistically to whether or not energy-efficiency features are built into new homes. Of the more commonly used sources, only the use of suppliers appears to be strongly related to eventual adoption of energy-conservation features. Many of the information sources that are used by fewer builders, on the other hand, appear to be much more strongly related to adoption by those who do use them. These sources include the National Association of Home Builders' handbook on designing, building and selling energy-conserving homes (National Association of Home Builders, 1978), as well as books and architects. The NAHB handbook is also one of the three sources most relied upon by builders in making their final decision about whether or not to use an energy-conserving feature.

Overall, it seems that the design of a simple, yet efficient informational program to reach builders is made difficult by their use of such a wide range

Exhibit 9

**BUILDERS' USE OF SOURCES OF INFORMATION
ABOUT ENERGY CONSERVATION FEATURES AND
THEIR INFLUENCE ON ADOPTION**

Independent Variable: Source of Information about Energy Conservation Features	Percent of Sample Using the Information Source[a] (N = 100)	Dependent Variable: Number of Features Used[b] (N = 80)		
		Energy Efficient Equipment	Thermal Efficiency Features	Total Number of Features
*Electric utility company	79	NS	NS	NS
Suppliers	67	NS	.23	.24
Popular magazine, TV, newspaper	67	NS	NS	NS
*Consumers	61	NS	NS	NS
Trade publications, generally	60	NS	NS	NS
Seminars and meetings on energy	59	NS	.24	NS
Subcontractors	59	NS	NS	NS
Federal government publications	53	NS	NS	NS
*National Association of Home Builders' handbook on energy	51	.29	.27	.34
State governmental officials	50	.28	NS	NS
Local governmental officials	46	NS	NS	NS
Other builders	45	NS	NS	NS
Consultant engineers	45	.27	NS	NS
Books	43	.23	.34	.36
Architects	41	.37	.22	.36
Gas utility company	40	NS	.21	.22

NS = Not statistically significant at .05 level.

*Sources builders relied on most in making final decisions about the use of energy-conservation features.

[a]QUESTION: Here is a list of possible sources of information about energy-conservation features. HAND CARD H. For each, I want you to tell me whether or not you used it during the past year in finding out about energy conservation features you would include in the houses you have built?

[b]The figures in the table are Kendall's Tau$_c$.

of information sources and the fact that the ones most widely used are not the ones whose use is most strongly related to the observed use of energy-conservation features. A well conceived handbook by a credible source (like the NAHB handbook) seems to carry weight for those who use it, but only about half the builders used that particular source. To be effective, the use of the handbook by a credible builder's organization would probably have to be supplemented by a shotgun approach in numerous other sources, perhaps calling attention to the handbook.

Energy Policy Implications

State and local policies directed toward energy conservation in the residential sector can provide a strong supplement to those outlined in the president's National Energy Plan. Making energy conservation and energy efficiency the cornerstone of federal energy policy, the former energy plan included several policies particularly affecting energy consumption in the residential sector—energy performance standards for new buildings; tax credits for homeowners for retrofitting existing housing; minimum efficiency standards for appliances; direct subsidies to low-income persons for retrofitting, and the establishment of rate structures to encourage energy conservation (U. S. Department of Energy 1978; Council on Environmental Quality, 1979, pp. 35-39).

These major policy pronouncements were reinforced by those set forth in state and local energy-management plans and by a number of other federal programs. Broadly, these former policies and other potential policies may be categorized in several policy areas—price strategies; supply restriction or allocation strategies; regulatory strategies; incentive strategies; and information strategies (see Healy and Hertzfeld, 1975, p. 7). This section outlines potential additional energy-conservation policies for the residential sector, concentrating on policies that our analyses suggest will be effective in affecting the energy efficiency of new housing. Since price strategies and supply or allocation strategies are more generally the province of the federal government, here the emphasis is placed on the role of the states and local government in instituting information, incentive and regulatory strategies.

PROPOSED INFORMATION AND EDUCATION POLICIES AND PROGRAMS

Ten percent of the builders we interviewed thought that information and educational material provided by public agencies was the one public policy most responsible for their use of energy-conservation features. We also asked builders what kinds of information they needed in order to make better decisions regarding the adoption of energy-conservation features in the homes they built. The most frequent responses were: (1) information about building product characteristics, including their reliability and effectiveness (mentioned by 34 percent); (2) information about the cost implications of energy-conservation features, including added cost from using a feature and the cost and amount of energy saved by the feature (its overall cost effectiveness) (34 percent); (3) information on consumer demand for energy-conservation features (9 percent); (4) information about solar energy (9 percent); and (5) the provision of one reliable source of information about energy in housing (4 percent). Only 10 percent of the builders volunteered that they already had all the information they needed about the energy efficiency of housing.

1. *Expand Workshops, Seminars, Educational Material.*—Builders' interest in obtaining additional information about energy conservation features suggests that state and local governments could play a very useful role in assembling, packaging and distributing information about energy conservation in home building. Although energy-related information is available from the major national trade association—the National Association of Home Builders (NAHB)—a sizable proportion of builders (at least a third in North Carolina, for example) are not members and are unlikely to be aware of the NAHB publications. As reported above, information related to energy conservation in home building is currently being issued by a vast array of sources, ranging from federal agencies to craft unions to building materials suppliers and associations of suppliers. A continuing workshop and seminar series, backed up with easily digested summaries of the latest developments in energy and home building, would provide a means for all builders in a state or locality to keep pace with this rapidly changing field.

2. *Establish State Builders' Institutes.*—In addition to making more information more easily available to home builders, there is a need for an organization that can assemble information about energy and other aspects of building and construction, process the information in terms of conditions in each state and the needs of builders in the state, and issue authoritative advice and recommendations about the effectiveness and dependability of new building and development materials, processes and techniques. At the conclusion of our home-builder interview, we asked builders whether they thought there was a need for a state builders' institute

and whether they would be willing to financially support such a service. Almost half of the builders (48 percent) thought the need was important and said they would pay $100 a year to support such an effort. Another 27 percent thought the need for a builders' institute was important, but were not willing to pay $100 a year for such an institute's services. Only a quarter of the builders felt the need for a builders' institute was not very important at this time. Given the need for such a service and builders' expressed interest in it, state governments should actively explore the establishment of individual state building institutes.

3. *Establish a Market Research Service.*—Providing better information about energy conservation features is one method of reducing the risk involved in their use and speeding up builder adoption of the features. Our research indicates, however, that builders are also often concerned about the market for energy-conservation features and are not likely to proceed solely on the basis of what they know about energy conservation. As another means of promoting the construction of energy-efficient homes, state and local government could provide a market information service for home builders. Possibly beginning on a trial basis, "Housing Energy Market Reports" could be prepared for each county or other housing market area. The reports would be made available, at a nominal price, to home builders, subcontractors, building materials suppliers, architects and others involved in the home building industry so that they could more accurately gauge consumer demand and willingness to pay for energy-efficient homes and housing products. The reports would be updated periodically. In addition to providing needed information to the housing industry, over time the Housing Energy Market Reports would enable energy officials, utility companies and others promoting the energy consciousness of households to gauge the success of their efforts to stimulate consumer demand for energy-efficient housing.

4. *Continue Research and Demonstration Program on Energy Conservation in the Residential Sector.*—For new ideas, including energy conservation ideas, to be adopted by the home building industry, their value needs to be clearly demonstrated. While a few builders will adopt a new feature or building technique based on what they have read about it, in many cases home builders need to see the feature in place or in use. Thus, in promoting energy conservation in the home-building industry, there is a clear place for continuing demonstration efforts. For demonstrations, both past and future, to have an effect on builders, builders must be aware of the demonstration. Any demonstrations funded should include steps to inform

builders of the demonstration, to show it to them, and to discuss its results in terms that are relevant to their concerns regarding cost, ease of installation, dependability and consumer reaction.

Based on the results of our research, the energy-conservation technology most in need of further demonstration is solar energy. As shown above, home builders have yet to use solar energy features in new construction in any great numbers. Demonstrations of solar housing should be designed not only to indicate the technical feasibility of particular solar energy systems, but also to show how major consumer reservations can be overcome. Those of particular concern to home builders include high initial costs (99 percent of the builders interviewed rated this as a major consumer reservation); the need for a supplemental heating system (77 percent thought this was a major consumer reservation to solar); and reliability of the system (71 percent rated it a major consumer reservation).

PROPOSED INCENTIVE POLICIES

Incentives are a second major class of policies that can be used to stimulate the production of energy-efficient housing. Because of the highly competitive nature of the home-building industry and its consequent sensitivity to the "first costs" of energy-conservation investments, incentives are a particularly appropriate class of policies to bring about change in the industry.

1. *Tax Credits for New Construction.*—Existing tax credit programs in the states could be expanded and modified to more directly address the new construction segment of the residential sector. For example, by making solar energy tax credits available directly to the builder (or, alternatively, providing solar builders with incentive grants) builders could lower the prices of solar housing to make them more competitive with housing equipped with conventional space and hot water heating systems. This might reduce some of the uncertainty involved with the tax credits so that they have a more direct effect on consumer demand.

2. *100 Percent Financing of Energy Conservation Features.*—Builders' concern about the cost of energy conservation features might be overcome if such costs did not add to the equity investment required to purchase a new home. For example, if builders could show that annual savings in the costs of operating the home were greater than the increased principle and interest payments required to finance the installation of an energy conservation

package of features, energy conservation features would make a positive contribution to home sales. In order to induce builders to take this approach, the states should consider establishing a revolving second mortgage loan fund applicable to the increased costs of new homes attributable to the use of selected energy-conservation features. Standard "extra" costs could be established for various energy-conservation features, with second mortgage funds made available to home purchasers to cover these costs. If interest rates were set at a rate equal to the cost of money to the state, plus administration of the program, it might be highly attractive to home purchasers and builders alike. Alternatively, of course, if lending institutions took home-operating costs into account in determining required loan to home-value ratios and other loan terms, a similar result might be obtained without the establishment of major new governmental programs.

3. *Raise Price of Energy Used in Housing.*—According to the Council on Environmental Quality (1979, p. 41), recent studies are showing that the elasticity of demand for energy is much more sensitive to price than previously had been thought. If so, then policies which used increased prices to promote energy conservation may be more effective than they have in the past. One means of promoting energy conservation in the residential sector, of course, is to further increase the costs of heating, cooling and other household uses of energy through the imposition of state taxes on electricity, fuel oil and natural gas. As the price of energy increased, households would make increased demands on builders for energy-efficient housing products.

PROPOSED REGULATORY POLICIES

A third approach to improving the energy efficiency of new housing is through mandatory regulations. When asked what policies were having the most effect on their use of energy conservation features, home builders rated building codes as the major policy factor inducing them to build more energy-efficient dwellings. Building codes could be modified to require the use of additional features that have been shown to be cost effective in reducing energy consumption. In addition, regulations could be adopted to improve the energy performance of other segments of the housing industry, such as subcontractors and developers, and to increase consumers' ability to exert an effective market demand for energy-efficient housing products.

1. *Add Additional Energy-Saving Ideas to Features Required in Building Codes*—Most states' building codes, as they apply to single-family detached houses, are like a cookbook—an energy-efficient house requires various ingredients, such as a certain amount of insulation, a certain amount of glass surface area and the like. Given this approach, it makes sense to add to the "recipe" any energy-conservation features that have been shown to be cost effective in most applications. For example, it was reported earlier in this paper that less than half of the builders interviewed were using each of ten features that appear to save more in energy costs than they cost to purchase (fabricate) and install. These include: (1) attic fan; (2) clock thermostat; (3) fireplace which uses outside air for combustion and provides supplemental space heating; (4) reduced glass area; (5) insulated hot water pipes; (6) 2 × 6 framing to allow extra insulation and tighter intersections; (7) insulated shutters; (8) reflective glass; (9) microwave oven; and (10) roof overhang for shade in summer. Other ideas that have been suggested as ways of reducing home energy consumption (see NAHB Research Foundation, Inc., 1979, pp. 51-57) include: (11) storm doors; (12) 24 inch o.c. wall framing; (13) appropriate roof colors; (14) insulated ducts; and (15) use of fluorescent lights for selected areas of the home. While all of these are obviously not appropriate for inclusion in a building code, consideration should be given to the use of the code as a device for promoting the widespread use of those that have a very high benefit/cost ratio.

2. *Address the Quality of Construction in the Building Code/Provide State Technical and Financial Assistance for Local Building Inspection.*—Studies have shown that air infiltration is a serious problem in energy conservation. According to Harrje (1978), for example, about one-third of the total energy in a house is lost as air moves into a house, is heated or cooled, and then moves out of the house. A key means of reducing air infiltration is through tighter construction. According to Seidel, Plotkin and Reck (1973, p. 55), "Exfiltration loss would be reduced 50 percent if regulations were written and enforced to include quality of installation as well as quality of material," and "Modifications to insure proper fitting of doors and window frames and caulking of leaks, could also result in major reductions of infiltration losses." Presumably, standards for acceptable rates of air infiltration could be established and made part of a building code. Or, using the cookbook approach, standards related to the performance quality of various operations needed to build a tight house could be developed, with simple indicators formulated so that the standards could be enforced by inspectors in the field. If more complex building code standards are adopted,

the state should explore methods of insuring that the standards are adequately enforced at the local level. These methods could range from state-provided training programs to state monitoring of the quality of local inspection to the provision of state financial assistance to enable localities to employ qualified persons for this vital task.

3. *Mandatory Disclosure of Home Energy Operating Costs.*—In order to increase consumers' awareness of the energy costs involved in home operation and to provide a means for households to consider energy-related operating costs when choosing a house to purchase or rent, state or local governments could require that energy operating costs be disclosed when housing ownership is transferred. Energy operating costs could be expressed in terms of actual or estimated fuel and electricity use for a preceding or future time period (the last or next 12 months) or in terms of estimated Btu's expended per gross square foot per year.

4. *Promote Adoption of Passive Solar Energy Principles.*—It is clear that relatively few builders or developers are incorporating passive solar energy principles in either the homes they construct or the subdivisions they develop. Careful consideration should be given to several means whereby the states and local government can promote accelerated use of passive solar energy in the residential sector. As a start, the states could amend their subdivision regulation enabling legislation to make energy conservation a valid purpose of subdivision regulation. Second, model subdivision regulations incorporating various passive solar energy and other ''neighborhood'' energy-conservation concepts could be developed and promoted among local governmental officials in a state, along with documentation of energy savings possible through passive solar design. Third, building codes could be amended so that builders were required to consider the potential for passive solar energy in the orientation of housing units, design of roof overhangs, size and location of windows, use of landscaping, and other characteristics of residential sites and structures.

5. *Consider Mandatory Licensing.*—In order to insure that solar and conventional heating, ventilating and air conditioning equipment is designed and installed properly, that insulation, caulking and weatherstripping is properly installed, and that other energy-efficient construction procedures are followed, licenses could be required of various installers, subcontractors, and craftsmen involved in the home building industry. Before a license was issued, the applicant could be required to demonstrate knowledge of the energy-related aspects of the particular job or procedure for which a

license was being sought. In addition, to keep licenses current, license holders and applicants could be required to take short-courses and other training to acquaint them with new developments and techniques.

BUILDERS' EVALUATIONS OF ENERGY POLICY OPTIONS

During the interviews with home builders, the respondents were asked to consider a range of public policies for new housing, including many of those discussed above, and to indicate which *one* policy they thought would be most effective in producing more energy-efficient new home construction. Although some policies drew more support than others, there was no consensus that any one policy would be most effective. The policies and proportions of builders who thought they would be most effective were: (1) bigger tax credits for homeowners (22 percent); (2) 100 percent financing of energy saving features (i.e., no down payment) (19 percent); (3) development of practical solar heating systems (18 percent); (4) better enforcement of building code requirements (15 percent); (5) higher prices for fuel and electricity (10 percent); (6) education of lenders to accept life-cycle costing principles (8 percent); (7) retraining programs for builders (6 percent); and (8) retraining programs for subcontractors (1 percent).

The variety of viewpoints about the *most* effective policy to promote energy conservation in new home construction reflects the diversity and complexity of the home-building industry. In addition, it may indicate that the states cannot rely on any one or even a small group of policies to achieve their residential energy conservation objectives. Instead, a mix of policies, spanning the entire range of policy types and focuses discussed above, seems appropriate. Finally, it seems worth reiterating that given the complexity of the home building industry, randomly selected policies for promoting improved energy efficiency in new housing will probably produce less than satisfactory results. Policies must be selected and formulated so that they combine into a coherent program of energy conservation for the home-building industry.

Conclusion

The trend toward building for energy efficiency in the residential sector has been explored in this paper by examining the current status of the thermal efficiency of existing and emerging housing and builder practices in new housing. Residential structures are becoming more energy efficient, as a result of both advances in building technology and consumers' demand for energy efficiency in their homes. However, the process of improving the thermal efficiency of residential structures may be hastened. A number of state and local policies for speeding the adoption of energy-conservation features in home building have been proposed.

Although this paper has focused on energy conservation in new owner-ship housing, an important topic for future study is the status of energy conservation and means of encouraging conservation in the rental sector. Such a study would focus on both the residents and investors of rental housing. Also suggested by the findings presented here is the more intensive study of the adoption of alternative technologies, such as solar energy, in the housing industry. A parallel study of solar builders and homeowners could provide useful information on means of increasing the rate of adoption of solar space or hot water heating. Finally, the fragmentation of the building industry brings to bear a number of key decision agents to the process of adoption of innovation in the industry. This fragmentation also necessitates the further study of subcontractors or investors who may facilitate or inhibit the process of building for energy efficiency.

Notes

This paper is based on research which was supported, in part, by the North Carolina Energy Institute. A complete report of the study findings, including information about household energy-conservation behavior and retrofitting, is available in Raymond J. Burby and Mary Ellen Marsden, eds., *Energy and Housing: Consumer and Builder Perspectives*, which is distributed by the National Technical Information Center. We would like to acknowledge the assistance of other members of the research team, including William W. Hill, Jeanne T. Hernandez, Duncan MacRae, Jr., Michael C. McKinney and David Orr.

References

Barrett, David, Peter Epstein, and Charles M. Haar. *Financing the Solar Home: Understanding and Improving Mortgage-Market Receptivity to Energy Conservation and Housing Innovation*. Lexington, MA: D. C. Heath and Company, Lexington Books, 1977.

Beyer, Glen H. *Housing and Society*. New York: The Macmillan Company, 1965.

Council on Environmental Quality. *The Good News About Energy*. Washington, D.C.: U. S. Government Printing Office, 1979.

Council of State Governments. *Energy Conservation: Policy Considerations for the States*. State Environmental Issues Series. Lexington, KY: The Council, November, 1976.

Division of Solar Energy, U. S. Energy Research & Development Administration. *Solar Energy in America's Future: A Preliminary Assessment*. Washington, D.C.: U. S. Government Printing Office, March, 1977.

Eisenstadt, Melvin M. and Albert E. Utton. "Solar Rights and Their Effect on Solar Heating and Cooling," *Natural Resources Journal*. Vol. 16, No. 2 (1976), pp. 363-414.

The George Washington University, *Solar Energy Incentives Analysis: Psycho-Economic Factors Affecting the Decision Making of Consumers and the Technology Delivery System*. Prepared for the U. S. Department of Energy Under Contract No. Ex. 76-G-10-2534. Washington, D.C.: U. S. Department of Energy, January, 1978.

Grier, Eunice S. "National Survey of Household Activities." Washington: The Washington Center for Metropolitan Studies, December, 1975.

Harrje, David T. "The Twin Rivers Experiments in Home Energy Conservation," in *Energy and the Community*. Raymond J. Burby and A. Fleming Bell, eds. Cambridge, MA: Ballinger Publishing Company, 1978. pp. 19-23.

Healy, Robert C. and Henry R. Hertzfeld. *Energy Conservation Strategies*. An Issue Report. Washington, D.C.: The Conservation Foundation, 1975.

Hirschberg, Alan and Richard Schoen. "Barriers to the Widespread Utilization of Residential Solar Energy: The Prospects for Solar Energy in the U.S. Housing Industry," *Policy Sciences*. Vol. 5 (1974), pp. 453-468.

Hirst, Eric and Janet Carney. "Effects of Federal Residential Energy Conservation Programs," *Science*. Vol. 199 (24 February, 1978), pp. 845-851.

Hittman Associates, Inc. *Residential Energy Consumption Detailed Geographic Analysis*. Summary Report, Report No. HUD-PDR-250. Washington, D.C.: U. S. Government Printing Office, January, 1978.

Leighton, Gerald S. "Statement." *Energy and the City*. Hearings Before the Subcommittee on the City of the Committee on Banking, Finance and Urban Affairs, House of Representatives, Ninety-fifth Congress, First Session, September 14, 15 and 16, 1977. Washington, D.C.: U. S. Government Printing Office, 1977, pp. 208-228.

Miller, Alan S., Gail Boyer Hayes, and Grant P. Thompson. *Solar Access and Land Use: State of the Law 1977*. Rockville, MD: National Solar Heating and Cooling Information Center, 1977.

Myers, Barry Lee. "Solar Rights in Residential Developments," *Practical Lawyer*. Vol. 24, No. 2 (March 1978), pp. 13-20.

NAHB Research Foundation, Inc. *Insulation Manual: Homes, Apartments.* Rockville, MD: The Foundation, 1979.

National Association of Home Builders. *Designing, Building, and Selling Energy Conserving Homes.* Washington, D.C.: The Association, 1978.

National Commission on Urban Problems. *Building the American City.* Washington, D.C.: U. S. Government Printing Office, 1968.

Office of Public Affairs, U. S. Department of Energy. *The National Energy Act.* Washington, D.C.: U. S. Department of Energy, November, 1967.

Schon, Donald. *Technology and Change.* New York: Delacorte Press, 1967.

Seidel, Marquis R., Steven E. Plotkin, and Robert O. Reck. *Energy Conservation Strategies.* Socioeconomic Environmental Studies Series, EPA-R5-73-021, 1973. Washington, D.C.: U. S. Government Printing Office, 1974, 1973.

Stanford Research Institute. *Patterns of Energy Consumption in the United States.* Palo Alto, CA: The Institute, 1972.

Thompson, Grant P. "The Law and Energy Conservation," in *Energy and the Community.* Edited by Raymond J. Burby, III and A. Fleming Bell. Cambridge, MA.: Ballinger Publishing Company, 1978. Chapter 8.

SECTION III

ALTERNATIVE LAND USE MEASURES TO ASSURE ADEQUATE ENERGY SUPPLY

The Future of Traditional Energy Sources

JON H. WEYLAND

Introduction

MUCH of the relationship between energy and land use is based on an assumption that a severe energy crisis is at hand. The advent of such a crisis would force changes in both patterns of new development and the use of existing land in order to accommodate new constraints. Conversely, new ways of using the land would provide opportunities to become more efficient and take advantage of certain opportunities to reduce consumption of energy. These effects will be felt, however, and these opportunities used, only to the extent that the availability of energy from traditional sources will diminish in the future. The purpose of this paper is to summarize the present view of the likely future in terms of its energy characteristics.

The forecasts which are summarized on the following pages are based on a single set of integrated demand and supply forecasts: those presented in the 1978 *Annual Report to Congress.* (It should be noted that there has usually been a 20-24 month delay in publication of the *Annual Report.*) The past political upheavals in Iran and ensuing steep price increases have, however, undermined the reasonableness of the *Report's* preferred projection series (Series C) since 1978. Consequently, the discussion of this paper is based entirely on Series B, a low-supply-high-demand scenario which has so far corresponded more closely to observed supply, demand, and prices than the authors' preferred projection Series C.

311

Many examples are available of reasonable forecast of supply and demand at the national level. These Energy Information Administration (E.I.A.) projections were selected because of their authority, reasonableness (at least series B), and detailed consideration of both supply and demand. The *Annual Report* includes comparisons of these projections to other recent projection series by several authors; consideration of these alternative projections is beyond the scope of this chapter and the interested reader is referred to the *Annual Report*. Some of the more recent and interesting work is listed in the references at the end of this paper.[1]

Definitions and Assumptions

It is assumed in economics that the supply of a commodity is a function of the price offered, a higher price stimulating increased discovery or production of the commodity. Likewise, demand for the commodity is assumed to vary as a function of price, a higher price reducing discretionary demand or stimulating investment in more efficient means of use or a substitute commodity. It is further assumed in economics that the commodity price which determines supply and demand is set in a free market meeting certain ideal conditions (Exhibit 1).

Starting with these assumptions, it is apparent that in any view of future fuel consumption, supply, demand, and price are critically linked. Exhibit 2 suggests a three-dimensional view of fuel supply and demand over time. Supply is usually broken down by fuel type and demand by sectors of the economy. Exhibit 2 also illustrates how any two dimensions can be considered separately if the third dimension can be held constant. The cells of this three-dimensional matrix contain *consumption*, the amount of fuel of each type supplied to each sector of demand. The matrix could as easily contain data on price, since consumption for each year could be considered to have taken place for an average unit price. In effect, the matrix in Exhibit 2 provides a practical way of tracing the interception of the supply and demand curves in Figure 1(c) over time. Both the level of consumption and price associated with this intersection can be included.

It may be noted that while the level of supply and demand may be dependent on many social and economic factors, the quantities included in the matrix are sufficient to describe completely the transactions which have occurred. Further, consumption and price, by fuel type, economic sector, and year, provide a complete, if general, description of the energy situation in a particular state or national economy.

EXHIBIT 1
SUPPLY AND DEMAND

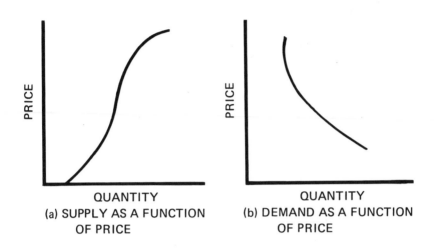

QUANTITY
(a) SUPPLY AS A FUNCTION
OF PRICE

QUANTITY
(b) DEMAND AS A FUNCTION
OF PRICE

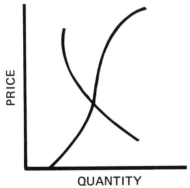

QUANTITY
(c) QUANTITY CONSUMED AT A PRICE—THE
INTERSECTION OF SUPPLY AND DEMAND

JON H. WEYLAND

EXHIBIT 2
CONSUMPTION BY CATEGORIES OF SUPPLY AND DEMAND OVER TIME

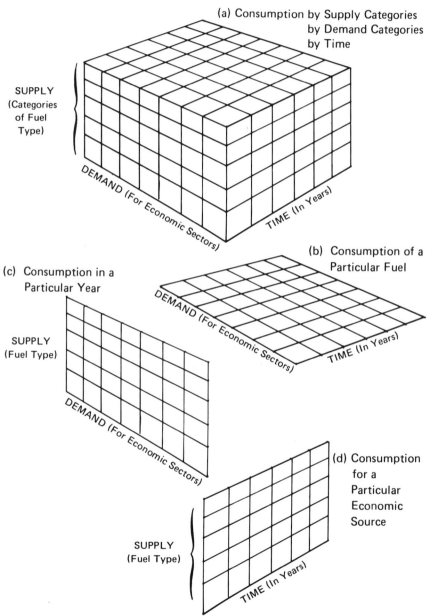

(a) Consumption by Supply Categories
by Demand Categories
by Time

SUPPLY
(Categories
of Fuel
Type)

DEMAND (For Economic Sectors)

TIME (In Years)

(b) Consumption of a
Particular Fuel

(c) Consumption in a
Particular Year

DEMAND (For Economic Sectors)

SUPPLY
(Fuel Type)

TIME (In Years)

DEMAND (For Economic Sectors)

(d) Consumption
for a
Particular
Economic
Source

SUPPLY
(Fuel Type)

TIME (In Years)

Source: Similar conceptual approaches have been suggested by Margaret Fels,
Princeton University, and Peter Meier, Brookhaven National Laboratory.

U.S. Demand By Land Use Category-Traditional Fuels

RESIDENTIAL

The residential sector consists principally of single-family houses and a variety of multi-family styles: duplexes (two family buildings), row houses, and apartments. All styles may be found either owner occupied or leased. Principal uses of energy in this sector are space heating and cooling, water heating, cooking, refrigeration, lighting, and small appliances.

U.S. residential demand has been characterized historically by an increase in consumption through 1972 with decreases thereafter. The rate of increase was approximately 3.1 percent per year, decreasing to about -0.5 percent annually after 1972. In absolute numbers, residential energy consumption was estimated to be 8,800 trillion BTUs out of a total national consumption across all sectors of 44,083 trillion BTUs in 1960; in 1977, the figures were 16,001 and 76,527 trillion BTUs respectively.

Major determinants of residential demand are thought to be population, *per capita* income (real), number of households, and the real price of each fuel used in the residential sector: electricity, natural gas, and oil. In the 1972-77 period these factors all showed real percentage increases, as opposed to the earlier 1962-72 period when many fuel prices fell in real dollars (Exhibit 3).

Exhibit 4 shows projected 1995 demand for each fuel used by the residential sector and total demand. Demand by the residential sector is expected to grow about 12 percent between 1977 and 1995, an average annual growth of approximately 0.65 percent per year. During the same period prices are expected to increase by about 86 percent in real dollars, an increase of approximately 4.8 percent annually. Since inflation is not expected to fall below 10 percent annually under any conditions in the near future and may be considerably higher, apparent price increases are expected to be well in excess of 4.8 percent annually under this scenario.

Demand for electricity is expected to increase from 21.7 percent of residential demand in 1977 to 32.6 percent in 1995. Oil will decrease from a 23.1 percent share to 15.7 percent, and the proportion of natural gas is expected to remain almost constant at about 48 percent under this scenario for the 18-year forecast period. Thus, there will be an increasing emphasis on electricity at the expense of oil in the residential sector. The proportion of all-electric homes is already increasing in many areas, with the heat pump and electric hot water heater tending to displace new oil burners and their built-in hot water heaters.

The residential sector is thus expected to be characterized by modest growth in energy demand over the forecast period, constrained to a significant extent by increased prices. Electrical use will increase about 68 percent, while demand for oil will decline to about 75 percent of its 1977 consump-

Exhibit 3

MAJOR DETERMINANTS OF ENERGY DEMAND
IN THE RESIDENTIAL SECTOR:
ANNUAL GROWTH RATES, 1962 - 1977
(percent)

	1962–1972	1972–1977
Population	1.2	0.8
Income per capita (real)	3.2	1.7
No. of households	2.0	2.3
Energy prices (real)		
Electricity	− 3.7	2.6
Natural gas	− 1.7	6.1
Oil	− 0.6	11.4
Total	− 0.6	7.6

Source: Energy Information Administration, *Annual Report* p. 280.

tion. Natural gas demand is expected to increase about 11 percent for the 18-year period.

COMMERCIAL

The commercial sector includes finance, insurance, and real estate; retail/wholesale trade; health/education services; office buildings, non-transportation-related government purchases of energy; use of asphalt and road oils for highway construction and maintenance; and other commercial activities. It is usually defined formally as Standard Industrial Classification codes A,B,C, and F through K. Energy demand in this sector is dependent upon economic conditions, which determine both the amount of energy-consuming activity and consumer willingness to absorb fuel-cost related price increases, net growth in commercial building stock, the energy consumption characteristics of building and equipment stocks, and the price-sensitivity of energy-consumption behavior.

Exhibit 4

**ENERGY DEMAND AND PRICES IN
THE RESIDENTIAL SECTOR:
1977 AND PROJECTIONS FOR 1985–1995
(10^{18} joules/year, 1978 dollars/10^9 joules)**

	1977	1988	1990	1995
Electricity	2.35	2.95	3.51	3.96
Price	11.11	12.79	13.19	14.72
Oil	2.51	2.14	2.06	1.90
Price	3.18	4.54	4.98	6.37
Natural Gas	5.25	5.40	5.62	5.86
Price	2.29	3.52	4.06	4.66
Total	10.86	11.06	11.67	12.12
Price	4.48	6.28	7.03	8.32

Source: Energy Information Administration, *Annual Report* pp. 284–287.

Commercial-sector demand increased 67 percent between 1962 and 1977, an average growth rate of about 4.4 percent annually over the fifteen-year period. Actually, however, the sector displayed a higher growth rate of 5.1 percent between 1962 and 1972, and a sharp dip and continuation of the earlier rate of growth after 1973, presumably reflecting the impact of the 1972-73 oil embargo. Weighted average real price of energy consumed in this sector for the 1962-72 period declined 12 percent, or 1.2 percent annually. After 1972, weighted average real price increased between 9 and 11 percent annually, marking a sharp change in price signals.

Growth in the stock of commercial buildings can be measured approximately by the increase in commercial floor space over time. This growth has been slow and constant, from a base total area of 6,328 million square feet in 1960 to 12,327 million square feet in 1975. Various studies have indicated substantial increases in the energy efficiency of commercial-sector heating, cooling, water heating, and lighting, and there is reason to expect that continuation of these increases will affect projected rates of consumption in this sector (the projections discussed below take these improvements into account). It seems likely that observed efficiency improvements have been stimulated primarily by the 9-to-11 percent weighted average increase in real prices mentioned above.

Exhibit 5 shows projected 1985, 1990, and 1995 demand for each fuel used by the commercial sector, and total projected demand. These projected figures correspond to the low-supply-high-demand scenario (projection B) discussed earlier. Demand by the commercial sector is expected to decrease about 1.1 percent between 1977 and 1995, an average annual decrease of less than 0.1 percent. During the same period prices are expected to more than double in real dollar terms, with 1995 fuel prices projected to be 218 percent of those in 1977, weighted by their relative demand within this sector. Apparent 1995 fuel price increases are expected to be well above this real increase because of inflation.

Demand for electricity is expected to increase from 23 percent of commercial demand in 1977 to about 40 percent in 1995. Oil will decrease from a 28 percent share to 11 percent, and the proportion of natural gas is expected to remain almost constant at about 33 percent under this scenario for the 18-year forecast period. Thus, like the residential sector there will be an increasing emphasis on electricity at the expense of oil in the commercial sector. The most tenuous aspect of this projection would seem to involve the relatively stable ratio of natural gas over total sectoral demand: recent evidence suggests widespread conversion from oil to natural gas facilities may be getting under way in commercial buildings, although a definite trend in this sector has not yet been verified.

The commercial sector is thus expected to be characterized by an almost constant annual rate of consumption over the forecast period, in the face of sharply higher fuel prices. Electricity use will increase about 69 percent, while demand for oil will decline to about 37 percent of its 1977 consumption. Natural gas demand is expected to drop slightly to 97 percent of 1977 consumption over the 18-year forecast period.

INDUSTRIAL

The industrial sector includes six industries which are particularly important for the study of energy demand: chemicals and allied products; primary metals; petroleum refining; paper and allied products; food and kindred products; and stone, clay, and glass products. The primary determinants of energy demand by this sector seem to be the overall level of national economic activity and the price of particular fuels used by the various industrial processes. More than any other sector, industrial demand appears to be price sensitive, perhaps because of the rational profit-loss calculations applied to most industrial purchase and investment decisions.

Exhibit 5

COMMERCIAL ENERGY DEMAND AND PRICES:
1977 AND PROJECTIONS FOR 1985–1995
(10^{18} joules/year, 1978 dollars/10^9 joules)

	1977	1985	1990	1995
Electricity	1.93	2.57	3.01	3.27
Price	11.02	12.84	13.28	14.87
Oil	2.31	1.27	1.03	0.87
Price	2.69	3.96	4.38	5.68
Natural Gas	2.72	2.42	2.51	2.64
Price	2.00	3.10	3.63	4.23
Total	8.30	7.53	7.91	8.21
Price	4.69	7.27	8.18	10.24

Source: Energy Information Administration, *Annual Report* pp. 301–2.

Industrial sector demand increased 29 percent between 1962 and 1977, an average growth rate of about 1.9 percent annually over the fifteen-year period. From 1962 to 1972 industrial energy demand grew at an average annual rate of about 2.8 percent per year, while from 1972 to 1977 the average annual increase was 2.5 percent per year. Between 1962 and 1972, however, value added in manufacturing grew about twice as rapidly as energy demand, or about 56 percent for the 10-year period. Between 1972 and 1977 value added grew at 16.7 percent or 3.3 percent annually, about 33 percent more rapidly than the 12.5 percent increase in energy demand for the five-year period (the 2.5 percent annual growth mentioned above). Thus, it would appear that the energy efficiency (changes in demand/changes in value added) of industrial development has been higher during the 1962–77 period than older industrial facilities as a whole, although the rate of improvement may have slowed somewhat in the latter part of this period. It should be noted, however, that the mix of diverse industrial activities may have shifted in unknown ways during this time, making these comparisons only suggestive. One conclusion they do suggest is that the apparent price-sensitivity of the industrial sector will result in the finding of significant future energy savings if prices rise steeply in the future.

Exhibit 6 shows projected 1985, 1990, and 1995 demand for each fuel used by the industrial sector, and total projected demand (E.I.A. projection B). Demand by the industrial sector is expected to increase by 78 percent between 1977 and 1995, an average annual increase of 4.3 percent. During the same period prices are expected to increase to 226 percent of 1977 prices in constant dollars, weighted by relative fuel demand within the industrial sector. This increase works out to 12.6 percent on an annual basis. Apparent price increases are of course expected to be much higher because of inflation.

Demand for electricity is expected to increase from 12 percent of industrial demand in 1977 to about 16 percent in 1995. Oil will increase from 10 percent share to 12 percent, and the proportion of natural gas is expected to fall from 37 percent in 1977 to 19 percent in 1995. Coal, generally not a significant factor in any of the sectors discussed above, is projected to increase from 17 percent share in 1977 to 28 percent by the end of the forecast period. Unlike the sectors previously examined, oil is projected to secure an increased share of the industrial sector, while coal demand is also projected to increase substantially, both at the expense of demand for natural gas. There are two reasons underlying these projections: prices for natural gas have increased at a 15 percent annual rate during the 1972-77 period (higher since then) and there were widespread curtailments of industrial deliveries during this time, resulting in continuing caution with respect to dependence on natural gas supplies; further, incremental pricing under the Powerplant and Industrial Fuel Use Act is resulting in significantly higher increases in the price of natural gas for customers in the industrial sector than in either the residential or commercial sector.

The industrial sector is expected to be characterized by a 78 percent increase in energy consumption over the forecast period, an average annual increase of 4.3 percent. The price increase for fuel over this same period would be 126 percent. Electricity demand will increase 149 percent, demand for oil will increase 105 percent, demand for coal 188 percent, and demand for natural gas will decrease less than 8 percent. In general, in this scenario, these shifts will be driven largely by relative fuel prices in upcoming years.

The Electrical Demand-Supply Bridge

ELECTRICITY DEMAND

Electricity bridges in some ways our understanding of demand and of supply. Simultaneously, electricity requires fuel for electrical generation

Exhibit 6

INDUSTRIAL ENERGY DEMAND AND PRICES:
1977 AND PROJECTIONS FOR 1985-1995
(10^{18} joules/year, 1978 dollars/10^9 joules)

	1977	1985	1990	1995
Electricity	2.72	4.22	5.50	6.78
Price	6.93	9.40	9.94	11.60
Oil	2.40	2.25	3.41	4.67
Price	2.63	3.82	4.15	5.35
Coal	4.03	6.18	8.57	11.63
Price	1.43	1.78	1.89	2.53
Natural Gas	8.69	9.21	9.07	8.05
Price	1.39	2.97	3.73	4.61
Total	23.38	28.92	35.16	41.54
Price	2.44	3.98	4.50	5.51

Source: Energy Information Administration, *Annual Report* pp. 312–313.

and supplies energy to the several economic sectors of demand. In addition, it serves as an important means of transporting energy and making it readily available to the various sectors of the economy. As the demand and supply characteristics and projections are summarized below, it should be kept in mind that the scenario being described coincides with the high-demand-low-cost assumptions of the particular scenario being considered. This description is intended to be consistent, but a range of interpretation is possible in such projections.

DEMAND

The electrical utility sector consists of electrical power plants needing one of several possible fuels: coal, oil, natural gas, nuclear fuel, and hydroelectric power. All are used to generate significant amounts of electrical power at present in the United States. In addition, several other sources of energy, such as solar, wind, geothermal, biomass, and others, are either used presently in small amounts or are possibly significant, though small, sources in the future. This section will examine the demand for the four most important conventional fuels by the electrical utility sector, leaving the bulk of the other papers in this volume the task of viewing the more exotic possibilities of alternative energy sources.

Energy demand on the part of the utilities is a function of the demand for electricity in a utility-company service area, which is usually a sub-state geographical area. In order to deal with planning, operating imbalances, and time-of-day imbalances, utility companies are usually affiliated with a regional power grid, permitting and systematizing exchanges of power among companies. "Peak-load" and "base load" are critical concepts in electrical planning, because demand for electricity varies substantially by time of day and season of the year. Idle generating capacity, unused in an off-peak period, is expensive to a utility because it is not being amortized by the profit from the sale of electricity; conversely, inadequate generating capacity during a period of peak demand can lead to power shortages unless electricity can be purchased from the grid. Grid sales and purchases are constrained somewhat, however, because most companies in a given region of the country tend to experience similar peak and off-peak periods, grid purchases tend to be expensive (although on the other hand, grid sales can be profitable), and in times when generating capacity may be short because of the difficulty of finding acceptable power plant sites or of constructing a power plant in the face of environmental opposition, there may be a strong reluctance to build excess reserve generating capacity above the necessary minimum which could be made available to other companies or regions.

Demand for electricity increased 150 percent between 1962 and 1977, an average annual growth rate of 10 percent over the fifteen-year period. From 1962 to 1973 the annual growth rate was 7.5 percent, and from 1975 to 1977 it leveled off somewhat to about 4.2 percent annually.

Exhibit 7 shows projected 1990 demand for each fuel used to generate electricity and total projected demand. Demand by the electric utility sector is expected to increase by 4.2 percent annually between 1977 and 1990. During the same period prices are expected to increase 1.6 percent annually in constant dollars. Inflation is expected to make these average increases appear more severe than the real-dollar figure would indicate.

Demand for coal for electrical generation is expected to grow to 58 percent of total capacity between 1977 and 1990 from 46 percent in 1977. Oil will decrease from an 18 percent share in 1977 to a 6.9 percent share in 1990. Natural gas is expected to drop from 15 percent of total generating capacity in 1977 to 1.4 percent in 1990.

Exhibit 7

**ELECTRICITY FUEL DEMAND
BY FUEL FOR ELECTRICAL GENERATION:
1977 AND PROJECTIONS FOR 1990**
(10^{18} joules/year)

	1977	1990
Coal	10.84	23.47
Oil	4.25	2.77
Natural Gas	3.47	0.57
Nuclear	2.77	9.66
Total input	23.75	40.28

Source: Energy Information Administration, *Annual Report*, p. 269.

Nuclear power is expected to increase from 11.7 percent of total input in 1977 to 24 percent in 1990. In general, coal and nuclear power are used in baseload facilities, while natural gas and oil are more often used for peak load stations, meaning that these generators are fired up only as needed, or in anticipation of peak demand.

The generation of electricity is expected to be characterized by an annual growth rate of just over 4 percent for the 1977-1990 forecast period. The average annual price increase, nationally, is expected to be about 1.6 percent. Annual growth rates for the 1977-1990 period for the several fuels are projected as follows: coal, 6.1 percent; oil, 3.2 percent; natural gas, 13.0 percent; nuclear, 10.1 percent.

ELECTRICITY SUPPLY

Electrical power is supplied to all sectors of the economy. Total sales in the 1962-1977 period have increased 150 percent. In 1962 electricity was supplied to the various sectors in the following proportions: residential, 29 percent; commercial, 22 percent; and industial, 48 percent. By 1977, these proportions had shifted to residential, 34 percent; commercial, 28 percent;

and industrial, 39 percent. Thus, the fifteen-year historical period is marked by moderate increases in the shares of the residential and commercial sectors and a decrease in the share of electricity supplied to the industrial sector.

During the historical period 1962–1977, electricity supplied to the residential sector increased 190 percent. The commercial sector received about 220 percent more in 1977 than in 1962, and the growth rate for the industrial sector was about 100 percent. On an annual basis for the 15-year period these figures work out to residential, 13 percent; commercial, 14 percent; and industrial, 6.8 percent. It is apparent that the industrial sector was being supplied electricity at a much less rapidly increasing rate during this period than either the residential or commercial sectors.

Exhibit 8 shows the projected 1990 supply of electricity for each sector and the total projected sales of electricity to be supplied. Supplies for the residential, commercial, and industrial sectors are projected to shift from 34 percent, 28 percent, and 39 percent of total supply, respectively, in 1977, to 29 percent, 25 percent, and 46 percent, for these three sectors in 1990. Average annual growth rates are projected to be residential, 3.1 percent, commercial, 3.5 percent, and industrial, 5.3 percent, for the 1977-1990 period. Thus, it is expected that electricity supplied to the industrial sector will grow at a more rapid rate than the supply to other sectors, producing a significant change in the use of electricity by that sector relative to the others. This projection portends a marked shift from recent historical trends toward slower growth in electrical supply to the industrial sector and a smaller relative sectoral share of total supply in the 1962 to 1977 period.

U.S. Supply By Fuel Type-Traditional Fuels

COAL

Coal is dangerous to mine. Its unprotected combusion can cause serious health hazards. Its transportation carries some environmental risks as well as creating physical and economic strains on a rail transportation system which has gradually lost its historical responsibilities for coal movement on a large scale.

Coal production declined for many years in the United States, reaching an apparent low point in the early 1960s and displaying a gradual increase since that time. Total national production shows a 41 percent increase in the 1962-1977 period, an average annual rate of 2.7 percent. In 1962, 45 percent of total coal production went to electric utilities, and 29 percent to the commercial and industrial sectors. By 1977 these percentages had changed

Exhibit 8

ELECTRICAL GENERATION SUPPLIED BY SECTOR:
1977 AND PROJECTIONS FOR 1990
(10^{18} joules/year)

	1977	1990
Residential	2.35	3.50
Commercial	1.93	3.01
Industrial	2.72	5.50
Total supply	7.03	12.01

Source: Energy Information Administration, *Annual Report*, p. 269.

to 69 percent and 9.7 percent, demonstrating a strong shift toward electrical generation (exports and production of various synthetics are not accounted for here). Growth for this period was 150 percent for the industrial sector, or 9.9 percent annually. The commercial and industrial sectors showed a decline of 46 percent, or about 3.0 percent annually, for this period.

Exhibit 9 shows the projected 1995 supply of coal to the commercial/industrial and electric utility sectors. The relative share of these supplies is projected to shift from 69 percent in 1977 to 61 percent in 1995 for the electric utility sector and from 9.7 percent to 16 percent for the commercial and industrial sectors. Net exports of coal and development of various synthetics are expected to take much of the remainder under this scenario. Average annual growth rates projected for these sectors between 1977 and 1995 are 9.7 percent for electric utilities and 22 percent for the combined commercial and industrial sectors. Perhaps surprisingly, the greatest increases are to be in coal supplied to the commercial/industrial, rather than utility, sectors. This projection deviates from the trend of the recent historical period in this respect.

OIL

Oil supplies are of critical national importance because of heavy reliance of all sectors of the economy on their continued availability. A major

EXHIBIT 9

COAL SUPPLY TO END-USE SECTOR:
1977 AND PROJECTIONS FOR 1985-1995
(10^{18} joules/year)

	1977	1985	1990	1995
Electric utility	12.76	20.29	28.86	34.92
Commercial/ industrial	1.80	4.37	6.97	8.90
Total domestic consumption	16.62	27.34	39.56	53.22

Source: Energy Information Administration, *Annual Report*, p. 234.

source of energy, crude oil and natural gas liquids account for approximately 50 percent of total U.S. energy consumption, and about 50 percent of this supply is imported. Important characteristics of this supply are its vulnerability to price increases set by the Organization of Petroleum Exporting Countries (OPEC) and its vulnerability to supply interruptions caused by an attempted use of political power, domestic instability in one or more Middle Eastern states, or a war which interferes with either production or transport of crude oil supplies.

Domestic oil production and imports together made up total supply in the proportions of 53 percent and 47 percent in 1977. Historically, the percentage of total supply which is imported has been steadily increasing. Table 10 shows the projected breakdown of domestic supply and imported oil for 1995. The relative share of domestic production is projected to increase to 58 percent, mainly because of the high prices for imported oil products on the domestic market. The overall rate of increase in total supply is projected to be extremely low, less than 0.5 percent annually. Domestic production is projected to increase 17 percent, or less than 1.0 percent annually. Imports of crude oil and products are projected to decrease 1.4 percent, a negligible decrease annually.

Exhibit 10

PETROLEUM SUPPLY FROM DOMESTIC
SOURCES AND IMPORTS:
1977 AND PROJECTIONS FOR 1985-1955
(10^{18} joules/year)

	1977	1985	1990	1995
Domestic				
supply	22.02	23.11	24.54	25.72
Imports (net)	19.14	17.28	18.13	18.87
Total supply	41.15	40.39	42.67	44.59

Source: Energy Information Administration, *Annual Report*, pp. 254–5.

NATURAL GAS

Historically, natural gas has been plentiful and inexpensive, relative to other available fuels. A 1954 court interpretation of the Natural Gas Act of 1938 resulted in a two-tier price system which has held the price of interstate natural gas under a regulated ceiling while allowing uncontrolled pricing intra-state. The Natural Gas Policy Act of 1978 is presently allowing the deregulation of prices in phases until 1985 when most types of gas will no longer be controlled. The first result of this phased deregulation has been a gas "bubble," an immediate release of increased supply to the interstate market which is expected to last for a few years. As the bubble in supply subsides and supplies come more into line with estimated reserves, supplies are expected to be much tighter and prices higher.

Domestic gas production and imports together made up total suply in the proportions of 94 percent and 6.2 percent in 1977: only small amounts of natural gas have been imported from Canada, Mexico, and the Middle East (particularly liquified natural gas). Exhibit 11 shows the projected breakdown of domestic supply and imported natural gas for 1995. The relative share of domestic production is projected to decrease to 83 percent because of limited domestic reserves. The overall rate of decrease in total supply of natural gas is projected to be 12 percent, or 0.6 percent annually, between 1977 and 1995. Domestic production is expected to decrease 21 percent, 1.2 percent annually. Imports of natural gas are projected to increase 140 percent, or 7.6 percent annually.

Conclusions

PRICE, SHORTAGE, AND REGULATION

It was noted in the introduction to this paper that economic modeling is usually based on assumptions of a free market: supply and demand are assumed to be equal, and both determined by response to price. Historically, these conditions have not been met in the marketing of petroleum products. Price controls on U.S. gasoline and natural gas and cartel-controlled pricing of crude oil have tended to distort the market, probably in different ways. Because crude oil is sold at prices which are set in international markets and is subject to politically-instigated supply interruptions, there have been rapid increases in constant-dollar fuel costs and a pervasive sense of supply insecurity. It is likely these rapid increases have contributed significantly to inflation and economic instability. Price controls on gasoline have probably allowed patterns of usage to develop which are higher than might have been observed under free-market conditions. It is also possible that retail prices of other petroleum fuels such as home heating oil and diesel oil have been raised by the oil companies to provide the wider profit margin that has been lost to gasoline price controls. There is also a good possibility that gasoline price controls have contributed to recurring shortages by inhibiting the normal working of market supply and demand mechanisms. Observation of such a supply breakdown in natural gas was behind the phased decontrol of natural gas prices in the interstate market enacted in the Natural Gas Policy Act of 1978.

THE SCENARIO APPROACH

Constructing a scenario employing the *Annual Report's* Series B Projections is useful because it allows construction of the current best guess about the future. Developing the detail of a "best guess" insures that a consistent projection has been constructed and that the implications of its many parts have been examined. Even if the scenario is not validated by future events, as will no doubt be the case to some degree, its usefulness to the present will

Exhibit 11

NATURAL GAS SUPPLY FROM
DOMESTIC AND IMPORTED SOURCES:
1977 AND PROJECTIONS FOR 1985–1995
(10^{18} joules/year)

	1977	1985	1990	1995
Domestic				
supply	20.22	17.90	16.36	15.89
Imports (net)	1.34	2.66	3.12	3.17
Total supply	21.56	20.56	19.48	19.06

Source: Energy Information Administration, *Annual Report*, pp. 243–44.

have been in construction and use of a coherent, detailed picture of the future; its usefulness to the future will lie in its historical record of how we saw our future in 1981.

A SUMMARY OF THE FUTURE

The scenario described suggests that demand for fuel will be high and supplies tightly constrained. Prices will be allowed to rise freely and will serve to dampen demand in many sectors. Residential demand will grow only moderately, with electricity displacing a substantial amount of residential oil consumption by 1995. Commercial sector demand is expected to level off in the face of sharply higher prices. Electricity is also projected to displace substantial oil consumption in this sector, with the consumption of natural gas remaining a constant share of sectoral demand during the forecast period. Industrial sector consumption is expected to grow moderately during the forecast period. Real price increases will be over twelve percent annually. Industrial-sector consumption of natural gas is expected to decline from 37 percent in 1977 to 19 percent in 1995, with a corresponding increase in coal demand. Oil demand will increase moderately and the demand for electricity will increase sharply. Prices of the various fuels will reflect this relative demand, with natural gas prices up sharply and steady increases in the price of oil as well. Electrical generation is expected to increase moderately over the forecast period. Use of natural gas to generate electricity is expected to decline sharply; oil consumption for electrical generation will be down moderately. Nuclear power is expected to increase rapidly, at a 10 percent annual rate for the forecast period, and coal will also increase substantially, at 6 percent per year. Of course, the political fallout of the nuclear accident at Three Mile Island and potential environmental, mine safety, and capital shortage (for mining equipment, pollution controls, and coal-banking-railroad reconstruction) problems may inhibit realization of these projections, but to a large extent they do reflect likely economic necessities in the face of short supplies and dramatically higher prices.

Notes

1 . U.S. Department of Energy, Energy Information Administration, *Annual Report to Congress, 1978 Vol. 3: Forecasts*. DOE/EIA-0173/3 (Washington, D.C.: U.S. Government Printing Office, 1979).

2 . *Phillips Petroleum Co. v. Wisconsin*, 347 U.S. 622, 1954.

References

Ford Foundation Energy Policy Project, *A Time to Choose: America's Energy Future* (Cambridge, Mass.: Ballinger, 1974).

National Research Council, Committee on Nuclear and Alternative Energy Systems, *Energy in Transition 1985-2010* (Washington, D.C.: National Academy of Sciences, 1979). Final Report.

Stobaugh, Robert and Daniel Yergin, ed., *Energy Future: Report of the Energy Project at the Harvard Business School* (New York: Random House, 1979).

U.S. Department of Energy, Energy Information Administration, *Annual Report to Congress, 1978, Vol. 3: Forecasts*. DOE/EIA-0173/3 (Washington, D.C.: U.S. Government Printing Office, 1979).

Coal and the
Residential Energy Future

GEORGE BENDA

Introduction

COAL once was the basic fuel of the modern age. But it is no longer. Coal once was the most widespread residential heating and cooking fuel in Europe and America. But it is no longer. The sky and ground were once blackened by the smoke and soot produced by home coal furnaces. But they are no longer.

The use of coal for residential energy supplies is an attractive option for the future, primarily because of the relative abundance of the resource. But many serious problems stand between the attractive option and the realization of it. To understand both the potential benefits, and the potential drawbacks to residential use of coal and coal derived energy, this paper examines some aspects of the history of coal use, some scenarios for reestablishing residential coal use, and some of the factors which can guide thinking and planning for land development based on these scenarios.

To prepare for this journey into the past and the future, it is necessary to take a look at some facts about coal and some facts about present residential energy use. The first fact about coal is that it is a finite resource. This is a fact that is often overlooked or dismissed, so I wish to emphasize it. The 200-500 year timeframe for exhaustion of the coal resource is really very short, put in the perspective of man's ten million years on earth so far. Coal is finite and that makes it precious.

On the whole, coal is thought of as an energy resource; it is thought of in terms of its heat value. But coal has a much greater range of values, tied to its chemical structure. The most well-known chemical constituent of coal is sulfur—which has a bad reputation for producing SO_2 when only the heating value of coal is extracted. But sulfur is a valuable and essential chemical in its own right. Coal contains many other substances, for example, zinc, in sufficient quantities to be useful. And through gasification, coal produces basic hydrocarbons which are primary feedstocks for many chemical products, such as plastics, fertilizers and solvents. Coal is precious not only because it is finite, but also because it offers several potential resource applications.

Matching coal resources to the energy needs of the residential sector requires some creative thinking. The residential sector consumes about 15 percent of the total energy budget for the U.S. (FEDS: July, 1979; See Exhibit 1). The direct burning of coal comprises less than 1 percent of the residential energy budget in the U.S. as we begin the 1980s (See Exhibit 2). While residential energy consumption increased over the past several decades, coal use has declined rapidly. In Illinois, for example, residential energy consumption increased at an annual rate of 2 percent between 1963 and 1975. Electric usage increased nearly 7 percent per year, and natural gas about 4 percent per year. At the same time, direct coal use in the residential sector dropped almost 18 percent per year (Weil and Bierman, 1977).

Principle residential energy uses are: heating and cooling of living space; cooking; heating water; powering appliances; and transportation. Gas, electricity and oil are the basic energy supplies for these uses. Nationwide, house heating is provided by natural gas in over 60 percent of the single-family detached houses; electricity in over 5 percent; oil in about 25 percent. Water heating is about 50 percent gas and 50 percent electric. Appliances and home cooling units are generally electrically powered, with a small proportion of refrigeration and clothes drying done by gas. Transportation is uniformly supplied by oil. Coal barely shows up in direct residential fuel use in the 1970s, providing about 5 percent of the space heating and virtually none of the other energy needs in single-family detached homes.

But the residential sector remains a major consumer of coal as an indirect energy source. Coal provides the fuel for about 45 percent of the nation's electrical production (Exhibit 3). In Illinois, coal provides nearly 70 percent of the fuel for utility electrical production (Wier and Bierman, 1977). Nationwide, residential electrical usage accounts for about a third of total electrical demand, and this demand has skyrocketed almost 400 percent since 1950 (Tansil and Moyers, 1974). The growth in this demand reflects the convenience and versatility of electricity as a fuel source, virtues which coal

Exhibit 1

U.S. RESIDENTIAL ENERGY CONSUMPTION

(in TRILLION BTUs)

Year	U.S. Total Consumption	Residential Total Consumption	Percent
1977	76,846.44	10,283.31	13.38
1975	71,294.15	10,004.03	14.03
1973	74,966.24	10,302.93	13.74
1970	68,065.78	10,075.76	14.80
1965	52,117.52	8,346.23	16.01
1960	42,357.66	7,183.19	16.95

Source: Federal Energy Data System (FEDS), Statistical Summary Update, DOE, July 79.

Exhibit 2

U.S. RESIDENTIAL COAL USE

(in TRILLION BTUs)

Year	Total Residential Consumption	Coal Consumption	Percent
1977	10,283.31	84.53	0.82
1975	10,004.03	107.75	1.07
1973	10,302.93	182.12	1.76
1970	10,075.76	312.87	3.10
1965	8,346.23	461.68	5.53
1960	7,183.19	673.02	9.36

Source: Federal Energy Data System (FEDS), Statistical Summary Update, U.S. DOE, July, 1979

Exhibit 3

U.S. ELECTRIC UTILITY COAL USE
(in TRILLION BTUs)

Year	Total Consumption	Coal Consumption	Percent
1977	22,515.25	10,271.96	45.62
1975	20,678.63	9,279.98	44.87
1973	19,401.59	8,313.53	42.84
1970	16,435.15	7,482.53	45.52
1965	11,064.73	5,875.01	53.09
1960	8,082.40	4,224.80	52.27

Source: Federal Energy Data System (FEDS),
Statistical Summary Update, July, 1979

itself lacks. Utilities are now the principle markets for coal and are likely to remain so in the near future.

Historical Perspective: King Coal and the Fuel Revolution

The recorded history of coal as a residential fuel stretches to a time when the notion of residential development was based on lords and serfs. And that history records more complaints about coal than praises of it. English nobility in the time of Edward I (1272-1307) complained of its foulness when burned, and bituminous coal use was generally banned or heavily controlled prior to the 17th Century. Nonetheless the use of coal grew as the demand for energy outstripped other energy sources such as wood. By the mid-17th and early 18th Centuries, a significant number of scientists were looking for ways to limit or eliminate the foulness of coal smoke. As coal use in England continued to grow, pollution control needs began to be recognized. A select Committee, appointed by the British Parliament in 1819, performed the first major study of smoke abatement (Goss *et al.*, 1915).

From the 17th Century to the middle of the Twentieth Century, coal was king. The industrial revolution was succorred on coal. It was the fuel for almost every conceivable use: home, business, industry and transportation. Technologies to use coal blossomed in the late 19th and early 20th Centuries. In the 1890s the first commercial coal-to-gas plants were cropping up in Europe, where the gas was used for municipal lighting. By the turn of the century electricity derived from coal was available. Coal had every appearance of being the energy source of the future.

A look at coal in Chicago during the period 1911-1915 illustrates just how significant coal was in that era. It is noteworthy that the source of information for the period comes from a book entitled, *Smoke Abatement and The Electrification of Railway Terminals in Chicago.* In 1912, 22.9 million tons of coal and coke were consumed in the Chicago area. Of that total, 4.1 million tons (18 percent) of coal were consumed in residential or small commercial furnaces. In addition, about 1.2 million tons (5 percent) of coal consumed in Chicago were used to manufacture gas and coke, much of which was consumed as a cooking fuel. Chicago had about 400,000 gas-fired stoves in 1915. Residential heating was also provided in some parts of the city through central steam or hot water plants. There were approximately 150 central heating and power plants in the U.S. in 1915—plants we would now christen with the modern name, cogeneration facilities.

Land use in Chicago was completely driven by the requirements of coal and coal-derived fuels. Transportation and coal-handling facilities dominated the landscape. The 23 million tons of coal passing through Chicago every year, handled by barge, train and wagon required an extremely large materials-handling system. The need for home delivery of the bulky fuel demanded a complex distribution system, as well. These factors add up to one conclusion: huge tracts of land had to be given over to King Coal, the fuel that made the city work, that made the city possible.

The futurists of 1915 would have no doubt bet upon the continued and exponential growth of coal use. No other fuel on the horizon—petroleum or natural gas—held the advantages of ready availability, low cost and multiple use. But a scant few decades later, these new fuels overtook coal, leading to the post-World War II petroleum and natural gas economy. King Coal was toppled; a rapidly growing, fuelish economy took part worldwide in the fuels revolution.

The revolution came about for some very good reasons, however. And these reasons are crucial to planning any possible future for residential use of coal. Coal lost out to oil and natural gas in the fuels revolution because:

• Coal is an inherently dirty fuel, especially compared to natural gas and oil.
• Coal is relatively inconvenient to use, especially in the home, where it brings dust and dirt problems, takes a long time to bring up to useful temperatures, cools slowly, and simply requires a lot of extra work.

- Coal is basically useful only for producing heat in the home; it cannot be applied to the modern appliances of refrigeration nor to mechanical mixers, washers and other devices.
- Coal is a bulky solid fuel, difficult to handle and transport, requires a great deal of available, useable land space for storage and distribution, and requires a great deal of floor space for home heat.

Air pollution regulations are not a recent bane to the coal industry. Coal is and always has been an essentially dirty fuel. As long as there were no real options, the dirt that came with coal was accepted as a price of modernity. But oil and gas burn much more cleanly than coal. The Chicago Committee on Smoke Abatement was a cog in the wheel of revolution. It concluded that there was (and still is) a significant case against widespread direct use of coal. The two greatest offenders were railroad engines and residential heating furnaces. Residential coal use was estimated to cause 3 percent of the smoke, 10 percent of the solid, 25 percent of the noxious gas, and 57 percent of the hydrocarbon pollution in Chicago in 1915. This proved to be a price which modern society chose not to pay.

Oil and gas were not only cleaner, but more convenient than coal, and could be adapted to a wider range of uses. In combination with electricity, oil and gas easily captured the residential market. Most residential units were switched from coal by the 1950s. No longer were the chilly Chicago mornings begun with a shivering trip to the basement to stoke-up the coal fire; no longer was the evening meal begun with a stoking of the kitchen coal stove; no longer was the winter's snow black within hours of its falling.

Land use changed rapidly with the switch away from coal. Coal yards and other handling facilities were removed and replaced with industry, housing and roadways for the new petroleum-powered transit mode, the auto. Remaining coal use, primarily in power generation, was fed by rail and barge. Smaller plants disappeared as utilities grew to realize economies of scale in electrical plants. The isolated large stations no longer infringed upon the space needs of the central city. Finally, in Chicago, nuclear energy captured nearly 40 percent of the electrical market. By the 1970s, King Coal was left only a modest stool in the democracy of modern power sources.

Scenarios: Coal and the Residential Energy Future

Four basic scenarios encompass the range of possible coal use for a residential energy source in the future:

- Maximum decentralization of coal use: coal as a basic, direct-use home fuel.

- Maximum centralization of coal use: coal as a feedstock for centralized, large-scale electrical and synthetic natural gas production for residential consumption.
- Community-based coal use: coal as a feedstock for moderate-scale gas and oil, electric and heat production (cogeneration).
- Coal as an element in community-based energy supply: coal as a basic resource in a mix of solar, waste or biomass, and other renewable resources.

Each scenario has a number of advantages and disadvantages related to the level of coal use, system efficiencies, economic costs and benefits to the individual, the community, and the nation, and environmental problems and benefits. Land use is impacted in various ways under each scenario, as well, including direct land requirements, transportation, storage and handling requirements, distribution networks, and other physical infrastructure considerations. These elements, added to some factors which influence individual choice, such as convenience and aesthetics, allow the prediction of the probability of realization of each scenario.

SCENARIO ONE: MAXIMUM DECENTRALIZATION OF COAL USE

Under this scenario a significant number of residential dwellings in the U.S. will be using coal directly as a fuel for all applicable purposes. Coal use on this level would require retrofit or new installation of coal-fired home furnaces, hot water heaters and cooking stoves. It is conceivable that, through technological development, each home could provide its electrical needs in conjunction with water heating and space heating. Mechanical feed systems might be developed to handle complex coal-stoking operations. A significant resurrection of the coal handling and distribution networks on a neighborhood scale would be required. Each neighborhood would need coal-handling and ash-disposal facilities; each home would also have to incorporate these elements.

A full home coal-use scenario would mean significant increases in the consumption of coal. About 15 percent of the total energy consumption in the U.S. is in the residential sector. This would translate into about 500 million tons of coal every year. But, realistically, perhaps a third of the residences of the U.S. could physically convert to coal, and perhaps 60 percent of their stationary energy use could be switched to coal. A more realistic estimate, then, would be an increase in coal consumption in the neighborhood of 100 million tons per year.

The overall system efficiency of the maximum decentralization scenario would be relatively low, but could be improved by technological development (See Exhibit 4). Based upon current technologies for home coal utilization—and these are essentially the same as they were fifty years ago—space and water heating might achieve average efficiencies of 40 percent, cooking is probably in the range of 20 percent efficient. The principal reason for the high waste levels is the relative inability of coal-fired appliances to follow the demand for heat. Long before, and long after the coffee is perked, the coal stove will be hot, but not hot enough to do anyone any good. These factors add together with transportation costs to reduce system efficiency for scenario one to an estimated 32 percent.

If new technologies, such as electronically-controlled mechanical feed systems and a small-scale cogeneration capacity were developed, it is conceivable that efficiencies could rise to as high as 64 percent. However, this would require each household having an interface with the electrical utility grid to buy power when heating needs are low and to sell power when heat is needed. Unfortunately, this would result in accentuation rather than reduction of utility-peaking problems, as electrical demand would generally increase during the summer for the cooling load.

Economic costs to the consumer and the community would be quite high, especially during the transitional phase. The requirements for transportation, handling, storage and distribution, as well as home floor space would provide a significant drain on economic resources. Land costs for handling facilities alone would be staggering, especially in older urban neighborhoods where easiest retrofit might occur. Facilities for a mile square of coal-fired residential units might take a city block, which if vacant in Chicago might cost $1-3 million. The current cost of residential coal with all of its inconvenience would be about the same as a convenient all-electric system.

Environmentally, scenario one would be perhaps the worst case for air quality and solid waste disposal. The 1915 Chicago study (Goss *et al.*, 1915) indicated that residential use of coal was the contributor of more than twice its share of pollution because of the inefficiencies of small-scale use. There are no solutions to particulate, sulfur and hydrocarbon emissions from home-scale furnaces or cook stoves. The sky and ground would again be blackened. Further, the environment and home would be filled with hazardous and toxic trace elements released in the combustion of coal.

Land use impacts would also be severe. As much as 20 percent of the urban landscape would have to be turned over to handling the bulky solid fuel. Transportation networks would be clogged by the traffic in coal. Heavier reliance on train and barge for coal transit to central depots, and on truck for local distribution would require a great investment of space for coal. And on the minute scale, new communities designed for coal use

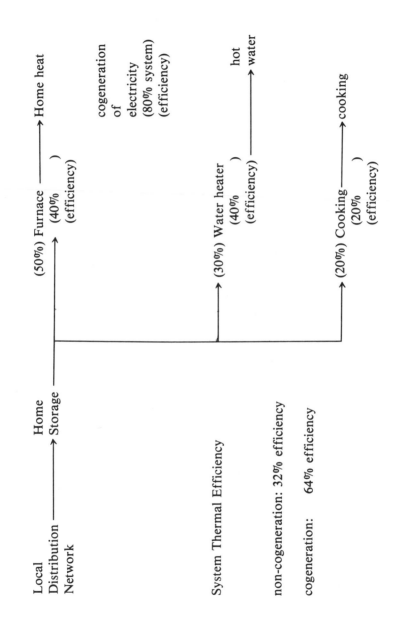

Exhibit 4
MAXIMUM DECENTRALIZATION
OF COAL USE

Local
Distribution ⟶ Home
Network Storage

(50%) Furnace ⟶ Home heat
(40%)
(efficiency)

cogeneration
of
electricity
(80% system)
(efficiency)

(30%) Water heater ⟶ hot
(40%) water
(efficiency)

(20%) Cooking ⟶ cooking
(20%)
(efficiency)

System Thermal Efficiency

non-cogeneration: 32% efficiency

cogeneration: 64% efficiency

would have to be designed around factors such as prevailing wind direction and average wind speed to minimize health concerns.

An evaluation of these general factors suggests that a return to direct coal use in the residential sector is unlikely. If the scenario were forced upon residential communities, it would require full recognition that all quality-of-life goals related to health and aesthetics would have to be ignored. The work involved in each individual dwelling would also cause significant resistance to implementation of this scenario.

SCENARIO TWO: MAXIMUM CENTRALIZATION OF COAL USE

This scenario presents the option which has been quietly, and perhaps somewhat unconsciously, chosen in America: development of large-scale energy production plants far removed from the urban environment. Since the beginning of the industrial revolution, the concept of economies of scale has held sway. From a time in 1915 when small cogeneration units were saddled beside residential and industrial customers, power plants have been increasing in size and in distance from the end user. The same principles which have been applied in electrical generation are being widely applied in coal gasification and liquefaction. Under this scenario, the rural areas of the U.S. will be dotted with huge coal-conversion facilities, producing electricity, synthetic natural gas, synthetic oil, and a range of chemical byproducts. These massive energy parks would include:

- Electric generating stations of one thousand to five thousand megawatt capacity, consuming 2-10 million tons of coal per year.
- Synthetic natural gas and oil production facilities, producing up to 90 billion cubic feet of gas and a million barrels of oil (or a somewhat different mix), and consuming upwards of 12 million tons of coal per year.

A dozen or two of these parks would generate ten-to-thirty percent of the nation's total energy needs, given current consumption patterns. The two principle residential fuels, electricity and gas, would continue in abundant supply, and oil would be avilable for both home use and transportation.

Adoption of this scenario would increase coal use, at the maximum, about 530 million tons per year (two dozen parks). If a heavy emphasis is placed on liquid fuels development, to replace imported oil, the number could go much higher. However, the efficiency of this option remains low. Gasification and liquefaction are generally in the range of 50 to 70 percent efficiency. Electrical production is usually about 30 to 35 percent efficient. Net system efficiencies are startlingly low: only about 38 percent for synthetic gas used at home; and only about 23 percent for electricity. The lion's

Exhibit 5

RESIDENTIAL SPACE HEAT DELIVERY
BY FUEL AND TECHNOLOGY

System	Delivered BTUs per 100 BTU fossil input	Fuel Cost
Direct Coal, current technology	32	$2.00-5.00 / $\overset{v}{m}$BTU[3] (residential)
Direct Coal, new technology	64	
Electric Conventional	75[1]	$11.75 / $\overset{v}{m}$BTU[2] (delivered cost)
Electric Heat Pump	200 - 300[2]	
Natural Gas Conventional	75	$2.50 - 3.00 / $\overset{v}{m}$BTU (regulated)
Natural Gas Heat Pump	200 - 300[2]	
Oil Conventional	70	$4.00 - 8.00 / $\overset{v}{m}$BTU (spot market)

[1] Hammond, Zimmerman

[2] Department of Energy, 1978

[3] OTA ($40-100/ton, residential)

share of the thermal value of the coal is released, unused, to the environment (See Exhibits 6 and 7).

Despite the advantages of economies of scale, the price of energy under this scenario is quite high. DOE (1978) estimates the real delivered cost of electricity to be $11.75 per million BTU. Synthetic gas is believed to cost $5 to 7 per million BTU to produce; delievered cost will be higher. Each of the energy parks described would require a capital investment of $4 to 6 billion. These investments and developments would fuel the economy, provide jobs and stimulate ripple developments. However, these developments would increase the real (not merely the inflated) cost of energy throughout the economy, which would probably worsen our inflation woes.

The basic premise of these rural energy parks is to isolate them from the urban environment—out-of-sight, out-of-mind. But the environmental impacts of this type of energy development will be wide-ranging and permanent. Major problems such as acid rain, produced by sulfur emissions to the atmosphere, and carbon dioxide build-up worldwide are already being seen and felt. To these problems will be added toxic tar and sludge problems from gasifiers and waste stream scrubbers. Agricultural land and natural areas will be paved and built over to develop the plants and their support facilities. Vast quantities of surface and groundwater will be consumed in the gasification process and used in cooling for the processes. The rural environment will be degraded significantly.

Land use examined narrowly for the residential community could be continued as it is today under this scenario. However, on a broad approach, land use will be widely and negatively impacted. The energy parks will consume land needed for production of food and for recreation. But these are impacts which our society has grown to accept. The question of scale, however, has never before been raised to that level.

A massively centralized, coal-based energy production scenario fits well with conventional wisdom about our energy future. Some rather strong opponents of the centralized concepts have emerged, led by Amory Lovins and others. Whether it is with glee or fear, most people believe that this scenario is inevitable. Recall, however, the futurists view of coal in 1915, and the fuel revolution of the following decades. The huge centralized plants may go the way of the dinosaurs, whose remains we now burn for fuel.

SCENARIO THREE: COMMUNITY BASED COAL USE

Cogeneration is the basic concept of this scenario. It is an old concept, as evidenced by the 150 or so cogeneration, district heating plants in the U.S.

Exhibit 6
COAL GASIFICATION,
MAXIMUM CENTRALIZATION
OF COAL USE

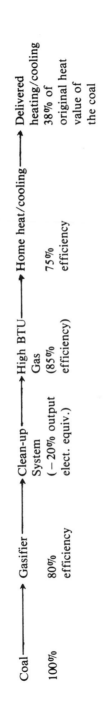

Coal ⟶ Gasifier ⟶ Clean-up ⟶ High BTU ⟶ Home heat/cooling ⟶ Delivered

100%   80%   System   Gas   75%   heating/cooling
        efficiency (−20% output (85%   efficiency   38% of
                   elect. equiv.) efficiency)   original heat
                                                                   value of
                                                                   the coal

System Thermal Efficiency

Conventional Gas Heat: 38% efficiency

EXHIBIT 7:
COAL GASIFICATION,
MAXIMUM CENTRALIZATION
OF COAL USE

Coal ———→ Utility ———→ Scrubber ———→ Electric ———→ Home heating/cooling ———→ Delivered
              Boiler   Power   heating/cooling
100%   35%   − 5%   75%   23% of original
              efficiency   output   efficiency   heat value of
                                                                                                              the coal

System Thermal Efficiency

Conventional Electric Heat: 23% efficiency

in 1915. But cogeneration, and most community-based energy supplies, disappeared under the pressure of economies of scale. Under this scenario, economies of distribution are balanced with economies of scale in moderate-scale plants which use coal as a feedstock to produce gas, oil, electricity and heat. Each plant would necessarily be directly tied to either a neighborhood or an industry. In general terms, such a plant might be made to fit community sizes ranging from 10,000 to 100,000 population. These plants would be relatively small by today's standards, consuming perhaps 100 to 500 tons of coal per day (compared to a 25 thousand ton per day synfuels plant). Ownership could be by private or municipal utility. Each plant would be within the boundaries of the service area.

It is difficult to estimate the potential for this type of development, but it can be assumed that somewhere between 1,000 and 5,000 plants could conceivably be built on this premise. The increase in coal use could range from 37 to 900 million tons per year. If these plants are built in areas where coal is used now to fire electrical power plants, they would displace current coal markets. The maximum use of coal under this scenario could supply all residential and many industrial or transportation needs.

Cogeneration has always been attractive because of its efficiencies. The option proposed here, using a medium BTU gasifier as the first stage in multiple-fuel energy production, is perhaps a bit less efficient than a straightforward electricity—hot water cogeneration system fired directly by coal. However, the inefficiency is compensated for through the generation of gaseous and liquid fuels and valuable chemical byproducts, production efficiencies reach near 80 percent, and residential delivery efficiencies, based on current home uses is about 60 percent. These thermal efficiencies are quite good. In addition, the entire coal resource can be available through this approach if associated industries use the chemical byproducts and the gas in chemical products.

Costs are difficult to assign in the context of the multiple-fuel plant. It is reasonable to assume that costs for gas and oil from these small plants would be about the same or a little higher than from major synfuels plants. Gas would probably range from $5 to 7 per million BTU, and oil would be $30 to 35 per barrel. The cost of electricity and district heating would depend greatly upon the geographical proximity of the plant and the end users. Since efficiency in the electrical process is twice-to-triple the average efficiencies, cost could be estimated at one-third-to-one-half the current cost of delivered electricity of $3.90 to $5.80 per million BTU. These prices do not account for sale of byproducts and other gains in efficiency which are tied to this concept.

Environmentally, the impacts of these plants would be proportional to the amount of coal they processed. Medium BTU gasification appears to be one of the least environmentally damaging ways to process coal, producing relatively little tar, and readily controllable sulphur emissions. Nonetheless, at the maximum coal usage for this scenario, the air pollution, solid waste and hazardous materials problems could become severe. It is more likely that the public would complain about this pollution because the plants would be in their own back yard, so to speak.

The land use impacts of this scenario would be extensive. As much as 10 to 15 percent of the urban landscape would be required for the plants and necessary handling and transport facilities. It also may be necessary to plan the location of the plant on the basis of wind direction in relation to living space, and to contain the plant with some sort of buffer. Community design would also have to be based upon district heating considerations to assure efficiencies.

The return of the town gas plant approach to energy is likely to meet significant resistance from people accustomed to a relatively clean living environment. For decades, residential communities have dedicated themselves to avoiding heavy industrial facilities within their environments. But that is precisely the character of a producer gas plant—noisy, smelly, dirty and troublesome. Yet the inherent efficiencies and potential for local energy stability make the cogeneration option attractive.

SCENARIO FOUR: COAL AS AN ELEMENT IN COMMUNITY-BASED ENERGY SUPPLY

In this scenario, coal becomes one of several sources of fuel for the community, but the community energy system is looked at as a closed loop, rather than a straight line flow. Coal, agricultural biomass and solar energy (broadly defined to include wind, hydropower and others), are the essential outside sources of energy for the community. In addition, the community's waste is recycled into the energy system and becomes an internal source. Coal can provide the basic resources of heat and chemicals for manufacturing processes required to sustain the community. The interface of the community with supplies of agricultural products is critical for liquid fuels production (alcohol) and for recycling wastes back into the land as fertilizer. Waste products also can be directly consumed or gasified for their energy values. Most of these technologies are available, but have not been substantially integrated before.

This scenario is a minimum coal consumption option, using the fossil fuel only where essential and using it for all of its possible resource values. It is not currently a widely applicable system, but assuming only 20 percent adaptability, and 50 percent of the total energy of the system deriving from coal, coal would provide about 2 percent of the residential energy needs. The difference is that this would be an application of coal as a total resource, rather than as only a heat source. Based on these assumptions, coal utilization under this option would increase only abouat 20 million tons per year.

The system output per BTU of fossil fuel input is estimated at 2. This assumes that coal is used to provide critical energy and chemical needs for the agricultural sector, as well as the residential sector. In addition, direct energy subsidies to agriculture can be minimized through use of waste materials for fertilizer and biomass-based alcohol for farm fuels (See Exhibit 9).

Environmental impacts of this scenario are minimized by use of the closed loop concept. Since 50 percent of the community's energy would still be coal-derived, a significant amount of air pollution would still be expected. Direct combustion of waste materials also would cause an increase in air pollution. However, offsetting these factors is the use of alcohol as the basic liquid fuel. Alcohol burns more cleanly than petroleum-based fuels used in the same situations. Simple replacement of combustion in 40 percent of the energy supply to the community implies a 40 percent reduction in pollutant loadings. Far and away, this scenario is the most benign from an environmental perspective.

Land use would come to have a new meaning in this scenario. The notion of integrating energy and material flows in and out of a community is fundamentally alien to the American urban mind set. The interface with agriculture would have to become a two-way connection, rather than the current one-way street. The interchange of fuels, food and waste also would present a space and handling problem. Land treatment of wastewater, as is being proven at the Muskegon project, requires significant acreage for proper treatment. But in an even broader sense, the institutional structures which underlie current energy and land use patterns in the U.S. would need to be shifted. The concept of this scenario reflects a structure nearer the ancient Greek polis than the contemporary American city.

Evaluation of this option suggests that several factors favor its realization, but several significant barriers relate to unknown cost factors and institutional complexities. The concept of integrating coal as an essential component in an otherwise reuseable energy resource scheme is one which neither the proponents of centralized facilities nor decentralized systems have really touched upon. Yet from the standpoint of industrial and chemical feedstocks and power generation, a fossil subsidy is still required under most renewable resource scenarios. Efficiencies and returns on fosssil

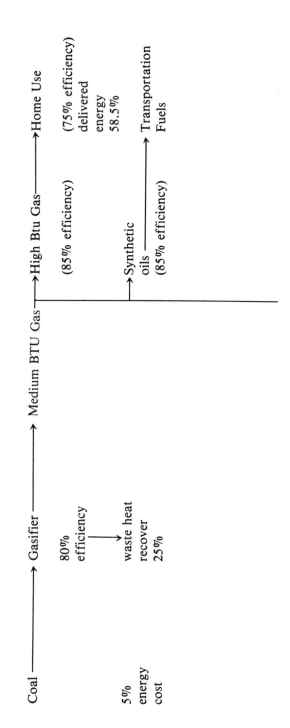

Exhibit 8
COMMUNITY-BASED
COAL USE

Exhibit 8 (Continued)
COMMUNITY-BASED
COAL USE

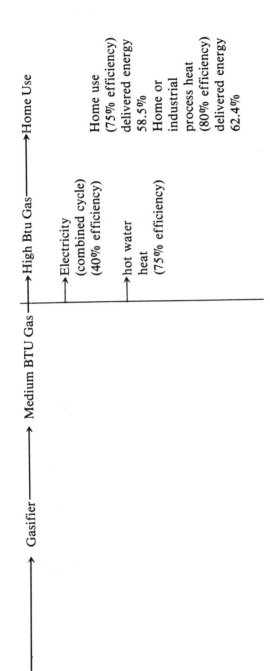

Coal ⟶ Gasifier ⟶ Medium BTU Gas ⟶ High Btu Gas ⟶ Home Use

Electricity
(combined cycle)
(40% efficiency)

hot water
heat
(75% efficiency)

Home use
(75% efficiency)
delivered energy
58.5%

Home or
industrial
process heat
(80% efficiency)
delivered energy
62.4%

System Thermal Efficiency

Production of Transportation fuels distorts systems efficiences

Production efficiency = 78% efficiency
System efficiency (residential only) = 59.8% efficiency

Exhibit 9

COAL AS AN ELEMENT IN COMMUNITY-BASED ENERGY SUPPLY

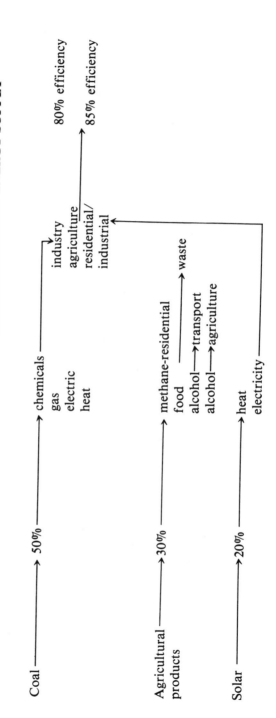

Coal ⟶ 50% ⟶ chemicals ⟶ industry   80% efficiency
                    gas   agriculture
                    electric   residential/   85% efficiency
                    heat   industrial

Agricultural ⟶ 30% ⟶ methane-residential ⟶ waste
products   food
                       alcohol ⟶ transport
                       alcohol ⟶ agriculture

Solar ⟶ 20% ⟶ heat
                electricity

investment are far better in the multiple-source system than in central facilities. It is conceivable that community-based energy systems will become experimental within a decade, and perhaps will come of age for new developments.

Thinking About Coal and the Residential Energy Future

The emphasis in this analysis so far has been on the question, how can coal be used in the residential sector. The remainder of the analysis will focus on this question with a slightly different emphasis: is the most appropriate use of coal the production of energy for the residential sector? Examined in the light of best use of resources, beyond the narrow perspective of energy demands, coal is a resource which should not be squandered on inefficient energy production or use systems. Coal is a precious resource, and should be used in applications which tap all of its potential chemical and heat values. The full range of social and ecological costs and benefits need to be weighed in making decisions about future coal use in the residential sector.

The case for increased use of native coal resources is often argued in terms of benefits to the nation's economy. Rather than dump our dollars into foreign oil—an investment on which the nation as a whole receives no return—our energy dollars should be recycled through our own economy. Then our energy dollars would create much needed employment, and growth in the money supply could be slowed without severe negative ramifications in interest rate and capital availability. Superficially, these are adequate reasons for a return to a coal-fired economy.

All coal-based scenarios, residential or otherwise, predict an energy price structure some level significantly higher than current oil or natural gas. In 1974, synthetic gas was predicted to cost $3.50 to 4.00 per million BTU; it is now predicted to be $5 to 7, and higher. Synthetic oil was predicted to cost $18 to 22 per barrel to produce in 1974; it is now predicted at $35 to 40 per barrel and higher. Clear benefit to the national economy would occur if coal-based energy prices were lower than foreign energy prices. But they are not. The price of synthetic fuels is sensitive to two primary factors: the cost of coal and the cost of capital. The 1974 assumptions were based on coal costs of about $3 to 10 per ton, and capital costs of about 10 percent. Current coal prices are 3 to 4 times the 1974 level, and capital costs have doubled. Synthetic fuels seem to be at a continuous price disadvantage to natural gas and oil, even though those fuels continue their upward price spiral. Exhibit 10 compares the rate of price increase for coal to imported oil; Exhibit 11 compares the price of coal to oil and natural gas prices.

Nevertheless, coal remains the largest potential fossil energy source in the U.S. Total recoverable reserves of coal in the U.S. are estimated at over 280 billion tons (OTA), though conservative geologists looking at current land use place the truly recoverable reserves at half that amount. At current levels of use, this reserve would last about 400 years. If coal alone were used to satisfy the current energy demand of the United States, these proven reserves would last less than 100 years.

One hundred to four hundred years of energy supply sounds like a long time frame to a society accustomed to five-year and twenty-year planning horizons. But when the subject of planning is a finite resource, normal planning time frames are meaningless. When our fossil fuel resources are gone, they are irretrievably and incontrovertably gone. The depletion of a finite resource brings disruption of the society dependent upon that resource long before it runs out. The cost implications of depletion are simple: every ton of coal is more expensive to mine than the previous ton. Every new ton requires a deeper shaft or pit, more washing, or some other increased effort to mine or use it. In Illinois, for example, over the decades the reserves of low-sulfur content were the first to go, each new ton comes from reserves of higher and higher sulfur content. From an air pollution perspective, each new ton of Illinois coal will be more expensive to burn. That is a significant cost of resource depletion.

The overall resource impacts of increased coal use include mining damage to agricultural land, acid rain, nitrogen build-up, carbon dioxide build-up from coal combustion. The range of short and long run environmental disturbances in the coal cycle are staggering (Exhibit 13). The problems begin with mining. Health and safety of miners has always been a problem, especially with underground mining. Deep mines present acid drainage problems, surface dust and soil deposit problems, and later on, subsidence problems. Surface mines disrupt or destroy large areas of the land surface. Early strip mining practices simply destroyed the land; current practices include land reclamation. Severe implications of strip mining occur when coal seams are under prime agricultural land, which is the case throughout much of the Midwest. While significant progress has been made in developing methods of restoring the productivity of mined lands, some portion of the coal value is lost. The question of long-run food needs must be set in balance with our need for fossil energy.

Acid rain is largely the result of the release of various sulfur compounds to the atmosphere through the process of coal combustion. The extent and intensity of this problem appears to correlate quite well to the total sulfur

Exhibit 10

AVERAGE DELIVERED PRICE OF COAL
AT UTILITIES (CONTRACT)

CRUDE OIL REFINER ACQUISITION COST
IMPORTED ($/barrel)
(includes transportation costs from well head to refinery)
(- average landed cost of crude oil to the refiner; $ represents
the amount which may be passed on to the consumer.)

Year	Average $/barrel	% Change
1974	12.52	
1975	13.93	11.3
1976	13.48	- 3.2
1977	14.53	7.8
1978	14.57	0.3
1979	18.45*	26.6

* average for first 7 months.
Percent increase during 67 months = 47.36%

emissions from utility plants (Exhibit 14, OTA p. 223). Increased coal use will increase sulfur emissions by some degree, depending upon the technology in place. A look at current geographical distributions of sulfur emissions (Exhibit 15, OTA p. 189), current sulfate concentrations (Exhibit 16, OTA p. 190), and projected 1990 sulfate concentrations under a high coal use scenario (Exhibit 17, OTA p. 217), suggests that by 1990, the acid rain problem would be as serious for the northeastern and midwestern U.S. as it is presently for northwestern Europe. Negative effects include destruction of trout and salmon fisheries, disruption of aquatic nutrient cycles, and possible damage to food crops, especially soybeans.

Nitrogen oxides are also the product of fossil fuel combustion, and part of the acid rain problem. But nitrogen oxide in the atmosphere also can be linked directly to negative impacts on plant and animal health (OTA, 1979). Coal is a significant source of nitrogen in the atmosphere, accounting for 24

Exhibit 10 (Continued)

Average Delivered Price of Coal at Utilities (Contract)

Year	$/Short Ton	% Change
1974	12.07	
1975	16.21	34.29
1976	17.90	10.42
1977	19.25	7.54
1978	21.41	11.22
1979	25.19*	17.65

* average for first 6 months.

Percent increase during 66 months = 108.69%

Source: DOE Monthly Energy Review

percent of the total and 49 percent of the stationary fossil fuel sources of the pollutant globally. The problem again primarily affects the northeastern U.S. (Exhibit 18, OTA p. 191), though emissions are significant nationwide (Exhibit 19, OTA p. 192). Aside from the acid rain complications, nitrogen oxides increase human respiratory infection, and in chronic doses may cause emphysema and other lung disfunctions.

Carbon dioxide is an inevitable product of combustion. But coal is among the largest contributing fuels per unit of heat value. The global impacts of carbon dioxide are being widely debated. The "greenhouse effect," the long-term increase of the world's atmospheric temperature, is the theoretically probable effect of increased coal combustion and resultant CO_2 emissions.

Large-scale increase in coal utilization, especially for synthetic fuels, also will impact water resources in the U.S. Water is consumed in the gasification process as a hydrogen donor. In addition, process çooling will account for higher water use. Major synfuels production centers are likely to be developed in the Midwest, where water and coal resources are both available. Water has always been recognized as a limiting factor in gasifica-

Exhibit 11

COST OF FOSSIL FUELS DELIVERED TO STEAM - ELECTRIC UTILITY PLANTS
(cents/mil BTU)
(national average)

Year	Coal	Residual Fuel Oil	Natural Gas
Jan. 76	80.2	194.1	86.5
July 76	85.7	187.0	106.2
Jan. 77	85.9	217.2	¹11.1
July 77	96.2	217.0	131.7
Jan. 78	99.6	211.3	133.3
July 78	110.2	205.0	149.8
Jan. 79	115.8	231.8	150.2
March 79	116.8	261.4	162.8
June 79	121.8	289.3	179.5
Percent change 1-76 to 3-79	45.6	34.6	88.2
Percent change 1-76 to 6-79	51.8	49.0	107.5

tion, and it will become more severely restrictive as the level of development increases.

These are the resource costs associated with developing coal for its full energy potential. Residential energy use accounts for about 15 percent of the nation's energy budget. Large-scale development of coal to supply residential and other energy needs would bring into effect these costs. And as coal use increases, the cumulative impacts of resource depletion will come into play. Both the economic and resource costs increase as larger and larger commitments are made to coal use.

Exhibit 12

COST OF FOSSIL FUELS SOLD IN RESIDENTIAL SECTOR
(national average)

Year	Natural Gas Sold to Residential Customers for heating use $/1000 cubic feet	Residential Heating Oil Prices (Average selling price) cents per gallon
Jan. 74	116.0	31.1
July 74	124.9	35.2
Jan. 75	141.2	37.4
July 75	154.7	37.2
Jan. 76	171.4	40.1
July 76	184.5	39.3
Jan. 77	213.8	44.4
July 77	229.9	45.8
Jan. 78	241.6	48.5
July 78	254.2	48.2
Jan. 79	297.7	53.7
March 79	305.5	58.5
July 79	N/A	73.9

Percent change

| 1-76 to 3-79 | 78.2 | 46.6 |

Exhibit 13.—Environmental Disturbances From Coal-Related Activities

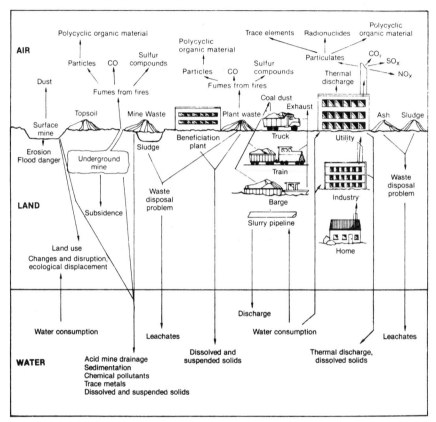

SOURCE: U.S. Office of Technology Assessment, *Direct Use of Coal,* 1979.

Exhibit 14.—Regional Impact of Acid Rainfall

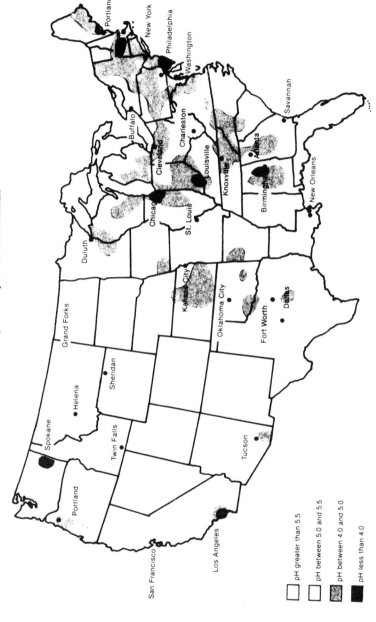

Portland · New York · Philadelphia · Washington · Savannah

Buffalo · Charleston · Cleveland · Louisville · Atlanta · Knoxville · New Orleans · Chicago · St. Louis · Birmingham

Duluth

Grand Forks · Kansas City · Oklahoma City · Fort Worth · Dallas

Sheridan

Helena · Twin Falls · Tucson

Spokane

Portland

San Francisco · Los Angeles

☐ pH greater than 5.5

☐ pH between 5.0 and 5.5

▨ pH between 4.0 and 5.0

■ pH less than 4.0

SOURCE: Environmental Challenges of the President's Energy Plan: Implications for Research and Development. Report of the Subcommittee on Environment and the Atmosphere. Committee on Science and Technology, U.S. House of Representatives. 95th Congress, October 1977.

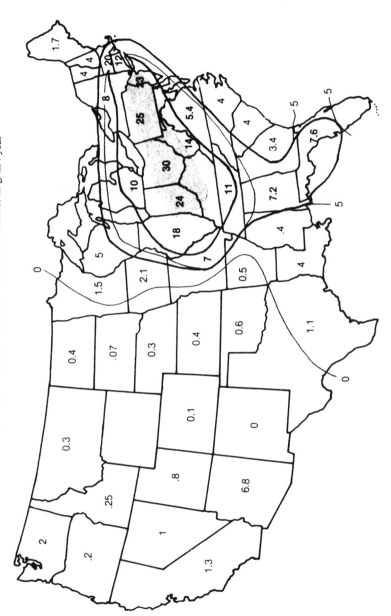

Exhibit 15.—SO₂ Emission Contours—Emission Densities in g/m²/year

SOURCE: Report on the International Symposium on Sulfates in the Atmosphere, in EPA-600/9-78-022, Energy/Environment III, U.S. Environmental Protection Agency, Office of Research & Development, October 1978

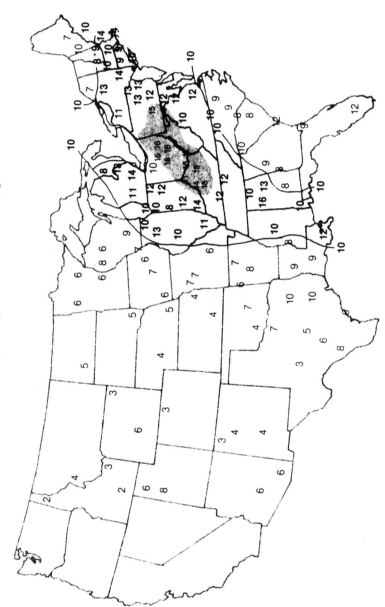

Exhibit 16.—Yearly Average Sulfate Concentrations, g/m³

SOURCE: Report on the International Symposium on Sulfates in the Atmosphere, in EPA-600/9-78-022. Energy/Environment III, U.S. Environmental Protection Agency. Office of Research & Development. October 1978.

Exhibit 17.—Annual Concentrations of Sulfate for a 1990 25.7 Quads Coal-Use Scenario
(Brookhaven National Laboratory, 1978)

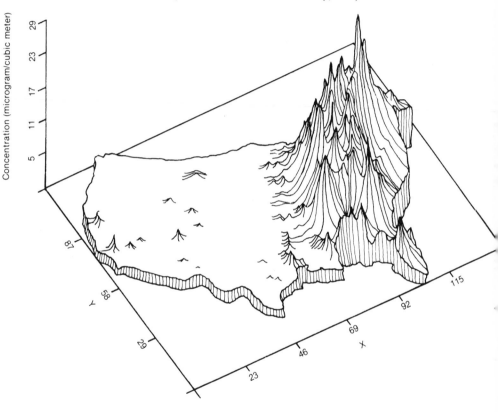

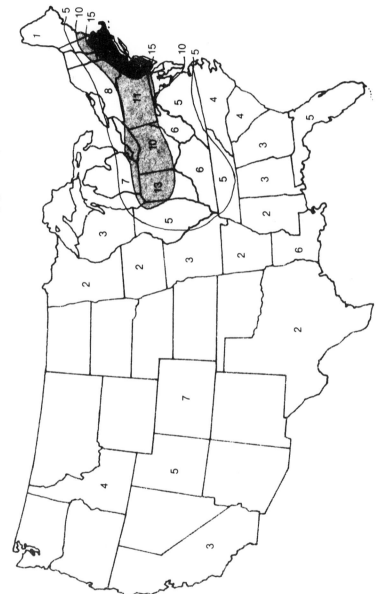

Exhibit 18.—Natoinwide Ambient NO$_3$ Levels, g/m^3

SOURCE: "Air Quality Data for Non-Metallic Inorganic Loss 1971-1974 from the National Air Surveillence Networks," Gerald G. Akland, EPA-600/4-77-003. Research Triangle Park. NC

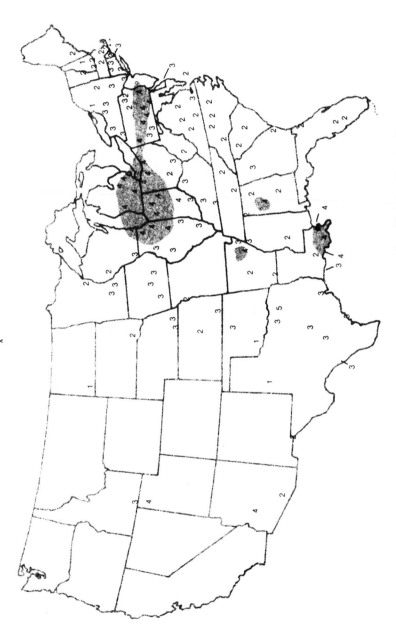

Exhibit 19.—National NO$_x$ Emission Densities, g/m^2/year

SOURCE:"National Emissions Report 1973," National Emissions Data System of the Aerometric and Emissions Reporting System, EPA-450/2-76-007, Research Triangle Park, NC

Confronting the finite nature of fossil resources, and focusing particularly on coal, future planning should incorporate a conservation ethic. Coal is too valuable—for both its chemical and heat constituents—to be wasted on inefficient modes of energy consumption. There is a wide range of modern needs that cannot be supplied without fossil fuels. But much, perhaps most of our energy needs can be supplied by renewable means. The limits of our fossil resources must teach us to budget our resources for their best and highest use—defined by wide social benefit and not new economics. The widespread use of coal for residential energy is a wasteful approach to our resource budget. As a policy option, residential energy supplies from renewable sources and from the gasification-cogeneration cycle based on coal is the best long run goal. Direct residential use and centralized power and gas production are the most wasteful approaches to the coal resource. We can no longer afford the waste; planning for the future must, for very practical reasons, be based upon a conscious budgeting of our natural resources, and a conscious application of the conservation ethic.

References

United States Department of Energy. *Federal Energy Data Systems (FEDS), Statistical Summary Update.* Springfield, VA: National Technical Information Service, 1979.

Weil, Keith and Wallace Bierman. *Illinois Energy Consumption: 1963-1975.* Illinois Department of Business and Economic Development. Springfield, VA: National Technical Information Service, 1977.

Goss, W.F.M. *Smoke Abatement and Electrification of Railway Terminals in Chicago.* Report of the Chicago Association of Commerce Committee of Investigation on Smoke Abatement and Electrification of Railway Terminals. Chicago: 1915.

Macrakis, M.S., ed. "Energy: Demand, Conservation, and Institutional Problems," in R.A. Grot and R.H. Socolow *Energy Utilization in a Residential Community.* for the Conference on Energy: Demand, Conservation, and Institutional Problems. Cambridge, MA: Massachusetts Institute of Technology Press, 1973.

Hammond, Ogden and Martin B. Zimmerman. "The Economics of Coal-Based Synthetic Gas," *Technology Review* (July/August 1975). 43-51.

Booz, Allen and Hamilton, Inc. *Evaluation of the Economics and Efficiencies of Heat Pump and Gas Furnace Space Conditioning Systems Using Coal as a Primary Source of Fuel.* prepared for U.S. Department of Energy under contract no. E(04-3) 1227. Springfield, VA: National Technical Information Service, 1978.

Office of Technology Assessment. *The Direct Use of Coal: Prospects and Problems of Production and Combustion.* Washington, D.C.: Government Printing Office, 1979.

United States Department of Energy. *Monthly Energy Review.* Washington, D.C.: Government Printing Office, 1979.

Verma, Arum. "From Coal to Gas—Part II," *Chemtech* (October 1978): 626-638.

Heddleson, F.A. *Fuels Used For Single-Family Detached Residential Heating in the United States.* by Oak Ridge National Laboratory. Springfield, VA: National Technical Information Service, 1975.

The Future of Nuclear Power

THE MORGAN GUARANTY SURVEY*

A stunning reversal in the fortune of nuclear power has taken place in this country within a relatively short time span. It was only a little more than half a decade ago that President Nixon, in announcing "Project Independence," confidently forecast that atomic energy would provide 30 percent to 40 percent of America's electricity by the end of the 1980s. Atomic plants now account for only 11 percent of electric power generation and the pace of future growth is quite uncertain. Indeed, utility officials see major challenges ahead if the nuclear industry is to survive and maintain the nuclear option for this country.

Many factors account for major problems where there used to be glowing promise. The quantum leap in oil prices charged by OPEC has triggered significant conservation efforts in power use by homes and industry. Slower growth in the use of electric power, now running at roughly 4 percent per year as against 7 percent annually in the earlier postwar period, has brought a scaling back in projected needs for both conventional and atomic power plants. All the while, opponents of nuclear power have impeded efforts of the nuclear industry, causing extensive and costly delays in construction programs. Adding to those obstacles have been government regulatory requirements that have stretched out completion time to build a nuclear power plant to more than twelve years, double the time needed in Europe or Japan.

Morgan Guaranty Survey June 1980 pp. 8–15.

Then, just over a year ago, came the trauma of Three Mile Island. The dangerous overheating of a TMI reactor, though it caused no casualties among workers or the public, has placed the burden of proof on those who say nuclear power is safe.

Taken together, the events of the past few years have sunk the domestic nuclear power industry in the doldrums. Last year there were no new orders by U.S. utilities for reactors and eleven cancellations of earlier orders. Meanwhile, many foreign nations—despite anti-nuclear protests—are pushing ahead on nuclear power. For example, the British government, after completing its own investigation of Three Mile Island, announced a stepped-up nuclear power program for the 1980s. Sweden, buttressed by a favorable public referendum last March, plans to double its nuclear power capacity over the next decade. So, too, does West Germany. France aims to supply 55 percent of its electric power from atomic plants by 1985 (Exhibit 1).

It was back in 1957 that the first reactor began producing electricity on a commercial basis. Expansion was swift in the next fifteen years. Since the mid 1970s, however, growth of nuclear power has slowed noticeably. As of May, there were 72 plants licensed to operate, another 88 under construction, and 17 reactors on order. The operating plants generated about 54,000 megawatts (1 megawatt equals 1,000 kilowatts) of electricity. They displaced the equivalent of about 500 million barrels of oil last year.

Even though existing nuclear plants account for only 11 percent of total electric power, some parts of the nation are heavily dependent on nuclear plants. For example, the area around Chicago, the Southern Atlantic states, and the New England states all rely on atomic plants for more than one third of their electricity.

Former President Carter was on record against closing the door on nuclear energy as some anti-nuclear groups urged. President Reagan is decidedly pro-nuclear.

Power for the People

Behind that judgment is the belief that nuclear power does not present unreasonable risks to public safety. Moreover, there is recognition that a growing nation and expanding industry will need enlarged electric-power capacity. Today's work force of roughly 90 million will climb to 113 million by the year 2000. It will take added power to provide the jobs, food, housing, and goods and services for that bigger work force. An official estimate of future power needs has been compiled by the National Electric Reliability

Exhibit 1

THE PUSH FOR ATOMIC POWER ABROAD

NATIONS abroad are pushing ahead with ambitious nuclear power programs, spurred by precarious oil supplies and rising oil prices.

There are nuclear units generating electricity in seventeen noncommunist nations outside the U.S. (accompanying table). Their combined generating capacity as of the second quarter of 1979 was 56,210 megawatts—about equal to U.S. nuclear generating capacity.

France has Europe's most ambitious program—with a nuclear commitment of $8 billion a year. Its 1,200-megawatt "breeder" reactor (Super Phoenix), when completed in 1982, is seen giving France a ten-year lead over the U.S. in the commercial application of fast-breeder technology. The U.S. "breeder" effort is contained in the Clinch River project in Tennessee.

Germany, Spain, and Italy recently have reaffirmed their goals of major nuclear expansion. West Germany has ten nuclear plants in operation and fifteen more in the planning stage. Spain, with three reactors, is planning to add fourteen more. Italy expects to more than double the number of its reactors in the decade ahead; four are now in operation. Belgium, Switzerland, and Sweden plan to produce 40% to 50% of their electricity with nuclear reactors by 1990. Japan, with twenty nuclear reactors, has a generating capacity of nearly 13,000 megawatts—second only to the U.S. among noncommunist nations.

Russia and other communist countries in Eastern Europe are forging ahead with nuclear power expansion. Comecon, the Soviet bloc's economic alliance, has a target of producing 25% of its electric power with nuclear reactors by 1990—compared with about 5% currently.

Nuclear Reactors and Generating Capacity

(Noncommunist nations; second quarter 1979)

	Reactors	Capacity (megawatts)
Asia		
Japan	20	12,840
India	3	620
Pakistan	1	140
South Korea	1	590
Taiwan	2	1,270
Europe		
Belgium	3	1,740
England	33	9,010
Finland	2	1,150
France	15	7,800
West Germany	10	7,050
Italy	4	1,490
Netherlands	2	520
Spain	3	1,120
Sweden	6	3,850
Switzerland	3	1,060
North America		
Canada	9	5,590
United States	71	54,180
South America		
Argentina	1	370
Total	189	110,390

Council (NERC). Even with conservation efforts, NERC estimates that generating capacity in the nation must be increased by 25,000 to 30,000 megawatts per year in the decade 1979–88 to meet rising power demand. That rise translates into a projected growth rate in electrical use in the ten-year period of 4.7 percent a year—down, it is true, from the annual growth rate of the past ten years, but still a very sizable increase in needed power-generating capability.

Large-Scale Options

All power sources will need to be tapped to meet energy needs. Most of them, however, offer little potential, for expansion in the near or intermediate term. Take solar power. The technology to use the sun to heat water in homes and to heat (and cool) buildings has existed for years, and solar water heaters are coming into more general use. But use of solar energy to generate large-scale electricity is quite another matter, given existing solar technology. Scientists estimate that it would take a field of solar collectors covering twenty to thirty square miles to produce the same amount of electricity as a typical nuclear plant. Scientists see the possibility of solar space stations that would collect and transmit the sun's energy to earth via a microwave system. That, of course, is far in the future. Solar energy, thus, is not counted on as a major source of power before the end of the century.

Hydroelectric power accounts for about 13 percent of total electric power production. All the best dam sites, however, are already taken. And new dams are vigously opposed by environmentalists. The possibilities for power expansion are very limited, too, in other fields. In fact, none of the other power sources—geysers, windmills, photovoltaics, electricity from waste products—is given any chance of meeting the nation's growing needs, at least not in this century.

That leaves only two nonoil power sources that offer any real capability of supplying the nation's needs for electricity in the next 25 years: coal and the atom. Both have good points; both have drawbacks. Yet scientists agree that expansion of coal and atomic power offers the only alternative to overdependence on unstable sources of high-priced foreign oil.

Coal is abundantly available in this country, with reserves sufficient to last several hundred years. Gasification of coal for delivery through existing gas pipeline systems may prove to be an added source of energy for space and water heating. An internationally sponsored "World Coal Study," found a bright potential for coal in the world power picture. To help meet world energy requirements, the study urged a tripling in coal use and stressed that such a rise could be accomplished without sacrificing health, safety, and environmental protection. The reason: oil is now so expensive that it is economic to clean up coal.

Nuclear energy, too, has many positive features. Supplies of uranium, the nuclear fuel, are sufficient to last well into the 21st century even without development of the "breeder" reactor, which produces more fissionable material than it burns. The nuclear industry's safety record (not a single radiation death in nearly three decades) ranks No. 1 in all of industry. And nuclear power is "clean."

Additionally, electricity from the atom generally has been more economic than that produced from the other major fuel sources. For example, an Atomic Industrial Forum (AIF) survey in 1978, covering 43 utilities with nuclear and other facilities, found the following cost per kilowatt-hour of electricity produced: 1.5 cents for nuclear; 2.3 cents for coal; and 3.5 cents for oil. These are costs of producing power at the "bus bar," the point at which electricity leaves the generating station. The costs include equipment, interest, fuel, operation and maintenance, insurance, and taxes.

The 1.5 cents per kilowatt-hour for nuclear power in 1978 compared with 1.2 cents in 1975, according to the AIF. Coal generating costs climbed from 1.8 cents in 1975 to 2.3 cents in 1978.

Coal plants generally take less time to plan, license, and build (six to nine years) than do nuclear plants (ten to twelve years). Nuclear plants have somewhat higher capital costs than coal plants, but the former have much lower fuel costs. In 1978, for example, fuel costs for nuclear power plants accounted for 22 percent of total production costs. For coal, fuel averaged out at 55 percent of total costs.

Projections of future costs of generating electricity—either by coal or atom power—are speculative and, indeed, controversial. The reasons are easy to understand. Forecasts on power costs require a whole series of assumptions—on interests rates, inflation, fuel costs, the regulatory environment, and so on.

While estimates of nuclear power costs in the future often show quite a variation (because of different assumptions about cost factors), nuclear analysts believe that nuclear power offers cost advantages over coal—particularly in areas where expensive pollution-control equipment must be installed. One estimate of future power costs—contained in a Ford Foundation study released in May 1979—placed the cost per kilowatt-hour for nuclear power in 1990, expressed in 1978 dollars, at 2.4 cents. For coal: 2.8 cents. Both figures are average costs for the U.S.

Three Problem Areas

All this is not to say, obviously, that nuclear power is problem-free. It clearly does have difficulties. Here are three major concerns:

SAFETY

The overriding public concern is the potential for catastrophe—a nuclear explosion or a "melt down" of the reactor core, the so-called China Syndrome.[1]

The former—an explosion such as in an atomic bomb—is not possible. The most important reason is that the reactor's fuel is a uranium dioxide ceramic material containing only 3 percent to 5 percent fissionable material. An atomic bomb requires a solid mass of nearly pure U-235 or plutonium.

The latter—a melt down—is possible, although it is significant that in well over 200 reactor years of commercial operation there have been no reactor-core melt downs. To melt the fuel requires a failure in the cooling system or the occurrence of a heat imbalance that would allow the fuel to heat up to its melting point, about 5,000 degrees Fahrenheit. In that event, the molten core might breach the outer containment building, releasing large amounts of radioactivity into the atmosphere.

The safety design of reactors includes a series of systems to prevent the overheating of fuel and to control potential release of radioactivity from the fuel. Such systems went into operation at Three Mile Island. Water was added to the system to cool the reactor. The TMI accident did result in a loss of cooling water from the reactor and the fuel was not covered with water during part of the period. However, while the extent of the damage to the fuel core is still uncertain, a melt down in the classic sense did not occur. And, in spite of serious damage to the plant, a Presidential Commission concluded that most of the radioactivity was contained and that the actual release will have a negligible effect on the physical health of individuals in the area.

What are the chances that one day a melt down actually could take place? The most comprehensive assessment of probabilities and consequences of accidents at nuclear power stations was conducted by 60 scientists and engineers in 1975. The three-year study was headed by Professor Norman Rasmussen of the Department of Nuclear Engineering at Massachusetts Institute of Technology.[2]

The consequences of a core melt would depend mainly on three factors: the amount of radioactivity released; the way it is dispersed by prevailing weather conditions; and the number of people exposed to the radiation. The Rasmussen study calculated the health effects and the probability of occurence for 140,000 possible combinations of radioactive release magnitude, weather type, and population exposed. The probability of a given release was determined from examination of the probability of various reactor system failures. The probability of various weather conditions was obtained from the weather data collected at many reactor sites. The probability of various numbers of people being exposed was obtained from U.S. Census data for current and planned U.S. reactor sites. The required thousands of computations were carried out in computer banks.

The results are shown in Exhibit 2. It compares the likelihood of a nuclear accident to nonnuclear accidents that could cause the same consequences. For example, the average probability of 100 or more fatalities from 100 nuclear plants is 1 in 100,000 years—the same as the chance of a meteorite falling from the heavens.

A related concern: If a major accident should occur, what would surrounding populations do to avoid the danger? As yet there is a general absence of effective evacuation plans. In some areas, rapid mass evacuations may be physically impossible. Nonetheless, efforts are being made to improve such planning among utilities and local governments.

WASTE DISPOSAL

Spent nuclear fuel contains many deadly elements—e.g., strontium 90, cesium 137, and plutonium. Such wastes remain lethal for very long periods. Clearly, disposing of them poses problems for government and industry, but the task is far from unmanageable.

Contaminated wastes from military programs (weapons, ships, research) account for more than 90 percent of total nuclear wastes. Over the past 30 years the government has generated enough wastes to cover four highway lanes ten inches deep for a hundred miles, according to the General Accounting Office. Those wastes are being stored in three government facilities. In addition, some 4,000 metric tons of commercial nuclear plant spent fuel are now being safeguarded in cooling pools at reactor sites.

The nuclear industry feels that waste disposal is a political, not a technical problem. Says the Electric Power Research Institute: "The technology for disposing safely of nuclear waste material is already available—and by several processes." The problem is said to be the lack of decision by government on which disposal system to accept.

Exhibit 2

FIGURING THE ODDS ON A CATASTROPHE

Catastrophe	Probability of 100 or more fatalities	Probability of 1,000 or more fatalities
Airplane crash	1 in 2 years	1 in 2,000 years
Fire	1 in 7 years	1 in 200 years
Explosion	1 in 16 years	1 in 120 years
Toxic gas	1 in 100 years	1 in 1,000 years
Tornado	1 in 5 years	very small
Hurricane	1 in 5 years	1 in 25 years
Earthquake	1 in 20 years	1 in 50 years
Meterorite impact	1 in 100,000 years	1 in 1,000,000 years
Reactors (100 plants)	1 in 100,000 years	1 in 1,000,000 years

Source: *Reactor Safety Study*, U.S. Nuclear Regulatory Commission, October 1975.

A government review group is on record with a relatively optimistic view about the waste-disposal challenge. Former President Carter established an Interagency Review Group on Nuclear Waste Management in March of 1978. The fourteen agencies in the review group reported a year later that: "Over-all scientific and technological knowledge is adequate to proceed with regions selection and site characterization . . . "

Energy Department officials think that the disposal system most likely to be settled on involves mixing the nuclear wastes into molten glass. The solid "glass" than can be sealed in welded stainless steel canisters. Such containers could be buried in deep salt beds. Salt beds, which underlie about a half million square miles of the U.S., are among the most stable geological foundations known.

The Edison Electric Institute calculates that all the wastes from the country's commercial nuclear energy program through the year 2000 would only make a stack eight feet deep between the goal lines of a football field.

PROLIFERATION

Problem No. 3 for the nuclear industry is the potential misuse of nuclear fuel. Plutonium is produced when uranium is irradiated in a reactor. That plutonium can be used as reactor fuel—or to make a nuclear bomb.

The concern is that expansion of nuclear power plants could abet the proliferation of nuclear weapons among nations. An added fear is that terrorists might steal plutonium and fashion a crude bomb.

The spread of nuclear weapons clearly should be prevented. To that end, the U.S. military has safeguarded the use and shipment of plutonium for 30 years. Commercial nuclear plants have tight security systems to protect plants and their fuel.[3]

Sale of nuclear power plant components abroad—where control over their use cannot be maintained—raises special problems. The U.S. and other nuclear nations have placed restrictions on their nuclear sales abroad. It should be noted, however, that other nations, if determined to produce nuclear weapons, can accomplish that objective (indeed, some already have done so) without having a nuclear power plant in operation.

Financial Fallout

Quite aside from such technical problems for the nuclear power industry, financial worries are sizable and on the increase. In an industry that is capital-intensive, costs have been climbing fast. And in the wake of Three Mile Island, refitting the nation's operating nuclear plants with additional safety features will cost millions of dollars. Add in the cost of redesigning plants under construction and the cost of added delay in putting these plants into service and the total may well reach in the billions. Fast-rising costs, regulatory uncertainties, and lower projections of future power needs recently caused the abandonment by a consortium of Northern Ohio utilities of construction of four big nuclear plants. For similar reasons, Detroit Edison recently cancelled plans to build two nuclear plants, explaining that continuing the plants would not be a "prudent allocation of the company's resources."

Soaring costs come at a time of serious erosion in the financial conditions of most utilities. Internal cash flow has decreased, forcing companies to raise a larger percentage of their construction funds externally. At the same time, interest costs of debt have only recently come down from historic highs and stock prices are still near their lows. Interest-coverage requirements mandate that fewer bonds can be sold when rates are high, and low stock prices mean more stock has to be sold to raise a given amount of money—diluting values for the present holders.

The question raised more and more often these days by industry analysts and utility managements: Has a commitment to nuclear power become too expensive and the payoff too uncertain for the investor to risk his money and for management to risk its company?

Surely, utilities will be exceedingly wary about pushing ahead with any reactors beyond those already under construction unless they receive a clear mandate from government to do so. Such a mandate would need to embrace changes in licensing and review procedures to shorten the present drawn-out construction time for nuclear plants.

Cloudy Future

Inflation, a mounting regulatory burden, the uncertain financial health of some utilities, opposition to nuclear expansion from environmental and consumer groups, erosion of public support—all those combine to make nuclear power's future cloudy. And yet the need for nuclear power in the nation's energy complex is clearly evident for decades into the future.

About 17 percent of the nation's electric power comes from oil-burning generators. Quite aside from reducing U.S. dependence on unstable and high-priced foreign oil, that liquid fuel is a diminishing resource. It makes good sense to use atom power to generate electricity and, thus, to free petroleum for use in gasoline, heating oil, medicine, plastics, and fertilizer.

Total energy demand in the U.S. will increase between 40 percent and 80 percent by the end of the century, according to the Electric Power Research Institute. The uncertainty in the projection is due to difficulties in estimating economic growth, labor productivity, and the success of conservation efforts. The estimate of electric power demand shows a similarly wide range—from two to three and one half times present consumption. Even the low end of the range, however, suggests an enormous need for added power facilities. Clearly, failure to move ahead with nuclear energy risks power shortages in this country in the 1990s and beyond, with all that would imply in terms of higher inflation, slower economic growth, continued energy dependence, and national insecurity.

What's needed is a stable political and institutional framework within which nuclear power can operate today—and plan for tomorrow. A broad-based consensus as to nuclear energy's safety and its essential role is vital to such a framework. For it is the public—much more than the technical experts—that will decide the future of nuclear power. Such a consensus may develop when there is more general recognition that all energy sources entail risks. Conservtion, solar power, and other noncoal-non-nuclear energy sources can not be counted on to close the gap between present electrical generating capacity and the real needs of a growing economy.

Notes

1. Despite the risks, the public seems opposed to abandoning nuclear power. A recent Harris poll disclosed that, while nearly three quarters of the public believed there was no guarantee against a castastrophic nuclear accident, the majority nonetheless felt that nuclear power was too important to abandon at this time.

2. A review of the Rasmussen study commissioned by the U.S. Nuclear Regulatory Commission found some deficiencies. Nevertheless, the Review Group strongly supported the objective of quantitative safety analysis, commended the Rasmussen group for a pioneering effort, and urged the NRC to make more extensive use of quantitative analysis in the regulatory process.

3. Theft of reactor fuel would be difficult. Scientists at the Edison Electric Institute calculate that a crude bomb using 25 pounds of plutonium would require the theft of about 1,500 pounds of radioactive fuel, removed from 12-foot-long reactor fuel elements transported in a shipping container weighing up to 100 tons.

Dr. Edward Teller, nuclear scientist, on nuclear terrorists: "Any terrorist who puts his mind to it can come up with ways to terrorize a population that are less dangerous to himself than handling plutonium. The answer is not to get rid of the reactors—let's get rid of the terrorists."

The Challenge of Electrical Generation

JOHN A. CARVER, JR.

Introduction

CENTRAL station generation of electricity is about a hundred years old. It developed rapidly from about 1879 to serve lighting loads based on Edison's invention of the incandescent light bulb. The entrepreneurial opportunities of the new invention were evident, but so were the risks: today's system standardization hadn't arrived and transmission distances were limited. Consequently many independent companies each serving comparatively small areas of metropolitan concentrations of population were formed to provide service to meet the new demand. Consolidated Edison was so named because it consolidated scores of previously separate utility companies.

If we imagine an 1879 conference called to consider today's subject, we might find some parallels. The electricity business was expanding to serve a night load, and it was possible to see that this load could and should be balanced to save costs, even though it was not conservation of fuel but spreading fixed costs over a larger number of output units which would be sought. It could have been foreseen that electricity would soon change the shape of the city by powering elevators, transit lines and industrial machinery; it also was predictable that its expansion would disrupt some businesses, and that new institutional arrangements for the control and regulation of this vital new source of energy would be necessary.

Economies of scale certainly could have been forecast in 1879, and it might even have been seen that the future of the electric power industry would be marked by shifting patterns of the siting of generation. Transformer technology was the key to bigger generating facilities and longer transmission distances, enabling generation to be moved closer to the water or rail terminal or even the mine mouth for coal supply. But it would not have been foreseen that natural gas would have an era during which generation with that clean burning fuel could move closer to load centers to realize economies by reducing transmission losses.

Although conservation of primary fuel resources would probably not have been a concern at an 1879 conference on the challenge of generation, we should remember that the conservation movement had in fact already begun. The Sargent Report, which forecast that the United States would run out of timber in a matter of decades, was already prepared, and would be an attachment to the 1880 Census Report. That in turn would lead to the movement to set aside national forests and to a philosophy of sustained yield management of that resource. The petroleum industry was growing rapidly, although it was only twenty years old in 1879. It, like the beginnings of the utility industry, was first seen as a source of illumination.

The subject, "The Challenge of Generation," is a subtitle of one of the larger questions of this conference, which is supply of energy. For at least a decade we have known that our country was exposed to great dangers by continuing to rely on foreign oil to meet its expanding needs for energy. Twenty years before 1879, the petroleum industry was born; twenty years before 1979, the United States was still a net exporter of oil. In 1973, before the Arab oil embargo, we got about one-fourth of our petroleum requirements from imports, and the figure barely six years later is double that. The price we pay for these imports, which will be about 65 billion dollars for 1979, is a transfer of claims on United States output and wealth, principally to OPEC countries, and a shift of domestic resources to produce definitionally costlier domestic supplies of energy.

During the Iranian crisis we learned a painful lesson in economics—Iran cannot be seriously damaged by any resolve on our part to forego taking Iranian oil, for the petroleum market is a world market which is shortage dominated. The key to present United States policy is the warning by our intelligence agencies that Middle East oil production physically cannot be increased much above present levels, and particularly Saudi Arabian production probably cannot be increased above the current level of 9.5 million barrels per day, even if the Saudis were willing to increase it. It cannot be lost upon oil exporting countries that Iran realizes as much from three million barrels per day as it did from five million barrels per day it produced prior to its revolution last spring.

Two years ago, the price of oil from abroad was about $13 or $14 a barrel, and we were speculating whether the OPEC countries would increase prices as much as 15 percent. Average prices have more than tripled since, as the percentage of oil being shifted to a spot market, which has been as high as $48 a barrel, increases.

What then is the challenge which these facts lay down to the generating phase of the electric utility industry? It has many parts, including technological, financial and institutional, but it is really the same challenge the industry has always faced, to minimize costs.

Over the past few years, we have learned to deal with costs which include externalities and much of our national policy of the decade of the seventies has been toward internalizing many external costs, particularly those related to the quality of our water and air. Unfortunately, one kind of nonfinancial cost we haven't been very good at figuring is the true cost of a depleting resource. In the case of the electric utility industry, the true cost of the primary energy which it consumes is higher than its financial cost. I must leave it to the economists to articulate the basic idea with clarity, but the mounting body of evidence is that in the case of energy, even the inflated prices being charged by the OPEC cartel or by its members who depart from it by charging even more than the cartel price, are probably below the margin. The margin is even harder to discuss when the dynamic situation is one of runaway domestic inflation, as it is today. We can't easily sort out how much of our inflation is directly traceable to the cost of energy, nor whether energy independence, if achievable, would increase or reduce it.

It was not too long ago that the Federal Power Commission perceived the challenge of generation with such sanguinity that it could forecast a growth in capacity and amount of electric energy supplied a percentage or two greater than the growth in the consumption of energy in the aggregate, in other words, a gradual shift to a largely electric economy. The key to such an optimistic view was technology which would maintain a trend to lower unit costs in real terms. This was expected to come through conventional technology improvements and particularly the development of the breeder and fusion in nuclear technology. It has been a challenge to readjust that thinking in several ways, among the more important of which is to scale down the extrapolations of the energy component of gross national product. By lowering our sights for the future, with respect to how much energy we really need, we have accomplished a narrowing of the gap between supply and demand.

There isn't a common understanding about how much of the country's energy should be delivered in the form of electricity, although there is a growing pessimism about how much will be. If we look only to supplies of domestic or indigenous primary conventional energy sources, we find that the United States has a disproportionate share of the world's total of coal, of shale, and of uranium. The world is not running out of energy, and neither is the United States, even if we leave out geothermal, and various forms of solar energy. Central station generation of electricity will continue to grow.

Of the three sources, we get the biggest numbers when the development of breeder technology is assumed, as Exhibit 1 developed by RFF shows. The United States has a quarter of the coal, converting to 35 thousand quads; it has 40 percent of the uranium converting into an incomprehensible 400 thousand quads of energy with the breeder, and 6500 quads with conventional light water reactors.

It is evident that on a long-term basis, the challenge to the United States is to solve the problems of nuclear waste disposal and the political problems of weapons proliferation which result from the production of plutonium as a byproduct of the nuclear reaction process. The reports of the findings of the International Nuclear Fuel Cycle Evaluation Conference[1] indicate that the latter problem is going to have to be faced or solved independent of actions taken domestically to achieve control by legislating against the enrichment process or use of the breeder.

The nuclear challenge is a long-range one, and one we may dispose of here without describing it. Exhibit 1 points to coal as the immediate challenge, and national policy in this respect is in phase. It is coal which is to do most to free us from dependence on foreign supplies of oil. Utilization of coal involves generation of electricity almost as much as nuclear energy involves generation of electricity.

Contemplation of increased use of coal in electric generation paradoxically will probably involve a certain amount of retracing of our technological steps. It is my opinion that downsizing and decentralization of generation is the trend we can anticipate, and that this in part will be because of "energy sensitive land development."

The chain of reasoning for my conclusion is something like this: First, the pricing of primary fuel is rapidly changing use patterns, so that forecasts of growth rates of demand are being sharply reduced. If forecasts of higher demand are self-fulfilling prophecies, forecasts of lower demand are even more so.

Second, government policies are already in place which, when they take hold, will tend very strongly toward elimination of peaks in the load pattern, increasing the load factor within systems, and among interconnected systems. The program for improving the energy efficiency of structures by insulation uses tax incentives under present law, and ultimately will use interest-subsidized loans, grants and other methods.

Of even greater significance are initiatives which foster cogeneration, because this definitionally will involve matching electric power output with associated heat output. This will be easiest with smaller, decentralized generating units, whether industry or utility operated.

Measures to integrate both small and medium-sized solar facilities with central station generation to improve the efficiency of the fossil-fired system already in place are being expanded beyond the tax incentive measures of the National Energy Act of 1978. A Solar Energy and Conservation Bank (12 USC 3601) to provide interest subsidies for loans and mortgages to finance solar energy systems, administered by a board composed of the Secretaries of HUD, Commerce, and Treasury, has been provided for in the Energy Security Act of 1980, and larger industrial firms will be encouraged to install such facilities by a 10 percent additional investment tax credit.

What can we say about enhancing our supply of domestic energy through our actions at the generation phase of the electric energy cycle? Obviously conservation accomplishments can be translated into equivalents of barrels of oil. Some of the numbers can be very big, but it is well to remember that we get big numbers saved only where we have big numbers used, and the focus of conservation, therefore, naturally is upon comfort heating, upon transportation, and upon efficiency in industrial applications, including generation of electricity itself.

All of these are directly tied to urban land use considerations. Ten percent of the total energy consumption in our country is for residential space heating, and about the same amount for industrial process steam. Thirteen or fourteen percent is for automotive transportation.

The program for improving the energy efficiency of structures is locked into place, a recognized objective of national policy. The principal policy tools have moved from the purely educational approach, through more elaborate tax incentives, into a combination of an institutional and a subsidy approach which seems fairly certain of enactment, whereby federal policies will push regulated utilities into a one-stop type of energy service which could extend to financing of conservation projects by interest-subsidized loans, to behind-the-meter expenditures being included in rate base for rate purposes, and others.[2]

Another generation-related aspect of current energy policy initiatives is a provision which facilitates various forms of cogeneration, which essentially is a return to the decentralized, small-unit generating pattern of

Exhibit 1

World Recoverable Reserves and Resources of Conventional Mineral Fuels

Region or Nation	Coal (billion metric tons coal equivalent)		Oil (billion barrels)		Gas (trillion cubic feet)		Uranium (thousand metric tons U)	
	Reserves	Resources	Reserves	Resources	Reserves	Resources	Reserves	Resources
United States	178	1,285	29	110–185	205	730–1,070	643	1,696
Canada	9	57	6	25–40	59	230–380	182	838
Mexico	1	3	16	145–215	32	350–480	5	7
South and Central America	10	14	26	80–120	81	800–900	60	74
Western Europe	91	215	24	50–70	143	500	87	487
Africa	34	87	58	100–150	186	1,000	572	772
Middle East	—	—	370	710–1,000	731	1,750	—	—
Asia and Pacific	40	41	18	90–140	89	a	45	69
Australia	27	132	2	b	31	500	296	345
Soviet Union	110	2,430	71	140–200	910	2,850	n.a.	n.a.
China	99	719	20	a	25	a	n.a.	n.a.
Other Communist areas	37	80	3	a	10	a	n.a.	n.a.
Total	636	5,063	642	1,450–2,120	2,502	8,710–9,430	1,894	4,288
Quintillion (10¹⁸) Btu[c]	17.7	140.6	3.7	8.4–12.3	2.6	8.9–9.7	7.4 (LWR) 443.2 (FBR)	16.7 (LWR) 1,003.4 (FBR)

Exhibit 1

World Recoverable Reserves and

Resources of Conventional Mineral Fuels (Continued)

Note: All resource figures are cumulative. They include reserves. The figures for international coal reserves and resources in this exhibit are given in metric tons of fixed heat content rather than actual metric tons.

Sources: Coal—World Energy Conference. *Coal Resources* (Guildford, England: IPC Science and Technology Press, 1978). A recoverability factor of 50 percent is assumed for resources.

Oil and gas—Reserves: *Oil and Gas Journal* vol. 76. no. 52 (25 December 1978) pp. 102–103; U.S. resources: U.S. Geological Survey estimates as reported in table 7-3; Canadian resources: Department of Energy, Mines, and Resources as reported in *International Petroleum Encyclopedia 1978* (Tulsa, OK.: Petroleum Publishing Co., 1978), p. 39; Mexican resources: PEMEX data as reported in *Mexico's Oil and Gas Policy: An Analysis*, Committee Print, Joint Economic Committee, 95 Cong. 2 sess. (Washington, D.C.: December 1978), p. 13. An undertainty margin of 20 percent for possible resources is introduced; other oil resources, Richard Nehring, *Giant Oilfields and World Oil Resources*, Report R-2284-CIA (Santa Monica, CA.: Rand Corporation, 1978), p. 88; other gas resources, Joseph D. Parent and Henry R. Linden, *A Survey of U.S. and Total World Production, Proved Reserves, and Remaining Recoverable Resources of Fossil Fuel and Uranium* (Chicago, IL.: Institute of Gas Technology, January 1977), p. 15.

Uranium—International Atomic Energy Agency/OECD Nuclear Energy Agency, *Uranium: Resources, Production and Demand* (Paris, Organisation for Economic Co-operation and Development, 1978), pp. 20–21. The 300,000 metric tons of Swedish uranium have been excluded from the reserves.

[a] Included in Soviet Union.

[b] Included in Asia and Pacific.

[c] Heat contents: coal, 27.78 million Btu per metric ton; oil, 5.8 million Btu per barrel; gas, 1,025 Btu per cubic foot; uranium, 390 billion Btu per metric ton U (without breeder) or 23.4 trillion Btu per metric ton U (with breeder).

Source: "Energy in America's Future: The Choices Before Us," *Resources for the Future* (1979), pp. 242–243.

the early days of the industry. Some or most of these programs relate to the development of solar energy facilities capable of backing out at least part of the energy costs involved in central-station generation of electricity. For example, the Land Conservation and Solar Energy Bank provides interest subsidies (up to 6 percent below market rate) for loans and mortgages to finance solar energy systems. Larger industrial firms are encouraged to install such facilities by a 10 percent additional investment tax credit for expenditures for solar, recycling, shale, wind, liquefaction equipment, and other equipment designed to supplant the burning of oil to generate electricity. The Energy Security Act of 1980 also expands the tax credits for installation of qualifying renewable energy source equipment in residences or commercial buildings.

The quantification of these steps in terms of reduced load upon generating capability is imprecise. For example, assumptions must be made as to the relative numbers of new and retrofitted structures and about the energy efficiency of existing heat sources (direct firing of fossil fuels, resistance-type electric heating, or heat pumps), as well as on the availability and cost of technology. The calculations made by RFF[3] suggest that the present aggregate amount of energy used for residential comfort heating (8.6 quads) will be sufficient to meet the heating requirements for the projected increase in population and number of dwelling units anticipated by the end of the century. This leaves us with "savings" calculations which are essentially hypothetical, in the sense that an equivalent shortfall of energy supply in the aggregate would have exactly the same effect upon generation. However, it also tells us that the shape of future generation will not be bigger and bigger generating stations.

One major omission in the "grand design" of current energy policy is any special push for the utilization of rejected heat from electric generation for district heating and cooling. There are problems as to the storage of heat and its transmission from the point of electric generation to the point of use, but the amount of energy capable of being saved warrants more discussion. A great deal of work is being done on heat storage in aquifers, which would release the useful output from the existing peaks and valleys characteristic of current generation patterns. If storage of heat is mastered so that a dependable flow of heat for district heating and cooling is assured, the patterns of electric generation will change exactly as they will with cogeneration. The size of units may come down, the configuration of the relative production of heat and electricity could change. The impact on urban land uses could be significantly altered.

Conservation measures which provide for making current levels of either capacity or output serve the increased requirements of a larger population are difficult to quantify in "supply" terms, although conservation is clearly a "source" of energy. In the case of utilization of rejected heat, which is equal to roughly 70 percent of the primary energy inputs for electric generation, a 1 percent improvement, that is from 70 percent to 69 percent, equates to 24 million barrels of oil. In the case of rejected heat, its use would amount to internalizing pollution control costs as a credit, thus adding to the savings. Since past technology improvements involved principally our ability to manage higher and higher levels of heat and pressure, much of the rejected heat is well within useable ranges, even accounting for considerable loss in storage and transmission—after all, all of its costs is sunk before its recovery begins, whether recovery is undertaken or not.

Other links in the chain of my argument are financial and institutional. Financial considerations are a complex matter, but as I see it, we have already moved sharply away from reliance upon traditional capital markets for energy projects. Some of the load has been assumed by government through tax incentives, directly or through using the credit of public agencies with its associated tax exemptions for pollution control and other projects. All-events tariffs or government guarantees have been looked to for the financing of the Alaska Natural Gas Pipeline. The Energy Security Corporation (42 USC 8711) currently advances front-end money for synthetic fuel projects, many of which develop fuels used for generation of electricity. Current ratepayers are paying for future projects through inclusion of construction work in progress in rate base. Finally, extraordinary events such as the Three Mile Island disaster can chill financing. If the costs of that event are thrown entirely upon stockholders, the economic viability of the utility itself could end, and just one bankruptcy of a utility company will make it an order of magnitude more expensive for the rest of the industry to finance expansion. The result could be smaller, and to some extent more numerous, utility entities.

Financial and institutional considerations are intertwined, although as a former regulator I feel much more comfortable talking about the institutional characteristics of the generation of electricity, and trends thereof. At the outset, we know that the institution of federalism has been considerably strained in the matter of generation and transmission of electricity. When a utility seeks to locate a plant for generation of electricity near the source of its fuel in another state, offsetting transmission losses by solving pollution related obstacles, it soon becomes apparent that an impasse between two sovereign states about siting can be resolved only by a preemptive federal statute. Those statutes are not common, but a common variant is for the federal government itself to take the lead role, as is being proposed in the Pacific Northwest in a pending bill to award the Bonneville Power Administration authorizing legislation.

The structure of the electric utility industry is pluralistic. At the same time the private, investor-owned segment was being tied together by holding companies (later to be broken up by the Public Utility Holding Company Act of 1935), municipalities perceived the advantages to them in owning and operating electric utility systems, paralleling their role with water utilities. As a result, for more than fifty years something over two thousand municipalities have owned and operated facilities for the generation and distribution (or only distribution) of electric energy within their boundaries and adjacent areas. (The largest of these systems is the Los Angeles Department of Water and Power.) In the aggregate they account for less than 13 percent of electric energy sales, and about 9 percent of total electricity production.

If the municipal and other nonfederal public systems which get power from federal dams are excluded, in other words, if we consider only municipalities which generate their own energy or most of it, we find a pattern to facilitate matching generation to the particular characteristics of urban load. By definition, the tradeoff between efficiency and environmental standards can be more readily agreed upon where the generating entity is a governmental body serving a customer group which is coextensive with its constituency. To put it most boldly, I think it is easier for a municipality to build a new generating plant in or near the city itself than it is for an investor-owned utility serving the same kind of load.

The missing part of the energy package—utilization of rejected heat for district heating and cooling—offers its own set of institutional problems. One of the reasons we have such a bizarre pattern of thermal efficiencies for structures is because we have left to the customer, or to the builder, to choose the technology based on fixed and variable costs (the furnace and the fuel), which have been giving him false signals. The customer should not choose the level of insulation to be installed, or the kind of heating system, based upon assumptions which are contrary to common sense. The worst of these assumptions has been the one that natural gas is cheaper than coal or oil; another is that natural gas is *not* cheaper than electricity used for resistance-type heating, whatever the fuel being used by the generating company.

Viewed broadly, we have no institutional way to assure that what the customer really wants, and really intends to pay for, is not Mcf of gas, or kilowatt-hours of electricity, or gallons of fuel oil—it is BTUs of heating or cooling service. If what was regulated by the PUC was the latter, and if insitutionally the utility supplier could be allowed to make an economic choice as to the source of that service without the customer's price-dictated preference entering into the equation, then the option of utilizing rejected heat would be a positive one for the supplier. His generating load would be reduced accordingly. This kind of reform, if it is to come, will come by small-scale trial and error experiments, and some of these will be by municipalities.

It is possible to identify other developments related to the current energy shortage and higher energy prices which work in the same direction as the ones discussed above. The two biggest targets of opportunity for energy improvement are almost identical in size, by coincidence. Both transportation and electric generation rejected 14.4 quads of energy in 1976, although in the case of transportation the rejected amount was a much higher percentage of the total energy input. Furthermore, transportation is almost totally reliant on petroleum, while electric generation uses a great deal of coal, less nuclear, and even less hydro, so that direct backing off of oil is not so likely. However, the population patterns of the city seem likely to change, toward more centralization. It is possible that this will facilitate such things as cogeneration, district heating, and also the number and kind of cars that people use. Perhaps a very light, limited-range vehicle powered by off-peak generated electricity could make a net contribution to energy efficiency.

In summary, the challenge of generation must be seen as a challenge to the human beings who run our companies and cities, who make and administer our laws, and who finance our undertakings. That challenge is to make what we have go farther, to match our plans to a new kind of future, and above all, to think *conservation*.

Notes

1. New York *Times*, Nov. 4, 1979, p. 1.
2. Cf., Titles IV, V, and VI of the Synfuels Bill, S.932, 96th Congress, 1st Session.
3. Schurr, *Energy in America's Future*, Resources for the Future, Johns Hopkins, 1979.

Cogeneration and Recycling as Alternate Energy Sources

STEPHEN A. MALLARD, WILLIAM WOOD,
ARTHUR W. QUADE

Introduction

COGENERATION and recycling have traditionally been regarded as two separate and distinct areas of activity. Time and energy problems have tended to blur the distinction and it is now reasonable to consider some aspects of cogeneration and recycling jointly. This paper will discuss cogeneration and recycling as alternative energy sources, especially in an urban setting.

Cogeneration, per se, is a relatively new term for an old concept and there are several definitions which apply, depending on the party defining it. The electric utility industry and the federal government[1] define cogeneration as the production of useful thermal energy as a by-product of electric generation. The chemical and petroleum refining industries take a broader perspective and say that cogeneration is the "sequential generation of electrical or mechanical power and useful heat from the same primary energy source or fuel."

Regardless of the definition selected, it is evident that cogeneration refers to the simultaneous production of shaft horsepower and thermal energy with a given quantity of fuel which would be less than the fuel required to produce that shaft horsepower and thermal energy independently. This concept then does appear to be a means of increasing the efficiency of energy utilization.

The broad term "recycling" also has many definitions, depending on the perspective of the definer. Recycling in the paper and paper products industry refers to the collection of used paper products, processing that collected material, and producing a product that has a specific positive use. Recycling in the wood and wood product industry can apply to such disparate activities as chopping up waste products into pieces of various sizes and reassembling these pieces into particle board and flake board or burning that waste in a boiler to produce process steam.

Again, just as with cogeneration, while the specific definitions depend on perspective, they all fall into a single broad category. Recycling of material refers to the collection, processing and conversion of scrap into new raw materials or products.[3]

While the preceding comments referred to industrial recycling, the same thinking can be applied to municipal wastes. However, the problem is somewhat more complex since the separation of the wastes frequently must be considered before the recycling can be carried out.

History of Cogeneration

Cogeneration is a relatively new term, but the concept has been in use for many years. At Public Service Electric and Gas, we use the utility definition of cogeneration, i.e., the production of industrial process steam as a by-product of electric generation located at the industrial plant.[4] In the late 1950s and early 1960s, the "Total Energy" concept was promoted by many of the nation's gas companies and some equipment suppliers. There are many similiarities between Cogeneration and Total Energy (and some differences which will be discussed shortly), but both concepts have their roots even further back in history.

Early in this century, at the time electricity first became widely available, many large industrial or commercial enterprises maintained large steam boiler plants for space heating and/or process heating. It was a relatively easy matter to boost the boiler steam pressure somewhat beyond the process requirement, insert a turbine which was connected to an electric generator, and then use the steam for its original requirement after the added energy was used to generate electricity. This first "Total Energy" concept was practical for several reasons:

1) The cost of purchased kilowatthours was high.
2) The cost of running electric distribution lines, which the customer had to bear, was high.
3) The cost of boiler fuel was low (most of it was coal which did not have to comply with Air Quality Standards).
4) Because of the low cost of energy, the equipment did not have to operate at maximum efficiency for economic justification.
5) The electrical requirements were not great.

As the electric utility industry matured, many of the above advantages of on-site power production were eliminated, and many private generation facilities were shut down. As electric utilities sold more electricity, the economies of scale resulted in a decrease in electric rates and distribution facilities became more widespread. By 1950, cogeneration provided only 17 percent of the total electric energy produced.[5] By the mid-to-late fifties, the majority of these private plants were abandoned.

In the 1950s, the natural gas industry grew at a rapid pace as interstate pipelines were constructed. The gas distribution companies had an exceedingly poor seasonal load factor with tremendous quantities of natural gas available on a "take or pay" basis year round. The "Total Energy" concept was born. It produced an almost constant demand for natural gas year round to operate a combustion turbine or gas-fired engine which produced the electric energy required by the facility. And as a by-product, the exhaust could be used to produce process or space heating (or cooling). The unfortunate aspect of a Total Energy installation was that a very good load balance was required to make the installation pay for itself (the electrical demand curve had to coincide with the heat energy demand). It was this aspect of the Total Energy concept that was strongly attacked by the electric utilities, because they did not want to lose the electric business. Consequently, an adversary position developed—the electric utilities fighting the Total Energy concept as uneconomical because of poor load balance and the gas utilities promoting it as a means of ending dependence on those "unreliable" electric utilities (remember the Northeast blackouts). Total energy grew, albeit quite slowly, until 1971 when the bottom fell out of the natural gas supply.

Finally we arrive at cogeneration. Initially, cogeneration was promoted by aggressive equipment manufacturers and in New Jersey, at least, by "anti-big" groups as a means of decreasing the "power" of the "huge, all powerful" electric utility industries. However, calmer voices soon entered the discussion with the suggestion that rather than develop competition between the two forces, cooperation be developed between cogenerators and electric utilities. And herein lies the difference between Total Energy and

Cogeneration—the appeal of total energy was based on complete independence from the electric utility, whereas cogeneration is based on both parties participating in, and sharing the benefits of, cogeneration.

PREVIOUS COGENERATION EXPERIENCE

The most successful large-scale cogeneration arrangements involving electric utilities have been with the chemical and petroleum industries, large users of process steam and electricity.

Since 1928, Gulf States Utilities and the Exxon Corporation have been cooperating in a cogeneration project where Gulf States produces steam and electricity for Exxon who in turn provides the fuel for the Gulf States power station. This arrangement has been expanded several times. Similarly, Pacific Gas and Electric has been furnishing steam and electricity to three oil companies from three different generating stations and one of those has been doubled in size.

Closer to home, the Atlantic Electric people have been selling steam and electricity to the DuPont Company at its Deepwater plant since 1928. They completed a similar arrangement in 1953 in Gibbstown.

The year 1957 signaled the beginning of PSE&G's largest and most successful cogeneration experience with the start-up of Linden Generating Station as a steam supplier for Exxon's Bayway refinery.

This arrangement is similar to the previously mentioned Gulf States—Exxon effort; we provide steam to Linden and they return the equivalent energy to us in the form of fuel oil.

Both PSE&G (and its customers) and Exxon benefit significantly from the arrangement. Because of the fuel-for-steam swap, kilowatthours are generated with much lower heat rates (higher efficiency) than from any other fossil generating station and this keeps the price of energy down for our customers. Exxon, on the other hand, benefits because it doesn't have to maintain a large boiler plant to supply its steam.

The above mentioned cogeneration installations represent a sampling of the very large applications throughout the country. In addition, there are a multitude of other smaller cogeneration arrangements, some of which involve utility participation, but most of which do not. Regardless of the size of the cogeneration operation or the identity of the participants, the vast majority have been established because of economic reasons (as opposed to political or sociological reasons).

LEGISLATIVE ENCOURAGEMENT

As a result of the national and worldwide energy crisis, the New Jersey Legislature passed and Governor Byrne signed, the Department of Energy Act. Amendments to this act imposed on the newly created Department of Energy (NJDOE) the responsibility of investigating the issue of cogeneration.

At approximately the same time, the United States Congress was wrestling with the proposed National Energy Act. This act was passed in several parts, two of which, the Power Plant and Industrial Fuel Use Act of 1978 (PL95-620), and the Public Utilities Regulatory Policies Act of 1978 (PL95-617) impact directly on cogeneration. The Energy Tax Act of 1978 (PL95-618) provides for additional investment tax credit for portions of individual cogeneration equipment.

The Department of Energy Act— The Department of Energy Act of 1977 (N.J.S.A. 52:27F-1 et seq) did not specifically mention cogeneration as a separate item to be considered.[7] However, an amendment, Chapter 80, specifically adds cogeneration to the definitions of terms; requires reporting of any steam use of 50,000 lbs. per hour or more; provides for the design, implementation and enforcement of energy conservation programs which include cogeneration; and requires the consideration of the establishment of cogeneration facilities to simultaneously produce electricity and steam to conserve fuel. In addition, the original act empowers the Commissioner of the Energy Department to conduct research studies, hold hearings, and consider any matter relative to the production, distribution, consumption, or conservation of energy.[8]

Powerplant and Industrial Fuel Use Act— At the same time the NJDOE was conducting its Cogeneration Hearing, former President Carter signed the Powerplant and Industrial Fuel Use Act, PL95-620, a part of the National Energy Act. This act specifically mentions cogeneration in several different contexts.

First, in the definition section, a cogeneration facility is not considered as an "electric generating unit," if less than half of the power generated is sold or exchanged for resale.[9] Also, "cogeneration facility" is defined as an electric power plant or major fuel burning installation which produces (A) electric power and (B) any other form of useful energy (such as steam, gas or heat) which is, or will be, used for industrial, commercial, or space heating purposes.[10]

Second, whereas (1) gas or petroleum shall not be used as a primary energy source in any new power plant, and (2) no new electric power plant may be constructed without the capability to use coal or any other alternate fuel as a primary energy source, cogeneration may be permanently exempted from these requirements after an appropriate petition is filed with the Secretary of the Department of Energy, provided the cogeneration saves energy and the same benefits cannot be derived using another fuel.[11]

Public Utility Regulatory Policies Act— This act (PL95-617) requires a public utility to connect to a private cogenerator, wheel the energy for the cogenerator and sell energy to and/or purchase energy from the cogenerator.[12] Of course, there are the appropriate exemptions within the law and the utilities are protected to a small degree. But the overwhelming fact is, cogeneration is being strongly promoted by this law.

In effect, the federal and state governments have both taken the position, through appropriate legislation, that cogeneration should be given special treatment due to its inherent energy-saving characteristics. But neither agency has acknowledged that the energy savings are appropriate only if the source of energy being displaced is oil or natural gas.

ACCEPTANCE BY INDUSTRY

Prior to the preceding legislation, PSE&G had begun its own investigation into the potential for cogeneration. We developed a listing of all industrial customers estimated to have 250 boiler horsepower (bhp) or more. This PSE&G list consisted of approximately 220 customers. The listing was then refined to those 182 customers with 500 bhp or more. A second refinement was conducted and approximately 50 percent of the 182 customers were eliminated because they had steam demands too small to be realistically considered for cogeneration. A third iteration narrowed the list to 78 customers with a demand of 44,000 pounds of steam per hour or more. Twenty-three customers had a use of 100,000 lbs/hr and from those, PSE&G selected 18 customers with 100,000 lb/hr and a minimum use of 50,000 lb/hr for 5000 hours or more for a detailed specific analysis.[13] Steam profiles were prepared for these customers to aid in the feasibility analyses.

CUSTOMER ECONOMICS

In addition to a very detailed study conducted on the 18 customers singled out above (of which only 12 were marginally economical with cogeneration) special visits were made to several large industrial steam users. Key points which surfaced at these visits were:

1. If the cogeneration scheme proved economic from the customer's point of view, they would be interested.

2. Most industries would be reluctant to establish a long-term contract with anyone for a cogeneration system.

3. Reliability is essential to many customers. Most would require backup steam supplies for a cogeneration system.

4. Cogeneration would not significantly reduce customer manpower, maintenance, or plant equipment requirements.

5. The operation of existing steam plants as standby or topping to a cogeneration system would cause several problems. Most prevalent are: increased boiler maintenance due to low circulation; and the necessity of keeping boilers at a reduced firing rate for standby.

6. Most of the customers would incur an additional insurance expense with a cogeneration facility on their property.[14]

OTHER FORMS OF COGENERATION

In addition to cogeneration that would be used for individual customers, there are two other applications of the cogeneration concept which are currently being studied.

District Heating— For years, many European countries have been using the district heating concept for providing space heating and domestic water heating for their cities. District heating refers to the distribution of heat energy in the form of high temperature hot water or steam from a central production plant to individual residential, commercial or industrial users which are located close to the production facility. This district heating can be accomplished by the production of heat energy alone or by the cogeneration of heat energy and electricity. New York City has long been involved with district heating using both approaches.

PSE&G is involved in a study of district heating which would involve extracting some of the steam from existing generating stations and converting the energy to high temperature hot water for distribution to neighboring industrial and commercial customers. The thrust of the study is to determine the economic feasibility of replacing oil as a fuel used by these customers.

Preliminary indications are that retrofitting district heating to existing facilities is difficult to justify economically. On the other hand, new development, such as that proposed for the Hackensack Meadowlands could prove to be economically attractive.

ICES— Integrated Community Energy System is an expansion of the district heating concept which is specifically designed for urban, residential and commercial areas. Production of electricity is a significant part of an ICES and carried to its extreme an ICES also could incorporate space heating and cooling and domestic water heating which could all be produced by a refuse burning boiler.

PSE&G has been involved in the study of an ICES for Trenton, NJ. This ICES would use combustion turbines which would burn light oil and produce part of the electricity and all of the space heating and domestic water heating for a portion of the city. Based on our studies, the ICES, as the plan was presented to us, would not be economically feasible. The energy balance (relationship of heat energy production and electrical production) is poor and the cost of retrofitting the system in the existing city is high. The City of Trenton is currently looking at alternate ownership options and possible DOE funding. One plan is to sell the electric energy to the City of Vineland which has its own distribution system. PSE&G has agreed to continue to work with Trenton on this program.

SOCIETAL FACTORS

Proponents of cogeneration claim that cogenerators would have a positive impact on the environment. They claim that central generating stations use #6 fuel oil or coal which are much more polluting than the #2 fuel oil, diesel fuel or natural gas which would be used in a congeneration facility. On the other hand, the cogeneration facility would be located adjacent to the point of use of electricity and heat energy, which is generally a built-up area, whereas the central station would be located possibly in a remote area with tall stacks (as opposed to low level exhaust from the cogeneration facility). While there might be less pollution injected into the total atmosphere with cogeneration, the central station pollution would be remote and widely dispersed as compared with the local injection of pollutants from the cogeneration facilities.[15]

Another cost to society that is sometimes charged to cogeneration is the loss of tax revenue. If energy is produced by a utility, the revenue derived is subject to a gross receipts tax. This tax would not be attached to cogenerated energy. Similarly, the franchise tax the utility must pay as a percentage of its investment in a community is not applied to cogeneration facilities.[16]

Some proponents of cogeneration are pushing very hard for special incentives for cogeneration, i.e., additional investment tax credits, low cost loans, subsidies, etc. In addition to these incentives representing a cost to society, they would tend to place cogeneration in a more favorable competitive position than economics would dictate. It is felt by some that if cogeneration is to be a real contribution to the solution of this country's energy problem, it must contribute on the basis of real economics, not artifically induced advantages.[17]

Where a utility has a mix of generating units, such as PSE&G with coal and nuclear in addition to oil, one other factor must be considered. Cogeneration, which is generally light-oil fueled, would not displace only oil-fired central station electric production but would probably displace coal and even possibly nuclear generation during the off-peak periods. Our Nation's dependence on foreign oil would then be increased rather than decreased as the proponents of cogeneration claim.

History of Recycling

While cogeneration has had a fairly long history (at least fifty years), recycling, or resource recovery, has been an area of concern since the mid 1960s. Initially, the recycling concept was promoted to slow the depletion of precious natural resources. A secondary purpose was to reduce the amount of waste that was returned to the environment.[18]

Lately, however, there has been a great deal of thought given to the conversion of waste into energy. Methods of converting waste into energy will be discussed shortly. However, it is necessary to take a fast glance backward to look at the methods used for handling waste in the past.

GENERATION OF REFUSE

This country is the most prolific producer of refuse in the world. We have become obsessed with disposables - cans, non-return bottles, disposable diapers, paper towels and napkins, disposable eating utensils, etc. In 1963, we produced 48 billion cans, 26 billion bottles and jars, 65 billion metal and plastic caps and crowns plus one and a half billion dollars worth of miscellaneous packaging material that must be disposed of.[19] In other terms, we produced over 35 million tons of packaging material in 1958 and over 60 million tons in 1971.[20] In 1971, the total waste stream amounted to 125 million tons of municipal solid waste compared with 65 million tons in 1958.[21]

DISPOSAL OF REFUSE

With such a large, ever increasing stream of refuse, it is necessary to find acceptable means for its disposal. Most of the traditional methods of refuse disposal are no longer acceptable.

Open Dumping— It was not too many years ago (as recently as 1953) that the garbage man would come by our homes once or twice a week and dump our banged-up metal garbage cans into his open dump truck in the early morning hours. He would then drive to a nearby vacant piece of land, leaving a trail of debris and odor, and dump his load onto a pile to give off odors and attract vermin and sea gulls. Periodically the dumps would catch fire causing noxious and unhealthy odors as well as hazardous smoke. Public pressure finally grew so strong that open dumping was banned in just about all areas.[22]

Sanitary Land Fills— When open dumping was banned, there was still a great deal of open, unusable space that was suitable for waste disposal. The sanitary landfill concept was developed which could make use of these former dumps and also reclaim other marginal "waste" land.

As in the past, the refuse trucks would pick up the residential (and some commercial) refuse, but they were now closed compactor trucks rather than open dump trucks. The refuse trucks would drop their loads, as in the past, but rather than remain open to the environment, each day's refuse dumping would be covered with two feet of clean dirt. This approach solved the major problems of the open dump, but a heretofore overlooked situation became evident.

As layer upon layer of refuse decomposed, various undesirable by-products, especially acid compounds, were formed. As long as these by-products remained dry, there was no problem. But with periodic rain and tidal action (since many of these landfills were in tidal estuaries), those by-products turned active and migrated through the layers of soil and refuse to pollute adjacent waterways. Sanitary landfills then were placed under stringent controls and became somewhat more expensive to operate. Many were shut down.[23]

Incineration— At about the same time that open dumping was flourishing in the open areas of separated cities and suburbs, many congested cities found themselves with insufficient open space to dump all their refuse. Incinerators were constructed to burn their surplus refuse. But with the developing concern over air pollution and the Clean Air Act, these incinerators gradually shut down rather than undergo the expense of retrofitting particulate removal and flue gas scrubbing equipment.[24]

Ocean Dumping— Another alternative available to many major cities was the concept of refuse dumping in the oceans. Trash would be loaded onto barges and when the barges were full, they would be towed to sea and their contents dumped into the ocean. This practice will be stopped in 1980 as the result of legislation passed in the late 1960's.

Resource Recovery— Due to the pressure exerted by society to reduce or eliminate the problems created with the conventional methods of waste disposal mentioned above, thought has turned to recycling. Much of the material in the refuse generated by people is still in a form that, if separated, can be reused directly or converted into another form which is useful. This recycling could accomplish several beneficial things:

1) The amount of refuse to be ultimately disposed of would be reduced.

2) Our natural resources could be saved by reusing previously used materials.

3) Energy would be conserved since a lot of material could be recycled using less energy than necessary to use virgin materials.[25]

RECYCLING AND CONVERSION

As mentioned above, whereas the standard approach to refuse disposal in the past has been to "dump," "throw away," or discard without regard to potential reuse, current thinking is directed at making the greatest use possible of all refuse.

There are two ways to make use of refuse—recycling in the same basic form, or conversion to another, completely dissimilar form.

Recycling— When a refuse material is collected, processed and converted into a product made of the same material as the refuse, this is known as recycling. Materials such as steel, aluminum, glass, lead, copper, zinc and dry paper are commonly recycled. In this context, the refuse material is processed into a raw material and then reassembled into a product similar to that which was disposed of. For example:

> Scrap automobiles are shredded and melted and formed into steel ingots which then became raw material and formed into steel ingots which then became raw material for pipe, construction reinforcing rods, or automobile parts.
> Alumunium beer and soft drink cans are melted, formed into sheets or bars and extruded or formed into such things as structural or ornamental shapes, or beverage cans.
> Old newspapers are chopped up and pulped and newsprint is made from the pulp.[26]

Conversion— When refuse material is received in one form and through some process, is converted to a completely dissimilar but useful form, it is considered to be converted. This conversion results in such things as compost from the bacterial degradation of food scraps and cellulose material (scrap paper, cardboard, and wood), and energy. Energy can be produced from refuse using many different methods and the following sections will discuss those methods in detail.

Recycling Problems— While recycling has generally positive results, it is not without problems. There is generally a great deal of political pressure to keep landfills open as a means of recovering land of marginal utilization. The costs associated with recycling are generally quite high and in many cases the revenue from recovered material and/or energy do not provide the return necessary to amortize the investment in facilities. To increase the revenue, it is many times necessary to increase the refuse tipping fee far beyond that which is necessary for landfill dumping.

Environmental concerns are greater with recycling, especially energy conversion, than with landfill operations. Whereas there are low-cost, accepted methods for treating landfill to minimize environmental impacts, the burning of refuse must be done in such a way so as to minimize air pollution as well as water pollution. Many times this requires the use of precipitators and even scrubbers in some instances.

Sound pollution also must be dealt with. The noise that comes from a recycling facility is much greater than from a passive landfill operation.

Energy from Refuse

There are many methods for deriving energy from refuse and they can be classified into three general categories which can then be subdivided further into specific systems:

Commercially Operational Technology
Developmental Technology
Experimental Technology

For the past several years, PSE&G has recognized that refuse is New Jersey's most significant (and probably only) raw material. As such, the Company has been looking at many different ways to make use of refuse in satisfying our customers energy needs. We feel that if this "domestic fuel" source has potential, it should be used providing that its use does not require subsidization by electric and gas customers.

COMMERCIALLY OPERATIONAL TECHNOLOGY

Existing refuse-to-energy facilities which are operating continuously fall into the category of Commercially Operational Technology.

Water Wall Combustion— This sytem uses an almost standard industrial boiler with a water walled furnace incorporating other boiler tubes to recover the maximum amount of heat. Most systems burn unprocessed solid wastes on mechanical grates which move it through the furnace. A recent modification utilizes shredded solid waste which is blown into the furnace and burned on semi-suspension spreader stokers. Generally, no additional fuel is used in this type of boiler.

Discussions are currently under way between PSE&G and parties who are interested in building and operating a refuse-fired boiler. The steam generated from this refuse burning would produce electricity which would be sold to PSE&G for resale to its customers. If the arrangement proves technically and economically feasible, there could be as much as 9000 tons/-day of refuse burned to produce energy in New Jersey.

DEVELOPMENTAL TECHNOLOGY

Developmental Technologies have been proven in pilot operation or in related but different application. There is sufficient experience to predict full-scale system performance success, but the performance has not been confirmed.

Refuse Derived Fuel (RDF)— In RDF waste processing systems, the waste is classified into a heavy fraction (glass, metal, rock, etc.) and a combustible fraction. The combustible fraction is further processed into either a fine dry powder or into a "fluff". This RDF is then used to supplement fossil fuels in large existing boilers—powder in oil fired boilers or fluff in coal fired boilers.

PSE&G's first venture into energy from refuse activity has been in the area of RDF. We have a contract with an engineering and construction firm to provide us with powdered RDF. It is expected that when the powdered RDF is available, and after necessary test burns have been completed, it could provide as much as 20 percent of the energy input for an oil fired boiler. In addition, we have participated in studies that have considered fluff as a supplement in some of our coal-fired boilers.

Pyrolysis— Pyrolysis is the term generally applied to the high-temperature combustion of a material in an oxygen-deficient atmosphere. The products of combustion in a pyrolysis system are combustible in themselves and are either in a liquid or gaseous form, of either a high or low BTU content.

Biological Conversion— Biological conversion involves the decomposition of solid waste by bacterial action to produce combustible gases. These gases could be burned immediately to generate steam or transported elsewhere for use. Biological conversion occurs naturally in landfills and can be accelerated in controlled vessels. (This latter application falls into the next category—Experimental Technology.)

Just recently, PSE&G drilled the first gas wells in New Jersey into a South Jersey landfill. The gas recovered from this landfill is being piped to an existing industrial customer who uses the gas to fire a boiler. The gas recovered has a heating content of 600 BTU per cubic foot as it comes from the ground. This is acceptable for the customer's use, but if the gas were to be mixed with natural gas (1020 BTU), a great deal of CO_2 and water vapor would have to be first removed.

EXPERIMENTAL TECHNOLOGY

This category includes new technologies that are still being tested at the laboratory and pilot plant level. There is insufficient data available to predict technical or economic viability.

Biological Conversion— As mentioned above, this is the action of bacteria on refuse to produce a combustible gas which would take place in an enclosed reactor vessel digestor.

Waste-Fired Gas Turbine— This technology involves the conversion of solid wastes into a combustible material (gas or liquid) either through pyrolysis, conversion, or some other method, and then burning that material in a combustion turbine to drive a generator.[27]

Fuel Cell— A fuel cell produces electricity through the chemical reaction of oxygen and hydrogen. Research is being conducted to determine if a fuel cell could successfully operate using the hydrogen-rich waste gas produced by a sewage treatment plant.[28] If the fuel cell could be successfully operated from sewage waste gas, there should be no reason why gas generated through Biological Conversion could not similarly energize a fuel cell.

PSE&G is currently involved in a study with Gilbert Associates and the U.S. Department of Energy to determine the feasibility of using off-gas from a sewage treatment plant to fuel a large fuel cell which would cogenerate electricity and process heat.

Conclusions

While there is much to be said for the large-scale development of cogeneration and energy from refuse facilities, practically speaking, there are many impediments to this development.

Energy Balance

The most significant requirement for the successful operation of cogeneration is an effective energy balance, or the simultaneous demand for heat energy and electrical energy. This heat balance is achieved most effectively with a large industrial customer. However, most of those large industrial customers are moving out of the urban areas of New Jersey to other parts of the country, which significantly limits this application of cogeneration.

ECONOMICS

The second most important item to effect the widespread acceptance of cogeneration is that of economics. Cogeneration should be required to stand on its own economic feet. However, there are some that point to the supposed energy-saving nature of cogeneration and say that it should be supported by tax and revenue incentives provided by all other tax and ratepayers. That proposition is wrong—those who gain the advantages of cogeneration, where they exist, should pay for those advantages. The burden should not fall on the shoulders of others.

INDUSTRIAL INERTIA

As indicated above, the most effective user of cogenerated energy (regardless of the fuel source) is large industry. To effectively realize an economic benefit from cogeneration would require a commitment to remain in a given location and produce at a given output level for many years. Many industries in New Jersey, and especially older industries in the urban areas, are unwilling to make such a commitment.[29]

SPACE

Any energy-production facility requires a great deal of physical space, regardless of fuel and independent of whether or not the energy output is in one form or several forms, as with cogeneration. There must be a space for the production equipment, but even more space for the fuel storage and processing facilities, regardless of whether the fuel is oil, coal or refuse. In addition, there must be means of accommodating traffic, either water, rail, or roadway. It is often very difficult to meet such requirements in any of the built-up urban settings in New Jersey.

Recycling and Conversion

As has been indicated, refuse is New Jersey's most abundant natural resource. And energy is our scarcest and most expensive resource. It would appear logical for a large part of the refuse generated in this state to produce some of the energy used here. If it is determined by society to encourage the development of the refuse from energy concept, state action would be logical in the following areas:

Encouragement of industry to make long-term commitments to remain at present urban sites.

Regulatory bodies to agree to expedite the granting of appropriate permits.

Governmental agencies to decide on land use priorities vs. low-cost refuse disposal. Artificially low prices for dumping municipal waste in landfills will not allow the refuse-to-energy plants to be built.

ENVIRONMENTAL

While many of the foregoing comments in this section appear negative, there are still some very strong positive aspects of cogeneration and refuse disposal which are developing.

Many existing landfills in or near New Jersey's urban areas provide excellent potential sources of:

Landfill gas which could be burned in a boiler
Space for constructing refuse-derived energy facilities
Land for economic development that is being promoted by the states and the federal government

This economic development could then encourage the development and expansion of cogeneration and energy from refuse.

Notes

1. *Public Utility Regulatory Policies Act of 1978*. P. 195-617 (HR.4018).
2. Manufacturing Chemists Association. *A Proven Way to Save Energy—Cogeneration.* (Washington, DC,: 5/78).
3. Testimony of M. J. Mighdahl, Executive Vice President, National Association of Secondary Material Industries, Inc., before the Subcommittee on the Environment of the Senate Commerce Committee during hearing on Senate Bill S1122, 3/6/73.
4. PSE&G. *Cogeneration Evaluation* (January, 1977).
5. John, W. L., "Cogeneration in the Chemical Industry," *Energy The New Rules.* (San Francisco, March 22, 1979).
6. PSE&G. *Cogeneration Evaluation* (January, 1977).

7. *Department of Energy Act* (NJSA 52:27-F) New Jersey Legislature, effective July 11, 1977.

8. *Department of Energy-Cogeneration Facilities-Aid In Developement* (NJAA 52:27-F Amended) New Jersey Legislature, effective July 13, 1978.

9. *Power Plant and Industrial Fuel Use Act of 1978*, PL 95-620 [H.R.5146] definition, effective November 9, 1978.

10. Ibid, definition.

11. Ibid, Subtitle A-Prohibitions.

12. *Public Utility Regulatory Policies Act of 1978* PL 95-617 [HR 4018], effective November 9, 1978.

13. PSE&G. *Cogeneration Evaluation* (January, 1977).

14. Ibid, p. 12.

15. "Oil Fueled Cogeneration in Urban Areas," paper presented by Bertram Schwartz, Senior V.P., Consolidated Edison Corporation, October 3, 1978.

16. Ibid.

17. Hearing conducted by the NJDOE, "In the Matter of the Investigation of Cogeneration in the State of New Jersey", November 9, 1978, Testimony of Sheldon Steiner, Flack and Kurtz Consulting Engineers.

18. Boggs, J. Caleb. *Planning for Solid Waste Management* (St. Louis, MO: September 9 1969).

19. Ibid.

20. Testimony of M. J. Mighdahl, Executive Vice President, National Association of Secondary Material Industries, Inc., before the Senate Subcommittee on the Environment, Senate Commerce Committee, regarding Bill S1122, 6/11/73.

21. Ibid.

22. American Public Works Administration. *Municipal Refuse Disposal*, p. 64, (Chicago: Public Administration Survey 1966).

23. *Solid Waste Management Symposium*, ASME, Teaneck, NJ 1971.

24. Ibid.

25. Shipelsky, Alan and Robert A. Lowe. *Resource Recovery Plant Implementation: Planning and Overview* (Cincinnati: USEPA, 1976).

26. Testimony of M. J. Mighdahl, Executive Vice President, National Association of Secondary Material Industries, Inc., before the Subcommittee on the Environment of the Senate Commerce Committee during hearings on Senate Bill S1122, 3/6/73.

27. Levy, Sten J. & Gregory H. Rego. *Resource Recovery Plant Implementation: Technologies* (Cincinnati: USEPA, 1976).

28. Gilbert Associates, PSE&G, USDOE.

29. PSE&G. *Cogeneration Evaulation* (January 1977).

The Status of Solar Energy Development

ROGER L. JOHNSON

Introduction

The purpose of the article which follows is to discuss the basic nature and provide a brief summary of the developmental status of the various technologies lumped under the category of solar energy. In addition to definition, each component of solar energy will be assessed as to its current level of required technology, locational application (site specific or centralized location), and time horizon for useful implementation (short/long range). Other articles on solar energy found in sections one and four of this monograph deal almost exclusively with one segment of solar, thermal application, including its legal hurdles and implications for metropolitan form. This article views solar energy as broadbased and diverse, reflecting an encompassing definition of energy derived from the actions of the sun, i.e., that body's ability to heat/light, sustain plant growth, and affect the force, direction and temperature of both wind and water bodies.

Definitions of the Components of Solar Energy

Under the definition just stated, a number of energy-producing technologies may be grouped. In an effort to provide some standarization and order to this grouping, the Department of Energy has divided solar energy into three groups which are composed of eight individual categories. These are:

1. *Thermal (heating and cooling) applications*
   Heating and cooling of buildings—including hot water heating, agricultural and industrial process heating

2. *Fuels from biomass*
   Plant matter, including wood and waste

3. *Solar electric*
   Solar thermal electric
   Photovoltaics-solar cells
   Wind-windmills
   Ocean thermal electric
   Hydropower-hydroelectric dams

THERMAL APPLICATIONS

Within the group of heating and cooling (thermal) applications, the category of devices which draw principally on the motion of the sun for heat intensity is called *passive* solar energy; those which employ mechanical moving parts are called *active* systems. Both require relatively low technology and may be used directly at the site of energy demand. Passive systems have most impact on new construction as double glazed windows and walls which accommodate heat storage are expensive to retrofit on existing buildings. Active systems have more broadbased application as they may be retrofitted on the standing stock or put in place as integral components of new structures. Both may be used for heating a structure and for a structure's hot water. The difference between active solar hot water provision and active solar heating is one of scale; the latter may require three to four times the number of solar panels as the former.

FUELS FROM BIOMASS

Fuels from biomass are derived from the burning, conversion or decomposition of organic matter. Wood and plants may be burned to produce energy, alcohol can be extracted from plants (gasohol), bacteria may be used to convert organic waste into energy-laden, methane gas (recycling). Wood is the principal source of biomass energy; organic wastes also contribute to this source of energy, yet extraction and transmission are more difficult and involved than the basic cutting and harvesting of wood. Wood burning is a low technology, site-oriented form of energy; gasohol and methane products are moderately sophisticated energy technologies usually requiring centralized locations and derivative energy transmission.

SOLAR ELECTRIC

One of the most technology-prone, solar energy applications is found in the use of a parabolic mirror. The sun's rays are transmitted to a mountain-sited, parabolic mirror from other mirrors in a collector zone on the side of an adjoining hill. The sun's rays are focused, via the parabolic mirror, on a boiler which converts the incident light to steam. The steam maybe used as a direct heating source or further converted to electricity via turbines. Whereas most of the *principal* types of solar discussed previously (thermal applications, wood burning, etc.) are site oriented, this form of solar energy is usually centrally located, and requires further transmission to the location of individual users.

Another category of solar electric is the area of photovoltaics or solar cells. Photovoltaics are most often commercially made from silicon and exhibit the property of electric voltage generation when light-exposed. Current technology sees application for both on-site and centralized use although centralized applications appear to have the most promising, short-run future.

The conversion of wind to useful energy has a long history of application. In Europe, mid-western and eastern United States, wind has been used to drive grinding wheels, pump water and later to produce electricity. A more sophisticated type of wind machine is being reintroduced in the United States, typically in the presence of others, on what is termed wind farms. A plan for hundreds of windmills, aggregated in central locations near the coasts, will take advantage of day/night, off-shore/on-shore breezes to produce electricity for transmission to potential regional electric users. They involve only moderately sophisticated technology and are capable of being replicated in most coastal zone locations.

The final categories of solar electric, ocean thermal electric and hydroelectric dams, rely on water temperature flows to generate electricity. Ocean thermal energy relies on the principal that energy can be extracted from the temperature differences maintained between different layers of sea water as a result of the absorption of solar heat. This is a high technology centralized application that is barely beyond the experimental stages of development. Hydroelectric power is a well established technology in which water is dammed to make an impoundment and intake structures are built so that a flow of water can be run through a turbine generator. Location is centralized and resulting energy is transmitted to on-site users.

Status of Solar Energy Applications

The concept of most people regarding solar power in both its direct and indirect forms is that it is "Year 2000 Technology." This view is incorrect: moderate-technology, solar systems are here today, albeit technical improvements are and will continue to be made. The coming of age of solar energy is evident from the following brief listing of its applications.

THERMAL APPLICATIONS

Solar hot water systems are being purchased across the nation. These installations are clearly competitive against electricity and fossil fuels—expensive energy sources dominant in the Northeast and North Central regions of the country. Other areas have also sought solar hot water systems because of climate and other factors. The state of Florida is one example; portions of the Southwest are another.

Solar space heating installations are fewer in number than solar hot water systems but the former are no longer novelties. They are being installed in all types of facilities—large and small, residential and nonresidential, urban and rural, etc. To illustrate, one of the largest solar systems in the country is now supplying all the heating/cooling demands of the Georgia Power Company headquarters in Atlanta.

FUELS FROM BIOMASS

Important progress has also been made in the development/utilization of *indirect solar energy sources—wood, other plant matter, waste, etc.*

Wood energy has been with us for many years and has become quite popular of late, in some instances assuming dramatic energy supply proportions. The state of Vermont is experiencing a very dramatic downturn (about 20 percent) in oil consumption and a commensurate increase in the utilization of wood for heating purposes. Nationally the sale of wood stoves has skyrocketed—from 100,000 units per year to over one million.

Methane is being produced from waste materials. Some jurisdictions are tapping central garbage collection points for methane which in turn is being used as an energy source for generating electric power and steam. A pilot facility at the U.S. Department of Agriculture's Hastings (Nebraska) Center utilizes anaerobic digesters to produce methane from manure. A similar process is being used by some sewage treatment plants; in some cases these facilities have combined the use of solar energy to power the methane conversion apparatus.

SOLAR ELECTRIC

New/expanded applications of solar energy are also being readied such as solar thermal electric power plants. One such facility is being constructed by General Electric in Albuquerque, New Mexico. This installation will ultimately contain about 20,000 tracking mirrors generating heat/steam (by reflecting sunlight onto a central tower receiver) for a 100 Mw electric generator.

Another important development concerns improvement in the basic technology for producing solar cells (photovoltaics). The traditional Czockralski single crystal growth process is being challenged by more time and cost-efficient methods. Examples include the Mobil-Tyco/Westinghouse techniques and the amorphous silicon cell process developed by RCA Laboratories at Princeton, New Jersey.

There is also experimentation with *wind and ocean thermal* resources. Wind farms—massing wind-energy equipment in strategic locations—are nearing the operational stage. Single wind-powered generators are already in place. A 2 Mw tower is being operated in Boone, North Carolina, a 200 Kw machine in Clayton, New Mexico. The latter was evaluated by engineers from the Electric Power Research Institute (EPRI) who concluded that the Clayton facility was able to supply electric power to the local grid system at an efficiency of over 90 percent.

A pilot OTEC (Ocean Thermal Energy Conversion) facility has been installed off the shores of Hawaii. Similar projects are planned for other locations.

As the above discussion indicates the *direct* solar systems—hot water and space heating—are much further along as far as usage and installation are concerned; some of the *indirect* solar technologies such as wind farms and

OTEC are mainly at the pilot stage. The point which must be emphasized, however, is that solar technology is available today and is developing a proven track record. The future is even more promising, as existing direct solar systems will be improved and indirect solar technologies will move from experimental to proven field level application.

Restraints and Opportunities

While solar energy's technological problems are being resolved, it is important to recognize the *financial* and *institutional* restraints to further usage and to consider/develop public policies to overcome these drawbacks.

Cost considerations are an important hurdle. Conventional fossil sources and even electricity generated from coal/nuclear remain relatively inexpensive, thus muting solar energy's attractiveness. This conventional energy price advantage is largely due to a price control policy which, until recently, has protected most American fuel sources from "real world" cost pressures. A second factor is the capital cost of solar installations and equipment—an investment requirement which runs counter to the practice of years past where capital investment has been at the point of energy supply and provided by energy producers, not at the point of consumption and paid for by energy consumers.

Regulatory hurdles are also problematical. Local planing and zoning boards have sometimes delayed, if not prohibited, solar installations because of aesthetic and other considerations. Building code officials, not familiar with solar equipment, have posed certification problems. Unfavorable property tax assessment practices, significantly raising the value and therefore the tax burden of homeowners opting for solar, is still another restraint. These regulatory problems all translate into added delays, frustration and costs—factors inhibiting conversion from conventional fuels to solar.

It is comforting to note that many of these hurdles are being overcome. Numerous changes have improved the financial attractiveness of solar energy. Its competition—conventional fuels—has become much more expensive as price controls have been lifted from both oil and gas. Gradual decontrol was provided by the *National Energy Act of 1978*. Accelerated decontrol is taking place under the Reagan Administration. Furthermore, the 1978 Energy Act also provided generous tax incentives for solar energy systems. It allowed an income tax credit for installation of solar equipment up to a total maximum grant of $2,200. In addition, the 1978 Energy Act authorized a $100 million Solar Energy Loan program to facilitate the financing of solar heating and cooling equipment in residential units. While this program has been cut by the Reagan Administation it remains the basis for significant commercialization of solar equipment.

Regulatory restraints are also being addressed. Numerous planning bodies are encouraging, and in some instances requiring, solar installation in both new and retrofit situations. Even more are paying serious attention to removing *restraints*. The growing effort to protect solar access is the most notable effort in this regard. Other actions include quick, "one-stop" certification of solar equipment by local building inspectors and the removal of property tax disincentives to solar upgrading. Arizona, Maryland, Massachusetts and New Jersey, for example, exempt solar devices from add - ed property taxation. Other states have gone even further, reducing the existing property tax obligation on homes where solar equipment is installed.

Conclusion

Solar energy is here today—it is not a next generation technology. Thousands of solar hot water systems are in place, and solar space heating/cooling has also become more commonplace. Other solar technologies such as energy from methane tapping, wind farms and ocean thermal sources are close to the take-off stage. They can become important energy sources by the early twenty-first century.

The progress of solar energy development is the consequence of the rising cost of conventional energy sources as well as directed efforts to provide tax, financing and regulatory relief to both the solar industry and the solar consumer. It is important not to be complacent, however. It is essential that we continue to advance solar technologies and secure the endorsement of local, state and federal public officials, financial institutions, and professional groups.

Energy Siting in Critical Environmental Areas

CARL N. SHUSTER, JR.

Introduction

THE subject of this paper constitutes an important issue: "Should energy facilities be sited in critical environmental areas?" This is a complex issue and one that is not easily resolved.

There is extensive literature and information available—so much that several books could be written on the subject. Much important material, unfortunately, has had a limited distribution, particularly within the federal bureaucracy or has been directed to a limited, usually technically-oriented audience. Further, the issue is so extensive, bringing together such a complexity of interactions between energy and environmental activities, that it is difficult for one person to deal effectively with all the details. The subissues that arise cut across economic, political and technological considerations.

We are dealing also with an issue that tends to be inflammatory, eliciting highly emotionalized and polarized viewpoints. It is an issue that is viewed differently by those responsible for environmental protection and management and those concerned with producing energy. As a result of the complex subject and the diverse viewpoints, decisionmakers dealing with permits for siting and operating energy facilities cannot help but be confused and frustrated by the extensive and often conflicting technical information.

Ultimately the issue will be resolved in the political arena and involve consideration of the degree of criticality of each situation—of specific energy/economic/environment scenarios—within the context of the national interest, regional needs and specific site problems. Thus, the choice between the merits of environmental protection and energy supplies may be based as much on political necessity and reality as on scientific findings. Nevertheless, environmental input to the decisionmaking process is essential.

Ever since the 1940s, the development and utilization of energy supplies and protection of critical natural environments have increasingly become issues, polarized against one another on a national scale.[1] So much has this been the case that it was pointed out in 1977, the time had arrived in the evolution of our environmental consciousness " . . . for both industry and environmentalists to stop engaging in the game of chicken, lest everybody lose."[2] It is obvious that both energy and environmental quality are of high, if not comparable, importance to our nation and all reasonable approaches toward accommodating the basic requirements of each must be made.

Although political events may have been the immediate cause of our energy crisis, other policies, such as regulatory controls that created cheap energy, undoubtedly contributed.[3] Whatever the cause and resultant relationships, this is but another example of the very complex system of interacting societal processes in our modern civilization. A change in one process causes an unexpected change elsewhere in the system. This observation leads to a counterpoint—that civilization is not as old or as complex as the natural environment. It is no wonder then that our difficulty in understanding or coping with the environment of our own making, i.e., civilization, is even greater when we are faced with the problems associated with the natural environment.[4]

Approach

The task here is to consider the assigned topic in its dual environment/energy context and reduce the record of governmental involvement and an extensive literature into a manageable discussion.[5] [6] [7] So, I will scope out some of the problems and remedies in siting energy facilities, at least to the extent of identifying information and concepts that might be useful in a planning context. Also, this paper can be considered an extension of the discussion begun at a previous conference in this series.[8]

There are three related themes that are intertwined throughout my discussion:

identification of the major environmental laws affecting energy systems planning; (2) sources of energy/environmental information; and (3) possibilities for accommodating energy facilities in environmentally "critical" areas.

The Legal Context and Implications

While there is no federal law specifically controlling the siting of energy facilities, a number of environmental laws, often implemented at the state/local level, mandate certain considerations affecting their siting, construction, operation and abandonment. This section identifies only some of the more important federal laws. These laws and regulations based upon them require, among other things:

> that the Federal Energy Regulatory Commission find a proposed hydroelectric power project be best adapted to a comprehensive plan for the development and utilization of a waterway for all beneficial purposes (Federal Power Act as amended, Section[9] (a).

Since 1964, "Planning Status Reports" and then "Appraisal Reports" have been issued on hydroelectric projects and related water resources in river basins.[9] In 1976, a methodology report was published to demonstrate the feasibility of adapting the water resource appraisal approach to a generic environmental impact statement process.[10]

> that federal agencies prepare an environmental impact statement (EIS) for all projects that constitute a major federal action significantly affecting the quality of the environment; utilize a systemative, interdisciplinary approach in planning and decisionmaking and recognize the worldwide and long-range character of environmental problems (The National Environmental Policy Act [NEPA] of 1969 [Public Law 91-190] as amended.]

Common with other federal agencies, the Federal Energy Regulatory Commission is implementing the Council on Environmental Quality NEPA regulations (43 *Federal Register* 55978-56007, 29 November 1978).[11]

> that federal actions, as in the licensing of hydroelectric power projects and certification of natural gas pipelines, be consistent with Department of Commerce—approved state coastal management programs (Coastal Zone Management Act of 1972 [Public Law 92-583]; Amendments of 1976 [Public Law 94-370]).

Since the regulatory activities of the Federal Energy Regulatory Commission begin when an application is made, the Commission has proposed that applications contain a certification that the proposed project complies with the state's approved coastal management program.[12]

> that the Secretary of the Interior consider available environmental information in making decisions on outer continental shelf exploration plans, drilling, permits, development and production plans, etc. (Outer Continental Shelf Lands Act of 1953 [Public Law 83-212]; Amendments of 1978 [Public Law 95-372]).

If sufficient supplies of oil and natural gas are discovered off the Atlantic coast, the Department of the Interior EIS on the OCS development plan will evaluate the impact of pipelines.[13]

> that certain users be prohibited on those federally-owned lands designated by Congress as "wilderness areas" unless authorized by the President (Wilderness Act of 1964 [Public Law 88-577]).

While the National Wilderness Preservation System has grown slowly, containing 16.5 million acres in 1978 (less than 2 percent of all public lands), there is a potential for great future expansion.[14]

> that power uses of rivers designated as "wild and scenic" be prohibited (Wild and Scenic Rivers Act of 1968 [Public Law 90-542]).

More than 1,650 miles of prime waterways had been designated as part of the National Wild and Scenic Rivers System by October 1976.[15]

> that water quality certificates be obtained prior to the operation of any federally-licensed or permitted energy facility affecting the navigable waters of the United States (Federal Water Pollution Control Act Amendments of 1972 [Public Law 92-500, § 401]).

Over the past decade water quality has slowly improved under the stimulus of the federal effort. For fossil-fired power plants, pollutant control has been mainly on thermal discharges and the biocides used in cleaning cooling systems. The unsolved radioactive waste disposal problem threatens the future of nuclear power in the United States.[16]

> that any licensed or permitted energy facility comply with national air quality standards and control techniques (Clean Air Amendments of 1970 [Public Law 91-604]).

Nearly two-thirds of all sulfur dioxide emissions and heavy concentrations of particulates come from coal- and oil-fired powerplants, yet approximately 74 percent of the nation's large power plants are in compliance with the sulfur dioxide emissions limitations.[17]

that all federal agencies cooperate with the Department of the Interior of the protection of endangered species and habitats (The Endangered Species Act of 1973 [Public Law 93-205]).

"Extinction of species is the surest way to convert a renewable resource into a nonrenewable resource."[18]

that federally-approved projects comply with the regulations governing the preservation of archaeological, historical, and cultural sites and objects (Historic Sites Act of 1935 [Public Law 74-292]; Reservoir Salvage Act of 1960 [Public Law 86-523], amended by the Preservation of Historical and Archaeological Data Act of 1974 [Public Law 93-291]; National Historic Preservation Act of 1966 [Public Law amended by the Land and Water Conservation Fund Act of 1976 [Public Law 94-422]; Executive Order No. 11593, Protection and Enhancement of the Cultural Environment [13 May 1971]).

In January 1978, the Heritage Conservation and Recreation Service was created within the Department of the Interior and the concept of a National Heritage Program approved.[19]

that federal agencies act to reduce the risk of flood loss, minimize the destruction, loss and degradation of wetlands, and preserve the natural and beneficial values of floodplains and wetlands (Executive Order 11988, Floodplain Management [24 May 1977] and Executive Order 11990, Protection of Wetlands [23 May 1977]).

Special attention by federal agencies is being given to floodplain and wetlands. The Federal Energy Regulatory Commission has proposed regulations that will carry out the Executive Order.[20]

SOME ENERGY/ENVIRONMENT AMBIGUITIES

In spite of the above laws and directives, the lack of a national land use policy is a serious flaw in environmental protection.[21] For example, we should be particularly protective of our prime farmland.[22] This situation is all the more curious due to the close linkage between food and energy supplies. They are the energy resources underpinning our national economy: one fuels humans, the other, their industries. The high productivity of modern agriculture is possible only through large commitments of energy; directly, as fuels for machinery and for drying grain; indirectly, in the conversion of natural gas into nitrogenous fertilizers. It would be a cruel fate to have attained an adequate, sustainable level of energy and then discover we have run out of productive farmland.[23]

The complexity of interrelationships among man's activities is such that the magnitude of the ultimate socioeconomic and environmental result of regulation is difficult to foresee without intensive planning. It may be more expedient, i.e., less wasteful of time and effort, to set a presumed meritorious action in motion then wait and see what happens and whether or not an artificial compensatory reaction is needed. It may be that the situation presently perceived as an imbalance favoring environmental goals over energy needs[24] is at least in part due to the earlier start made prior to the energy crises, on a campaign to decrease environmental impacts. Also, now that environmental impact costs are included as an expense of producing energy, energy is no longer cheap. If there really is a serious conflict, the situation probably is a type resulting more from lack of foresight and planning than from intent.[25] The course of a regulatory trend may be that of ultimately producing such significant conflicts between different activities that reactionary and compensatory actions are set in motion. Also, the momentum created by the earlier enacted environmentally-protective laws will tend to reenforce future federal regulatory action.[26]

Overall, the bulk power supply systems of the total electric utility industry in the United States and Canada are adequate at the present time. But, warns the National Electric Reliability Council (NERC), future adequacy and reliability of electric power supply is jeopardized by an imbalance of environmental over energy needs.[27] Mainly, the industry has not been able to cope with the unpredictable timetable in the permitting process due to some form of regulatory, litigatory or financial problem. Later, in the section on accommodating energy development in "critical" environmental areas, I will address the value of early, open planning as one approach to coping.

In looking toward the future, NERC sees a compelling need to restore a rational balance between the need for new electric generating facilities and the desire for an improved environment. Even with energy conservation providing an important brake on development, from 25,000 to 30,000 MW additional generating capacity per year is forecast by NERC for the 1980s. One cannot help but wonder on how this fits into the overall seriousness of national near- and long-term energy problems since the outcome will depend upon how well and how soon new technologies and expansion of resources within the United States will abate our energy supply problem.

Sources of Energy/Environment Information

This paper has been prepared from the viewpoint that it is important to first understand the larger relationships, as represented by generic considerations of power plant siting. This means that many detailed reports, particularly those describing the natural environment, have not been referenced. Neither has the vast number of environmental impact statements (EISs) been examined, nor the environmental assessments, reports and associated research.[28] During the first decade of NEPA (1970-1979), some 7,278 EISs were finalized and another 3,680 are still in the draft stage.[29] I hasten to point out that involvement in an actual planning/site selection exercise, in contrast to writing this paper, requires knowing the "territory" and the "players" as a prime requisite. Items associated with these requisites include the following:

KNOW THE "TERRITORY"

This demands the selection of the appropriate magnitude of the study—the proper geographic/environmental area and its associated socioeconomic and political parameters—that will best subsume the issues involved by the energy supply system(s) being considered.[30] This includes:

(1) identifying the problems and issues, knowing how to and then identifying "critical" and "sensitive" areas—special habitats and species,
(2) knowing the pertinent regulations and procedures,
(3) knowing the applicable methodologies, particularly the application of the systems analysis approach.

KNOW THE "PLAYERS"

This requires the early identification of who should know about what you are doing and who can contribute to the study/planning you are doing:

Who should be notified?
Who should be included in early planning?
Who are the key references (information sources) people/institutions/agencies/companies?[31]
Selection of consultants.[32]

Environmental Impact Study Model

An environmental impact study model (Exhibit 1) is a convenient framework in which to assemble pertinent information. The model is time dependent. For example, when assumed future alternatives become

EXHIBIT 1

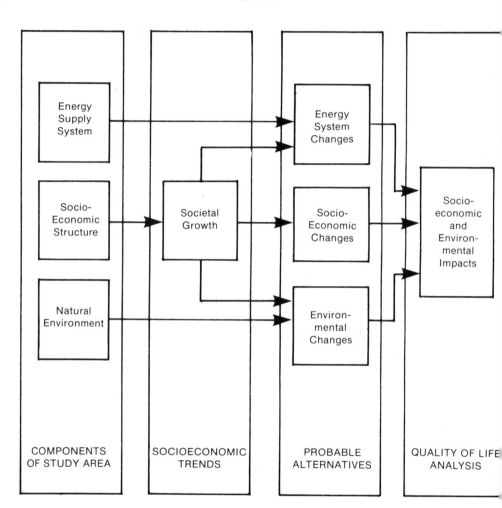

A model used in the study of significant interrelationships between energy
supply systems and natural and developed resources in appropriate regions.

realities, they move to the left and occupy the "existing" conditions module. As diagrammed in the figures, the study proceeds through four successive steps. The first step, the description of existing conditions, is followed by three integrative steps:

EXISTING COMPONENT PARTS OF THE REGION: Description of the energy supply system; socioeconomic structure within the area; chief physical aspects of the natural environment, including existing man-made alterations; and, impacts within the existing environment.

TRENDS IN SOCIOECONOMIC DEVELOPMENT: An examination of those aspects of societal growth which contribute significantly to changes in the environment (natural environment and socioeconomic structure, including the energy system).

ASSUMED FUTURE ALTERNATIVES: Consideration of probable alternatives: the need for energy changes in the energy system, and probable changes to socioeconomic structure and the natural environment.

IMPACTS ON THE QUALITY OF LIFE: The impact of the assumed alternatives upon the energy supply system, socioeconomic structure, and the natural environment: probable environmental impacts; significant probable adverse environmental impacts that cannot be avoided, and the extent of environmental impacts (with a summary of the possible environmental consequences of probable alternatives—the commitment of resources (natural and socio-economic, whether irreversible/ irretrievable or reversible)—evaluated against the duration of the impact (curtailment or enhancement of the short-term use or long-term productivity) upon the use.

The following three sections refer to information pertinent to:

energy supply systems,
natural systems, and
relationships between energy supply systems and natural systems.

INFORMATION ON ENERGY SUPPLY SYSTEMS

The juxtaposition of the proper size of the geographic territory commensurate with the proposed and alternate energy supply systems is a prerequisite to analyzing the parameters of a specific energy facility. The first attention then is toward that of bulk energy systems. For nation-wide assessments and reports on the reliability and adequacy of the bulk power supply of electric utility systems in North America, the starting place is the National Electric Reliability Council (NERC).[33] The nine councils of NERC have detailed regional data, and regional power pools are further information sources.[34] Governmental agencies provide data on energy supplies,[35] fuel use,[36] energy systems,[37] and the additional energy required to provide environmental controls.[38] Special resource studies[39] and trade journals are a further source, as for the development of newer methodologies or status reports.[40]

Energy processing (Exhibit 2), energy sources (types), and policy issues are elements of energy problems definition.[41] Here we consider only energy processing (the energy supply system: exploration/discovery, recovery/extraction, transport, conversion/processing, distribution/sales, and utilization). In Exhibit 2, the general interrelationships are depicted between the energy supply systems for five major energy sources (natural gas, oil, coal, uranium and hydropower). Only the steps in the natural gas supply system have been connected to aid the reader in visualizing the sequence of events. Steps in the other energy systems are suggested by arrows.

ECOLOGICAL/ENVIRONMENTAL INFORMATION

Basic biological principles and concepts pertinent to consideration of energy siting —e.g., biological organization and productivity, community metabolism, ecosystems, natural gene pools, species and habitat diversity and stability, etc.—are described in most ecological textbooks.[42] Special studies have been added to these: future North American environmentalists,[43] environmental integrity,[44] coastal ecological systems,[45] preservation of natural diversity,[46] and protecting natural gene pools.[47]

Another biological phenomenon is also important to ecological considerations—the interface between different media.[48] It pertains to the physiological activity at interfaces and to the high productivity of life at the edge of the continents.

ENVIRONMENTAL INTERFACES

The tremendous surface area (a little over half the playing area of a football field) of the human capillary system (some 60,000 miles long when the capillaries are placed end-to-end) is the surface through which oxygen and carbon dioxide are exchanged with 22 to 25 millions of red blood cells.[49] In other words, it is this interface in biological systems where the action is. It is one of the factors accounting for the high productivity of tidal marshes and estuaries where shoreline development is one of the indices of biological productivity.[50]

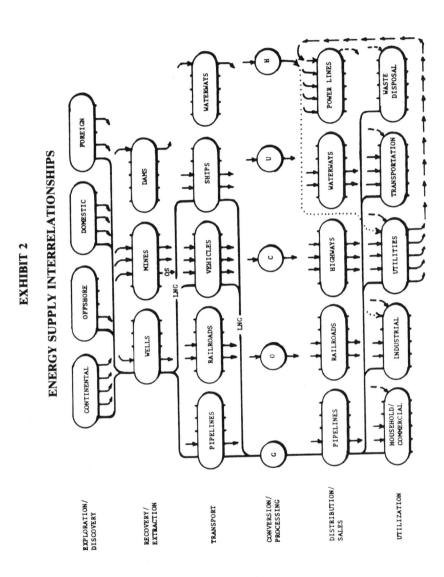

EXHIBIT 2

ENERGY SUPPLY INTERRELATIONSHIPS

A schematic showing general interrelationships between the energy supply systems for five major sources of energy: G = natural gas; O = oil; C = coal; U = uranium, and H = hydropower. Only the sequential events in the natural gas supply system are completely interconnected. The five lines projecting upwards and downwards from each capsule represent, in sequence (left to right), the respective sources of energy: G, O, C, U and H. Arrows, positioned appropriately on the capsules, indicate the lines of connection for the other supply systems. The other symbols are: LNG = liquefied natural gas, and OS = oil shale.

Generally the greater the shoreline development (see Exhibit 3) of a shallow body of water (its geochemical attributes being equivalent to another), the greater the potential for biological productivity. This physical relationship is enhanced even more in marine areas because the air-land-water interface is greatly increased by the fluctuations of the tides. The serpentine-dendritic pattern of water courses through a tidal marsh also greatly increases the shoreline development and appears to be an essential factor, among others such as nutrient cycling, in the high biological productivity of tidal marshes.[51]

Slightly more than half (70 million acres) remain of an estimated original 127 million acres of all inland and coastal wetlands of the United States. This is an extremely limited, valuable resource. Inland wetlands have been evaluated as potential registered landmarks.[52] A new handbook should aid planners and coastal program managers in better obtaining and utilizing maps.[53]

ENERGY/ENVIRONMENTAL INFORMATION

The bulk of the references reviewed for this paper were in this category. Only a few of these are cited here; but, of all of them, one deserves special attention.[54] It is the single most comprehensive environmental planning manual available, albeit it deals with coastal areas and problems. In fact, it is in the coastal area that consideration of interrelationships between energy and environment provide a lively debate[55] and a reasonable philosophy to apply in planning and management.[56]

The remainder of the selected, representative references fall into three general categories:

data and assessment reports[57-65],
siting studies and methodologies[66-75], and
other environmental impact considerations[76-87].

Harmonizing Energy And Environmental Concerns

As I have already intimated, the opportunities for accommodating energy supply systems should depend more upon the criticality of the environment than upon energy site needs. It is doubtful that any energy or similarly impacting activity should ever be permitted in a truly critical environmental area—one that has such unique features and scarcity that any impact upon

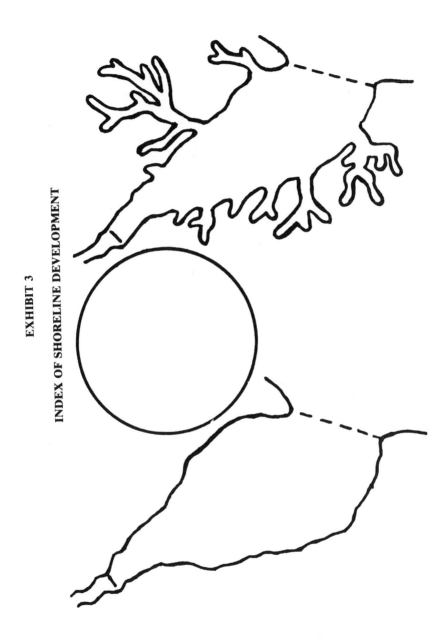

EXHIBIT 3

INDEX OF SHORELINE DEVELOPMENT

An index of shoreline development is obtained by comparing the length of the shoreline of a body of water with the circumference of a circle having the same area.[50] The greater the development of the shoreline, the larger the value of the index (ratio). This schematic compares two estuaries.

it would be a significant loss to the natural environment and to mankind. We are immediately guided in the identification of such areas of a number of laws, and the regulations based upon them, that place prohibitions upon the use of wild and scenic areas, historic sites, and endangered habitats, etc. Other areas, for which criticality has not yet been established, can be evaluated in terms of the relevant laws and by scientific assessment. It is assumed that areas amenable to the development of energy supply systems are not so critical that they are especially protected by law or that impacts upon them cannot be satisfactorily mitigated.

Three kinds of environmental impacts must be minimized to maximize energy system development. Of these—physical, chemical, and exploitation of renewable resources—physical destruction of habitat is the most disastrous because there are fewer and less effective means for mitigation. Habitat replacement/creation is difficult.

THE IMPACT OF DESTRUCTION

Avoidance of critical habitats may be the single most important consideration in energy siting. After considerable experience with an intensive study of pollution in the Mediterranean Sea[88] and worldwide oceanic exploration, Captain Jacques Cousteau[89] observed that coastal habitat destruction, not pollution, was the number one problem at the present time.

Given sufficient geologic time, physical habitats—the everchanging landforms and water bodies—are a renewable resource. But the sensitivity of living resources to habitat destruction is almost instantaneous when compared to geologic time (as in the celebrated case of the elimination of the Passenger Pigeon when the forests of the mid-continent were decimated in the 19th Century). Thus we must evaluate the sensitivity of living resources to habitat destruction. To do so we must have answers to such questions as the following: What is the relationship of the habitat to the quantity and productivity of the living resource? Is the habitat necessary to maintain natural "gene pools"? How long does a habitat remain relatively stable? What is the length of time required to replace a natural habitat? Is it possible for us to create a habitat with a quality equal to a natural habitat? In quantity as well? What do we know about the quality and quantity of a habitat? Have we or can we evaluate degrees of habitat quality? How much of a specific habitat exists in the world, in the United States, regionally, and in lesser geographic areas. How dependent, ultimately, is man upon each habitat type? None of these are easily answered, but in the following sections some approaches are suggested for answering these and other problems.

It is imperative, therefore, that decision-makers and persons planning and siting energy facilities are provided with reliable estimates of the environmental consequences of their actions. The proper frame of reference for these actions is an examination of environmental systems to the same extent and detail as that which goes into the planning and engineering details for the siting, construction, operation and retirement of energy facilities.

SOME APPROACHES TO IMPROVING SITING CONDITIONS

The systems approach to siting energy facilities, early participation by the public in the planning process, expeditious use of a third party contractor, and adoption of "standardized" procedures are suggested as ways to improve the technical, scientific and public interest inputs to the siting process.

THE NEED FOR ENERGY SYSTEMS PLANNING[90]

Against the recent history of delays in, and outright termination of, building bulk energy facilities, improvements in planning must be seriously considered. At the outset the proper context is at the level of regional energy systems planning.

It is no longer possible to justify a major new facility except as a requisite for the reliability and adequacy of a justifiable energy supply system. Although Congress has given considerable attention to a siting law, a concensus has not yet been reached. However, a regional plan is a common and generally accepted prerequisite for a siting decision.

Earlier a number of federal laws requiring some form of regional approach were cited (see text and references 5, 9, 10 and 12). In addition, these laws and court decisions related to them have created an era of open planning. The advent of NEPA opened the decision-making process to everyone, including the public and its self-appointed spokesmen. The what, when, where and how decisions on building major new energy facilities is no longer the sole responsibility of the owner; nor of government (local, state, and federal); nor of the owner and government together.

Public Participation. No doubt, problems associated with regulatory control of development are increased by public participation. And public participation may interfere with efficient planning and regulation. But so do overlapping local, state and federal controls. With "open" planning and decisionmaking, the challenge is to make adjustments in the planning and permitting processes. We believe that early participation in the public interest is a viable alternative, especially if appropriate steps are taken to keep the public well informed and continually involved in the major phases of the developmental process. A search for solutions to the problem of regulatory coordination led to a study of foreign experiences.[91] One positive aspect of the Dutch experience has been obtained through making resources available to citizen groups in an effort to encourage participation and to make it more efficient and competent.

Use of a Third Party Contractor.   One way to hasten the regulatory process and include public participation may be in the way environmental impact research is conducted. Regulations requiring a one-step study combining the environmental reports of an applicant and the environmental impact statement of the pertinent regulatory agency is possible.[92] While this procedure appears to provide an important step in research efficiency, acceptance of such an agreement by all parties does not insure that the project proposed by the applicant partner in the contract will in any other way receive prejudicial treatment or acceptance. In such a procedure the regulatory agency selects and monitors the contract work; the applicant finances the work through the regulatory agency and does not have direct contact with the contractor. The public interest could be served by having a "public interest" participant as a third member of the task force managing the contract.

The Possibility of Mitigation.   Imposition of structures upon the landscape and in water bodies create different environmental conditions. When this physical impact and the operational mode of the structure can be blended more closely with its environment, in form and function, its impacts will be lessened. A recent symposium on mitigation is illustrative of both probblems and solutions.[85] "Mitigation" was first used in connection with wildlife under the Fish and Wildlife Coordination Act of 1934 (amended in 1946 and 1958). The symposium report gives evidence of the great interest and scope of work in mitigation research. Among regulatory agencies the effectiveness of such measures can be assured only by a dedicated, knowledgeable staff with ample resources to conduct or contract necessary studies. We have not referenced agency EISs in this paper, but mitigatory procedures have been prescribed for most of the projects evaluated.

Unified Approaches.   With time, in any new endeavor where new methodologies are developed and tested, a concensus develops as to which is the best method for standardization. The need may be relatively straightforward—that ecological surveys conducted before, during, and after the construction and operation of an energy project be standardized.[93]

Where a multi-disciplinary approach is needed—as in most complex energy/environment problems, improvement in communication between the disciplines is a necessity. The energy/environment language developed in recent decades is a good example of a common denominator to the communication problem.[94] It is being used increasingly, and the students of this language are now geographically widespread and active in many types of environmental work. We hope that its adherents will multiply and their efforts make substantial contributions in resolving energy/environment problems.

Further Issues to Examine

As indicated in the opening remarks, much more information exists than could reasonably be reduced to a paper. Some very interesting questions, unanswered herein, should be considered elsewhere.

What is the degree to which each energy supply system has the capacity for impacting the environment? How are the different systems (energy sources) linked, especially in comparison of alternative choices?

When environmental issues are categorized according to energy systems, which of these systems, measured by degree of impact, show the greatest potential for successful siting?

Can the siting problem be reduced to quantification of the conflicts, energy versus environment? Of priorities? What relationships exist, as in a matrix considering the magnitude of an energy source (limited or large) versus the degree of conflict in different environmental/geographical areas.

Undoubtedly many more questions remain. It is in the nature of things that the further one probes a question for answers, the number of questions greatly exceeds the answers. But the quest for solutions, particularly in the energy-environmental conflicts, will at least resolve some conflicts, and open up further areas of inquiry.

Acknowledgements

The author is especially appreciative of the help of several persons and institutions in locating some of the material used in this paper.

Thomas Armentano, The Institute of Ecology. Herbert M. Austin, Virginia Institute of Marine Science. Robert W. Burchell, The Center for Urban Policy Research, Rutgers-The State University. John R. Clark, The Conservation Foundation. Clayton J. Edwards, Federal

Energy Regulatory Commission. Richard F. Hill, The Engineering Societies Commission on Energy, Inc. (Washington, DC). Betty Jankus, Office of Environmental Review, Environmental Protection Agency. Joyce Mann, National Fisheries Center (Kearneysville, WV), U. S. Fish and Wildlife Service. Howard T. Odum, Department of Environmental Engineering Sciences, University of Florida. John Robinson, Dames & Moore (Los Angeles, CA). Gustav A. Swanson, Department of Fishery and Wildlife Biology, Colorado State University. John Cairns, Jr., Virginia Polytechnic Institute and State University.

Notes

1. There were at least five key ingredients in the expansion of environmental awareness in the 1940s: 1) appreciation of the contribution of science to the war effort led to the emergence of national, post-war sponsorship of research; 2) the socioeconomic boom of babies, industry and the Gross National Product; 3) a cadre of well-informed, environmentally-concerned professors; 4) the G. I. Bill and the rapid expansion of student enrollments, particularly in graduate schools; and, 5) encouragement of basic science led to the establishment of sophisticated research centers and equipment (see Shuster, C. N., Jr. "Energy Systems Planning and the Coastal Zone," *Proceedings, Third Annual Conference: Energy Across the Coastal Zone.* Washington, DC: The Coastal Society, 1977, pp. 142-154.

2. This quote, from an address by William D. Ruckelshaus (Commonwealth Club of California, San Francisco, 20 May 1977), is the concluding statement in an excellent review of the impact of environmental management activities on industry (see Harrison, E. Bruce. "Environmental Trends: Part I - How Did We Get Where We Are?; Part II - The Environmental Age is Costly; Part III - Where Do We Go From Here?" *Hydrocarbon Processing*. Houston, TX: Gulf Publishing Co., 1977, 56 (10): 121-; 56(11): 425-; 56(12): 153-).

3. For example, it has been alleged, from an analysis of a complex situation, that regulated, depressed rates of natural gas led to greater consumption and acted as a disincentive to further exploration and production, helping to create shortages in supply (see Breyer, Stephen G., and Paul W. MacAvoy. *Energy Regulation by the Federal Power Commission.* Washington, DC: Brookings Institution, 1974, 196 pp.

4. This understanding is further hampered by our creation of a world-within-a-world through an increasing scientific capability to manipulate natural phenomena; by our becoming an enemy of the natural world while losing an understanding of what it is to be a part of it (Pirsig, Robert M., *Zen and the Art of Motorcycle Maintenance.* New York: Bantam Book, Inc. 1979, 373 pp; at 341-342.

5. The National Environmental Policy Act (NEPA) of 1969 (Public Law 91-190; approved on 1 January 1979) established the Council on Environmental Quality that reports annually to Congress. These reports, *Environmental Quality*, are an excellent source of information on environmental problems and what is being done about them at all levels of government and in the private sector. Section 102(2) (C) of NEPA mandated a new approach to federal decision-making by requiring an environmental impact statement process in all " . . . major federal actions significantly affecting the quality of the human environment . . . . " The Environmental Protection Agency was established on 2 December 1970.

6. Federal involvement in energy activities is reviewed in the "History of Federal Energy Organization," a staff analysis for the *Committee on Interior and Insular Affairs, U. S. Senate, Serial No. 93-19 (92-54).* Washington, DC: U. S. Government Printing Office, 59 pp.

In late December 1975, the Energy Policy and Conservation Act (Public Law 94-163) was enacted. Then, April 1977, the President's *National Energy Plan* (Executive Office of the President, Energy Policy and Planning. Washington, DC: U. S. Government Printing Office,

1977, 103 pp.) was submitted to Congress. It called for energy resource management, particularly energy conservation and fuel efficiency, to offset increasing demand for, but decreasing supply of, oil and natural gas.

The Department of Energy was established in October 1977.

The "National Energy Act," was comprised of five separate acts: Public Utility Regulatory Policies Act of 1978 (Public Law 95-617), Energy Tax Act of 1978 (P. L. 95-618), National Energy Conservation Policy Act (P. L. 95-619), Powerplant and Industrial Fuel Use Act of 1978 (P. L. 95-620), and Natural Gas Policy Act of 1978 (P. L. 95-621) (see *The National Energy Act*. Committee on Energy and Natural Resources, U. S. Senate, Publication No. 96-1. Washington, DC: U. S. Government Printing Office, January 1979, 598 pp.)

7. My viewpoint is circumscribed by my experience as an ecologist in an energy regulatory agency for seven years and a prior eighteen years in senior positions with academic and governmental programs that were principally oriented toward water-associated biological problems, particularly in the coastal zone.

8. The logical base for this paper is the session on "Energy Constraints and Growth," especially "Environmental Resources and Energy Supplies" (Seevers, Gary L., and Allan G. Pulsipher), "The Multiple Dimensions of Energy Policy" (Freeman, S. David), and "A Reply - Harmonizing Energy and Environmental Conflict" (Pulsipher, Allan G.) in Burchell, Robert W. and David Listokin. *Future Land Use*. New Brunswick, NJ: Rutgers, The State University of New Jersey, Center for Urban Policy Research, 1975, pp. 245-330.

9. These reports, in a series entitled *Water Resource Appraisals for Hydroelectric Licensing* (Federal Power Commission—since October 1977 the Federal Energy Regulatory Commission, Washington, DC: 1964-), are periodically updated. They provide a collation of available information on existing and possible future development of water and related land resources, including the status of hydroelectric projects within a river basin, pertinent to licensing considerations of the Commission.

10. This approach was examined because it coincided with the Commission's responsibilities under Section 10(2) of the Federal Power Act and NEPA; because a river basin EIS could provide an overall perspective for several licensing actions before the Commission, and consolidation of a number of possible project EISs could provide a more comprehensive understanding of the relationships affecting any single project and would be a more efficient utilization of staff resources (see Shuster, Carl N. Jr., Robert S. Restall, and Shou-shan Fan. *Report on Significant Interrelationships Between Electric Power Generation and Natural and Developed Resources in the Connecticut River Basin*. Washington, DC: Federal Power Commission December 1976, 288 pp.; Shuster, Carl N. Jr. "A Commentary on a Methodology for Assessment of the Environmental Impact of the Electrical Power System Within the Connecticut River Basin," *Energy Technology V*. Washington, DC: Government Institutes Inc., April 1978).

11. Federal Energy Regulatory Commission Docket No. 79-69 (Regulations Implementing the National Environmental Policy Act of 1969), Notice of Proposed Rulemaking (issued 20 August 1979). For a significant treatise on NEPA see: Rodgers, William H. "Chapter 7 - National Environmental Policy Act," *Environmental Law*. St. Paul, MN: West Publishing Co., 1977, pp. 697-834.

12. Federal Power Commission Docket No. RM76-38 (Certification of Compliance with Approved State's Coastal Zone Management Program in Applications for Authorization to Import or Export Natural Gas and Certification or License Applications, where applicable), Notice of Proposed Rule-making (issued 23 September 1976).

13. Council on Environmental Quality (CEQ). *Environmental Quality: 9th Annual Report*. Washington, DC: U. S. Government Printing Office, December 1978, p. 364.

14. CEQ, op. cit., p. 299.

15. CEQ, op. cit., p. 300. This act prohibited the Federal Power Commission from licensing the High Mountain Sheep hydroelectric project on the Snake River between Idaho and Oregon and has had the effect of terminating the license for the Blue Ridge project in North Carolina and Virginia on the New River, headwater of the Kanawha River.

16. CEQ, op. cit., p. 90 and p. 377.

17. CEQ, op. cit., p. 72.

18. CEQ, op. cit., p. 328. For an analysis of the development and application of federal wildlife law, see: Bean, Michael (Environmental Law Institute). *The Evolution of National Wildlife Law*, a report to the Council on Environmental Quality. Washington, DC: U. S. Government Printing Office, 1977, 485 pp.

19. CEQ, op. cit. p. 244. For historical background and other information see: Iroquois Research Institute. *Archaeological and Historical Investigations for Energy Facilities: A State of the Art Report.* A report to the Federal Power Commission (Washington, DC: 1977), 273 pp.

20. CEQ, op. cit., p. 245. Federal Energy Regulatory Commission Docket No. RM79-70 (Floodplain Management and Protection of Wetlands), Notice of Proposed Rulemaking (issued 20 August 1979).

21. In coastal management one alternative to new legislation on land use management is the coordinated use of traditional mechanisms to influence land use through government regulation, taxing, and acquisition (see Schoenbaum, Thomas J., and Ronald H. Robsenberg. "The Legal Implementation of Coastal Zone Management: The North Carolina Model," *Duke Law Journal* (March 1976), 1976 (1):1-37.

22. Fields, Shirley F. *Where Have the Farm Lands Gone*? National Agricultural Lands Study. Washington, DC: U. S. Government Printing Office, 1979, 20 pp. The Soil Conservation Service is conducting a farmland inventory on the 1,200 high priority counties, to be completed in 1981, based upon a nine-point definition of "prime" farmland (CEQ, op. cit., p. 269).

23. The historical dependence of the world food supply upon an essentially solar energy-powered, traditional agriculture contrasts with the heavy dependence of modern agriculture on energy. This relationship is examined, worldwide, in terms of the impact of population growth when food production is dependent upon limitations of land and fossil energy resources: Chancellor, W. J., and J. R. Goss. "Balancing Energy and Food Production, 1975-2000," *Science* (1976), 192(4236):213-218. Federal Energy Administration and U. S. Department of Agriculture, *Energy in U. S. Agriculture: Compendium of Energy Research Projects.* Energy Conservation Paper No. 37A, Washington, DC: U. S. Government Printing Office, January 1976, 176 pp.

24. Seltz-Petrash, Ann. "Regulations—Enough is Enough," *Civil Engineering*, ASCE, 49(9):99-103.

25. Freeman (7, op. cit.) characterized an earlier energy crisis as resulting more from stupidity than cupidity; that widespread, even global disaster results more from sheer stupidity than from some design—"It is a lack of planning that is getting us into trouble, rather than too much planning" (at p. 305).

26. Harrison's review (1, op. cit.) is dated by newer issues, but except for Congressional consideration of such bills as H. R. 4985, "Priority Energy Project Act of 1979," the inertia still remains with the perpetuation and revision of environmental regulations by federal agencies, as noted by 11, 12 and 20, op. cit.

27. National Electric Reliability Council. *Ninth Annual Review of the North American Bulk Power Systems.* Princeton, NJ: Research Park, 1979, 37 pp.

28. One cannot help but wonder how useful all of this information would be if it could be easily accessed.

29. Environmental Protection Agency, Office of Environmental Review (Personal communication, February 1980). *The 102 Monitor* (Washington, DC).

30. The Soil Conservation Society of America " . . . believes it is essential to manage water resource systems according to their physical rather than political boundaries." (see Frere, Maurice H. [Chairman, The SCSA Water Resources Task Force]. *Water Resources Management in North America: A Position Statement by the Soil Conservation Society of America.* Ankeny, IA: SCSA, August 1979, 11 pp). Torrey, Irina Perlis, "Book Review: Marsh, *Environmental Analysis for Land Use and Site Planning.* U.S.A.: McGraw-Hill, 1978, 292 pp; Simonds, *Earthscape.* U.S.A.: McGraw-Hill, 1978, 34 pp, *"Journal of the American*

Planning Association (January 1979), 45(1):95-96, [Describes seven areas of information necessary to accomplish environmental planning" 1) the nature of the natural and cultural environments, 2) interactions between components of the natural environment, 3) significance of identified phenomena, 4) relation of natural areas to traditional planning (land use, transportation, urban design), 5) relation of natural to socioeconomic conditions, 6) mitigation measures to offset adverse impacts, and 7) synthesis of relevant environmental information for presentation to decision-makers and the public.]

31. Environmental Protection Agency, *U. S. Directory of Environmental Sources (3rd Edition): National Focal Point of the United Nations Environment Program/International Referral System for Sources of Environmental Information (INFOTERRA)*. Washington, DC: EPA-840-79-010, January 1979), 861 pp., currently being revised.

32. Johnson, B. G. "Choose Environmental consultants carefully," *Hydrocarbon Processing* (Houston, TX: October 1977), 56(10):101-103.

33. The National Electric Reliability Council (NERC, Princeton, NJ 08540) was formed voluntarily by the electric utility industry in 1968 and incorporated in 1975 to augment the reliability and adequacy of bulk power supply of electric utility systems in North America. NERC consists of nine regional councils whose membership is comprised essentially of all major electric utility systems in the United States and Canada. Recent publications include: *The Coal Strike of 1977-78: Its Impact on the Electric Bulk Power Supplies in North America* (May 1978); *1978 Annual Report, with Reports from its Regional Councils and NAPSIC* (April 1979); *Assessment of the Overall Adequacy of the Bulk Power Supply Systems in the Electric Utility Industry of North America During the Summer of 1979* (June 1979); *1979 Summary of Projected Peak Load, Generating Capacity, and Fossil Fuel Requirements for the Regional Reliability Councils of NERC* (July 1979); 27, op. cit.; *Review of Major Power System Interruptions in the North American Bulk Power Systems Between July 1, 1976 and December 31, 1978* (August 1979); *Assessment of the Overall Adequacy of the Bulk Power Supply Systems of the Electric Utility Industry of North America, Winter of 1979/80* (November 1979).

34. For example, in the New England area, see Shuster, Restall, and Fan (10, op. cit.: pp. 3-28 and 3-29). An electric demand model is described on pp. 4-26 through 4-30 in Northrop, Gaylord M., William V. McGuinness, Joseph C. Reidy, and Harvey M. Katz (The Center for the Environment & Man, Inc.). *A Framework for Environmental Impact Evaluation for Electric Power Systems in a River Basin*. Washington, DC: Federal Power Commission, December 1975, 323 pp.

35. U. S. Department of Energy, Energy Information Administration. *Energy Information: Quarterly Report to Congress* [required by Public Law 93-319, amended by Public Law 94-163]. Washington, DC: U. S. Government Printing Office, quarterly: A quantification of United States resource development, production, consumption, and net imports/exports, including estimates of proved reserves, of major energy supplies (coal, petroleum, natural gas, nuclear, and electric power). *Energy Information Data Bases*. Washington, DC: U. S. Department of Energy, Technical Information Center, January 1978, 24 pp.

36. Seidel, Marquis R. *State Projections of Industrial Fuel Use*. Washington, DC: Federal Power Commission, August 1976, 34 pp. + 94 tables.

37. Federal Power Commission, *Principal Electric Facilities in the United States* [map, in color: scale, 1 inch equals approximately 55 miles; a set of 11 regional maps are available in black & white], (Washington, DC: 1977); *Principal Natural Gas Pipelines in the United States* [map in color: 1 inch equals about 55 miles], (Washington, DC: 1974); *The 1970 National Power Survey* [in four parts]. Washington, DC: U. S. Government Printing Office, 1971-1972, 1,659 pp; *The National Power Survey* [6 reports]. Washington, DC: 1973-1974, 905 pp; and the *National Gas Survey* [5 volumes]. Washington, DC: USGPO, 1973 and 1975, 2,241 pp.

38. U. S. Department of Commerce, *Energy Consumption of Environmental Controls: Fossil Fuel, Steam Electric Generating Industry*. Washington, DC: USDOC, Office of Environmental Affairs, March 1977.

39. Schurr, Sam H., Joel Darmstadter, Harry Perry, William Ramsay, and Milton Russell, *Energy in America's Future: The Choices Before Us*. Resources for the Future.

Baltimore, MD: The Johns Hopkins University Press, 1979, 555 pp.

40. Shaw, T. I. "The Status of Tidal Power," *International Power & Dam Construction* *(1978)*. 30(6):29-34.

41. Greeley, Richard S. *Meeting Energy Demands and Environmental Quality: A Case Study.* McLean, VA: Mitre Corporation Report MTP-362, 21 pp.

42. For example: Odum, Eugene P. *Fundamentals of Ecology.* 3rd edition, New York: Holt, Reinhart & Winston, 1971, 574 pp, being revised.

43. Darling, F. Fraser and John P. Milton, *Future Environments of North America.* Garden City, NY: Natural History Press, 1966, 767 pp.

44. Maryland Department of State Planning, *Integrity of the Chesapeake Bay.* Publication No. 184, Baltimore, MD: July 1972, 52 pp.

45. U. S. Department of the Interior, Fish and Wildlife Service, *National Estuary Study* [7 volumes]. Washington, DC: U. S. Government Printing Office, 1970; see especially: Volume 1 (Main Repot), 96 pp; Volume 4 (Appendices C and D), 216 pp; Volume 5 (Appendices F and F), 216 pp. Odum, H. T., B. J. Copeland, and E. A. McMahan (ed.), *Coastal Ecological Systems of the United States* [4 volumes]. Washington, DC: The Conservation Foundation, June 1974, 1,977 pp.

46. U. S. Department of the Interior, *The Preservation of Natural Diversity: A Survey and Recommendations.* (prepared by The Nature Conservancy, Contract No. CX0001-5-0110, 1975), 212 pp. + app.

47. Franklin, Jerry F. "The Biosphere Program in the United States: A Program has Been Developed to Select Key Sites for Environmental Research and Monitoring," *Science* (1977). 195:262-267. Lovejoy, Thomas. "The Conservation of Evolution," *Parks*, 4(2):23. Deacon, James E., et al. [members of the Endangered Species Committee, American Fisheries Society], "Fishes of North America—Endangered, Threatened, or of Special Concern: 1979," *Fisheries* (1979). 4(2):29-44.

48. This principal derives from form and function relationships and the general surface area law, see Davson, Hugh. "Basal Metabolism of Homiotherms," and "Surface-Area Law," *A Textbook of General Physiology.* 2nd edition. Boston, MA: Little, Brown & Co., 1959, pp 171-177.

49. Kahn, Fritz [translated from the German and edited by George Rosen], "the Blood-Vessels" and "The Blood," *Man in Structure & Function.* NY: Alfred A. Knopf, 1947, pp. 189-193 and 206-209.

50. Shuster, Carl N. Jr. "Biological Evaluation of the Delaware River Estuary," *State of Delaware Intrastate Water Resources Survey.* Dover, DE: 1959, pp. 21-1 through 21-73.

51. Shuster, Carl N. Jr. "The Nature of a Tidal Marsh," *The Conservationist* (State of New York Conservation Department, Albany, NY: August-September 1966), 21(1):22-29, 36. Gosselink, James G., Eugene P. Odum, and R. M. Pope. "The Value of the Tidal Marsh," *Louisiana State University Publication No. LSU-SG-74-03.* Baton Rouge, LA: Center for Wetland Resources, May 1974, pp. 1-30. Shabman, Leonard A. and Sandra S. Batie, "Economic Value of Natural Coastal Wetlands: A Critique," *Coastal Zone Management Journal* (1978). 4(3):231-248.

52. Goodwin, Richard H. and William A. Niering. *Inland Wetlands of the United States: Evaluated as Potential Registered Natural Landmarks.* National Park Service Natural History Theme Studies No. 2. Washington, DC: U. S. Government Printing Office, 1975, 550 pp.

53. Ellis, Melvin Y. (ed.). *Coastal Mapping Handbook.* Washington, DC: U. S. Government Printing Office, 1978, 200 pp.

54. Clark, John R. *Coastal Ecosystem Management: A Technical Manual for the Conservation of Coastal Zone Resources.* NY: John Wiley & Sons, 1977, 928 pp.

55. Hargis, William J., Jr. (Chairman, CSO Executive Committee). *Development of the Coast: Facing the Tough Issues (Final Proceedings).* Gloucester Point, VA: Virginia Institute of Marine Science, 1980, 48 pp.

56. Bardin, David J. "Remarks," (55, op. cit.), pp. 31-34.

57. Beall, S. E., et al. *An Assessment of the Environmental Impact of Alternative Energy Sources.* Oak Ridge National Laboratory. Springfield, VA: National Technical Information Service, September 1975, 125 pp.

58. Council on Environmental Quality [and other federal agencies], *Energy Alternatives: A Comparative Analysis.* prepared by The Science and Public Policy Program, University of Oklahoma. Washington, DC: U. S. Government Printing Office, May 1975, 666 pp.

59. Anderson, Ernest C. and Elizabeth M. Sullivan (eds.). *Impact of Energy Production on Human Health: An Evaluation of Means for Assessment* [Proceedings, LASL Third Life Sciences Symposium], ERDA CONF-751022. Springfield, VA: National Technical Information Service, 1976, 144 pp.

60. U. S. Department of Energy, Office of the Assistant Secretary for Environment, *Environmental Readiness of Emerging Energy Technologies: Summary Report.* Springfield, VA: National Technical Information Service, January 1979, 73 pp.

61. U. S. Department of Energy, Office of the Assistant Secretary for Environment. *National Environmental Impact Projection No. 1.* prepared by The Mitre Corporation (McLean, VA). Springfield, VA: National Technical Information Service, February 1979, 99 pp.

62. U. S. Department of Energy, Office of the Assistant Secretary for Environment. *An Assessment of National Consequences of Increased Coal Utilization.* Executive Summary, in 2 Volumes (ID-29425. Springfield, VA: National Technical Information Service, February 1979, Vol. 1, 79 pp; Vol. 2, 316 pp. The major environmental impact of replacing oil and natural gas as fuels by coal will be to air quality. On a nation-wide basis, there will be little effect on water quality. However, availability of water will be a limiting factor in coal extraction (mainly at western sites) and as cooling water (in Midwest areas).

63. Council on Environmental Quality, *Environmental Statistics.* Springfield, VA: National Technical Information Service, March 1979, 264 pp.

64. U. S. Department of Energy, Office of the Assistant Secretary for Environment. *Regional Issue Identification and Assessment: First Annual Report.* Washington, DC: September 1979, 822 pp.

65. U. S. Department of Energy, Office of the Assistant Secretary for Environment. *Environmental Data for Energy Technology Analysis.* Vol. 1, Summary, prepared by The Mitre Corp., McLean, VA: August 1979, 2nd edition.

66. Goldberg, Murray D. (ed.). *Energy, Environment and Planning: The Long Island Sound Region.* Brookhaven, NY: National Laboratory, Associated Universities, Inc., February 1973, 171 pp.

67. Dames and Moore. *A Report on Data Management for Power Plant Siting: Delmarva Interface Study.* Cranford, NJ: D&M, April 1975, 68 pp [an illustration of the use of computer mapping techniques—described by Dennis R. Smith and John H. Robinson. "Computer-Aided Siting of Coal-Fired Power Plants: A Case Study." Los Angeles, CA: D&M, undated manuscript. 24 pp].

68. *Energy Facility Siting in Coastal Areas.* prepared for the Committee on Commerce and the National Ocean Study Policy, U. S. Senate. Washington, DC: U. S. Government Printing Office, 1975, 126 pp.

69. Washington Public Power Supply System, *Siting Study* [re suitable sites in Pacific Northwest for thermal electric power generating stations]. prepared by Woodward-Clyde Consultants (December 1975).

70. Eberhart, Russell C., and Thomas W. Eagles. "Regional Energy Facility Siting Analysis," *Coastal Zone Management Journal* (1976). 3(1):71-82 [evaluates the advantages/disadvantages of two approaches to siting: area ("cell") versus specific site analysis. The "cell approach is useful in a national siting context, to screen out undesireable areas; specific studies provide in-depth analyses].

71. Morrell, David, and Grace Singer. *State Legislatures and Energy Policy in the Northwest: Energy Facility Siting and Legislative Action.* Brookhaven National Laboratory. Springfield, VA: National Technical Information Service, June 1977, 263 pp.

72. Bishop, Richard C., and Daniel L. Vogel. "Power Plant Siting on Wisconsin's Coasts: A Case Study of a Displaceable Use," *Coastal Zone Management Journal* (1977), 3(4): 363-384.

73. Maryland Department of Natural Resources, Energy and Coastal Zone Administration. *Maryland Major Facilities Study: Executive Summary.* prepared by Rogers & Golden, Inc. and Alan Mallach/Associates, January 1978, 85 pp.

74. U. S. Nuclear Regulatory Commission, Office of Standards Development. *Early Site Reviews for Nuclear Power Facilities: Procedures and Possible Technical Review Options* (DRAFT). February 1978.

75. Pennsylvania Power & Light Co. *A Model Process for Site Selection of Power Facilities.* report of the PP&L Permanent Siting Advisory Committee. Allentown, PA: September 1978, 10 pp.

76. Goodland, Robert (ed.). *Power Lines and the Environment.* The Cary Arboretum, New York Botanical Gardens. (Box 609, Millbrook, NY, 12545), 170 pp.

77. National Science Foundation. *Energy/Environment/Productivity: Proceedings of the First Symposium on RANN: Research Applied to National Needs.* Washington, DC: U. S. Government Printing Office, 1974, 251 pp.

78. Saila, Saul B. (ed.). *Fisheries and Energy Production: A Symposium.* Lexington, MA: Lexington Books, 1975, 300 pp. [papers on responses of ichthyoplankton and young/small fish to thermal, entrainment and hydrostatic stresses; use of electrical barriers and other by-pass/deterent structures; and, multiple discriminant analysis of benthic communty responses to environmental variation].

79. Environmental Protection Agency, Office of Research and Development. *Energy/Environment II.* EPA-600/9-77-012, Washington, DC: November 1977, 563 pp; *Energy/Environment III.* EPA-600/0-78-022. Washington, DC: October 1978, 386 pp.

80. Van Winkle, Webster (ed.). *Proceedings of the Conference on Assessing the Effects of Power-Plant-Induced Mortality on Fish Populations.* New York: Pergamon Press, 1977, 380 pp. [case histories; methods for estimating abundance, production and mortality rates of young fish; compensation and stock-recruitment relationships; monitoring programs and data analysis; use of population models; and, conclusions and recommendations].

81. Cairns, John Jr., K. L. Dickson, and E. E. Herricks (eds.). *Recovery and Restoration of Damaged Ecosystems.* Charlottesville, VA: University Press of Virginia, 1977, 531 pp. Cairns, John Jr. (ed.). *The Recovery Process in Damaged Ecosystems.* Ann Arbor, MI: Ann Arbor Science Publishers, Inc., 167 pp.

82. Shreeve, Daniel, Charles Calif, and John Nagy. *The Endangered Species Act and Energy Facility Planning: Compliance and Conflict.* Brookhaven National Laboratory, Associated Universities, Inc., May 1978, 51 pp.

83. Crabtree, Allen F., Charles E. Bassett, Lynn E. Fisher. *The Impacts of Pipeline Construction on Streams and Wetland Environments.* Lansing, MI: Michigan Public Service Commission, 1978, 164 pp.

84. Pister, Edwin P. "Endangered Species: Costs and Benefits," *Environmental Ethics.* 1(4):341-352.

85. Swanson, Gustav A. (technical coordinator). *The Mitigation Symposium: A National Workshop on Mitigating Losses of Fish and Wildlife Habitats.* Rocky Mountain Forest and Range Experiment Station, General Technical Report RM-65. Fort Collins, CO: U. S. Department of Agriculture, 1979, 684 pp.

86. Electric Power Research Institute. "Water: Pinch on Energy Development," *EPRI Journal.* Palo Alto, CA: October 1979, 4(8):23, 6-13.

87. U. S. Department of the Interior, Fish and Wildlife Service. *Publications of the Biological Services Program.* Washington, DC: Office of Biological Services, Information Transfer Coordination, FWS/USDOI, 20240. The Biological Services Program has produced a number of pertinent publications, from several research teams: National Coastal Ecosystems, National Energy and Land Use, and National Power Plant.

88. Keckes, S. "The Co-Ordinated Mediterranean Pollution Monitoring and Research Program (MEDPOL) in the Framework of the Mediterranean Action Plan, "*4th Journal on Studies of Marine Pollution in the Mediterranean.* Monaco: International Commission for the Scientific Exploration of the Mediterranean Sea, 1978, pp 17-25.

89. Capt. Jacques Cousteau. Luncheon address at the Coastal States Organization "Conference on Development of the Coast," reference 55, op. cit. (Charleston, SC: 28 September 1978).

90. This section is borrowed, liberally, from: Hill, Richard F. "Regional Systems Planning," *Energy Technology III: Commercialization*. Washington, DC: Government Institutes, Inc., 1976, pp 307-314. This important paper provides the rationale for, the philosophy of, and some guidelines for, energy systems planning.

91. Noble, John H., John S. Banta, and John S. Rosenberg (eds.). *Groping Through the Maze: Foreign Experience Applied to the U. S. Problem of Coordinating Development Controls*. Washington, DC: The Conservation Foundation, 1977, 165 pp.

92. The Environmental Protection Agency has adopted regulations that encourage, at the option of either applicant or agency, such a procedure (CFR 40—Section 6.908(e)92). It is interesting that representative of energy interests also have expressed an interest in the advantages of the third party contractor (Rennie, Sandra M. [Director, Office of Environmental Assessment, U. S. Department of Energy]. "Delegation of EIS When Consultants Are Used to Prepare the EIS," a speech in a course on *The Environmental Impact Statement Process Under NEPA* (Washington, DC: 8 November 1977). Harmon, Robert W. [American Electric Power Service Corporation], comments on FERC RM79-69 (see reference 11, op. cit.), letter (New York, NY: 2 October 1979), 5 pp).

93. Austin, Herbert M. "The Need for Standardization in Pre- and Post-Operational Ecological Surveys," *New York Fish & Game Journal* (1976). 23(2):180-182.

94. Odum, Howard T. *Environment, Power, and Society*. New York: John Wiley & Sons, Inc., 1971, 331 pp; Gilliland, Martha W. (ed.). *Energy Analysis: A New Public Policy Tool*. American Association for the Advancement of Science, Selected Symposium 9. Washington, DC: 1978, 110 pp.

SECTION IV

IMPLEMENTATION OF ENERGY-SENSITIVE LAND USE MEASURES: LOCAL, STATE, AND FEDERAL CONTROLS

Energy Conservation Implementation Through Comprehensive Land Use Controls

DANIEL R. MANDELKER

Introduction

THE use of time and space clearly affects energy conservation. Different residential dwelling types have different energy needs that reflect land use densities and dwelling size. People must also move through space from one area to another, and movement through space always consumes energy in some form. The density and arrangement of land development in turn affects travel times and the amount of energy travel consumes. These interdependent uses of time and space provide opportunities for energy conservation through land use controls.

The Containment-High Density Strategy

The accepted land use strategy for energy conservation rejects the low-density dispersed development pattern typical of American metropolitan areas. Critics claim that this development pattern wastes energy. Lower-density development has higher operating costs and requires longer trips to work and other destinations.

Energy strategists call for a radically different land use and density pattern. They favor an urban containment strategy that prohibits urban dispersal and contemplates concentrated development at higher densities. This strategy also reduces travel by locating employment opportunties and commercial facilities close to higher-density residential areas. Energy strategists argue that the concentrated land development pattern saves energy by reducing residential energy use and by shortening work and other trips.

Containment need not mean the incremental, continuously outward expansion of urban areas. Alternative containment patterns contemplate a restriction on urban area expansion and the dispersal of new development to distant, self-contained urban centers. The British adopted a decentralized land use strategy of this kind in their Green Belt and New Towns program, though for different reasons.

Several studies have estimated the energy savings that can be expected from contained, high-density development.[1] These estimates vary, but suggest savings up to 30 percent. These savings are marginal, and have an even smaller impact on total energy consumption because they are possible only for new development. Pre-existing development continues its energy use level, except when it is modified to save energy and when sites are redeveloped at higher densities.

The marginal energy savings achieved by contained, high-density development provide a compelling but not a sufficient reason for land use controls that can achieve this land use pattern. Policy-makers must weigh the energy benefits produced by this land use pattern against its costs. They must also consider the limitations inherent in an energy-oriented land use strategy that affect its ability to achieve energy conservation.

Costs and Limitations

1) The energy use split.   Society uses energy for a variety of purposes including automobiles, residential and nonresidential uses. An energy balance survey conducted for the Washington, D.C. metropolitan area indicated that residential use accounted for about one-half, the automobile for about one-fifth, and nonresidential use for the rest of energy use in this area.[2] These figures may not be typical, since the Washington area is not industrialized, but they do indicate that no single use dominates energy consumption.

Additional complications arise because work and nonwork trips divide energy consumption by automobiles. One statistic indicates that work trips account for about 30 percent of all trips under 30 miles,[3] although these percentages may vary by metropolitan area. The work and nonwork component of travel may require different land use strategies.

These statistics suggest that land use strategies to achieve energy conservation may affect only a comparatively small part of energy use unless they achieve concurrent reductions in more than one use sector. The containment-high density development pattern may achieve this objective. It also requires a centralized and comprehensive land use control system that can influence energy use over a wide geographic area.

2) Urban form and travel mode. Metropolitan areas assume a variety of forms and their residents use a variety of travel modes. Metropolitan areas also vary in the distribution and concentration of employment centers. The mix of urban form and travel mode significantly affects energy consumption by transportation. A concentric urban area with dispersed employment and population uses more energy for transportation than a linear city which concentrates population and employment. Travel mode also affects energy consumption. Urban areas in which people travel primarily on public transportation use less energy than urban areas in which they travel by automobile.

Concentrated urban areas may consume less energy but concentration imposes other environmental and personal costs. Pollution increases as density increases.[5] Travel by public transportation decreases accessibility because travel by this mode is less convenient and consumes more time.[6]

Deconcentration in small urban centers is an alternative that can mitigate these costs. Some studies indicate that the dispersed, deconcentrated urban development pattern holds the most promise for energy conservation.[7] Because it loses the economies and amenities of scale, this urban pattern also imposes costs on its inhabitants. A deconcentrated dispersal policy also requires a centralized and comprehensive land use control system. The British land use control system fits this model; the American system does not.

3) The costs of high density. High-density development, whether concentrated or not, also imposes personal costs. An increase in pollution is one, and there are others. Higher densities often contemplate multi-family apartment development. Proponents of higher-density multi-family development as an energy conservation measure often overlook the reduction in housing amenities this development can impose. Dwelling units in higher-density developments are often smaller than single-family residences, especially in high-rise structures. Higher-density development also provides less on-site amenity. Residents may not view common recreation areas as an adequate substitute for private, on-site open space. Urban planners, whether energy-oriented or not, decry dispersed, single-family living patterns and generally favor higher-density residential development. This preference may be laudable, but society may not be willing to accept the reduction in housing amenity that higher density brings if marginal energy savings are the only benefit.

A St. Louis study[8] indicates that consumers of housing may decide to accept alternative costs to reduce energy use. That study showed that carpooling can achieve substantial transportation energy savings. Even though carpooling imposes time and convenience costs on travelers, housing consumers may prefer these costs to the amenity costs of higher-density living.

The tendency of higher densities to increase pollution of all kinds also complicates the choice problem. Housing consumers may prefer high-energy consumption costs as an alternative if the land use pattern necessary to save energy increases pollution. The energy use split also affects policy-making on this cost-benefit trade-off. Higher densities increase pollution from residential and nonresidential uses, but can reduce automobile pollution by reducing travel journeys. Conversely, lower densities can reduce pollution from land use through dispersal but will probably increase energy use by transportation.

Society has placed a high priority on pollution control. It must decide whether to curtail this environmental priority for an alternative environmental priority that emphasizes energy conservation that may increase pollution levels. This decision requires the careful calibration of land use, energy conservation and pollution control programs. It also requires the further centralization of planning and land use control systems and more extensive control of private choice in the land market.

This discussion does not imply a rejection of the containment-high density land use pattern. Quite the contrary. Compelling reasons support this land use strategy. The point to make is that a social decision to accept this strategy cannot be based solely on the energy-conservation objective. To structure choice in this manner may lead to a rejection of this strategy should society decide that its energy benefits are not worth its other costs.

Neither does this discussion imply that land use controls have no place in the energy-conservation effort. It does imply that energy-oriented land use controls must be aware of the leverage points it can use to reduce energy consumption through a modification of land use patterns. The energy use split and the influence of urban form and travel mode on energy suggest some of these leverage points.

Energy-Oriented Land Use Control Systems

The discussion so far has concentrated on the containment-high density land use strategy for achieving energy conservation. This land use strategy requires a highly centralized, highly restrictive land use control system. Society must consider alternative systems for achieving energy efficiency in land use if it finds this system intolerable. How much centralization and control society will accept becomes another variable to consider in selecting a land use control system to reduce energy use.

A tentative hypothesis is possible: Land use controls for energy conservation become more centralized and comprehensive as more of the energy interdependencies in land use and transportation are brought within the control system. The dispersed, deconcentrated land use strategy illustrates this point. As in the British system, this strategy requires controls to restrict the incremental growth of urban areas. It also requires controls and major development programs for dispersed, newly developed urban centers. Control over urban growth throughout entire metropolitan regions is necessary.

The deconcentration strategy is not easily accomplished. Even the British dispersed only a relatively small part of new urban growth to their New Towns. In this country, the failure of the federally sponsored New Towns illustrates the frustrations a deconcentration and dispersal program can encounter.

The first land use control system described here, which can implement the containment-high density strategy, requires comprehensive control over land use and transportation needs that affect energy consumption. It requires a high degree of centralization and regulation as well. The next two systems require less centralization and regulation, but also achieve less control and conceivably a smaller reduction in energy use. They do not necessarily contemplate a containment-high density strategy throughout metropolitan regions.

1) *The highly centralized, highly regulated system.* This land use control system contemplates a centralized agency, either at the state or regional level, that can adopt and implement land use plans. The system would include all energy users. It would also include planning and implementation powers that can structure land use and transportation needs to maximize energy savings.

The powers included within the system would extend beyond traditional land use planning and regulation to include transportation planning and the use of transportation facilities. The centralized agency should also have access to federal and other governmental assistance to strengthen and improve public transportation. Control over automobile use is also necessary. Few state and regional agencies presently possess all these powers, although the Metropolitan Council in the Twin Cities, Minnesota, metropolitan area comes close. A structure for combined transportation and air quality planning exists in the federal air quality and highway programs, but this planning process concentrates on transportation planning to improve air quality. As noted earlier, pollution control may conflict with energy conservation.

A few planning and land use control systems incorporate energy-conservation or related development criteria in their land use regulation programs. The Oregon state planning system is an example.[10] Under the Oregon planning law, a state Land Conservation and Development Commission adopts statewide planning goals to guide planning by local governments. The goals are not a state land use plan. They contain planning criteria local governments must follow in their local land use plans. The state Commission reviews local land use plans for consistency with the statewide planning goals.

One of the statewide planning goals contains an energy conservation directive:

> Land and uses developed on the land shall be managed and controlled so as to maximize the conservation of all forms of energy, based upon sound economic principles.[11]

This planning goal is elaborated in additional guidelines for local government planning. The guidelines include the following directive:

> Land use planning should, to the maximum extent possible, combine increasing density gradients along with capacity transportation corridors to achieve greater energy efficiency.[12]

This planning guideline appears to contemplate the containment-high density strategy. It encourages high density development and selects the linear urban form as its energy-saving development option. This urban form is not necessarily the most efficient for energy consumption.[13]

The Hawaii Land Use Law[14] also authorizes a state land use program that incorporates the containment-high density strategy. That law created a state Land Use Commission with authority to divide the entire state into three non-urban and one urban land use district. The Commission exercises control over new urban growth through its power to reclassify any of the non-urban districts to an urban district classification.

The legislature enacted an Interim Statewide Land Use Guidance Policy to guide the Land Use Commission in its decisions to reclassify non-urban land to the urban district. This policy does not encourage energy conservation, but does contain the following guidance that contemplates a contained urban development pattern:

> Maximum use shall be made of existing services and facilities, and scattered urban development shall be avoided.
> Urban districts shall be contiguous to an existing urban district or shall constitute all or a part of a self-contained urban center.[15]

These policies control the location of new urban development but do not directly control density. Hawaii local county governments control density through their zoning ordinances and the Land Use Commission usually considers density when it reviews applications for urban district reclassifications. A legislatively adopted state land use plan has now incorporated the Interim Land Use Guidance Policy in its planning guidelines.[16]

Mandatory comprehensive planning is necessary in a system that includes all energy users and that takes advantage of land use interdependencies to reduce energy consumption. The Oregon legislation mandates local planning and the Hawaii state planning legislation likewise contemplates local plans. Only planning can produce the comprehensive land use policies necessary to achieve energy conservation. As in Oregon, the energy-oriented land use plan can provide the basis for land use controls that implement its energy policies. A requirement that land use controls be consistent with the plan can accomplish this objective.

The Hawaii and Oregon legislation illustrate the point that comprehensive land use strategies for energy conservation require centralized planning and land use control systems. These systems run risks common to all attempts to centralize governmental power. They can be comprehensive only if they are all-inclusive, but attract political opposition just for this reason. They can reduce opposition if they include fewer powers but they may then be ineffective.

The Oregon goals and guidelines also illustrate the risk in energy-oriented land use policies whose energy-saving potential is untested or whose costs are not fully known. Their preference for linear, transit-oriented development as the land use pattern best able to reduce energy use is arguable. The Hawaii Interim Land Use Guidance Policy permits more flexibility.

Flexibility in the development of an energy-oriented land use policy is necessary. Centralized planning agencies could undertake a "balancing process" in which they consider a variety of costs and benefits in the formulation of an energy oriented land use policy. A "balancing process" permits discretionary decision-making by land use control agencies that can have major consequences for land development. This extensive delegation of power is another cost society may have to accept if it pursues energy conservation through highly centralized decision-making systems.

2) *A less centralized, less inclusive system.* A review of the leverage points for achieving energy conservation through land use control suggests that alternative systems may be possible. The point can be made by considering the transportation component of energy use, which is split between work and nonwork trips. In a mobile society land use controls cannot easily arrange residential and employment locations to minimize work journeys because workers commute to jobs throughout metropolitan areas. Work journeys are also shorter in American metropolitan areas than many assume. The potential for more reduction in work journeys is problematic.

Land use controls can more easily influence nonwork journeys. These journeys include journeys to shop, journeys to school, and journeys for recreational and social purposes. Residential uses are more easily linked with commercial, educational and recreational facilities to reduce nonwork journeys. More important, the planning and control powers necessary to link these uses may not require centralized agencies. Nonwork journeys often take place within a geographically limited area that falls within a single governmental jurisdiction. Controls over land uses affecting nonwork journeys can also be less comprehensive than controls that include all energy users in the regulatory program.

Growth center policies common to many local land use planning programs can encourage land use linkages that reduce nonwork journeys. These policies locate facilities attracting nonwork journeys at important transportation nodes, such as stations on public transit lines or highway interchanges. These facilities can be scaled to fully serve the commercial, educational and recreational needs of the surrounding area. A planning policy that concentrates high-density residential uses adjacent to the growth center also can reduce nonwork trips.

The Montgomery County, Maryland, ordinance authorizing high density, mixed-use zones near transit stations illustrates a growth center planning policy.[17] It requires a plan of development for these zones that contemplates multi-family residential densities "for use in locations within walking distance of the transit station," together with associated commercial uses.

Local planning and land use control powers are usually adequate to implement a growth center policy of this kind. Controls must be tightened to prevent high-density residential and commercial development at points outside these growth centers, but this problem is political. High-density growth center policies also contemplate intensive land uses that impose negative external costs on adjacent lower-density residential areas. Residents in these adjacent areas may then object. This problem is also political.

Planners should not dismiss political problems casually. They indicate that even a more limited land use control program to reduce energy consumption requires a trade-off between costs and benefits. Local governments may not be able to overcome the political opposition these programs provoke. A centralized agency can be more resistant to local pressures. It can take the energy needs of an entire urban area into account, and can more fairly distribute costs and benefits because its jurisdiction is comprehensive.

3) *Site-specific review for energy impact.* A final land use control system avoids many of these problems. It dispenses with a land use policy and provides for the review of individual developments to assess their energy impact. This system assumes that site-by-site development review will in the long run achieve an energy efficient land use pattern.

This system has many advantages. Local governments can review the energy impacts of new development under statutory criteria that guide the development approval process. A land use plan that designates future land uses is not necessary. The system can include but does not necessarily require decisions on land use locations.

Several models exist for this kind of site-specific review system. The National Environmental Policy Act and counterpart state legislation require the preparation of environmental impact statements. These statements disclose the environmental impact of major governmental actions significantly affecting the environment. California and a few states have expanded the environmental impacts discussed in impact statements to include energy impacts.[18] A few states, including California, also have extended the impact statement requirement to private land development that requires zoning and other local approvals. California environmental impact statements must therefore discuss the energy impacts of private land development.

Impact statements are not really a land use control. They merely disclose environmental impacts at the development site. While the issue is not settled, a land use control agency (except in Washington) cannot rely on negative environmental impacts discolosed by an impact statement to disapprove a development project. Energy impacts disclosed by impact statements can still affect land use. Impact statement legislation usually requires the impact statement to discuss alternatives, including alternative locations. These alternatives in states that require discussion of energy impacts presumably include alternative locations that reduce energy use.

Environmental impact statement legislation that requires discussion of energy impacts also can compel attention to energy conservation measures at the development site. The California impact statement legislation provides an example. It states that the impact statement must contain—

Mitigation measures proposed to minimize the significant environmental effects including but not limited to, measures to reduce wasteful, inefficient and unnecessary consumption of energy.[19]

A few states have enacted legislation creating state land use control programs that require the review of environmental impacts. They explicitly provide authority to disapprove development projects. One example is Florida's program authorizing state level review of Developments of Regional Impact (DRI). This legislation is based on the American Law Institute's Model Land Development Code.[20] In Florida, DRI is a development control that has an effect on more than one county.

Under the Florida legislation, the developer files an application for approval of a DRI with the local government. The regional planning agency then prepares a "report and recommendations" on the DRI which includes consideration of a number of impacts, some of which are environmental. In part, the regional planning agency must determine—

. . . the extent to which the development would create an additional demand for, or additional use of, energy, provided such criteria and related policies have been adopted by the regional planning agency.[21]

An appeals from a local DRI decision can be taken to a state adjudicatory agency.

The Florida legislation may authorize disapproval of a DRI if its location or density increases energy demand. Conceivably, a residential development qualifying as DRI can be disapproved if its density is too low, on the assumption that higher-density development can save energy. In this sense, Florida DRI review may be a land use control.

Another state land use control program that provides for an environmental impact review is the Vermont state development permit law, known as Act 250.[22] This law provides for the review of all major development in the state by a state environmental agency, the District Environmental Commission. One of the criterion to be applied in the review process states:

A permit will be granted when it has been demonstrated by the applicant that, in addition to all other applicable criteria, the planning and design of the subdivision or development reflect the principles of energy conservation and incorporate the best available technology for efficient use or recovery of energy.[23]

Like the California energy impact requirement, this criterion appears limited to energy impacts associated with the site.

A District Environmental Commission has now given the energy impact criterion this interpretation. In the *Pyramid Mall* case,[24] a developer applied for an Act 250 permit to build a large shopping mall several miles from the City of Burlington. Opponents urged the Commission to deny the permit because traffic generated by the mall would induce substantially more fuel consumption than would occur if the mall were not constructed.

The District Commission rejected this interpretation of the statute:

> Although the "planning" of a development might well be deemed to include its location as affecting the amount of energy it will consume, we do not believe that the legislature intended us to consider the effect of energy consumption by customers.[25]

Energy impact review can be useful in energy-conservation programs because it provides a method for assessing site treatment and other development features that affect energy use. Local governments also may implement an energy impact review program without the advance planning and controls for energy-efficient land use that can attract controversy. Energy impact review also has disadvantages. It is less easily applied to modify development density and location to improve energy consumption. It is also applied on a case-by-case basis, and may not produce a satisfactory land use and density pattern that achieves the greatest energy savings.

Conclusion

It is seductive to treat each new and urgent social policy objective as a mandatory imperative. Environmental quality protection commanded this kind of social priority not so long ago. Environmental protection, like energy conservation, also attempted land use and transportation strategies to meet its objectives. Policy-makers quickly learned that social preferences and social inertia represented by prevailing land use patterns change slowly, if at all. A congressional delegation to the Federal Environmental Protection Agency to require land use controls in air quality programs was quickly withdrawn.

Energy-oriented land use control programs can avoid these mistakes if they properly assess the complex trade-offs these programs require. The energy use split problem and the marginal savings that land use control programs can achieve in energy use suggest caution. Trial-and-error experimentation is necessary. Limited controls such as site-specific energy reviews and growth center policies that local governments can implement with minimal resistance should receive priority.

One encouraging finding is that energy efficiency requires a containment-high density land use strategy many policy makers have urged all along. Energy conservation provides a powerful but not the only argument for this strategy, which also serves many other worthwhile social objectives. Energy-oriented land use controls will succeed best if they give energy conservation a preferred but not a dominant position among the many objectives an informed land use policy must consider.

Notes

1. Harwood, C. *Using Land to Save Energy.* 1977, p. 11.
2. Roberts. "Energy and Land Use: Analysis of Alternative Development Patterns," *Environmental Comment.* September, 1975, pp. 2, 4.
3. Wallace, N. and A. Kahn. *Conservation of Fuels in the St. Louis SMSA: Alternative Residential Land Use Patterns and Transportation Strategies for 1995.* 1978, p. 21.
4. Edwards. "The Effect of Land Use on Transportation and Energy Consumption in Energy and the Community," R. Burby & A. Bell eds., 1978, pp. 47, 51.
5. Windsor. "A Critique of 'The Costs of Sprawl,' " *Journal of the Amer. Planning Ass'n, 45.* 1979, pp. 279, 289.
6. Edwards, p. 59.
7. Edwards, p. 52; Van Til. "Spatial Form and Structure in a Possible Future, *Journal of the Amer. Planning Ass'n 45.* 1979, p. 318.
8. Wallace and Kahn, p. 26.
9. D. Mandelker. *Environmental and Land Controls Legislation.* pp. 55-60 (1976) and 14-19 (Supp. 1978).
10. Oregon Land Conservation and Development Commission. *State-Wide Planning Goals and Guidelines No. 13.* 1975.
11. Id.
12. Roberts, p. 7.
13. Mandelker, Ch. VII.
14. Hawaii Rev. State § 205-16.1 (1976).
15. E.g., id., § 226-104(c) (2) (Supp. 1978).
16. Reproduced in Harwood, pp. 266-77.
17. Harwood, p. 194.
18. Calif. Pub. Resources Code § 21100(c) (Supp. 1979). See *People v. County of Kern*, 62 Cal. App.3rd 761, 774, 133 Cal. Rptr. 389, 398 (1976), ("The energy mitigation amendment is substantive and not procedural . . .").
19. Mandelker, pp. 110-17 and pp. 23-27 (Supp. 1978).
20. Fla. Stat. Ann. § 380.06(f) (Supp. 1979).
21. Mandelker, Ch. VIII.
22. Vt. Stat. Ann. tit. 10, § 6086(F) (Supp. 1979).
23. In re: Pyramid Company of Burlington, Vermont Dist. Envt'l Comm'n #4, Application #4C0281, Findings of Fact and Conclusions of Law—Order Denying Land Use Permit (Oct. 12, 1978).
24. Id. p. 37.

Managing Energy Conservation Under Planned Growth

ROBERT H. FREILICH

Background: The Process of Growth in the United States

THE process of growth in America in the post-war era, the so-called urbanization phenomenon, is composed of simultaneous *centrifugal* and *centripetal* forces. At the same time that moderate-to-middle income, white families developed the economic capacity to seek single-family suburban homes, the cities suffered increased inmigration of poor, minority households from rural areas. The net effect of these population shifts was to leave the central city with severe housing, educational and fiscal problems, while fostering environmental and economic pressures in the rapidly growing suburbs.

Governmental actions are associated with both metropolitan phenomena. In the past three decades, federal and state programs have attempted to address the dysfunctions of older urban centers. To date, these have not succeeded in creating a viable urban core—Urban Renewal, Model Cities, NDP, Neighborhood Preservation, CDBG, and UDAG strategies have experienced serious shortfalls insignificantly improving the quality of the urban environment.

Governmental action has been more intimately associated with; in fact, was an important catalyst to, the suburban metropolitan dynamic. The policies of the Federal Housing Administration (FHA) during the 1930s and 1940s provides a case in point. The single greatest racially discriminatory program in the history of the United States was that of Federal Housing Administration during the period 1934 to 1948. One could count on the fingers of two hands the number of black families that actually obtained FHA loans during that period. In fact, it is well known that the FHA insisted on racially restricted conveyance in most of the developments which it helped to finance. This practice, curtailed in 1948 when the Supreme Court declared this type of action unconstitutional, was an important government action fostering the movement of white, middle class families out of cities to subsidized, suburban locations in the post-war era.

The same type of suburban push and urban bypass characterized the Federal Interstate Road system. Choice industrial and commercial establishments were pulled from the urban core via circumferentials and connectors to major new highways, while city road systems lost their users and ultimately their function. A federal income tax system which encouraged new construction as opposed to rehabilitation contributed to the abandonment of aging urban housing stocks and nonresidential plants. The fragmentation of governments into COGs, counties, regional school systems and special districts also facilitated outward movement by making it fiscally beneficial (lower taxes) to seek peripheral locations.

Favorable *energy* factors were a further prop to metropolitan sprawl. The real price of energy, both for space heating and transportation, declined for most of the post-war period. In addition, energy availability was never an issue—the different energy producers competed aggressively for the consumer's dollar. Households/jobs could move to suburban and even rural locations, despite the concomitant higher energy consumption of such decentralization, because energy was both inexpensive and plentiful. In contrast, the urban population and economic base declined because the city's energy advantage meant little in an era of energy abundance.

The Consequences of Metropolitan Decentralization

It would be comforting to think that the new energy realities of rising costs and uncertain supply would reverse the metropolitan development pattern of suburban push and urban bypass. This energy-land use linkage ignores, however, the fact that *energy is but one of six major crises in American society* which have come to the fore in the past two decades as a result of insensitive urbanization policies. These are discussed in chronological order below:

1. *Central city decline.* The riots of the mid-1960s sparked attention to the problems faced by central cities. Their housing stock was old/deteriorating and growing portions were becoming abandoned. The flight of the cities' white, middle-class population continued, thereby aggravating divisions along racial/social lines. Crime and other social pathologies assumed alarming proportions. Cities also faced a loss of economic functions as their traditional industrial, warehousing and retail activities became functionally obsolete. These urban ills evoked the Great Society and subsequent urban revitalization programs of the 1960s and 1970s.

2. *Environmental degradation.* The late 1960s saw growing sensitivity on the part of the public and government to ecology and the harmful consequences of despoiling the environment. Rachel Carson's *Silent Spring* and other early environmental studies were rediscovered. Federal legislation was enacted protecting against air, water, and noise pollution and harmful development in the coastal zone, flood plains and other fragile areas. The federal initiatives were followed by state and local legislation adding further environmental safeguards.

3. *Energy shortfall.* The United States became an oil importing nation in 1958 and OPEC was formed in 1960. It was not until the 1973 OPEC oil embargo, however, that a sense of energy crisis took hold. Ensuing shortages over the next few years confirmed the nation's energy vulnerability and evoked a rash of federal, state and local efforts to reduce energy consumption and encourage expansion of domestic/renewable energy sources.

4. *Fiscal insolvency and tax revolts.* The mid 1970s saw New York, Detroit, Cleveland, Oakland and other major cities experience near or actual financial bankruptcy. Their financial problems were the result of overspending and a decline in the cities' economic fortunes as a result of rampant metropolitan decentralization. Suburbs also experienced a growing financial burden as a result of having to rapidly provide expanded public services and new capital facilities. Local financial problems, in cities and suburbs, were aggravated by a concurrent taxpayer revolt. Taxpayer resistance to rising expenditures and public charges resulted in the adoption of Proposition 13 (California), Proposition 2½ (Massachusetts), and other public revenue/spending curbs.

5. *Agricultural land consumption.* In the late 1970s attention turned to the significant loss of agricultural land—an important natural resource, especially in an era when United States' agricultural exports were needed to offset its energy imports. The 1977 Soil Conservation Society Conference in Omaha, Nebraska revealed a shocking statistic—approximately one percent of the nation's total agricultural land was lost each year. The nation's farmland was being encroached by continuing urban sprawl.

6. *Housing costs and inflation.* The most current crisis is an inflationary

spiral affecting the price of all goods but having an especially severe impact on the cost of housing. In years past, only the poor were denied homeownership. Today, inflation has priced out much of the middle-class as well.

In short, energy is but one of the six crises which together vie for the attention of American society. Addressing the energy shortfall must therefore be done in a manner complimentary to these other concerns. Promising energy sources/techniques, such as surface coal mining, coal gasification, and shale-to-oil conversion, must be developed without significantly degrading the quality of environment, destroying agricultural land, or driving up the price of housing. Evaluation of alternative energy sources/techniques must also consider urban impacts: namely, how adoption of a specific energy strategy will affect the relative ability of cities to attract new households and economic activities. Energy conservation standards for housing should be developed by weighing the standards' energy saving potential against their effect on housing cost/affordability. Other energy-land use linkages must also be considered.

The same sensitivity is necessary in addressing the other crisis areas. Environmental standards should not unnecessarily add to housing prices nor compound the threat to agricultural land. Belt-tightening to reduce public expenditures must not leave stillborn, urban revitalization programs. Such complementarity is fostered by implementing *growth management*—an approach considering and dealing with the many forces affecting metropolitan development.

Development of Growth Management Strategies

Planning and zoning in the United States since the 1920s, have primarily been two-dimensional. These activities were essentially a process allocating residential, industrial, and commercial opportunities and then allowing development of these opportunities, regardless of facilities, regardless of timing, regardless of sequence. In short, traditional planning and zoning ignored the third dimension of *time*—these controls did *not* govern *when* development would be most opportune.

Growth management, in contrast, incorporates this missing dimension. Its scope is illustrated in the first such effort—the Ramapo Growth Control Program instituted in 1970 in Ramapo, New York. This community was a prototypical, rapidly growing suburb that was developing faster than its capacity to provide public services/facilities. The Ramapo Growth Control

Program was a comprehensive approach to deal with this service demand-capacity shortfall. It developed the concept that by preparing both a comprehensive plan as well as a long range capital improvement program, and tying development timing and sequencing to the constraints/opportunities identified in these basic documents, there would be achieved a rationally paced, sequence of development.

The Growth Control Plan was challenged as an unconstitutional taking of property. Ramapo's strategy was upheld in a landmark decision by the New York Court of Appeals in *Golden v. Planning Board of the Town of Ramapo*. The significance of the decision is not the specifics of the Ramapo strategy itself, but primarily the affirmation of a comprehensive planning approach to controlling development.

GROWTH MANAGEMENT APPLIED AT THE REGIONAL, STATE AND FEDERAL LEVELS

Ramapo was a *local* effort at comprehensive growth management. The strategy has subsequently been extended to a *regional* level. Examples include growth management schemes developed in San Diego, California; Lexington, Kentucky; Baltimore County, Maryland, and the Twin Cities area. To illustrate, in the last region, the Metropolitan Council developed a growth policy designed to stem the increasingly costly sprawl in the Minneapolis-St. Paul Region—a dispersion evident from the following figures. The first million population in the area consumed 180 square miles of land. The second million settled an additional 600 square miles of land. And the third million, which they were into in 1970s at current zoning, would have consumed another 1,600 square miles of land. The Twin Cities Growth Management Plan attempted to reverse such sprawl by encouraging growth in the urban center and close-in suburbs and discouraging continued spread to the outer, agricultural counties of the region.

In addition to these regional planning efforts, some growth management initiatives have been attempted at both the *state* and *federal* level. Massachusetts has issued a policy statement, *City and Town Centers: A Program for Growth* emphasizing the need for metropolitan compaction as opposed to decentralization. A similar approach was followed by California in its Urban Development Strategy. Other states, including Colorado, Hawaii and Oregon, have adopted important land use control/review procedures. Oregon, for example, established a State Land Conservation and Development Commission (LCDC) empowered to establish land use goals to be adhered to by local communities. These goals include: "preservation of agricultural lands," "energy conservation," and an urbanization goal requiring the "establishment of growth boundaries separating rural from urbanizable land."

Certain *federal* actions regarding comprehensive planning strategies also bear mentioning. Numerous federal reports have recommended comprehensive regional planning to deter further metropolitan sprawl. The 1968 Douglas Commission urged that "the prevention of urban sprawl should qualify as a valid public purpose justifying the use of valid zoning and timing regulations." A decade later, the President's Urban and Regional Policy Group (URPG) *Report* similarly urged actions encouraging development in settled, urban areas as opposed to the metropolitan fringe. The 1978 URPG *Report* was followed by numerous implementation measures to foster the revitalization of the urban core. The actions regarding the placement of shopping centers received the greatest notice. A federal review mechanism—Urban and Community Impact Analysis*—was established to evaluate the urban effects of planned, regional, suburban shopping centers. These shopping facilities were singled out for analysis because of their "magnet" influence in attracting retail, service and other important economic functions, and ultimately jobs and people, from urban areas.

The Carter Administration took another action to deter sprawl. In 1979, the Farmers Home Administration (FmHA) changed its underwriting criteria to take into account impacts on agricultural land. Henceforth, FmHA would not insure/grant loans for projects involving the loss of such lands. The important point in considering these and the other federal actions described above is not that they are so far reaching, but, rather that they represent a new federal awareness of the need for and the benefits of metropolitan growth management. This new attitude is a marked departure from the activities of the FHA, Interstate Highway System and other federal actions which were described previously as instrumental in fostering sprawl in the post-war era.

Components of Growth Management

Growth management, at whatever level of government, has the *common goal of correlating all aspects of growth-related policies comprehensively*. This, in turn, involves the following activities:

---

* Subsequently curbed under the Reagan administration.

1. Differentiating the various parts/functions of the metropolitan area. Most metropolitan regions can be functionally distinguished into four geographic areas or tiers: Working from the urban center outward: Tier I consists of the retail, office, governmental and cultural uses in the urban, central business district. Tier II includes the residential neighborhoods in the central city and older suburbs. Tier III encompasses the suburban fringe where most development is occurring in a leap frog fashion. Tier IV consists of rural and agricultural areas which are premature for active, intense development.

This is the stage of planning in which the energy efficient nodes of the metropolitan area could be designated. Depending upon the level of metropolitan development this designation could fall in Tier I (free standing cities) or between Tiers II and III metropolitan-oriented cities. Energy efficiency designation directly affects the growth management policy to be employed in each of the tiers.

2. Developing appropriate management strategies. Growth management sets different policies for inclusive metropolitan locations. It *encourages* growth in Tiers I and II—areas serviced with water, sewer, road and other public facilities and having substantial amounts of land (especially in Tier II) for development because many serviced, choice locations were bypassed by the outward leap frog development. The potential for further development in the inner metropolitan rings is illustrated in San Diego. This community adopted a *Growth Plan* in 1972 to encourage growth near its center. At the time, the *Plan's* detractors argued that such concentration would stifle growth. This fear proved groundless. In 1979, San Diego experienced a record building year—16,000 building permits were issued, over half of which were located in the downtown area.

In contrast to compaction near the metropolitan center in Tiers I and II, growth management *discourages* development in the outer rings. The Metropolitan Council's Growth Policy, for example, attempted to significantly reduce the pace of construction in the outer counties of the Twin Cities region (Tiers III and IV).

The tier designation might also affect or be affected by energy supply characteristics. Tier I might be an appropriate site for land use designations that emphasize cogeneration or recycling forms of energy, or those that provide storage and transportation sites for coal; Tiers II and III might prove ideal locations for rooftop and land-sited solar energy installations; whereas Tier III could provide remote and protected locations for central station electric generation or for nuclear facilities.

3. Utilizing multiple implementation tools. Growth management directs the locus and pace of development through various tools and strategies.

Growth can be fostered in the inner rings by offering special development *incentives*. Density bonuses could be accorded to developers to attract them to prime in-fill sites. In addition, or as an alternative, capital improvement extractions could be reduced/eliminated in promising Tier I and II areas. In contrast, growth can be discouraged by posing *development restraints/disincentives* in the form of large minimum lot sizes and extensive subdivision infrastructure requirements. Where appropriate, such as in Tier IV, fragile agricultural land could be removed from development pressures by the transfer of development rights or other compensation techniques.

This activity of growth management is where the legal hardware of energy sensitive land use can be integrated and coordinated with existing controls. Zoning modifications, creative subdivision siting controls, solar access laws, energy supply-enhancing ordinances could be planned for zones of differing development intensity. Energy considerations could serve as an element of the rationale as to why certain lands may be developed currently while others must await a future use.

4. *Promoting intergovernmental coordination.* Growth management typically entails coordination by local, county regional and other levels of government. For instance, the county might review the conformity of local land use controls to the land use goals adopted by a regional or state planning board. The literature on growth management has viewed such coordination as desirable in theory but problematical in practice given differing interests and constituencies. The experience of the growth management plans which are in force belies these apprehensions. Implementation of growth controls in the Twin Cities, and other areas shows that the various governmental bodies have been able to work together to achieve a balance of regional objectives while allowing sufficient flexibility for local input and variation.

Here, perhaps most importantly of all, are demonstrated the interrelationships of energy considerations and growth management. Successful growth management is dependent on intergovernmental cooperation. So too is energy-sensitive land development. The interrelationships of governing bodies generating development goals, environmental quality standards, building codes, and zoning/subdivision regulations is critical to the development of energy conservation measures that save energy rather than enhance intergovernmental bureaucracies.

Conclusion

Growth management requires a significant planning commitment. It necessitates: recognition of the varying needs and functions of different areas within the metropolitan region, development of appropriate growth strategies for each area, provision of effective implementation tools, and coordination by different governmental bodies. These activities, involving decisions regarding ends and means, in turn, raise dilemmas of their own. How are different values—energy conservation, housing costs, environmental preservation, urban revitalization etc.—to be balanced in deciding upon a regional development strategy? How should long-range planning be effected in the face of uncertainties regarding the future? The one factor of energy for example, involves unknowns concerning future energy *supply* from renewable and non-renewable sources, future energy *costs* and the actual energy *savings* to be realized from different land use patterns.

While one must recognize the inherent difficulties and dilemmas of long-range, comprehensive growth management, the benefits make it compelling. It affords growth opportunities while protecting the environment and endangered agricultural land. By encouraging compaction, the strategy acts as a positive support to urban revitalization efforts. Growth management also has a fiscal benefit. Numerous studies considering the growth management systems in San Diego, Baltimore, Lexington and other areas indicate very clearly that there are substantial fiscal savings in the form of public services and capital improvements achieved by managed growth as compared to trend, sprawl development. Last, but not least, growth management also provides an energy benefit. While there is some uncertainty concerning the exact energy implications of different land use patterns, there is little doubt, based on both the studies to date and intuitive knowledge, that growth management—encouraging development compaction in multiple locations in the metropolitan area—can reduce energy consumption. Comprehensive, growth controls challenge the planner's skills and knowledge but offer the promise of ordered, efficient and well balanced metropolitan development.

References

1) Einsweiler, Freilich et al., *The Design of State, Regional and Local Development Management Systems,* National Science Foundation 2 vols. (1978).
2) Freilich. "Development Timing, Moratoria and Controlling Growth." *Institute on Planning, Zoning and Eminent Domain* 147 (1974).
3) Freilich and Greis. "Timing and Sequencing Development: The Ramapo Plan." in *Future Land Use,* eds. Robert Burchell and David Listokin, Center for Urban Policy Research, Rutgers (1975).
4) Freilich and Stuhler. *The Land Use Awakening,* American Bar Association (1981).
5) Freilich and Ragsdale. "Timing and Sequential Controls - The Essential Basis for Effective Regional Planning." 58 *Minn. L. Rev.* 1009 (1974).

Energy-Conserving Development Regulations: Current Practice*

DUNCAN ERLEY AND DAVID MOSENA

Introduction and Findings

ALMOST every aspect of land development has an effect on energy use, from minute architectural details to broad considerations of urban density. Energy efficiency depends in part on how development is planned and carried out. Conventional development regulations, such as zoning ordinances and subdivision regulations, can be adapted in many ways to promote energy conservation at the community level.

This report is about energy-efficient site and neighborhood design. It examines recent experiences of local governments that have adopted new development regulations or amended existing ones to promote energy conservation, more efficient generation and distribution or a switch to alternative, renewable energy sources. Although much has been written in recent years about saving energy through community design, our actual experience in applying these new ideas is still limited. To date, most communities have focused their efforts on studying the problem, documenting consumption

* This chapter is based on research conducted under the aegis of The Community Energy Program, Community Systems Division, Office of Conservation and Solar Energy, U.S. Department of Energy.

patterns, and writing reports and plans. But a handful have amended their land use controls for the express purpose of saving energy.

This "first generation" of energy-conserving development regulations offers a wide range of examples and approaches. Some are quite ambitious; others address energy conservation in terms of minor, simple adjustments. They illustrate how energy use is reflected in virtually every aspect of land use, from regulations on clotheslines to street widths. These early cases are important to the rest of the country, because they offer initial guidance and direction for other local governments to follow.

This study has identified 13 communities which have adopted energy-conserving development regulations, after undertaking a survey of 1,426 local, regional and state planning agencies.[1] It takes a look at their experiences to learn what they have done, how well it has worked, and what problems they have had.

PURPOSES OF REGULATIONS

The regulations identified addressed four major areas of energy use: space heating and cooling, transportation, energy consumed in construction materials and processes ("embodied energy"), and alternative energy sources and systems.

Reducing heating and cooling needs is the most common objective of the regulations. This is accomplished primarily through energy-efficient site design practices such as solar orientation or landscaping. Seven communities adopted such regulations.

Promoting the use of alternative energy was the objective of adopted regulations in six communities - four communities promoted the use of solar energy, one encouraged the use of geothermal energy, one promoted use of clotheslines.

Reducing the dependence on automobile transportation was the objective of adopted regulations in three communities. This was encouraged through the provision of incentives for energy-efficient location of development, and through encouraging expanded use of home occupations.

Reducing the consumption of energy in construction material and processes was encouraged by one community by reducing street width standards. Two other communities have proposed such width reductions.

TYPES OF PROVISIONS

Most of the regulations represent adaptations of conventional development controls. Rather than designing radically new regulations, communities incorporated a variety of energy objectives into existing zoning, subdivision, site plan review, landscaping regulations and incentive provisions. The regulations represent three levels of action:

Removing regulatory barriers to energy-conserving practices. The reduction of street-width standards is an example of this level of action.
Encouraging energy-conserving practices. Providing incentives such as density bonuses for protecting solar access or encouraging voluntary energy-conserving practices by developers are examples.
Requiring energy-conserving practices. One community, for example, mandated solar water heating in new development.

Half of the communities with regulations also have more comprehensive programs or plans for conserving energy. The other communities adopted regulations on an ad hoc basis, in response to specific interests or problems, rather than as a result of detailed analyses of issues and problems.

ADOPTION AND ADMINISTRATION

The initiative for energy-conserving regulations came from a variety of sources. Planning staff, developers, local energy interst groups, public officials, and local professional groups are examples of those who initiated regulations or the ideas that led to them and that worked to develop the regulations themselves.

Few of the communities reported major difficulties in getting energy-conserving regulations adopted. Developers, local utilities and citizens groups cooperated closely with planning departments in preparing regulations.

No major problems have been encountered in administering regulations. This situation may be attributed to the fact that most energy conserving regulations represent minor adjustments to conventional regulations already used in the communities. Experience however is still limited.

Few of the communities attempted to predict the energy savings that could result from changing their development regulations. The most comprehensive cost and energy analysis was done by the community with the most controversial and stringent regulations (mandatory solar water heating). The data, while not definitive, suggest that "low cost or no cost" actions have needed little technical support, but those requiring substantial investment by developers and consumers may need to prove they save energy and dollars in order to be adopted.

Using Development Regulations to Conserve Energy: An Overview of the Options

There are many techniques for conserving energy through development regulations. They can be broken into four functional categories:

Reducing the energy required to heat and cool buildings,
Reducing the energy used in transporting people and goods,
Reducing the embodied energy consumed in materials and construction,
Using alternative energy sources and more efficient generation and distribution systems.

This section is an overview of: (1) development practices that can save energy in each of these categories, and (2) how development regulations can be changed to encourage or require these energy-conserving practices.

DEVELOPMENT PRACTICES THAT SAVE ENERGY

Energy-efficient development techniques are wide-ranging in scope, cost and effectiveness. Some options are small and easy to use, requiring only minor changes in current development practices. Passive solar orientation, for example, is a relatively simple, low-cost way to reduce the heating and cooling needs of a new building. Other options are more complex to design and implement, such as using mixed-use development to cut down on the number and length of automobile trips. Where major changes in development patterns are involved, public receptivity and political reactions can vary. So can development costs and, of course, the energy savings vary. Finally, some options yield results that can be predicted or measured relatively accurately. For others, informed judgment tells us they save energy, but the actual amounts depend on consumer behavior.

As an overview of these techniques, the following list briefly describes the range of options that fall into the four categories mentioned above.

Reducing Heating and Cooling Needs. The energy that is required to heat and cool buildings is determined in part by how buildings and their sites are designed with respect to climate. Sun, cold winds, warm breezes, landscaping and topography affect a building's heating and cooling needs and can be taken advantage of to reduce them. The orientation and arrangement of buildings with respect to the sun and wind and the use of landscaping are examples of actions that can be taken to moderate climatic extremes, make a living environment more comfortable naturally, and thus save energy.

Housing types also affect the energy needed to heat and cool buildings. Housing with common walls usually needs less energy for space heating and air conditioning than detached buildings (i.e., townhouses, apartments and other kinds of housing that have shared walls). Also smaller housing units further reduce energy requirements.

Natural Solar Heating

Design developments so that buildings are oriented to the sun. This means designing streets to run from east to west; lots to run from north to south; and the long axes of buildings to run from east to west.

Develop south-facing slopes first. They are warmer in winter than slopes facing other directions.

Natural Cooling

Use landscaping to shade buildings, parking lots, streets and other paved areas. This prevents overheating of buildings in summer and lowers summer air temperature near pavement.

Design developments to take advantage of cooling breezes. The placement of vegetation and the arrangement of buildings can be used to channel breezes through buildings. Developments can be located (especially in hot climats) in places that receive the strongest breezes, often near hilltops and bodies of water.

Wind Protection

Use windbreaks (trees, hedges, fences, earthworks) to protect buildings from winter winds. Windbreaks reduce the infiltration of cold air into buildings.

Avoid developing low areas of topography where cold air collects.

Avoid developing locations (especially in cold climates) where winds are strongest.

Arrange buildings so that they protect each other from the wind. Often, such an arrangement is compatible with taking advantage of summer breezes too—when winter and summer wind direction is different.

Housing Types

Build housing with a lower proportion of outside surface to interior space (e.g., more common walls, multi-family housing, more compact buildings forms).

Reduce the size of dwelling units.

Encourage housing design innovations which save energy (earth-sheltered housing, for example).

These principles, while generally holding for all areas, need to be specified and applied differently in different climatic zones (for example by choosing appropriate types of vegetation). There are also many aspects of building design—windows, overhangs, construction materials, etc.—that are appropriate which are not covered here.

Reducing Transportation Needs. The energy used to move people and goods in a community is determined in part by patterns of development. The spatial relationships of individual buildings and entire neighborhoods—their density and the degree to which different kinds of uses are integrated—determine in part how far and by what means people travel. Compact development with a mixture of different land uses permits less travel and perhaps more opportunity for public transportation than low density, sprawling development when goods, services, jobs, residences and recreation are closer together.

Density

Develop at increased densities, especially near activity centers, mass transportation stops, and in areas with existing sewer, water and street capacity to handle it.

Use clustering, even at low net densities, to shorten distances within developments.

Develop skipped over parcels that are located within existing development (urban infill).

Integrating Uses

Combine different kinds of land uses within development projects and neighborhoods.

Develop multiple-use buildings. Large urban development complexes with residential, hotel, entertainment, office and commercial uses under one roof are examples. Can also be done on a small scale—an apartment building with a few shops for example.
Allow increased use of home occupations.

Provide convenience shopping and service facilities in otherwise residential neighborhoods. Convenience food stores in residential areas provide an alternative to driving long distances for minor purchases.

Bicycling/Walking/Mass Transit

Provide facilities to encourage bicycling and walking. Pathways, parking facilities, landscaping and other amenities in projects can encourage biking and walking.

Develop at densities that support mass transit. Seven dwelling units to the acre appears to be a minimum density capable of supporting mass transportation (bus system).[2]

Locate highest-density development near mass transit lines.

Provide amenities and facilities to encourage mass transit use. Connect residencies to mass transit stops with pathways. Provide shelters for waiting.

Efficient Traffic Flow

Design street systems to reduce overall lengths.

Design street systems that facilitate traffic flow. Reduct the number of intersections; make efficient connections with existing street systems.

Reducing Embodied Energy Needs. Compact, higher-density development has less energy tied up in streets, utilities and other infrastructure than low-density, detached development. Some things can simply be built smaller. In addition, excessive design standards waste embodied energy. For example, requirements for parking spaces that are based on older, larger automobile sizes, waste asphalt and energy for construction as well as money.

Develop at higher densities. More people can be served by sewers, water lines, streets and so on.

Develop areas that are already served by streets and utilities.

Use narrower streets where possible.

Use fewer and smaller parking spaces and lots where possible.

Cluster buildings together to reduce the length of streets and utilities.

Reuse existing buildings.

Using Alternative Energy Sources and Systems. This category includes energy that could be supplied from alternative energy sources which may include: 1) solar, wind, and use of available heat from power plants; 2) more efficient generation, conversion and distribution systems; and 3) small-scale power generators, district heating systems, and others. The ease with which many of these systems can be used depends on how land is developed. The practicality of using solar energy, for example, is affected by the amount and location of shadows that are cast by buildings and landscaping. When a district heating system (a system that supplies heat to buildings from a central source) is being considered, the density of development must be high enough so that the system is economical. Integrated energy systems that use a variety of energy-conserving technologies are more practical for developments that have a mixture of uses—residential, institutional and light industrial, for example.

Develop at densities that will facilitate the use of district heating or other more efficient production and distribution systems.

Integrate different kinds of uses in new development to create the potential for using wasted energy.

Use solar energy systems (or others like wind generators) in development.

Design development to protect solar access for solar energy systems. The arrangement and height of structures and vegetation affects the amount and location of shadows that could block sunlight to solar collectors.

Develop locally available sources of energy and utilize them in development projects. Geothermal energy resources, energy created from agricultural wastes, ocean thermal and wind energy may be locally abundant and potentially tapped by new or existing development.

REGULATING ENERGY-CONSERVING DEVELOPMENT PRACTICES

There are three basic approaches to incorporating the types of development options described above into regulations, involving an increasing level of stringency. First regulations that stand in the way of energy-conserving practices can be changed to remove the barriers. Second, development regulations can be used to encourage energy-conserving practices either by providing incentives or by creating development settings in which they are easier to use. Third, regulations can require energy efficiency in new development.

Level 1: Removing Barriers to Energy Conservation. Some development regulations and design standards currently in use stand in the way of energy conservation, usually unintentionally. For example, the installation of solar collectors has been prohibited in some communities by aesthetic regulations. Yard and setback or lot layout specifications in subdivision regulations and zoning can limit flexibility in passive siting of buildings. In some cases design standards are excessive—for example, street width requirements that are greater than they need to be for adequate safety and traffic flow. Such standards can actually promote energy waste. Removing regulatory barriers allows developers and consumers to initiate energy-conserving design options without unnecessary hindrance.

Level 2: Encouraging Energy Conservation.   Planners and public officials are also in a position to actively encourage the use of energy-conserving development practices by providing regulatory incentives. In the past, local governments have offered developers incentives to provide desired public amenities such as open space, plazas and better design. The same can be done for energy-conserving design options. Density bonuses, priority processing of development proposals, even decreases in development fees are examples of carrots that can be offered to developers to make their projects more energy efficient. The incentive approach can be applied to almost all of the available practices for saving energy from landscaping of project sites to locating projects near mass transit stops.

Level 3: Mandating Energy Conservation.   The third and toughest approach is to require energy-conserving practices in development. Land use controls place numerous requirements on developers, many of which address issues that have important energy implications and offer opportunities to make development more energy efficient. For example, zoning imposes restrictions on building height and setback. This makes zoning a potential tool for requiring that development be designed to protect solar access or regulations can be imposed to require proper lot and building orientation, landscaping, etc. The police power inherent in development controls offers an excellent opportunity to require many energy-efficient practices in development.

Current Practice: Survey Results

APA surveyed over 1400 local, regional and state planning agencies in Fall, 1979 and conducted an extensive literature review to identify communities that had adopted development regulations designed to save energy. The survey questionnaire sought only regulations that met three criteria: 1) they had been legally enacted and were not just under consideration;[3] 2) they dealt with land use and development and excluded building or architectural design; and 3) they were designed exclusively to save energy. This excluded many provisions that may have saved energy but only as a side effect (e.g., clustering).

The results of the survey were diverse. The regulations identified addressed many different objectives and covered a wide range of approaches. Thirteen communities met all the criteria. All four of the categories of energy-efficient design options were covered. The regulations involved zoning, subdivision, site plan review requirements, landscaping and development incentives.

This section presents a list of the 22 different energy-conserving development regulations identified in the 13 communities (presented in Exhibit 1). Regulations of communities that have adopted more than one type are considered separately.

REDUCING HEATING AND COOLING NEEDS

The majority of regulations identified in the survey addressed the heating and cooling needs of buildings, primarily in the residential sector. The regulations covered include several kinds of energy conservation techniques including passive solar orientation; arranging buildings and using vegetation to control natural ventilation; the use of underground housing; and use of shade control devices; and the use of common wall, townhouse development.

1. Pt. Arthur, Texas: Subdivision Requirements for Passive Solar Orientation

The city amended its subdivision code to require passive solar orientation in new development to prevent overheating. Section C of the amending ordinance states:

> Streets shall be designed so that at least eighty (80%) percent of the buildings in the subdivision can be oriented with their long axes parallel to nine (9°) degrees south of west with a possible variation to six (6°) degrees north of west or to twenty-five (25°) degrees south of west.
> (Ordinance No. 79-89 Port Arthur,
> Texas 9.4.79, 2, 1979)

The ordinance requires developers to orient streets and lots according to these standards. But the burden of compliance is eased somewhat by rewar-

Exhibit 1.

COMMUNITY EXAMPLES AND CHARACTERISTICS*

Category and County	Type of Regulation	Date Adopted	Provision
REDUCING HEATING AND COOLING NEEDS			
1. Port Arthur, TX	Subdivision requirements for passive solar orientation	Sept. 1979	Mandatory
2. Sacramento County, CA	Resolution and administrative procedure for passive solar orientation	1977	Voluntary
3. Dade County, FL	Site plan review criteria for energy-efficient site design	1975	Voluntary
4. Boulder, CO	Incentives for energy-efficient site design	Aug. 1977	Incentive
5. Douglas County, KS	Zoning amendment to permit underground housing	March 1979	Removes Regulatory Barrier
6. King County, WA	Regulations to permit and encourage townhouse development	Dec. 1979	Removes Regulatory Barrier/ Encourages
7. Davis, CA	Zoning amendment to permit flexible siting of fences and hedges for solar heating	1979	Removes Regulatory Barrier
8. Davis, CA	Zoning amendment to permit greater use of shade control devices	1979	Removes Regulatory Barrier
9. Davis, CA	Landscaping requirements for energy conservation	1979	Mandatory
REDUCING TRANSPORTATION NEEDS			
10. Boulder, CO	Incentives for energy-efficient location of development	Aug. 1977	Incentive
11. Windsor, CT	Incentives and requirements for energy-efficient location of development	1976	Incentive/ Mandatory

---

* Within each category, the examples are not listed in any particular order, except that similar techniques are grouped together.

Exhibit 1. (Continued)

COMMUNITY EXAMPLES AND CHARACTERISTICS*

Category and County	Type of Regulation	Date Adopted	Provision
12. Davis, CA	Zoning amendment to expand use of home occupations	Apr. 1979	Removes Regulatory Barrier
REDUCING EMBODIED ENERGY			
13. Windsor, CT	Reduced subdivision standards for street width	1974	Removes Regulatory Barrier
14. King County, WA	Reduced subdivision standards for street width	Proposed	Removes Regulatory Barrier
15. Davis, CA	Reduced subdivision standards for street width	Proposed	Removes Regulatory Barrier
USING ALTERNATIVE ENERGY SOURCES AND SYSTEMS			
16. San Diego Cty, CA	Mandatory use of solar water heaters in new development	1979	Mandatory
17. San Diego Cty, CA	Protection of solar access in new development	1979	Mandatory
18. Albuquerque, NM	Zoning provisions to protect solar access	1976	Mandatory
19. Los Alamos, NM	Zoning provisions to protect solar access	1977	Mandatory
20. Lincoln, NB	Incentives for protecting solar access	Oct. 1979	Incentive
21. Imperial County, CA	Overlay zoning provisions to manage geothermal energy development	1972	Manages and Facilitates
22. Davis, CA	Deregulation of clotheslines "solar dryers"	1977	Removes Regulatory Barriers

ding proper orientation with a reduction in required street width standards.

2. Sacramento County, California: Resolution and Administrative Procedure to Encourage Passive Solar Orientation

The county adopted a voluntary resolution and administrative procedure to encourage developers to design streets and lots so that buildings in new developments could be passively oriented for natural heating and cooling. The resolution was adopted in 1977 and establishes that:

> new lots and new residences be oriented to make maximum effective use of passive solar energy.

(Board of Supervisor's Resolution 77-987)

A memo that serves as a guide for planning staff in reviewing proposals for passive orientation defines proper orientation. It states that the long axis of each lot should run from north to south with a possible variation of 22-1/2 degrees in either direction. In this way, lots will be designed so that a conventionally-sited building will have its long axis running from east to west and with a major wall facing south.

Actual implementation of the resolution's provisions occurs when developers and planning staff sit down to review project plans. Although the developer is not required to meet the provisions of the resolution, the planning staff suggests changes to provide for better orientation of the greatest possible number of lots. If a developer does not make an effort to comply with the resolution, the planning staff can recommend that changes in the plan be made in its report to the planning commission.

3. Dade County, Florida: Site Plan Review Criteria for Energy-Efficient Site Design

The county added energy conservation to its development code criteria used in reviewing development proposals. Energy-related review criteria apply to four types of development: office park districts, clustered, planned and planned area developments (all types of "PUDs"). The intent in each district's provisions is to promote energy-efficient site design, although the specific language differs from one district to the next. For example, the provision for the planned development district states:

> *Plan review standards.* The following criteria shall be utilized in the site plan review process.
> 10. Energy Conservation—Design methods to reduce energy consumption

shall be encouraged. Energy conservation methods may include, but not be limited to natural ventilation of structures in relation to prevailing breezes and sun angles, insulation of structures, use of landscape material for shade, direction of breezes and transpiration.

(Dade County Code, 33-284.21, p. 1024.24)

As part of the plan review process, planners negotiate the provision of these practices with developers.

4. Boulder, Colorado: Incentives to Encourage Energy-Efficient Site Design

The Boulder growth management system awards bonus points for energy-conserving amenities to competing developers. In general, projects are ranked on a scale of one to ten, depending on the level of amenities provided. To simplify the awarding of points for energy-saving amenities, a developer may receive only ten, five, or no points, rather than one-through-ten points. The provisions are clear-cut on what developers must include in their project design to obtain the bonus points. Ten points (the maximum) are granted, subject to the following:

Conservation

10 pts. - Maximum use of conservation measures by incorporating improved insulation within the site and building design (Public Service Company Minimum Insulation Recommendations for Residential Structures, Revised 3/1/76) and any two of the following items as a significant aspect of the project: (1) design or landscaping techniques for energy conservation including building orientation to take maximum advantage of the sun, (2) use of solar or other non-fossil fuel sources, (3) recycling of an existing building or buildings, (4) water conservation efforts including recycling.

(City of Boulder, Ord. No. 4208, 1977.)

Five points are awarded if the project includes: (a) improved insulation (defined above); and (b) one of the other options. The project receives *no* points if no conservation efforts have been made.

5. Douglas County, Kansas: Zoning Amendment to Permit Underground Housing

The county amended the zoning ordinance to include a definition of "Underground Structures," and to redefine basements and dwellings, to permit earth-sheltered dwellings. The new definition reads:

> Underground Structures. Any completed building that was designed to be
> built partially or wholly underground. A completed structure which was not
> intended to serve as a substructure or foundation for a building.
>
> (Douglas County Zoning Code, Article 3-1.7a)

In addition, the definitions of "basements" and "dwellings" were revised
to be compatible with the definition of "underground structures."

   6. King County, Washington: Zoning Revisions to Permit and Encourage
      Widespread Use of Townhouse Development

The county recently adopted a comprehensive package of zoning and
plan amendments to encourage and permit increased use of townhouse
development on a wide scale. Energy conservation was a major impetus for
this action. The regulations to promote townhouse development are quite
extensive and are summarized below:

> Ordinance 4687—Makes townhouses a permited use in RS (one and two-
> family dwelling) zones. Average density of the zone must be met.

> Ordinance 4688—Adds new residential development and design policies to the
> comprehensive plan in order to accommodate townhouse developments in the
> county.

> Ordinance 4689—Creates a floating townhouse zone (RT) which includes com-
> prehensive development standards for townhouses. Allows for range of den-
> sities from one unit per 3600 feet to 1 per 1600 square feet. RT zone is applied
> to the land subject to a rezoning. Density to be determined subject to com-
> prehensive plan, environmental concerns, proximity to transportation and ex-
> isting utilities. etc.

> Ordinance 4667-Raises maximum allowed density in RS (one family dwelling)
> zones. This change was made to increase density of detached structures and
> also allow for higher density use of townhouses within the zone.

> Ordinance 4304-Permits flexible siting and zero lot line siting in subdivisions
> for townhouses and detached structures.

The sum of these provisions is that: (1) townhouses are now permitted uses
in any residential zone (subject to specified densities); (2) townhouses may
be constructed without the use of a planned unit development through ap-

plication of the RT zone; (3) the maximum density of single-family zones has been increased to one unit per 5,000 square feet; and (4) townhouses and any other detached structure can be sited flexibly for energy-efficiency.

7. Davis, California: Deregulation of Fence and Hedge Setbacks For Passive Solar Heating

Davis has modified its fence and hedge setbacks so that they do not interfere with passive solar heat collection. The revised hedge and fence ordinance allows for more flexible siting. It permits fences, walls and hedges in a variety of locations, subject to the following provisions:

If the fence is under four feet it can be located in any of the building's yards, except for a corner lot.

If the fence is under seven feet it can be located in any rear or interior side yard, subject to a building permit for fences over six feet.

On a corner lot, a fence over three feet cannot be located in a triangular area measured 25 feet along the inside face of the sidewalk in each direction from the sidewalk intersection.

In front yards, fences under six feet can be located within 10 feet of the lot line on an interior lot and 15 feet on a corner lot.

On a corner lot a fence six feet high and set back five feet from the lot can be located along the street side yard.

(City of Davis, California, Ordinance No. 1024, June 1979)

8. Davis, California: Deregulating Sun Shades, Trellises, Arbors and Other Climate Control Structures

The city has adopted an ordinance that allows greater use of sunshading devices and their energy-saving potential, while still preserving sufficient open space on a given lot. The new provisions are more liberal than the original provisions, allowing projections as follows:

To within 10 feet of the front property line;

To within 3 feet of the interior side property lines;

To within 10 feet of the side street property line; and

To within 5 feet of the rear property line.

(City of Davis, California, Ordinance No. 1025, 1979)

9. Davis, California: Landscaping Requirements for Energy Conservation

The city has adopted an ordinance that requires increased levels of landscaping for parking lot and commercial development. The purpose of the ordinance is illustrated in its section on "Intent":

1. Deciduous trees, vines, and other landscaping can provide excellent cooling during the summer. This cooling reduces heat load on buildings and reduces energy use.
2. Air temperature and heat radiation affect human comfort equally.
3. In Davis, unshaded walls and paved surfaces can reach surface temperatures in excess of 140 degrees fahrenheit. Such high temperatures make walking and bicycling unpleasant and thereby encourage automobile use.
(City of Davis, California, Ordinance No. 920, November 2, 1977)

REDUCING TRANSPORTATION ENERGY NEEDS

Current practice in this area was limited in comparison with the last section. The survey identified three examples of regulations or incentives to promote more compact development and greater integration of land uses.

10. Boulder, Colorado: Incentives for Energy-Efficient Location of New Development

The city encourages the location of new development close to a variety of existing facilities and services by granting bonus points for such practice to developers who compete for a limited number of development permits. Like the bonus points awarded for energy-efficient site design, the provisions for locating new development are set forth in the city growth management ordinance. The regulations are specific about the distances between new projects and existing development that must be provided to obtain the points.

The maximum number of points is awarded for the following:

A project is within 3/4 miles of a developed neighborhood park.

Half of the project is within 1/4 mile of existing bus service.

A project is within one mile of three of the following facilities: library, recreational center, day care center, and neighborhood shopping facilities with at least a grocery store.

A project is within four minutes of fire department response.

A project is within safe walking distance of two of the three schools that serve it (distance unspecified).

(City of Boulder, Colorado, Ordinance No. 4208, 1977, pp7-12). NOTE: The energy provisions were recently revised and amended.

11. Windsor, Connecticut: Requirements and Incentives for Energy-Efficient Location of Development

The town of Windsor offers density bonuses to converters of one and two-family residential structures (to office and multifamily uses) if the projects are near convenience shopping and mass transit. The town requires that residential units in Planned Unit Developments be close enough to schools for children to walk. The requirements and incentives are set forth in the town's zoning regulations. Excerpts of the ordinance provisions follow:

Location and site standards for planned unit developments state that all residential units be within 1 and 1/2 miles of one elementary school and within one mile walking distance of an existing bus route. (Sec. 11.2)

Location and site standards for developments in the central business district state, that if the location of a development results in sufficient generation of pedestrian, bicycle and mass transit trips, required parking may be reduced up to ten percent. This is subject to the discretion of the Planning Commission. (Sec. 11.3.33)

Zoning provisions for conversion of residential buildings to other uses state that if the conversion is within 1/4 mile of a grocery *or* drug store *or* mass transit route, the regular density may be doubled. If it is within 1/4 mile of both a grocery or drug store and a mass transit line, the density may be tripled. (Sec. 4.5.1)

12. Davis, California: Expanding the Use of Home Occupations

The city adopted a home occupations ordinance to provide for greater use and smoother management of home occupations. It developed and adopted a new, flexible ordinance governing home occupations. This ordinance is a performance standard—it sets minimum levels of performance for home occupations rather than listing specific types. The criteria used to determine a valid home occupation include the following:

Work only by family members and one non-family member;

No external use of material or equipment not normal to the residential district;

No direct sales from home;

No generation of excessive traffic (pedestrian, auto);

No signs other than a nameplate;

No use of commercial vehicles.
                                    (City of Davis, California, Ordinance No. 875, 1977)

Instead of the original conditional use permit which required approval by the planning commission, the applications for home occupations are handled by the planning department under the terms of the ordinance.

REDUCING EMBODIED ENERGY NEEDS

The survey found that changing development regulations to reduce energy that is consumed in construction materials and construction itself was limited to communities that have reduced street width standards. Such reductions allow developers to use narrower streets in projects—something which also reduces construction costs significantly. One community has actually reduced street standards. Two others that have proposed such changes are also included here because of the lack of implemented examples.

13. Windsor, Connecticut: Reducing Street Width Standards

The town reduced subdivision code standards for the width of cul-de-sacs and one-way-loop local streets. Under the original ordinance, standards for both cul de sacs and one-way loop local streets were 30 feet (pavement width not right-of-way). The standards in the new Windsor Subdivision Code, 1975, reduce the widths, when parking is restricted to only one side of the street, as follows:

Cul-de-sacs: 24 feet.

One-way-loop local streets: 20 feet.
                                    (Windsor, Connecticut, Subdivision Code, 1975)

These reductions are granted at the discretion of the Planning Commission, which considers recommendations by the Departments of Public Works and Public Safety.

14. King County, Washington: Proposed Revision of Street Classification to Allow Width Reductions Tailored to Neighborhood Conditions

A report commissioned by the county recommends four categories of streets and a flexible formula for determining their widths, to replace the present two-category classification system that requires some streets to be wider than necessary. The study suggests replacing the county's two-tier classification with the one spelled out in a publication by the American Society of Civil Engineers, *Residential Streets*.[4] Briefly, the report identifies five street types, based on the number of average daily trips (ADT) each carries: places, lanes, subcollectors, collectors and arterials. An ADT range is suggested for each category. For example, a road classified as a "place" carries from 0 - 100 trips per day; arterials receive over 3,000.

The King County report also suggested a formula for determining right-of-way widths that takes into account such factors as: pavement width, size of utility easements, planting strips and space for future street widening. If right-of-ways can be tailored to specific neighborhood needs, less land will be used for streets, and development costs should decrease.

15. Davis, California: Proposed Reduction of Subdivision Standards For Street Widths

Reductions in width standards for collector streets, local streets, and cul-de-sacs are proposed. The proposed standards, would reduce street widths as follows:

Type of Street	Existing Width (Feet)	Proposed Width (Feet)
Collector	52	38
Local	34	28
Cul-de-sac	28	25

ALTERNATIVE ENERGY SOURCES AND SYSTEMS

Several communities have adopted development regulations to promote the use of alternative energy technologies—specifically solar and geothermal energy. The solar energy systems addressed by the examples in this section are active as opposed to passive systems. One community has passed an ordinance mandating the use of solar water heaters in new residential development. Several others have adopted regulations to protect the access to sunlight that is required for efficient operation of solar energy systems (solar access protection). One community has developed a zoning classification (an overlay zone) to manage the development of geothermal energy reserves. (This example is somewhat different in character from the others in this report—it does not directly affect development but it is included here as an example of how planners can facilitate the use of locally available alternative energy resources.) Finally, one community passed an ordinance to prohibit restrictions on the use of clotheslines ("solar dryers").

The survey form also requested enacted regulations dealing with the use of energy production, conversion, or distribution systems such as district heating, cogeneration or more complex integrated systems. No examples were provided that specifically allowed or mandated such systems.

16. San Diego County, California: Mandating the Use of Solar Water Heat in New Development

San Diego County adopted an ordinance that requires the use of solar energy systems for heating water in all new development in unincorporated parts of the county. The ordinance, itself, is very simple. It states that:

> no permit shall be issued by the administrative authority for a new residential building . . . unless said building includes the use of a solar energy system as the primary means of heating water.
>
> (San Diego County Ordinance
> No. 5324, Section 53.119,
> December 12, 1978)

The ordinance is to be implemented in stages. For the first year, beginning in October of 1979, only projects in areas that are not served by natural gas must have solar systems. This covers about 20 percent of the unincorporated land in the county. By Year III all new development must comply with the regulations.

17. San Diego County, California: Solar Access Protection Ordinance

The county of San Diego also adopted an ordinance requiring protection of solar access for subdivision lots. The ordinance states:

No tentative subdivision or parcel map . . . shall be approved unless each lot within the subdivision can be demonstrated by the subdivider to have unobstructed access to sunlight to an area of not less than 100 square feet, falling in a horizontal plane 10 feet above the grade of the buildable area of the lot.

It goes on to define access to sunlight:

When a specific area of not less than 100 square feet has an unobstructed skyview of the sun between azimuths of the sun at 45 degrees to the east and 45 degrees to the west of true south on December 21st.

(San Diego County Code, Ordinance No. 5589)

In other words, the subdivision has to be designed so that after homes are built, each lot will have at least 100 square feet of space, raised ten feet above the ground that receives direct sunlight between the times of day when the sun is 45 degrees east and west of due south (which probably corresponds to between 8 or 9 in the morning and 3 or 4 in the afternoon.)

18. Albuquerque, New Mexico: Zoning Regulations to Protect Solar Access

The city amended its height regulations for certain medium and high-density zoning districts to protect access to sunlight for solar collectors. The solar access provisions are set forth in the zoning ordinance section on height for higher density residential and some non-residential districts. The regulations specify an automatically allowed height for a building at any point on the buildable area of a lot (i.e., within regular setbacks). Any part of a structure that is over the allowed height must lie within a pyramid defined by 45 degree planes that rise from the four sides of the lot. To allow some flexibility in height on the south, east and west sides of the lot where the impact of shadows is least critical and also to avoid absolute centering of buildings on their lots, a second set of planes, at 60 degrees is established for all but the north side of the lot. A building with its height and form established under the 45 degree pyramid can then be shifted laterally (except to the north and closer to buildings it would shade) to the limits set by the 60 degree pyramid.

19. Los Alamos County, New Mexico: Provisions for Protecting Solar Access in Existing Development

The county amended its zoning code to provide for protection of solar access in existing development. The provisions protect solar access to a collector that has been installed as long as the collector is registered with the county clerk. The ordinance states:

II. *Solar Energy Collection System*

   a. When a solar energy collection system is installed on a lot, accessory structures or vegetation on an abutting lot shall not be located so as to block the solar collector's access to solar energy. The portion of a solar collector that is protected is that portion which:
      (1) is located so as not to be shaded between the hours of 10:00 a.m. and 3:00 p.m. by a hypothetical 12 foot obstruction located on the lot line; and
      (2) has an area not greater than one-half of the heated floor area of the structure, or the largest of the structures, served.

The ordinance thus, defines the level of solar access to be protected and establishes that if a collector is larger than one-half of the floor area that it serves, the excess collector area is not guaranteed protection. This places a reasonable limit on the restrictiveness of the ordinance on neighboring parties. Again, the collector must be registered with the county clerk to be protected.

20. Lincoln, Nebraska: Incentives for Protecting Solar Access

The city of Lincoln offers density bonuses to developers who design planned unit development projects with a variety of energy-conserving practices. The section on design standards for the zoning regulations of the city authorize a density bonus of up to 20 percent for dwelling units for community unit plans (CUP). In order to qualify for the density bonus the plans for a CUP proposal must include:

A plan showing the shading effects of trees;

A plan showing the shading effects of buildings;

A copy of covenants to be used in protecting solar access.

The plan is also evaluated against the following criteria for energy-efficient site design:

Highest densities on south-facing slopes;

Streets oriented in east/west manner (varying up to 20 degrees);

Lots oriented with greatest dimension north/south;

Buildings sited with longest dimension east/west;

Buildings sited as close to north lot line as possible to create larger yard space to south where shading could be caused and placing it under control of owner;

Tall buildings sited north of shorter ones;

Building form designed to maximize solar heating and reduce exposed exterior surface;

Landscaping designed to provide shade where needed but also to protect access for solar collectors;

Natural ventilation shall be considered.

The density bonus is determined in the CUP negotiation process as plans are reviewed. The ordinance states:

> The bonus . . . shall be determined by the eligibility of solar access criteria as illustrated in the shadow plans and review items. Shadow plans will be reviewed as to the number of building envelopes with clear access to solar energy. (The site design criteria) . . . will be reviewed against the proposal and their implementation and effectiveness in accomplishing energy objectives (by planning department review staff)
>
> (City of Lincoln, Nebraska, Design Standards for Zoning Regulations, October 1979, p. 40)

Although the provisions indicate that the plans will be reviewed for energy-efficient site design, the basic objective of the incentive is protecting solar access; shadow plans for trees and buildings and the provision of convenants to protect solar access are the real requirements for obtaining the density bonus.

21. Imperial County, California: Regulating Development of Geothermal Energy Reserves

The county developed special overlay zoning provisions to manage exploration, testing and production of geothermal energy reserves. The regulation of geothermal exploration and development in Imperial County is done in three phases. First, a conditional use permit is required for drilling an exploratory well. If the permit is granted, the applicant must comply with a set of terms, conditions and procedures that include specifications for insurance, cleanup, waste disposal, noise and vibration suppression, air pollution, odor and others.

The next phase is associated with the testing of exploratory wells that have yielded positive results. Another conditional use permit is required, but for this permit, the terms and conditions are negotiated with the applicant based on the conditions of the site and its reserves and the proposed testing procedures. The list of conditions serve as a contract that the applicant must comply with during the testing procedure.

The final phase is construction of a power plant to tap a geothermal reserve that has been identified and tested. The county has developed a special geothermal (G) overlay zone to regulate this phase of development. The G zone may be applied over any land use zone subject to a third set of terms and conditions for construction and operation of a power plant. The zoning ordinance states:

Geothermal (G) Zone

    a) Overlay classification. Land classified in a particular zone may be classified as being with a Geothermal (G) Zone;

    b) Uses permitted. The production and development of geothermal resources may be permitted in a Geothermal (G) Zone under a Geothermal Production Permit, authorizing particular uses and specifying terms and conditions. (See terms, conditions, and applicable procedures adopted by the Board of Supervisors.)

                Imperial County, California, Land Use Code, s83226, Ord. No. 410; eff. June 1, 1972 repealing Ord. No. 369 and enacting a new section 83226)

22. Davis, Calfornia: Promoting the Use of Clotheslines ("Solar Clothes Dryers")

Davis has promoted the use of clotheslines in two ways: (1) by requiring their installation in new multi-family developments; and (2) by banning deed restrictions and covenants that prohibit clotheslines in single-family neighborhoods. The city passed an ordinance prohibiting any agreements that restricted the use of clotheslines. The ordinance states:

It shall be unlawful . . . to establish any private covenant or restriction which prohibits the use of a clothesline in any residential zone . . .

                (Ordinance No. 876, City of Davis, California, 1977)

The city went a step further and required that space be provided and facilities for drying clothes on lines be provided in all new multi-family developments:

. . . all multi-family developments . . . shall require suitable space or facilities . . . to enable residents to dry their clothes using the sun. Such clotheslines shall be convenient to washing facilities and oriented so as to receive sufficient sun to dry clothes throughout the year.

(Ordinance No. 876, City of Davis, California, 1977)

Analysis of Current Practice

This section of the report compares and analyzes the collective experiences of the 13 survey communities to see what lessons can be learned from their energy-conserving development regulations. (Exhibit 2 presents a tabular overview of this section.) The survey results presented in the previous section are important for these reasons:

The actions of the 13 communities, although tentative and preliminary, indicate a step toward incorporating energy conservation into local development regulations and planning. Even the more modest techniques show the willingness to use local police powers to promote energy efficiency in land development.

The techniques illustrate a wide variety of technical, energy-efficient design options. They deal with building space conditioning, efficient land use patterns, transportation, embodied energy and alternative sources and systems.

The case studies suggest three important ways that development regulations can be tools for conserving energy. Some of the examples entail minor changes to remove barriers. Others offer economic incentives for energy-conserving practices. Still others mandate energy conservation.

The examples represent the major geographic and climate regions of the country, as well as a variety of government types at the local level—large cities, small towns, urban counties and rural communities.

It is not known if these 13 communities have or will realize substantial energy savings. As with all land-use controls it will take time to see what the effect really is. As development occurs it will become more obvious which options lead to the greatest energy savings. In addition, some of the regulations address long-term solutions—changing development patterns, for example—that may not produce results for some time. Although it may take a decade to realize the ultimate yield in energy savings from the efforts of these communities, today they are laying the foundation for other local governments to build and expand upon.

Exhibit 2

Preparation and Administration of Energy-Conserving Development Regulations by Community

Community	Type of Regulation(s)	Comprehensive versus Ad Hoc Approach	Estimated Energy Savings?	Source of Initiative	Experience to Date	Have You Evaluated Results?
Albuquerque, NM	Zoning provisions to protect solar access	Ad Hoc	No	Local architects/planning staff	Many buildings constructed under regulations/many solar collectors in use	No
Boulder, CO	Incentives for energy-efficient site design and location of development	Comprehensive	No	Planning staff	Several projects have received bonus for energy-efficient design	Plan to evaluate entire growth management system
Dade County, FL	Site plan review criteria for energy-efficient site design	Comprehensive	No	Planning staff	Most projects now use increased landscaping and overhangs	No
Douglas County, KS	Zoning amendment to permit underground housing	Ad Hoc	No	Interested citizens/planning staff	None	No
Davis, CA	Zoning amendments to permit flexible siting of fences and hedges for solar heating; greater use of shade control devices; expanded use of home occupations; reduced street width; deregulation of clothes-lines; landscaping regulations	Comprehensive	No	Interested citizens/planning staff	Substantial	No
Imperial County, CA	Overlay zoning provisions to manage geothermal energy development	Ad Hoc	No	Planning staff/geo-thermal industry	Two geothermal plants in construction	No

Exhibit 2 (Continued)

Preparation and Administration of Energy-Conserving Development Regulations by Community (Continued)

Community	Type of Regulation(s)	Comprehensive versus Ad Hoc Approach	Estimated Energy Savings?	Source of Initiative	Experience to Date	Have You Evaluated Results?
King County, WA	Reduced subdivision standards for street width; townhouse regulations	Comprehensive	Yes	Planning staff/county executive	None	No
Lincoln, NB	Incentives for protecting solar access	Comprehensive	No	Energy council/planning staff	Three projects in preliminary review	No
Los Alamos, NM	Zoning provisions to protect solar access	Ad Hoc	No	Local scientists/planning staff	50 solar systems registered and protected	No
Port Arthur, TX	Subdivision requirements for passive solar orientation	Ad Hoc	No	Interested developers/planning staff	Four projects approved but not built	No
Sacramento County, CA	Resolution and administrative procedure for passive solar orientation	Comprehensive	Yes (for other options)	Planning staff/local interest groups	Few subdivisions have been built with proper orientation	No
San Diego County, CA	Mandatory use of solar water heaters in new development; protection of solar access	Comprehensive	Yes	Local interest groups/county brd./planning staff	Several projects have been reviewed	Plan to keep track of homes with solar systems for future evaluation
Windsor, CT	Incentives for energy-efficient location of development; reduced subdivision standards for street width	Ad Hoc	No	Planning staff	Several projects have used narrower streets or energy-efficient locations	No

Choosing Energy-Conserving Options: Comprehensive Vs. Ad Hoc Approaches

The 13 communities arrived at their adopted energy-conserving development regulations from two directions. Six adopted their regulations within the framework of a comprehensive plan or program to conserve energy. This may have involved developing a set of policies and objectives for conserving energy, or preparing and adopting a special "energy element" for the community's comprehensive land use plan. The other seven adopted their regulations in an ad hoc fashion—in response to specific concerns or interests of planners or public officials, for example, or requests by developers or citizens to use energy-conserving practices.

COMPREHENSIVE APPROACH

A comprehensive approach to saving energy through development regulations and practices means simply that a community has taken a look at a range of conservation options. For the communities discussed in this report (quite a few other communities have made a comprehensive consideration of conservation options but not adopted any regulations) the comprehensive approach may encompass many facets of community operations or be limited to community planning and development. For example, some of the communities' comprehensive programs address such things as energy use in government buildings, schools, and municipal automobile fleets, as well as land use.

Sometimes too, the communities discussed in this report included an analysis of community energy use as part of a comprehensive approach. A community energy audit was conducted to determine where energy comes from, where it is used and where the greatest potential for conserving it lies. Others simply identified and organized conservation options into a coherent base from which to work and proceeded with implementation. Two of the counties, Dade and San Diego, provide examples of the comprehensive approach.

Dade County— Planners in Dade County examined and presented options in a 1978 report on energy conservation in urban development. The report combined background information on energy/land use relationships in the county with recommended policies, goals and strategies for saving energy. The first four chapters present: an overview of trends in supply and use of energy in the county; a breakdown by community sectors of actual energy use; a review of existing policies and programs on energy from the local to federal levels; and a discussion of several aspects of the relationship between planning and energy (transportation and travel patterns, density, mixed use, site planning, and alternative energy technologies). The fifth chapter contains the proposed policies that were developed out of the review of Dade County's energy picture and its options for saving energy.

The report is a good illustration of translating energy audit findings into policies and objectives for solving local energy problems. The policies and objectives in the report focus on reducing the need for air conditioning, promoting the use of mass transit, and reducing the frequency and length of automobile trips. The policies respond directly to the energy problems identified for the community. Although the site plan review regulations discussed in this report were adopted before the set of policies was developed, the example illustrates how one community has considered a range of options, many of which are tailored to local conditions.

San Diego County—The San Diego County approach resembles Dade County's in that it spells out objectives, policies and strategies for conserving energy. But in San Diego County, energy conservation findings and policies were incorporated into the general plan as a separate energy element, adopted in late 1977. The energy element is more detailed than the Dade County study, and covers aspects of community energy use that go beyond land development (e.g., conservation in county buildings and auto fleets, public education on conservation, intergovernmental cooperation and coordination, energy shortage contingency plans, and others). The energy element is organized into three sections: urban and site design, transportation, and supply policies. And land use practice cuts across all three. The policies and actions most directly related to land use and development include encouraging innovative building design and orientation techniques, encouraging higher development densities, promoting land use aimed at minimizing transportation requirements, promoting safe and practical walking and bicycling routes, and promoting accelerated market penetration of solar equipment and technology. The county board established an energy office within the planning department to oversee and coordinate implementation of the energy plan, although county line agencies are directly responsible for carrying out some of the particular actions.

AD HOC APPROACH

The ad hoc approach is represented by the communities in the survey that prepared and adopted energy-conserving regulations as a single issue. Such regulations were developed in response to specific local problems or interests. Four examples from the survey communities are described below.

response to an interested developer.

The amendments to the Douglas County zoning code to permit underground housing were made in response to requests by individual citizens to build this type of housing.

The solar access regulations in Albuquerque and Los Alamos were developed out of a strong local potential for and interest in solar energy use.

Imperial County's overlay zone provisions for geothermal energy development were developed in response to the abundance of this alternative energy source in the county.

These examples illustrate how planners and local governments have responded to individual requests by developers and citizens; local groups interested in a specific aspect of energy conservation or to an especially good local potential. This is in contrast to the other communities that considered many kinds of options in anticipation of adopting energy-conserving development regulations. The communities that took an ad hoc approach really began with implementation. These communities focused on a particular problem and developed a regulatory solution in relatively little time. Although they did not have a comprehensive program to save energy, each still took a significant individual action to meet the same objective.

LESSONS LEARNED

Should energy-conserving development regulations be developed on an ad hoc basis, as needs arise, or in a comprehensive fashion? The communities surveyed more often took the former approach. This is to be expected in the early stages of regulating development for energy efficiency. There are yet few models on which to base a comprehensive program—especially ones that involve significant implementation—and many of these communities have narrowly focused approaches.

There are arguments in favor of and against both kinds of approaches.

What are the advantages of the comprehensive approach and corresponding disadvantages of an ad hoc approach?

1. Some design options have a relatively low potential for saving energy by themselves. A program that combines a variety of them can produce cumulative energy savings that are more significant. An ad hoc approach that focuses on only one kind of option may have only limited results.
2. Many energy-conserving principles and practices are interrelated—development density and the use of district heating, for example. Considering options in concert can prevent overlooking certain opportunities that may have highly beneficial results, something that could occur under an ad hoc approach.
3. Comprehensive energy planning illustrates to developers, decision-makers and courts that a community takes energy seriously. This is especially important when mandatory regulations are enacted and new restrictions are imposed on developers and the public. A court decision on mandatory regulations may hinge on whether it is based on a comprehensive plan. An ad hoc approach may not produce sufficient support for adoption of regulations or their defense in court.
4. Planning can eliminate potential conflicts among energy-conserving practices. For example, encouraging the use of landscaping to provide summer shade could conflict with the objective of protecting solar access. An ad hoc approach could overlook such conflicts.
5. Fiscal austerity and rising costs mean that communities must pick and choose actions carefully, selecting only those that are truly energy and cost-efficient. Comprehensive programs and plans provide a means of weighing these aspects while an ad hoc approach could fail to differentiate on the basis of cost effectiveness.
6. Finally, a comprehensive approach, like all planning, allows for consideration of a broad range of actions, rather than a more limited set developed in response to narrow concerns. It opens up opportunities for innovation and initiative and the potential for greater overall payoffs.

What are the advantages of an ad hoc approach and the corresponding disadvantages of a comprehensive approach?

1. An ad hoc approach allows a local government to proceed on a limited basis with energy conservation if technical aids are not available or a commitment to energy conservation is not fully developed. It can be an appropriate way to lead into more comprehensive activity as broader interest increases and more information on how to save energy becomes available.
2. An ad hoc approach may lead to action more often than a comprehensive approach if a community responds to a specific interest by making regulatory changes. With a comprehensive approach there is the possibility that a study or set of policies could be considered to be enough effort, stifling further implementation of energy-saving options.
3. An ad hoc approach is likely to cost less than a comprehensive approach. Less staff or consultant time is likely to be required to evaluate a single option than a range of options.
4. An ad hoc approach can take immediate advantage of public interest in energy conservation if planners act quickly. It is conceivable that public in-

terest could wane during a lengthy process of developing a comprehensive program.

The two approaches can certainly be combined. A community that is developing a comprehensive program also can be open to specific opportunities as they arise and may decide to put aside the broad based effort in order to respond. Another community may wish to respond to individual issues as a primary focus but develop a broader plan over a period of time, as interest develops and technical aids become more available.

It is not yet clear whether a community energy audit should be included as part of a comprehensive conservation program, and if so, to what degree it should be done. Although a few communities have conducted energy audits relatively easily and cheaply, comprehensive audits can be expensive. Where there is a strong political base of support for energy conservation, a detailed audit may not really be needed. The Comprehensive Community Energy Management Program (CCEMP) should yield a great deal of data on the feasibility and value of community energy auditing, as it proceeds. The program has just moved out of the audit stage and into the stage of developing energy objectives, strategies, and the plan itself. For the time being, however, the decision to conduct an audit depends on the factors of time, budget, available expertise and political support.

Building Technical Support: Estimating Effects of Proposed Regulations

Three of the communities surveyed made an effort to calculate the effects of proposed energy-conserving development practices (as indicated in Exhibit 2). Sacramento County attempted to estimate potential energy savings from a variety of energy-conserving development practices. San Diego County prepared a cost-benefit analysis of mandatory use of solar water heaters. King County investigated the energy and cost implications of townhouses. The studies done by the latter two communities are specifically related to the regulations each has adopted and described in this paper. They are described in greater detail below.

The other 10 communities did not feel the need to predict potential energy savings or investigate other effects of proposed regulations in great detail. Rather, they assumed to varying degrees that the options they were considering were effective. In many cases, planners involved in developing the regulations knew of the general benefits of a particular technique and accepted this as adequate justification. It was not important to them to know exactly *how much* energy could be saved, but rather *if* energy could be saved. And some options entailed such minor changes to regulations that they could be justified and adopted without supportive information. This was the case, for example, with many strategies to remove regulatory barriers to energy-efficient practices. The pros and cons of evaluating versus not evaluating the effects (especially energy savings) of regulations are discussed at the end of this section.

San Diego County—The San Diego County energy element called for a feasibility study of solar energy systems before an ordinance requiring them could be adopted. The task was assigned to the county energy office (within the planning department). The study addressed several impacts of adopting the ordinance: the energy and cost savings possible; the cost of initial installation; maintenance and insurance costs; impact on labor, the housing industry and the local electric utility; consumer protection; and solar access.

Perhaps the most important part of the study was the evaluation of the life-cycle costs of solar versus conventional water heating systems. Staff in charge of the study adopted a comprehensive methodology that had been developed by the California Energy Commission to determine and compare the costs for purchasing, installing and operating different types of water heating equipment. The methodology considered 41 variables supplied from local sources of information, many of them within the planning department itself.

The cost analysis determined that solar hot water systems are more cost-effective than conventional natural gas systems. The study convinced both opponents and board members, and was a key in getting the ordinance adopted. The study methodology could be used by other communities across the country to make their own comparisons. It cost about $7,000 in staff time. The staff member who conducted the study is a planner by profession, with a background in public administration and systems ecology. His familiarity with programmable calculators helped him with the life-cycle costing methodology but he felt that most planners could make a similar study without great difficulty.

King County, Washington—This county commissioned a study of the cost, energy and regulatory implications of townhouse development, before enacting development regulations to permit and encourage townhouse development on a wide scale. Available data indicated that townhouses were more energy-efficient than detached structures, but the planners felt that a concise summary of the data would help in enacting the necessary regulations.

An architectural firm made the analysis without sophisticated modeling methods. It involved a comparative analysis of single-family detached and single-family attached (townhouse) dwelling units with respect to energy consumption for heating, annual heating costs, construction costs and life-cycle costs. The study concluded:

> . . . townhouses are better conservers of energy than single-family dwellings based on energy use for heat only. The most significant difference occurs between the one-story detached unit and the two-story attached unit with two common walls (interior unit); the annual energy use for heating is 40% less for the attached unit. The two-story attached with one common wall (end unit) uses 20% less energy than the one-story detached unit.

> *Land Use Regulation and Energy Conservation:*
> *Revised Working Paper on the Townhouse Element,*
> p.i.

The study also found that construction costs were slightly lower for the townhouses (same size as detached unit) and that operating costs (including mortgage payments, energy costs, painting and roofing) were about 15 percent less for the attached structures.

LESSONS LEARNED

Analysis of feasibility of specific techniques can be important in two ways. First, it can help set priorities among options. When a technique promises only limited energy savings but adds to government costs or the price of housing it will likely be given a low priority for implementation. Second, an estimate of potential results can also be important in getting it adopted. San Diego County is a case in point.

There are cases when it is not necessary or possible to evaluate energy savings. When it is obvious that energy is being wasted—streets that are wider than necessary for example—exact figures are not important. In other cases, energy savings may be difficult to quantify. Estimating energy savings from mixed-use or higher-density development, for example, would require sophisticated methods and a great deal of staff time. It is certainly easier to rely on the several existing theoretical studies on these topics than to make predictions of local savings.

There may be times when it is not necessary or possible to evaluate energy savings or other effects of proposed regulations. The savings and effects from some options may be difficult to quantify. There may be many variables to consider or energy savings may be dependent on consumer choice. Mixed-use development as an option for saving energy is an example. There are many complex questions that have to be considered in evaluating potential energy savings. Will people who live close to certain industries actually be employed in those industries? Will people who live closer to grocery stores walk to them for small purchases or just use their cars more often? Questions like these make it difficult to assess the effects of some conservation options.

Sometimes, too, planners may be able to do without quantitative calculations of energy savings. Regulatory changes that are not potentially controversial, that have obvious energy benefits or that are relatively passive (removing barriers for example), may be supportable without detailed data on energy savings, cost effectiveness or other effects. For example, the Davis ordinance that permits greater flexibility in siting fences for passive solar heating removes a regulatory barrier to encourage a relatively benign energy-conserving practice. Although issues such as traffic were considered in developing the ordinance, precise data on energy savings that could result were not needed. In contrast, the San Diego ordinance, as mentioned earlier, had to be carefully analyzed for its effects.

The decision to evaluate energy savings and other effects of energy-conserving regulations hinges on the kind of option and the regulatory approach in question, and on local conditions like the political climate. Certainly, there may be times when the backers of energy-conserving development regulations are able to rely on existing general information on savings from a proposed option. However, it is important to consider the need for more detailed supportive information before going to the community or public officials with information they may consider inadequate.

BUILDING COMMUNITY SUPPORT: RESPONDING TO INITIATIVES

No single group of actors seems to be behind changes in local energy-conserving actions. Exhibit 2 indicates that regulations were initiated by planning staff, developers, private citizens, public interest groups, and local professionals such as architects or scientists. In many cases, planners indicated that interest originating outside the planning department was important—both in preparing effective, realistic regulations and in generating the support needed to adopt them. The three following examples illustrate the variety of actors and roles in building community support.

A developer in Port Arthur, Texas approached the planning department to help determine the solar orientation to prevent overheating buildings in a new town that he was developing. The planning department took the opportunity to act. The planning staff not only developed the necessary technical information, it also suggested that additional units in the project be oriented properly. The staff also prepared an amendment to the city's subdivision ordinance that would require proper orientation in other projects as well. As the effort progressed, other sources of support were tapped. The local utility, concerned about summer peak loads from air conditioning, testified in support of the ordinance. In addition, staff showed a videotape on the benefits of passive orientation developed by the Texas Governor's office at a public hearing. The involvement by outside parties was a deciding factor in the passage of the ordinance.

There are two other examples of how planners have taken advantage of existing interest in energy conservation in preparing and adopting energy-conserving regulations. The solar water heating ordinance in San Diego County imposed stringent requirements for an expensive technology in new homes. The planning staff built on the strong local interest in solar energy and worked with a variety of interested and affected parties in writing and adopting the ordinance. Manufacturers of solar equipment, developers, the local public utility, and public interest groups were consulted and involved in the adoption process. One staff member felt that this was a key to getting the ordinance adopted.

Likewise, in Albuquerque, an ordinance for protecting solar access was drafted by the planning staff in conjunction with the local chapter of the American Institute of Architects, a group already interested in the use of solar energy systems. There was a double benefit in this case, because many of the buildings affected by the proposed regulations would be designed by architects (commercial and apartment buildings in medium and high-density zones). Experience in Albuquerque has shown that there are few problems in interpreting and complying with the solar access regulations.

LESSONS LEARNED

The interest in energy conservation is growing, but much of it is still unfocused—people don't know which ways to save energy work best. There is evidence that developers are beginning to incorporate energy-conserving passive site design techniques and occasionally active solar technologies into otherwise conventional projects. This interest can be led, nurtured and supported through studies and incremental action. The political dimension in land-use regulations necessitates that groups with competing interests cooperate. Energy-conserving development regulations are no exception and outside help may result in more realistic and effective regulations in the long run.

Administration

Few of the communities surveyed have encountered or anticipate any major problems in administering energy-conserving regulations. Planners report that procedural adjustments were minor, and added administrative costs were minimal.

One reason for the apparent ease in administering energy-conserving regulations is that many of the regulations are so recent that few development proposals have been reviewed under them. The real problems may only become clear with the passage of time. A second reason is that many of the communities made relatively minor changes to existing, conventional development regulations. In fact, several of the regulatory changes have actually reduced administrative burdens—the deregulation of shade devices in Davis and townhouse zoning provisions in King County, for example.

LESSONS LEARNED

Even the limited experience with energy-conserving regulations has raised some general administrative issues. Problems or bottlenecks are apt to center around two issues. First, reviewing proposals for their energy impact adds a new dimension to the review process and thus increases the extent of discretion that planners exercise. Second, it requires more expertise on the part of planners—new skills in assessing energy implications and the application of conservation practices in development proposals and projects. Each of these factors is discussed in the following sections.

Increased Regulatory Discretion—Many energy-saving development practices are site specific. District heating systems, for example, vary in design and application from one site to another. Planners are not directly concerned with conventional heating systems contained within each dwelling unit, of course. But district heating systems or other alternative technologies for central heating and distribution do have implications for density, the arrangement of structures and the integration of different land uses. Planners must address these and many other concerns in reviewing such proposals. Likewise, landscaping for energy conservation will vary from one site to the next, depending on topography, wind flow, the location of parking lots, and so on. This, too, adds another dimension to landscaping previously reviewed only for aesthetic purposes.

This means that staff exercises discretion over more issues. The discretionary quality of staff review will be especially pronounced in the early years, until more prescriptive guidelines can be developed. The costs and benefits of increased discretion in development review are widely discussed. Does more discretion result in development that is better tailored to community needs and environmental objectives? And if so, does it justify increased processing time and planning costs for developers? Does it open up opportunities for abuse?

There are no easy answers to these questions. Energy-related regulations cannot escape being drawn into the debate. The communities surveyed did not anticipate major problems in this area, but most of them are only considering a few aspects of development related to energy. None is yet looking at the full range of items in development proposals that affect energy consumption. As more details of a project are reviewed for their effects on energy, delays from discretionary review may emerge as a more pressing problem.

The Need for New Skills—Reviewing land development for energy-efficiency means looking at it from a new perspective and requires a range of new skills. The experience in Dade County, Florida offers an example. The planning department found that building inspectors were not familiar with the correct tree species or the significance of 10-foot sapling minimums in the landscaping ordinance. The county has requested the addition of an inspector trained in forestry to correct this situation. The importance of climate is another area where many planners need more education. From which direction are winter winds? How do the arrangement of structures affect wind flow? What is the optimal orientation for solar heating in a given location?

Planners also will have to learn more about how the density and integration of land uses affect the use of decentralized community energy systems. What densities are optimal for use of district heating? Do any local industries produce useable waste heat, and are they compatible with non-industrial uses that could use it?

To summarize, the following administrative concerns are grouped according to the four categories for energy-conserving regulations described previously:

1. Reducing Heating and Cooling Needs:

Passive orientation-requirements for *street* and *lot* orientation do not always guarantee *building* orientation. This means an additional review when building plans are submitted.

Natural ventilation and protection from prevailing winter winds are highly site specific. Thus, regulations that address them must be flexible. Further, planners will need new information about local wind conditions and wind-flow patterns.

Landscaping requirements for shading, wind protection or solar access protection imply certain species and locations for vegetation. If species are not outlined in an ordinance, planners must be familiar with them to review plans effectively.

2. Transportation Needs:

Projects proposing mixed uses are more complex than single-use proposals. This can mean more time for reviewing and analyzing plans.

Infill development of substandard lots requires flexible development standards and a case-by-case approach to development review.

Performance standards for the expanded use of home occupations may require more review time than regulations that simply list permitted home occupations.

3. Embodied Energy:

Reducing street width or other development standards may have to be done on a case-by-case basis, as appropriate with other needs (e.g., safety, access), with an associated increase in review time.

4. Alternative Energy Sources:

To administer solar access regulations, planners must learn how to assess shading impacts of development proposals—developing shadow patterns for trees and buildings.

Developers may not always be able to meet requirements for using alternative energy sources. Planners must be familiar with how site constraints could limit a developer's flexibility.

Monitoring and Evaluating Results

It is too soon for most of the communities surveyed to evaluate the performance of their energy-conserving regulations. The few jurisdictions that *have* had significant development occur under the regulations have not assigned a high priority to evaluating the results.

One community, however, San Diego County, is making plans to evaluate energy savings from its solar water heating ordinance. The energy office is keeping track of the homes built with solar water heating systems through a specific code in the county's computer filing system. This will enable the planning department to identify these homes in a few years when enough development has occurred to make an evaluation. The staff plans to compare individual utility bills from the homes with solar water heaters with the bills from homes with conventional systems to estimate the actual energy savings.

LESSONS LEARNED

The virtual absence of local experience makes it difficult to discuss whether to monitor and evaluate the results of energy-conserving development regulations. One obvious consideration is the level of effort required. Some conservation practices—space heating and cooling of individual buildings for example, may be relatively easy to monitor. But practices that are applied on a community-wide scale, such as increasing development densities, may be all but impossible to evaluate. In addition, some options are so thoroughly analyzed before they are approved, that post-installation evaluation would be superfluous. The potential energy savings from a district heating system, for example, would have to be well-documented before its use because of the cost involved.

If regulations and practices are to be evaluated it may be necessary to keep track of the development that is affected by them from the start. San Diego County is a case in point. Planners there have not yet designed the actual method to evaluate the solar ordinance—this would be premature. But they are making sure that a future evaluation will be possible by logging the buildings constructed under the new ordinance.

Where monitoring and evaluating makes sense for a particular option or regulation, it can offer two benefits. First and foremost, an evaluation indicates how well a regulation is working—is it having a significant effect on energy use? If so, should its use be extended to other parts of the community? Should it be required, instead of encouraged? Information on energy savings can help to modify or fine tune regulations. For example, a community that requires passive solar orientation may discover that specific provisions save more energy in one type of housing than another. It may consider developing different regulatory standards to improve performance in different housing.

Second, it is important to generate information on the actual performance of energy-related regulations. Data gained through practical experience will very likely begin to replace the hypothetical results obtained from computer models, especially for more restrictive controls. Communities experimenting with these regulations right now are pioneers. More and more communities will be considering energy-conserving practices, and they will benefit from any information obtained from earlier experiences.

Conclusions

Energy-conserving development regulations are still in their infancy. Most of the communities that have adopted them did so within the last year and few have had much experience with administering them. Many of the regulations focus on a single issue—addressing one aspect of site design like passive solar orientation—rather than a range of conservation options. Justifiably, many of these communities that are pioneering energy-conserving regulations have addressed the less difficult issues, encouraging site design practices that involve fewer risks to developers and make fewer changes in the character of conventional development.

Many have started by removing regulatory barriers to developers and others already interested in energy-conserving development practices. In communities where energy-conserving practices are actively promoted, voluntary approaches are often used. Only a few communities have adopted truly mandatory requirements for energy-efficiency in new development.

These first steps must be characterized as evolutionary, not revolutionary. This may disappoint those who believe that sweeping changes must be made in one or two years, in the face of the energy crisis. But experience has shown that land use, as one of the value-laden and emotional issues at the local level, changes only gradually. The whole history of zoning is only sixty years old. Ten years ago there were few communities with any type of environmental regulations. Rising housing costs, legal rights, and administrative costs and political acceptability must all be considered in designing energy-conserving development regulations.

However limited and tentative the activity in this area, the trend is clear. Energy conservation is quickly becoming a major concern for planners. The question is, how fast will this concern grow, and what form will it take? The most important variable is the price of energy. As heating and electricity take a larger portion of household operating budgets, people will look for more energy-efficient homes. The current interest in proper insulation and storm windows will likely expand to include site design practices.

The answer also depends on our willingness and ability as consumers to make tradeoffs between energy conservation and changes in lifestyle. For example, as the price of gasoline rises and its supply becomes more critical, it would be effective to plan more compact communities with jobs and homes close together. This is a departure, though, from the suburban ideal, and it is quite possible that we will find other ways to reduce our dependence on the automobile instead.

Finally, the growth of our concern for energy will be dependent on how fast our ability to conserve it progresses. The level of technical development of energy-conserving practices and technologies currently limits their use. The state of the art with site planning practices, community energy production and distribution schemes, wind generation systems and others is not far advanced. While a great deal of work is being done to refine energy-conserving practices and technologies, we must have demonstrated proof of their performance before they will be considered for wide use.

As the pressure, willingness and ability to conserve energy increase, we should see other communities consider and implement the kinds of options and regulations described in this report. All three regulatory approaches—removing barriers, encouraging and requiring energy efficiency—will be considered in programs to save energy. Communities will examine regulatory barriers or establish incentives for energy-efficiency first. Mandatory regulations may take longer to be implemented because of their potential controversiality.

There may also be a shift or focus from small, project-scale solutions to larger, community-scale planning and development practices for saving energy, addressing overall community land-use patterns rather than just individual development sites. Conventional planning techniques that are inherently energy-conserving should be used even more as this happens. Urban infill, increasing development densities and the use of multi-family housing on a broad scale, greater use of mixed-use districting, and other techniques that have been advocated by planners in the past also make good sense from an energy standpoint and can be used to increase community energy efficiency.

The communities described in this paper are beginning to experiment with energy-conserving development regulations. Over time they will learn how much the design of development projects, neighborhoods and entire communities affects energy consumption. These communities have proceeded on their own, without many technical aids. So far, only a few lessons have been learned but over time these communities will gain experience and greater numbers of others will refine and adopt energy-conserving regulations.

There also may be much greater interest by homebuyers and developers in the details of site and community design that affect energy use. The public purpose of development regulations and community planning techniques that save energy should become stronger and more evident. As this happens energy will likely become a major focus of development regulation and indeed community planning as a whole.

List of Communities with Energy-Conserving Development Regulations/Publications List

LIST OF COMMUNITIES

Albuquerque, New Mexico
Solar Access Provisions (Comprehensive City Zoning Code, 1977, Sections 12c 13c, 14c, 20c, 22c, 24c, 25c, 30c, 31c)
Contact:
>Zoning Enforcement Officer
>P.O. Box 1293
>Albuquerque, New Mexico 87103
>505/766-7595

Boulder, Colorado
Permit Allocation System with Incentives for Energy Conservation (Ordinance No. 4208 as amended by Ordinance No. 4251, 1977)
Contact:
>Director
>Energy Office
>City of Boulder
>P.O. Box 971
>Boulder, Colorado 80306
>303/441-3270

Dade County, Florida
Site Plan Review Criteria for Energy-Efficient Site Design (Dade County Code, No date, Sec. 33-284 (.9, .21, .26, .38))
Contact:
>Chief
>Development Division
>Planning Department
>Suite 900, Brickell Plaza
>900 S.E. First Avenue
>Miami, Florida 33131
>305/579-2880

Douglas County, Kansas
Zoning Code Revision to Permit Underground Housing (County Zoning
Ordinance Sec. 3-1.06, and 3-1.25, 3-1.70a, 1979)
Contact:
> Director of Planning
> or
> Current Planner
> Lawrence/Douglas County Planning Commission
> 910 Massachusetts
> Box 708
> Lawrence, Kansas
> 913/841-7700

Davis, California
Flexible Location of Fences (Ordinance No. 1024, 1979)
Yard Projection Amendment (Ordinance No. 1025, 1979)
Deregulation of Clotheslines (Ordinance No. 876, 1977)
Landscaping Requirements (Ordinance 920, 1977)
Contact:
> Director
> Community Development Department
> 226 F Street
> Davis, California 95616
> 916/756-3740

Imperial County, California
Geothermal Overlay Zone, County Land Use Code, Sec. 83226, (no date)
Contact:
> Director of Public Works
> County of Imperial
> Courthouse
> El Centro, California 92243
> 714/352-2851

King County, Washington
Ordinances 4689-creates townhouse zone; 4687-townhouses in one- and
two-family zones; 4667-density increase in one- and two-family zones;
4304-flexible siting; 4688-policies
Contact:
> Director
> Growth Management Section
> Department of Planning and Community Development
> W217 King County Courthouse

516 Third Avenue
Seattle, Washington 98104
206/344-7550

Lincoln, Nebraska
Incentives for Solar Access and Energy Efficient Site Design (Design Standards for Zoning Regulations, Section 4, 1979)
Contact:

City Planning Division
City Hall
Lincoln, Nebraska 68508
401/475-5611

Los Alamos County, New Mexico
Solar Access Ordinance (Zoning Code, Article V-C. Sec. 11, 1977)
Contact:

Planning Director
County of Los Alamos
3200 Trinity Drive
Los Alamos, New Mexico 87544
505/662-4122

Port Arthur, Texas
Ordinance for Proper Building Orientation (Ordinance No. 79-78, 1979)
Contact:

Planning Director
P.O. Box 1089
Port Arthur, Texas 77640
713/983-3321

Sacramento County, California
Resolution for Proper Lot Orientation (Board of Supervisor's Resolution No. 77-987)
Contact:

Director
Planning and Community Development Department
827 Seventh Street
Sacramento, California 95814
916/440-6221

San Diego County, California
Ordinance for Mandatory Use of Solar Water Heat in New Construction (Ordinance No. 5324, 1978)
Solar Access Ordinance (Ordinance No. 5589, 1979)

Contact:
> Director
> San Diego County Energy Office
> 1600 Pacific Coast Highway
> San Diego, California 92101
> 741/236-2173

Windsor, Connecticut
Incentives for Energy-Efficient Location of New Development (Zoning
Regulations, Sec. 4.5.1, 11.2.3, 11.3.3)
Contact:
> Town Planner
> Windsor, Connecticut 06095
> 203/688-3675

Annotated Listing of Publications by Communities with Energy-Conserving Development Regulations

APRA/Morse Stafford Partnership. *Land-Use Regulation and Energy Conservation: Working Paper on the Subdivision Element*. Seattle, WN. King County Department of Planning and Community Development, 1978 (Draft).
Report on energy conservation considerations in subdivision development. Also contains comparative cost analysis of neighborhood and cluster development patterns and suggests ordinance provisions.

———. *Land-Use Regulation and Energy Conservation: Revised Working Paper on the Townhouse Element*. Seattle, WN. King County Department of Planning and Community Development, 1979.
Report on proposed townhouse zoning which includes energy and cost comparisons of detached versus attached dwelling units and a review and analysis of legal regulations affecting townhouse development.

———. *Land-Use Regulation and Energy Conservation, Working Paper on Revisions to the Zoning and Subdivision Codes and Guidelines for Climatic Design*. Seattle, WN. King County Department of Planning and Community Development, 1979.
A report on reconnemded energy-conserving revisions to land-use codes in King County. Includes ordinance language for townhouse district and development standards. Also has climatic design guidelines.

Living Systems, Inc. *Davis Energy Conservation Report—Practical Use of the Sun*. City of Davis, CA: April 1977.

Metropolitan Dade County Planning Department. *Energy Conservation: Proposed Goals and Policies for Urban Development.* Published by author. 909 S.E. 1st Avenue, Suite 900, Miami, FL, 1978.
Report on energy consumption and recommendations for conservation in Metropolitan Dade County.

Pulliam, Eric, *et al. Solar Ordinance Feasibility Analysis for San Diego County.* San Diego County, CA: December, 1978.
As directed by the San Diego County Energy Element Action program, this study was made in preparation for adoption of the mandatory solar water heating ordinance. The report examines and evaluates many aspects of the use of solar water heating in the county including economics, initial costs, maintenance and insurance, energy savings, impact on labor and others.

Sacramento County Planning and Development Department. *Sacramento County General Plan, Energy Element, Action Program.* Published by author, Sacramento, CA: 1979.

———. *Sacramento County General Plan, Energy Element, Policy Plan.* Published by author. Sacramento County, CA: February 1979.
A set of policies, objectives and strategies, adopted as an energy element, covering energy conservation and the use of renewable resources within Sacramento County.

———. *Sacramento County General Plan Energy Element—Research Report.* Advance Planning Section, Sacramento County, CA: January 1978.
A background study prepared for development of the county energy element. Contains information on energy use and conservation alternatives.

San Diego County Board of Supervisors. *San Diego County General Plan Part XI, Energy Element.* Published by author. San Diego, CA. October 1977.
A policy plan and action program for energy conservation and use of renewable resources in San Diego County. Lists major goals and objectives and policies and actions to meet those goals and objectives.

The Elements. *The Davis Experiment.* Washington, D.C: Public Resource Center, 1747 Connecticut Ave, 1977.
A summary of the energy-conservation actions taken by the City of Davis, California. Written for mass distribution, the report discusses the Davis building code, solar houses, clotheslines, solar swimming pool heat, fence and hedge setbacks, street widths, shade trees, and others.

Energy, Land, and Values:
The Davis Experience

PAUL P. CRAIG

Introduction

IN every age there are a few issues which loom especially large and which come to be seen, from the perspective of the sweep of time, as critical. From a short-term perspective, identification of major themes is controversial. Still, the exercise of identifying the important is useful, for it helps us to separate out the trivial and transitory. Energy, seen from this vantage point, meets all tests. Available information provides abundant evidence that ours is a period of transition; that we are in the brief couple of centuries during which oil and gas are key energy forms; and that it is up to us to figure out how to operate a rapidly advancing technological society in the post-oil/gas era. The imminence of this transition, and the powerful political, social and economic forces surrounding it, provide the matrix in which our discussion is embedded.

The prospect of a decline in global oil has been extensively documented (e.g., WAES, 1978). The dominance of political factors in the international oil scene finds eloquent support in an analysis by the head of the Venezuelan national oil company (Coradi, 1979). The need for the United States to develop alternative strategies has been the focus of several federal energy plans (e.g., NEP II, 1979). The Harvard Energy Project (Stobaugh, 1979) is the most recent in a series of studies pointing out the importance of developing a national strategy which deemphasizes use of oil and gas and heavily emphasizes demand moderation (conservation) and solar energy. This viewpoint also has been extensively advocated within both ethical and economic contexts by Amory Lovins (1978) in his Soft Energy Path strategy. The case for large, centralized technologies is well represented in a series of papers at the Oak Ridge Associate Universities Conference on Strategies for Energy Development (ORAU, 1976, especially Haefele and Sassin, 1976). U.S. policy has in practice (though often not in rhetoric) been consistently oriented toward centralized supply strategies.

Many types of analysis and arrays of perspectives have been used to examine energy issues. These range from the purely technical to the philosophical. The perspective which plays the dominant role in national policy discussion emphasizes the role of economics and is couched in the language of benefit cost ratios, payback periods, etc. Other analyses (e.g., Lovins, 1978, and Coradi, 1979) are value and power oriented, respectively. The issues are so complex and involve so many aspects of the society that virtually no perspective can be deemed irrelevant.

This paper focuses on the desirability of treating energy as an organizing principle for social decision-making. It argues that energy-induced stresses are already causing severe dislocations in the society and that these are likely to increase in severity. There will be a great deal of uncertainty as to how events are treating us and little confidence that any one decision is the best one. Under such circumstances it may be best to err on the side of conservatism and move so as to increase resiliance—the ability to withstand those shocks which, though difficult to anticipate, seem virtually inevitable. Resiliance is compatible with other goals, especially the redesign of urban regions so as to decrease the growing feeling of anomie which characterizes so many settings and which has become an omnipresent feature of life in modern America.

In exploring this theme, the paper moves quickly from a review of changes in perspective on energy futures to a discussion on the role of centralized, as compared to decentralized, authority and concludes with some discussion of local planning which includes energy as one among many determinants of action.

The paper focuses on the energy conserving city of Davis, California. This city has developed extensive conservation ordinances and has a solar development, Village Homes, in which solar equipped buildings are combined in an environment which places heavy emphasis on amenities and lifestyle, thereby showing through example that energy can be used as an organizing principle to assist in accomplishing other, equally important objectives.

Forecasts

In 1975, the Rutgers Center for Urban Policy Research published a report on a conference that was the antecedent of this one (Burchell, 1975). A series of articles on energy contain many insightful remarks. They also include a number which, only five years later, are remarkably out-of-date. Energy forecasters, like economic forecasters, are prisoners of their tools and their biases. Both can be misleading. There is a tendency in our society to believe experts. The experience of the post-embargo era shows that prognostication in this area is some ways from perfection and that our perceptions of what is possible can be turned upside down, if not overnight, then surely in the few years since the OPEC oil embargo. Consider the demand for energy in the United States. Sean Wellesley-Miller ("Fuel Prices and Energy Technology: A Look at the Future") reported energy forecasts for 1979 of close to 100 quads versus the 77 quads or so we will actually use this year. His oil imports for 1979 were about 12 million bbl/day, a full 50 percent greater than we will actually import. The forecast for 1985 was about 120 quads, a figure rather close to a current conventional estimate for the year 2000.

What has changed? In 1975, most forecasters could not bring themselves to take seriously the idea that the oil price jump of 1973 was simply the first of many shocks. They did not believe prices would jump again. The oil prices considered for 1980 in President Nixon's "Project Independence Blue-print" (the source of Wellesley-Miller's forecasts) were in the $7-to $11 range per barrel. Even taking into account inflation since that time, which would raise those prices by about 30 percent, the Project Independence oil prices are grossly out of line with the $20 per barrel we are paying today.

While the forecasts of 1975 were far too high on the demand side, they erred even further in underestimating the costs of new energy supplies. Prices for synthetic natural gas, synthetic oil from oil shale or coal, and methanol were all low compared to present day estimates by factors of two-to-four (even after correction for inflation). Oil shale is a particularly interesting example because its estimated cost of production has been a few dollars above the costs of conventional imported oil for many years—from the time when imports cost $2 per barrel to today, when imports cost $20 per barrel. There surely is some price at which oil shale can be produced, but experience shows that this price has an uncomfortable way of staying just ahead of the alternatives.

Forecasts of electrical generating capacity have shifted even more strikingly. Wellesley-Miller quotes estimates for installed capacity of 1500 GW in 2000 (versus about 400 GW today) and 800 GW of nuclear power by 2000. Yet, demand for electricity has slowed drastically, and reactors are in limbo. A plausible present-day estimate is 600 GW of installed capacity by 2000, of which at most 200 GW might be nuclear.

Wellesley-Miller's report reflected the thinking common in government and industry in 1975. Perceptions have changed strikingly in the past four years in three important directions:

1) The costs of supply replacements for oil are turning out to be much higher than had been expected, and the lead times for starting up new technologies are much longer than anticipated.

2) The opportunities for demand reduction through conservation are substantially greater than most people believed in 1975.

3) There is growing recognition that energy decisions affect the society at every point and that energy is a key locus for social decisionmaking.

Demand Moderation — Technical Considerations

The technical capabilities for moderating energy demand are enormous. The National Academy of Science's CONAES study (1978) provides technical analysis of energy futures ranging, in the year 2010, from 74 quads to 136 quads (versus 71 quads in 1975). Yet this analysis is conservative in many ways. To understand something of the potentials for technical conservation, it is useful to examine briefly the major sectors.

BUILDINGS

In the buildings sector, the CONAES results are based on conservation measures which, at their maximum, approximate the new building energy performance standards (BEPS) being proposed by the federal government for implementation now. Solar systems play only a minor role. Yet the Davis, California, experience (duplicated in many other places) shows that for an investment of just a few thousand dollars in solar design features, it is possible to reduce space heating needs to levels approaching hot water needs. At these levels, heating energy requirements are so low that residential heating needs become relatively unimportant, even with substantial fuel cost increases (McGregor, 1977; Ridgeway, 1979; Bainbridge, 1979). Similar conclusions have been obtained in the commercial sector, as evidenced especially in the Department of Energy conservation building in Manchester, New Hampshire (Fazzolare, 1979).

INDUSTRY

The industrial sector is more difficult to analyze since it is composed of a very large number of different processes, many of which are quite technical. The Harvard project (Stobaugh, 1979) reviewed the experience of a number of industries and concluded that 30-to-40 percent savings in energy use per unit of product can readily be obtained. In the long run, though, it appears likely that the industrial sector will prove more difficult to modify than other sectors, so that this sector may grow in relative importance.

DISTRICT HEATING

District heating systems are arrangements whereby heat is centrally produced at a large facility and distributed (usually as hot water or steam) to buildings to provide either space heating or (through absorption air conditioners) cooling. A major advantage of district heating systems is that the central source of heat can be a co-generation unit which provides heat and electricity simultaneously, thereby leading to significant improvements in efficiency (Williams, 1978).

Co-generation systems should be analyzed in terms of their efficiency measured by the second law of thermodynamics. The overall efficiency for production of heat and electricity can be as high as 48 percent, as compared with efficiencies of about 34 percent for electricity generation or 5 percent for space heating using a conventional gas furnace (Williams, 1979 [a]). Thus, there are substantial advantages in terms of fuel use efficiency and, under some circumstances, also major advantages in terms of economics.

Some decades ago district heating played a large role in many U.S. cities. Today, it is important in only a few (New York City is an important example—Consolidated Edison company still obtains a large portion of its business from steam distribution, as is apparent to anyone who has walked around New York City on a winter evening and seen the steam leaking from the streets). Some European cities are relying heavily on new co-generation (Lovins, 1978).

District heating is an attractive technology for areas which have relatively high energy use densities. However, there is competition between district heating and energy conservation. To the extent that energy conservation reduces the needs for space conditioning energy, the cost of district heating systems will rise. Even under relatively high density conditions, district heating appears competitive only if there are substantial fuel cost increases. A good example of a comprehensive analysis is the work of Root (1979) on the Conners Creek area around Detroit. This territory is served by Detroit Edison. It comprises an area of 13.6 square miles, and the total energy use for space conditioning is estimated at 1.7×10^{10} BTU by the year 2005, corresponding to 1.2×10^9 BTU/square mile.

This energy density is just about ten times as high as that of the developed portion of Davis (excluding agricultural and undeveloped land). The higher energy density in Detroit comes about both because of a higher density residential use and also because of a considerable amount of industry. Further, the region is particularly well suited for co-generation because of an existing generating facility which could be converted to district heating use. Despite this, the economic analysis is discouraging for district heating in Detroit, unless there are substantial increases in energy costs, or substantial subsidies. For Davis, with its low energy use per unit area, the economics of a large-scale district heating system are even less favorable.

The situation is even more discouraging if the promise of energy conservation in buildings is taken into account (Williams, 1979(b)). Average energy use for residential space heating in the U.S. is 22 BTU/square foot per degree day. This is the kind of energy use which was used in carrying out the Detroit study. Yet it is possible to reduce energy use for space conditioning to a small fraction of this amount through relatively simple changes in construction, much of which is already coming into use. Thus, new U.S. residential buildings use about 10 BTU/square foot per degree day; the current California standard is 6 BTU/square foot per degree day (the same as the current Swedish standard); and a townhouse in Twin Rivers, New Jersey, was retrofitted to bring the use down to 2.7 BTU/square foot per degree day.

To the extent that conservation can reduce the need for space conditioning, systems which have expensive distribution requirements (like district heating) will be at an economic disadvantage. This is not to say that there will not be uses for district heating. The uses will be in areas of high industrial demand for heat and in urban areas with high population densities, especially with apartments and high-rise buildings. But it will be important in planning for such systems to take into account reductions in heating and cooling demand which may result from improved building practices.

Finally, there are technical opportunities for small-scale district heating systems for particular applications such as integrated systems to provide heat and power to shopping centers and integrated industrial complexes. With high reliability power plants and the use of computer control systems, this kind of installation could play a significant role.

The key to the success of district heating systems in small communities like Davis will hinge on the development of economical systems which do not require extensive maintenance and which can be counted on to operate for long periods with the kind of reliability we expect from present oil, gas and electric systems. As such systems come into existence, it will become possible for local planners to insist on compact building clusters which can take advantage of the economics of district heating. There are many different technological choices, relying on conventional fuels, on biomass, or on solar power. All of these need to be explored, and the small size and relatively low prices of the systems justify a major federal program to assure that the needed experience is developed quickly.

TRANSPORTATION

The transportation sector uses about 25 percent of the nation's energy. Within this sector the automobile uses about 55 percent of all transportation energy, about equally divided between urban and intercity use. The actual importance of transportation is much greater than these numbers indicate, however, since there is a great deal of energy required to manufacture and maintain vehicles and roads. Hirst (1974) estimates that a full 40 percent of all the energy used in the nation can be attributed to direct and indirect requirements of the automobile.

The technical potential for conservation in automobiles is substantial. Federal regulations require that fuel efficiency rise to 27.5 miles per gallon by 1985, though these efficiencies are measured using the EPA procedure which underestimates actual fuel requirements substantially (by as much as 25 percent in some instances). Very much higher efficiencies are achievable. The Volkswagen diesel Rabbit already gets 40 mpg, and 60-to-70 miles per gallon economy is within view in vehicles with performance and comfort similar to that of today's cars. Trucks use about half as much fuel as autos, and this usage may be difficult to modify. Our society has intensified its specialization over the years, and this means that goods are increasingly manufactured in one place and consumed in another.

The airplane is a great unknown in future energy use. While planes use only about 7 percent of total energy, mostly for passenger transport, the opportunities for growth are almost limitless. The growing trend toward business and recreational travel suggests that airplanes may well absorb a growing portion of our disposable income, even if fuel prices increase substantially (CONAES, 1978). Cutbacks in commercial airline service are attributed to increases in fuel costs. The effect is probably transitory since the desire of people to travel has been growing dramatically for a number of years and shows little sign of stopping.

The transportation sector will certainly prove one of the most difficult to deal with in the coming years of oil shortage. Transportation is the one sector which is almost entirely "hooked" on liquid fuels. It is the sector most sensitive to oil import shortages. The replacement time for cars is, however, short (about ten years), and this means that major improvements in efficiency can occur quite rapidly. Indeed, the most striking energy use change since the 1973 OPEC oil embargo is in the fuel efficiency of new cars, which has improved from about 14 mpg in 1973 to near 18 mpg today.

The transportation sector is one which will require a number of major national policy decisions. The direction these will have to go is, I believe, toward:

1) improvements in efficiency of cars going well beyond the mandated 27.5 mpg standard for 1985.

2) improved mass transit for inter-urban transportation (which accounts presently for half of all automobile mileage).

3) changes in land use patterns to reduce the need for urban transport.

4) new concepts of urban mass transit based on light rail and computer directed rubber-tire systems.

5) improved efficiency of airplanes.

Each of these directions raises major problems, though these problems are likely to be less than the consequences of not pursuing them. We will not explore these issues further here.

Given pressure, change can occur rapidly. In 1973, the City of Los Angeles experienced severe oil shortages, and a major conservation program was undertaken. Per capita energy use was decreased by 4.5 percent and commercial energy use by 10.5 percent. A similar result occurred in New York City, where in 1974, gas use decreased by 6 percent, electricity for heat by 12 percent, and gasoline use by 4 percent. Within the entire OECD, total energy use declined by 3.3 percent in 1974-1975 (OECD 1977). The prospect for energy futures based on much lower energy use, while economic growth continues, has been documented extensively for the U.S. (CONAES, 1978). For a doubling of GNP between now and 2010, energy use was found to be variable over a range of about two—from virtually constant to doubling, depending on energy prices and on policy.

Energy and Lifestyle—"The Davis Experience"

Technical improvements can lower energy use substantially and provide much flexibility for planning a low-oil future. But does the "tech fix" approach go far enough? If the objective is purely to meet the demands of a society evolving in the future just as it has in the past few decades, end use improvements are complemented by a vast array of emerging supply technologies. These include both fossil and nuclear based systems, plus a spectrum of renewables using sunpower directly (solar collectors) or indirectly (biomass, wind, waves). Taken together, it appears likely that significant portions of the U.S. could, by sometime toward the middle of the next century, be energy self-sufficient and fueled almost entirely by renewable resources (CEQ, 1979; Craig, 1978, 1979).

But this optimistic picture is hardly the whole story. There are two major issues:

1) What do we want our society to look like in the next century?

2) How do we get through the next few years, when we will, at best, be painfully vulnerable to energy supply disruption?

These issues are intimately linked. The way we proceed to get through the next few years will largely determine what our society will look like in the next century, and our long-term goals ought to influence our day-to-day decisions.

If we fail in our efforts to keep our energy appetites in harmony with supplies, there is no end of trouble in store. The warnings from OPEC are clear, perhaps nowhere better stated than by Coradi (1979) in a somber warning about the prospects of global economic collapse if the U.S. fails to curb its oil appetite. The Iranian revolution led to a relatively minor cutback in oil which was, however, quickly converted into major gasoline lines by confusion and an excess of governmental intervention.

The future is uncertain always and probably more so today than at most times in the past. In times of uncertainty, it often pays to play it safe, and this strategy may have important ramifications for land use. The Davis experience offers some insight about what kind of directions can work.

Davis is a small (30,000) university town. It was once the university farm and is still the University of California "Aggie" school. Davis is located in the central valley of California, about 60 miles northeast of San Francisco and 15 miles southwest of the state capital, Sacramento. A few years ago the city decided to go in heavily for bicycles, energy conservation and solar energy (McGregor, 1977; Ridgeway, 1979). The reasons this occurred have been documented in part (Vine, 1979), but it remains something of a mystery why one city should decide to do so much while hundreds of similar towns are doing little.

Davis is known as a town of bicycles. There are almost as many as there are people. There is a vigorous conservation ordinance (Davis, 1976, 1977) which has led to natural gas usage about 60 percent of that of neighboring communities and electricity bills which are about 80 percent of those of adjacent communities. Solar systems abound, and the nation's first solar subdivision, Village Homes, is a resounding success (Bainbridge, 1979).

The Village Homes development is especially important in helping to understand what is possible. While virtually every home in the Village Homes has some solar component, the energy aspects of the community may be secondary in importance to the ambiance achieved through careful attention to planning, large common space areas, bicycle paths, and a community recreation hall and swimming pool. A small commercial center is just about to be constructed, and this will make it possible for some people to walk to work and many residents to walk for at least some of their shopping needs. (Of course, almost all needed services in the small town of Davis are within a 15-minute bike ride).

Energy consciousness played an important role in the complex decisions to move the town in these unusual directions, but concern over values was even more important. Recurring themes in the debates leading to the Davis ordinances dealt with amenities and quality of life. This concern is clearly seen in the process by which developers are given approval for new construction. There are only limited opportunities for new development between now and the end of the century. (The Davis Master Plan [1977] calls for increasing the town population from its present 30,000 to 50,000.) Permits for new developments are thus difficult to obtain, and concepts which have high amenity levels are likely to win out in the competition. This same kind of situation may be occurring with solar systems elsewhere in the state, with preferential or accelerated permit approval going to proposals which include solar elements (and with extra incentive provided by the 55 percent solar tax credit which California has had for the past two years).

The changes which have occurred within Davis in the past half decade have affected the attitudes of most citizens. There is a pride in living in the bicycle city, and there is surely a pride associated with living in Village Homes. To those of us who are privileged to live here, it is something of a mystery that the changes which Davis is experimenting with are not being picked up immediately throughout the nation. Perhaps this is not surprising, though. Excepting in crisis situations, change takes place slowly. Our cities and towns are built to last for many decades, and revolutions which occur overnight are based on antecedents that may cover decades.

Yet signs of change abound. Throughout the nation experiments are going on. Some are successes; others are not. Almost every city has its share of examples of innovation, and it is likely that as energy prices increase and awareness of the prospect for shortage becomes embedded in the public consciousness, these centers of nucleation will expand.

Energy Densities, Sunlight and Distributed Technologies

The energy conservation measures implemented in Davis have begun to take effect, and substantial reductions of use have been found relative to neighboring cities. Within California as well, energy use is well below the average for the U.S. California uses 280 million BTU per capita per year, about 85 percent as much energy per capita as the U.S. as a whole. Energy for transportation, surprisingly, is equal to the U.S. average, while the benign climate and emphasis on service and entertainment industries lower the energy use in the residential/commercial and industrial sectors to 68 percent and 65 percent of the U.S. average, respectively (Ruderman, 1979).

Another way to view energy use is in terms of energy density. Energy density is important for two reasons: a low energy density leads to the possibility that solar energy sources may meet a significant portion of the total need, while a high density is advantageous for use of integrated energy systems (some of which are listed below). Of course, the requirements need not be, and should not be, incompatible. Low average density can be combined with high localized density, leaving space for green belts and other amenities.

The city of Davis has a land area of 43,000 acres, of which about 80 percent is presently in agriculture or agriculture reserve. Based on the total area, the population density is 470 persons per square mile, substantially below that of San Diego (2,200), Los Angeles (6,100), or San Francisco (16,000). This low population density is highly advantageous for solar systems of all types. Expressed as a percent of the incident solar radiation on the entire city per year, the total commercial energy used is: Davis 0.9 percent of insolation; San Francisco 4 percent; Los Angeles 12 percent; and San Francisco 30 percent. Solar energy conversion systems operate at a maximum efficiency of about 20 percent, and more typical numbers are 5-to-10 percent for non-biological systems and 1 percent for bioconversion. Thus, Davis is low enough in population density that an appreciable portion of the city's needs might be met with renewables, whereas this is out of the question for the higher population density cities, even if they practice extensive conservation.

This comparison, while interesting, is misleading in several important regards. There was no effort made to distinguish among energy needs for space conditioning, for industrial processes, and for transportation. Yet needs of industry for high temperature heat and the needs for liquids in the transportation sector place special burdens on any solar system. Further, there is no compelling reason to limit solar systems to within city limits (see Craig, 1978, especially the report by R. Nathans therein, and Craig, 1979).

If we expand our horizon, there are many additional choices, especially in the desert regions. Indeed, on a national basis the land requirements to meet our total national energy needs are quite modest. Only one-tenth of one percent of the Southwest receives as much energy as the entire nation uses, and even at a conversion efficiency of 10 percent, national needs could be met by an area less than 250 miles on a side.

High population densities have significant advantages for energy distribution. Urban core regions are characterized by quite high densities. In Hanover, Germany, half of the total energy used by the city is concentrated within a radius of 1.5 km of the city center, and 80 percent of the energy lies within 5 km of the center (Haefele, 1976). An extensive analysis of German energy use by Düring, et al. (1979), has emphasized the importance of energy density as a guide to planning. The most frequent energy load in Germany for space conditioning is about 30 megawatts per square kilometer (2.8×10^8 BTU/mi) which corresponds approximately to space conditioning energy use in the low-density cities of California (e.g., San Diego). Even at these relatively low energy densities, the study found that decentralized energy systems were close to economically competitive. Some of the systems examined were district heating and cooling, integrated thermal storage, diesel generators with heat recuperators, and various co-generation schemes.

If these results are transferable to the U.S.—and there is a growing body of data which suggests that they are—then an important element of new design is the linkage of energy systems among buildings to provide economies of (small) scale together with much improved efficiency.

These considerations suggest that there will be increased opportunity to achieve both economic savings and increased system resilience by examining integrated energy planning for new land use development. Such planning will not be accomplished easily but will become increasingly attractive as, or if, energy shortage becomes an increasingly important element in our lives.

Levels of Decisionmaking

The discussion above made little reference to activity at the federal level, save for the energy forecasts which were distinguished chiefly by their rapid rate of obsolescence. This omission is not by accident. Neither the federal nor the state government played any role in the Davis developments, and there was no federal support provided to Village Homes (excepting a small HUD grant received only after the success of the community was proven). There is a lesson here, and it is worth examining. It is a lesson of self-reliance and of the problems of large bureaucracies.

Energy, as I have noted, is serving as a focal point for a large array of conflicts involving technology, values and national priorities. The most complex of the issues come to focus, inevitably, at the national level. This focus can, in time of concensus, lead to rapid national commitment. Before concensus is achieved, however, the national focal point can lead to stagnation, and innovation is more likely to occur at a lower level. But the matter is far worse than that, for there are fundamental conflicts between state and local versus national objectives.

There are many reasons for this. At the federal level there has long been a desire to simplify energy issues so that they become comprehensible by the federal bureaucracy. Because our nation is so large, this leads inevitably to over-simplification. Issues must be rendered manageable. For some purposes this causes little difficulty, at least in principle. A decision to hold oil imports constant—a part of President Carter's energy plan—must be national in character. It is a statement from one nation to the world and must come from the nation's leader. Internally, the matter is far more complex. Who shall cut back on oil use? The northeastern portion of the country, which is heavily reliant on oil for heating; or the West which uses gas for heating? This kind of issue leads quickly into allocation strategies, an area in which the federal government has frequently demonstrated its incompetence—most recently during the Iranian revolution and its associated gasoline lines.

Natural gas pricing structure is another area where the federal government became involved in a complex process designed initially to hold gas prices up and later used to hold them down. The result has been a series of natural gas shortages (Winter of 1976) and surpluses (the current "gas bubble"), with the government oscillating wildly between policies designed to get users to shift off of gas, followed by policies to move them back (see, for example, the acerbic review in a recent Harper's [Bethell, 1979]).

The situation regarding technological choice is similar. It is far easier for a program manager to disperse large sums of money in a few giant contracts than much smaller sums to many small projects. Large contractors, be they aerospace firms, think tanks, or A&E firms, can readily build into their overhead structure the personnel to provide unlimited documentation, time flow charts, cost effectiveness analyses, and the like. These organizations find it easy to be responsive to bureaucratic demand, and, thus, we have evolved a system which discriminates, systematically, against small entrepreneurs, small businesses, and small units of government. Our technological armamentarium, as viewed from the stronghold of Washington, is based on centralized and massive technologies: coal gasification and liquifaction, light water and breeder reactors, solar power towers, fusion tokamaks, and (perhaps) even a satellite power system.

Redesign of cities exists as a program element, but one must search hard to find it. Conservation is finally beginning to develop a little muscle (to contrast with the lip service given it for many years, including even that moment on April 21, 1977, when it became "the moral equivalent of war" [MEOW]). But it lags far behind, and the concept of innovative or long-run R&D remains foreign to the Washington view of the matter.

At the state level perspectives are different. There is recognition of the environmental problems of all technologies. This sensitivity to destruction of the local environment gives rise to continuing state/federal antagonism: coal development in the Great Plains states, the California automotive emission standards, and the coastal zone act are illustrative. Even at the state level, land use planning has relatively little importance, but there are important exceptions. The California Energy Commission has responsibility for approval of siting of electric generating facilities and for establishing a wide variety of energy conservation standards. The Commission grew out of state need, as perceived by former state assemblyman Charles Warren (until recently head of the President's Council on Environmental Quality). Despite considerable agony and a great deal of political complexity, the Energy Commission is bringing a state perspective to bear on both of these issues, and the result will be more rational land use for energy systems and a great deal more conservation. (California's residential building standards are getting close to being as good as those of Davis, and it seems possible that the federal government will shortly pull ahead of both with its Building Energy Performance Standards—but it has taken a long time.)

The Davis experience resulted from a confluence of accidental factors. Davis is unique in the particular directions it chose to follow. It is far from unique in its concern over planning a future which will include energy as one important factor among many and which will serve well if times become more severe. What, then, are the considerations which may assist a community to adapt to rapid but often unknown change? The most important is to build for man, not against man. Witness Louis Mumford (1970), ". . . to be condemned to a devitalized megalopolitan habitat . . . is to unlearn and discard all the lessons learned in cooperation by living organisms during some three billion years on earth," or the architect Richard Stein (1978) on energy and architecture, "There is an appropriate, precise relationship between form and the controls that produce it."

Our built environment molds ourselves. In a period of instability, one designs for resiliance and cooperation. We will need both large and small systems, but the large ones should not make us captive. Buildings can become largely self-reliant in energy and designed to "fail-safe" (be habitable in time of energy shortage or outage). Urban areas can be designed to minimize transport needs, to provide recreation close at hand instead of at the end of hours of travel. Public transportation is all but ignored in public policy today. Yet, as the population ages, there will be ever more of us unable to rely on the car and yet seeking mobility.

Energy is and will remain a major issue. It is not our only national issue, perhaps not even our most important. But used as an organizing principle, it can provide additional leverage for structuring a more human oriented society, a society developed from humanity rather than from technical feasibility. Technology is flexible and can move us in a manifold of directions. The time has come to begin to make these choices.

References

Bainbridge, D., Corbett, J. and Hofacre, J. *Village Homes Solar Designs*. Rodale Press, 1979.

Bethell, T. "The Gas Pricefixers." *Harper's Magazine* (June 1979).

Burchell, R. W., and D. Listokin (eds.). *Future Land Use*. Rutgers University Center for Urban Policy Research, 1975.

California Energy Commission. *Second Bienniel Report* (draft), 1979.

———. "The Good News About Energy." Council on Environmental Quality, 1979.

Coradi, A. "Oil and the Exercise of Power." *Foreign Affairs* (Spring 1979).

Craig, P. P. and M. D. Levine. "Distributed Energy Systems in California's Future: Issues in Transition." Lawrence Berkeley Laboratory, DES-R-02, 1979.

———. Christensen, M., M. Simmons, M. Levine, and D. Mukamel. "Distributed Energy Systems in California's Future." U.S. Department of Energy HCP/P7405-01, 1978, (available from NTIS).

CONAES. "U.S. Energy Demand: Some Low Energy Futures." *Science* 200. 142-152 (1978).

Davis, City of. *Energy Conservation Building Code Workbook*. 1976.

———. General Plan, 1977.

During, K., H. Kullman, D. Oesterwind, D. Orth, and A. Voss. *Central versus Decentral Energy Supply Strategies for Industrialized Countries—Soft or Hard Energy Strategies*. Second Conference on Energy Use Management. Pergamon Press I, 137, 1979.

Fazzolare, R. A., and C. B. Smith. *Changing Energy Futures*. Second International Conference on Energy Use Management, Part III, Buildings and Dwelling. Pergamon Press, 1979.

Haefele, W., and W. Sazzin. "Contrasting Views of the Future and their

Influence on our Technological Horizons for Energy." Symposium on Future Strategies for Energy Development. Oak Ridge Associated Universities, 1976.

Hirst, E. "Direct and Indirect Energy Requirements for Automobiles." Oak Ridge National Laboratory, ORNL-NSF-EP-64 February, 1974.

———. "Energy Consumption for Transportation in the U.S." Oak Ridge National Laboratory, ORNL-NSF-EP-15, 1972.

Institute of Gas Technology. "Energy Topics," October 1, 1979.

Lovins, A. *Soft Energy Paths*. Ballinger Press, 1978.

McGregor, G. "Davis, California, Implements Energy Building Code." City of Davis, California, 1977.

Mumford, L. *The Myth of the Machine: The Pentagon of Power*. Harcourt Brace Jovanovich, 383, 1970.

BEP II, National Energy Plan II. U.S. Department of Energy, May, 1979.

OECD. Environment and Energy Use in Urban Areas. OECD, Paris, 1978.

ORAU. Strategies for Energy Development. Oak Ridge Associated Universities, Tennessee, 1976.

Ridgeway, J. *Energy-Efficient Community Planning*. Emmaus, PA: J. G. Press: Box 351, 18049, 1979.

Root, T. E. *Power Plant Retrofit for District Heating in Detroit*. International Conference on Energy Use Management, R. A. Fazzolare, and C. B. Smith, (eds.). Pergamon Press, 1979, p. A102.

Ruderman, H., J. Kay, F. Littermen, A. Vsibelli, and G. Welch. *An Energy-Environment Data Book for California, Hawaii, and Nevada*. Lawrence Berkeley Laboratory, LBL-7821, 1978.

Stein, R. G. *Architecture and Energy*. New York: Anchor Books, 1978.

Stobaugh, R., and D. Yergin. *Energy Futures—Report of the Energy Project at the Harvard Business School*. Random House, 1979.

Vine, E. L. *Planning for an Energy Conserving Society, The Davis Experience*. Distributed Energy Systems Group Report DES-R-09, Lawrence Berkeley Laboratory, 1979.

WAES (Workshop on Alternative Energy Strategies). McGraw-Hill, 1978.

———. "Industrial Co-generation." *Annual Review of Energy* 3. 1978, p. 313.

———. *Toward a Fuel Policy for Industrial Co-generation*. International Conference on Energy Use Management, Fazzolare, and Smith, C. B. (eds.). Pergamon Press, 1979(a) p. 1476.

———. "Use of Oil and Gas for Co-generation." Draft manuscript, October 15, 1979(b).

The Planning, Regulatory, and Design Implications of Solar Access Protection

MARTIN S. JAFFE

THE use of solar energy as an alternate energy resource is likely to have land use implications, both in terms of how cities develop and how they appear. On a superficial level, one can imagine that solar collectors (both active collector arrays and passive glazing) will become more prevalent as a feature of urban architecture. On a slightly more subtle level, many more features of the urban landscape will have to be modified to accommodate this new technology, as a result of the solar access and energy conservation requirements of these systems. These changes will affect how buildings look (their height and bulk), how they are positioned on lots (orientation and setback), and how they are landscaped. In a broader context, these changes will not radically alter the appearance or form of cities, but will offer new opportunities for increasing the visual diversity of the streetscape.

The Solar Access Issue

Solar energy systems require access to direct sunlight in order to operate most effectively. Although indirect sunlight (diffuse or reflected light) can provide an appreciable fraction of available solar radiation, and also can conserve energy by substituting natural daylight for artificial illumination, direct sunlight provides the greatest concentration of available solar radiation in a form most easily used by solar energy systems. Solar collectors—the components of active or passive solar technologies that absorb solar radiation and convert it to usable heat—must be located in places where they will not be shaded or deprived of this direct sunlight.

Although this seems to be a simple and easily achievable objective, it is actually quite difficult in a real-world setting. For example, a recent solar access study by the Central Naugatuck Valley Regional Planning Agency in Connecticut found, in a survey of solar installers, that fully one-third of all the solar installations in the region were either shaded or will be shaded in the next five years.[1] Most of the shading problems were caused by vegetation growing on site and on neighboring sites that blocked access to direct sunlight and shaded the collectors.

The shading of collectors by off-site development or vegetation is a problem because American property law does not recognize an inherent right to sunlight as a property interest. According to an old Common Law doctrine, a person owns from the center of the earth to the heavens.[2] But sunlight does not fall from directly overhead and must pass over adjoining properties before reaching a particular site.[3] Therefore, the traditional ownership of land in both a horizontal and vertical direction does not address the physical fact that sunlight enters a property obliquely and is thus subject to interference by actions beyond the property owner's control. The laws of England and Japan do recognize a limited property interest in sunlight, but these doctrines have been rejected in the United States.[4]

Collector shading also is affected by factors other than legal issues. For example, the type of solar collector used, its location on a lot or building, its daily and seasonal use requirements, and the extent of solar radiation blockage to a collector all enter into a determination of whether solar access is obstructed. These technical factors must be assessed before a decision is reached that solar access to a collector had been obstructed to the extent that the use of solar energy is technically or economically infeasible for a particular installation or location.

Private Approaches to Protecting Solar Access

Many of the legal issues of solar access protection can be settled between property owners by using private agreements. These can range from a simple restrictive covenant negotiated between landowners, where both contract not to shade each other's properties or collectors, to complex arrangements establishing limited three-dimensional property interests in adjacent parcels. These latter instruments are known as solar easements.

Twelve states presently have legislation affecting the preparation and recording of solar easements.[5] Generally, these statutes establish standards that must be met in order to record a valid solar easement. Their purpose seems to ensure uniformity among all such instruments, rather than specific authorizations that such property interests are valid. There is little doubt that they are valid since people can choose to restrict their development options in numerous ways, provided that the restrictions do not violate public policies.

These private approaches are limited, however. First, they are by definition voluntary. Both property owners must agree to abide by the covenant or easement provisions and both must agree to enter into the arrangement collectively. A recalcitrant neighbor, or an individual who does not wish to restrict his or her development options to protect another's solar access, can easily stymie a solar energy user's ability to secure continued solar access to a collector. Moreover, the enforcement of these instruments must be through litigation or declaratory judgment, an often costly and time-consuming process.

Public Approaches to Protecting Solar Access

Because of these limitations, it is often better to consider public approaches to protecting solar access. These public approaches can be categorized into two distinct strategies—lot-by-lot protection and area-wide protection.[6] Lot-by-lot protection strategies offer solar access only to a solar energy user. A person must have made the investment in a solar energy system before he or she can consider these approaches. The area-wide strategies, however, offer protection to both solar energy users and to persons who are not, at present, considering solar technologies.

These distinctions are important because solar access is lost for the life of an obstruction. For example, once a taller building is built which shades one or more lots to the north, then these lots will be shaded for the lifespan of the building. This can hinder future as well as present solar energy use on these shaded lots. While this shading may not affect a present homeowner using conventional heating fuels, it will prevent him or her from considering solar technologies as an energy option. Likewise, it will prevent future owners of the lot from considering this energy resource.

LOT-BY-LOT PROTECTION STRATEGIES

There are three lot-by-lot strategies that can be considered by local governments wishing to protect solar access for solar energy users. These are a public nuisance strategy, a prior appropriation strategy, and a public easement strategy.

The public nuisance approach is legislatively mandated in the state of California under the provisions of the Solar Shade Control Act of 1978.[7] Under this statute, communities can choose to exempt themselves from the provisions of the act; it is a discretionary rather than mandatory approach. The statute declares that collectors that are installed in compliance with the act are protected against the shading of more than 10 percent of the absorptive surface of the collector. If a collector is shaded beyond this amount, then injunctive or compensatory actions can be maintained under state public nuisance provision of the penal code. The person using the solar energy system must meet the burden of locating collectors in specified locations to qualify for the solar access protection provisions. Neighbors, also, must meet the burden of locating development on their parcels so as not to obstruct solar access to an existing collector.

The California approach has not been recommended in at least one legal study,[8] obstensibly because of its reliance on public prosecution under criminal statutes for its enforcement. On an administrative level, this may be a difficult program to enforce if collector installations become numerous and incidents of shading become common. Public expenditures for prosecution are likely to be costly, and the burden on the courts large.

New Mexico's Solar Rights Act establishes a different approach, called the prior appropriation strategy.[9] Under this legislation, sunlight is treated the same as water—it is a common resource that must be allocated among competing users. Under Western water law doctrines, the beneficial use of water is allocated to a "first in time, first in right" doctrine, and this principle has been expanded under the Solar Rights Act to sunlight and solar access as well. Under this approach, the person who first establishes a solar collector is protected against shading by neighbors; the person has obtained a prior right to direct sunlight that cannot be interfered with by subsequent actions of other individuals.

The planning implications of this approach are unclear. Provided that a solar energy user meets certain collector standards of size and beneficial use, a solar collector installation is protected at the expense of neighboring development actions on lots to the south. This means that a small collector array used for domestic water heating on a mobile home can prevent the construction of a multimillion dollar highrise to the south, if the construction would shade the collector.

Enforcement of this strategy is likely to have enormous impact on land use planning and development in areas undergoing both development and increasing solar energy use. Essentially, the prior appropriation rights seem to make zoning and other land use objectives more difficult to achieve and expensive to implement. A high density development that is both desired, planned for, and allowed under existing zoning but which obstructs the solar access of adjacent solar energy users can easily result in compensatory "taking" claims under state law, or may even be prohibited by injunction in some cases. The individual objectives of securing solar access could thwart the larger development objectives of the community as set forth in local plans and development regulations.

A variant of this approach is used in Los Alamos, New Mexico, which requires the filing of a solar access statement by a solar energy user to secure solar access rights under local regulations. The filing of the statement would restrict development on lots to the south that could shade the collector installation—the city will not issue a building permit for any development that breaches a light plane from the collector over adjacent properties. Essentially, this establishes a public solar easement created by local ordinance.

As of 1978, fourteen applications for these public easements have been filed at the city.[10] All were negotiated by the adjacent property owners prior to the establishment of the solar access statement, so there has been no litigation testing this approach. It is unknown whether this approach is being used or considered by other communities, but it seems likely that variants of this system could arise in California, under the provisions of the California Solar Rights Act of 1978.[11]

Under the California Solar Rights Act, communities may require solar easements as a condition of approving subdivision plats. Solar easements are authorized as a valid exaction under state law—communities requiring local plat approvals could require the developer to establish solar easements between lots prior to their approval by the community. This is analogous to a "public" solar easement requirement, although the instrument would be filed as a private restriction binding on all lot owners subject to the easement. Initial inquiry into how well this technique is working suggests that it may have some problems; San Diego County found that developers balked at meeting this dedication requirement, citing a per-lot cost of $400 for additional surveying expenses.[12]

The public easement approach also has its limitations as well as its strengths. Although solar access would be protected by public action, it would impose a considerable burden on both the developer and homeowner. The burden on neighboring lot owners or developers is fairly clear—increased development costs and decreased development options due to the solar access restrictions. What is unclear is the burden placed on the community as a whole.

The shadows cast by buildings to the southeast and southwest of a solar installation, and not necessarily by buildings on lots lying due south, will present the greatest solar access problems. So a public easement requirement may affect development activities on at least three lots to the south of a solar installation. Where there are a large number of solar installations, the public hearing and notice requirements can become expensive, particularly if courts perceive the public solar easement restriction as a rezoning action. If considered a rezoning, then publication, multiple hearings and revisions to comprehensive plans (in consistency states) may be required.

Areawide Public Strategies

The limitations of the lot-by-lot approaches suggest that an areawide solar access approach may be better. In an areawide approach, all lot owners would be given a certain degree of solar access protection. In exchange, all would give up some development options, based on these solar access objectives. How much solar access will be provided and how much development will be restricted become actions of public policy.

This focus on public policy means that planning—particularly land use planning affecting the location of buildings and vegetation—becomes preeminent. While the lot-by-lot approaches are basically incremental techniques that arise only upon a private action to install solar collectors, the areawide technique is comprehensive and can affect specified districts, areas, or even entire communities.

Four states already authorize local governments to plan for and regulate development for solar access protection or solar energy use.[13] This authorization generally takes the form of a phrase involving solar energy in the preamble to zoning or subdivision review enabling legislation, establishing the general purposes of the land use control technique.

It is unclear whether communities require this specific authorization in order to plan and regulate for solar access. Although the traditional zoning purposes of ensuring "adequate *light* and air," as set forth in the Standard Zoning Enabling Act, probably would not authorize a community to zone for solar access, it is likely that interpretations of the broader "general welfare" purposes of zoning would. Given the plethora of policy statements in state and federal energy legislation, it seems probable that courts will find solar access protection by local governments consistent with the stated public policies, meeting the burden of showing the promotion of the general welfare.

The areawide solar access strategies can all be carried out by conventional land use controls—traditional zoning, subdivision review and environmental regulations. They need not use any innovative techniques, such as public solar easements filed in a local agency, nor do they require the establishment of entirely new property interests, such as the prior appropriation approach. Moreover, they can be administered by existing local agencies as part of existing forms of regulation. These factors suggest that the areawide techniques may be more useful to communities than many of the lot-by-lot approaches. Therefore, they will be discussed in greater detail than the lot-by-lot techniques.

PLANNING IMPLICATIONS

It is unclear what level of planning is necessary by local governments to support a comprehensive solar access protection program. This extends not only to the question of what level of evidence will be needed by a court to sustain local solar access regulations, but touches upon the more subtle question of what kinds of data are needed to develop rational solar access standards and objectives.

To meet the evidentiary element of planning, communities will probably have to show that solar access protection offers opportunities to conserve conventional energy resources and that it will promote the use of renewable energy supplies to the benefit of the community. This should sustain the general welfare purposes in terms of interpreting the grant of authority offered by conventional state enabling legislation.

This is not a great burden to meet. Comprehensive energy audits are probably not necessary in most communities (although they can be useful in the next stage of establishing regulatory standards). Some estimates of conventional fuel use, availability and savings may be sufficient. This can be gathered from secondary data sources as well as from actual experience with solar energy use in the community.

Setting regulatory standards, however, becomes more complex. Decisions will have to be made concerning which technical choices among solar technologies are best for the community, where the collectors for such systems are most likely to be installed, and what degree of solar access already exists within the community. Essentially, communities will have to determine whether solar access to rooftops, walls, or south lots will be protected and whether enough solar access already exists in the community to justify public programs promoting solar energy technologies.

Obtaining this planning data and organizing it in a form to develop policies will require some administrative effort. But these efforts probably will not be onerous—for example, the assessment of solar orientation and shading for three smaller communities in the Central Naugatuck Valley region of Connecticut required only about sixty staff-hours to evaluate aerial photographs of the communities.[14] Based on this and other data, decisions can be made about possible solar energy options and collector locations. A rational solar access objective can be adopted, and standards derived from existing development patterns can be evaluated to meet this objective.

Care must be taken that the solar access planning component does not conflict with other planning objectives. For example, solar access is generally best on south-facing slopes, because of the incresed availability of solar radiation and the shorter shadow length in these areas. But a community must decide whether it wants increased development—even solar development—in these areas. Policies and regulations governing erosion and sedimentation control or hillside aesthetics may preempt a program emphasizing greater south-slope densities for solar development. Another example would be the obvious conflict between tree preservation programs and solar development, where access to sunlight might be obstructed by the preserved vegetation.

One conflict that will probably not arise is the one between the provision of low- and moderate-income housing and solar access. It is true that better solar access can be obtained on larger lots, merely because the lot owner has greater control over possible obstructions, but it is also true that south-wall access can be maintained at fairly high housing densities. Detached housing can probably have total access to the south walls of buildings at a density of eight to nine dwelling units per acre on level terrain.[15] This will be considerably higher on south-sloping terrain or if multifamily housing is considered. Solar access depends more on how the density is distributed on a site, than on the overall total density of the parcel.

ZONING IMPLICATIONS

Conventional zoning can provide an important component of a community's solar access program. Zoning will have to be evaluated or modified for two reasons: first, to remove barriers to the installation of solar equipment; and, second, to modify building geometries to minimize problems with the shading of collectors or collector locations.

Removing barriers involves changing zoning restrictions that prohibit the installation of solar equipment on or in buildings. For example, height or yard restrictions in many communities may prevent a homeowner from installing a solar collector on a lot. In California, under the provisions of the

Solar Rights Act of 1978, communities are prohibited from adopting regulations which unreasonably restrict the use or installation of solar energy systems, unless the restriction is specifically related to protecting the public health or safety. Communities in other states may have to reevaluate existing regulations to make sure that solar equipment can be used, or installed. This may involve such diverse actions as: expressly removing solar energy systems from the definition of "structure" in zoning ordinances (to allow collectors to be placed on a lot without violating setback restrictions); requiring a special permit for additions to roofs that are likely to cause substantial shadows across property lines (to protect against the shading of possible collector locations on adjacent lots); and revising aesthetic controls, to allow the installation of solar collectors in scenic view corridors and historic districts.[16]

Modifying zoning regulations in this manner can ensure that solar energy systems can be installed in existing buildings (retrofitted) as well as in new development. For stable communities, with little or no new development or infill projects, these modifications may be all that is necessary to develop a solar energy program (provided that vegetation standards and street tree programs also are reevaluated or modified). There are only a few changes that are possible with built-up communities to modify the solar access potential of existing development; removing barriers to solar installations would be an important consideration in these areas.

Modifying zoning provisions to actually protect solar access is a step most useful for communities undergoing substantial growth or infill development. Zoning actions generally affect only new development; even if zoning ordinances were to be changed to protect solar access, existing buildings would probably be exempt as nonconforming prior existing uses. Protecting solar access in nongrowth areas may affect additions or modification to existing buildings, though, and the solar access implications of these development actions may wish to be reviewed.

Height, setback, frontage and yard area standards can all be changed to minimize the shading of solar collectors or collector locations in developing communities. Each of these zoning techniques offer possible alternatives that can be considered and used to protect solar access.

Height restrictions. Height restrictions may have to be changed if an analysis of development patterns and shading problems indicates that shading can be minimized if buildings are shorter. For example, a community at 40-degrees north latitude on terrain sloping northward at a 5-percent grade may find that 35-foot tall buildings cast a shadow 140 feet to the north. If this interferes with solar access of buildings to the north, the height standard can be reduced to 28 feet, which will result in a north shadow projection of only 112 feet.[17] A 28-foot tall building can offer almost as many opportunities for various building styles as a 35-foot building, so it is doubtful that such a minor height reduction will affect the appearance or character of a neighborhood if adopted.

Height regulations also may be modified to provide a transition zone between high-density high-rise development, and shorter lower-density development to the north. Urban form and bulk gradients can be established to minimize shading problems between the high-rise and low-rise zones. The purpose of such transitional zones would be to concentrate high-rise development to only one section of a city, allowing unimpaired access to structures of lower density and height.

A transitional zone concept may not be necessary when shared community-level solar energy technologies are considered. In these systems, a large array of collectors provide heat to areas of a community that have inadequate area (or unshaded space) to accommodate on-site, dispersed systems. Locating these community-level installations along the south boundary of a high-rise zone would offer reduced shading problems from shadows cast by the buildings themselves, and would reduce heat transmission losses by locating the installation close to its point of use.

Setback restriction. Setback restrictions can be modified in two ways for solar access protection. First, buildings and collectors can be moved to north lot lines, so that the lot owner has maximum control over obstructions that can shade solar installations. This may require modifications to conventional zoning ordinances allowing such techniques as zero-lot-line siting, or permissive changes to prescriptive setback requirements using a special permit approach. Second, buildings or collectors may be moved closer south to east-west streets, so that the road right-of-way reservation helps buffer the installation against shading. But moving an installation closer to a street also means that street trees must be evaluated, so that solar access is not reduced by tree shading.

Frontage requirements. Zoning ordinances often establish minimum frontage requirements for all buildable lots. Changing the frontage standards can result in different lot shapes, and different solar geometries, minimizing shading problems. For example, on east-west streets, a lot with dimensions of 75 feet by 100 feet, and a lot of 60 feet by 125 feet have the same area of 7,500 square feet. But the narrower and deeper lot can move a building or collector further away from obstructions to the south, better protecting solar access. A 15-foot reduction in frontage corresponds to a 25-foot increase in depth, which can be important in achieving a particular solar access objective. It can mean the difference between using a passive system on a building's wall, or having to install an active collector (or skylight) on a building's roof.

Bulk plane and solar envelope techniques. A bulk plane provision is a zoning technique encompassing both a height and a setback standard. It establishes a building line running a distance above a property from a specified point and requires all development to be contained within the plane boundaries. If sloping upwards from a lot line, the technique means that a building may be taller the further it is set back from the lot line.

Albuquerque, New Mexico uses a bulk plane technique in several of its higher-density zones to protect solar access to neighboring districts. The solar access provision would apply only to buildings proposed on the boundary between the two zones. Although the angles of the solar access planes do not correspond to the solar angles required to assure unobstructed solar access to collectors, they may reduce shading problems to some extent.

Carrying this concept even further, studies underway between the City of Los Angeles, California, and the University of Southern California suggest that bulk plane provisions can be specifically tailored to provide solar access to adjacent properties by restricting building bulk in such a manner that all shadows cast by a structure will not fall on adjacent structures.[18] The specific geometries of the bulk planes—or solar envelopes—are computer-generated for each specific site being examined. They are to be applied to multi-acre land assemblies for commercial development in the city.

Initial studies indicate that the solar envelopes approach is feasible for these larger sites. Floor area ratios derived from the building shapes that would result from the plane restriction suggest that developments regulated in this fashion come close to allowable floor area ratios under existing zoning and are economically feasible.[19] It remains to be seen whether these techniques are feasible for smaller lots, areas at other latitudes with more restrictive solar angles, or areas with varying slopes.

The geometries of the solar envelope are largely determined by surrounding development patterns, as well as by solar angles. In some sites, the envelope shapes can become quite complex, resulting in buildings that look different than conventional high rise structures. For example, bulk plane provision in New York's zoning ordinance in 1916 resulted in the unusual "ziggurat" or "wedding cake" appearance of many of that city's major skyscrapers. It is likely that a solar envelope approach also will result in a great deal of visual and design diversity in buildings subject to such development restrictions.

The resulting changes in building height, setback, and facade slope may radically change the design of higher-density, concentrated districts. If buildings restricted under solar envelopes are infilled into existing high density areas, it is likely that the urban streetscape will lose a sense of uniformity. Contrasts in building height and setback between solar and conventional structures are likely to be sudden and dramatic. The urban "canyons" of Manhattan and many other of our cities are likely to be considered as an anachronism of earlier, energy-rich times. But these design changes may not be so prevalent if district-level solar technologies are adopted.

Incentive provisions. Another zoning technique that can be considered as an areawide solar access strategy is the use of incentive or bonus provisions. Under this strategy, developers would be given a development bonus if projects are designed to protect solar access.

The bonus can take several forms. The most common is likely to be a density increase. Lincoln, Nebraska, for example, recently amended its zoning ordinance to permit a 20-percent density bonus for residential subdivisions designed for solar energy use and solar access protection. Another alternative bonus is a reduction in open space dedications, or reduction of a payment in lieu of dedication. Since many of the "solar subdivisions" presently in existence were initially developed as negotiated planned unit developments, the additional negotiation required for a solar access bonus might be considered worthwhile. Developers proceeding under more conventional subdivision regulations might choose to forego the bonus over the certainty of a fast development approval.

SUBDIVISION REGULATIONS

Subdivision regulations may be modified to protect solar access and to promote solar orientation. Because east-west building orientation is optimal for most solar buildings, it is likely that a subdivision standard establishing an orientation criterion may be considered as a simple technique of promoting solar energy opportunities.

The optimal thermal characteristics of building form and orientation vary with climate and microclimate. Generally, if the long axis of a building lies within 45-degrees north or south of due east and west, a building will have good solar orientation for most climates.

The orientation of a building depends on street orientation, if subdivision regulations require that building lines run parallel to lot lines. East-west streets, therefore, provide the best orientation of buildings under these conditions, in suburban settings. In urban locations, where buildings are usually deeper than they are wide (to minimize utility costs along the lot frontage), a north-south street orientation will result in buildings with their long axes oriented properly. This concept probably applies to most mobile home developments, which orient their units perpendicular, rather than parallel, to the access road.

Sacramento County, California, established orientation guidelines in its subdivision regulations in 1978. The regulations allow a 22.5 degree variation in lot orientation from an east-west direction, assuming that buildings were sited with their longer axes running parallel to the street. This standard was used to minimize summer heat gain to buildings and to allow simple window shading strategies (especially for east and west windows). But this technique proved wrong in some cases; although lots were properly oriented, the buildings on the lots often varied from the regulatory standard. The country is now considering a building orientation, rather than a lot orientation, requirement.

The design implications of a street/lot/building orientation requirement are probably considerable. Buildings fronting east-west streets will probably look like conventional suburban development. The same buildings on north-south streets, however, will have their shorter sides oriented towards the street, rather than their longer sides, giving the neighborhood a more "urban"appearance. Buildings on streets running in intercardinal directions—southeast to northwest or vice versa—will have a radically different appearance than conventional development. The buildings will appear skewed on their lots, facing on a diagonal to the front of the lot.

For infill development in built-out communities, this unusual diagonal orientation can be compensated for by careful landscaping to maintain continuity with surrounding buildings. Garages can be placed in front yards and fencing or landscaping extended across the lot to carry the visual line established by neighboring building setbacks. In new developments, this orientation result can lead to interesting visual diversity, where different streets have completely different appearances or characters depending on their orientation.

The street and building orientation also can affect the design characteristics of the solar buildings themselves. On east-west streets, buildings on the north sides of the roads will have their solar collectors facing the street. The active or passive installations may be quite visible from a public perspective, and lend a specific appearance to buildings using these solar technologies. Buildings on the south sides of these streets will have their collectors facing their backyards and may present quite a different public appearance as viewed from the street. Moreover, since minimal north wall fenestration is common on solar buildings, an observer might notice that one side of the street is lined with buildings having extensive glazing, while the other (south) side is devoid of windows. Careful landscaping and site treatment may relieve this visual contrast, to some extent providing privacy to houses with glazing facing the street and breaking up the monotony of bare, windowless facades for buildings whose collectors face their own backyards.

Buildings using centralized, district-level solar energy systems may also present similar design issues as buildings using on-site installations. In these situations, buildings probably will use similar energy-conserving design features to reduce the load on, and the cost of, the central solar facility.

Although collector panels of active system arrays may not be visible on buildings using district-level systems, the centralized collector installation may be visible from some public vantages. The individual buildings themselves, if they use energy conserving building form and orientation, would have many of the same design features as solar buildings with decentralized, on-site solar energy systems. Window distribution, solar tempering by landscaping, and orientation would probably not differ extensively from buildings using on-site detached solar energy systems.

LANDSCAPING AND VEGETATION CONTROLS

Landscaping presents the greatest issue to solar access protection. Unlike buildings, whose shadows can easily be avoided once known, trees are living organisms and change form through time. The sapling next door may grow to shade a solar collector in five years time. Moreover, trees are a common feature of the urban and suburban landscape, both on private lands and on public right-of-ways.

Probably little can be done to minimize shading by trees in developed areas, except to locate a solar installation in places where it may not be shaded over the lifespan of the system. Because trees provide a variety of environmental and aesthetic assets, tree removal to assure solar access should only be considered as a last resort.

A better approach would be to use the energy-conserving features of trees to best advantage, in combination with solar technologies. Deciduous trees can provide useful shade and lower temperatures through evapotranspiration, and can successfully be used in conjunction with solar energy systems under many conditions. But even the bare branches of deciduous trees can block between 20 and 80 percent of the available solar radiation during the winter months.[20] It is, therefore, best to assess existing street tree maintenance and replacement programs, to select tree species that least obstruct solar access as replacement species.

In developing areas, more options are open. Where subdivision regulations have street tree requirements, the location and species of the trees can be selected to minimize shading problems. Similarly, landscaping requirements for developments and mobile home parks can be assessed for solar access needs, as well as resident amenity values. Clustering of trees (or the preservation of stands of trees) might be the most appropriate means of achieving both design and solar access goals.

Landscaping also plays an important design role as well as energy role for solar energy use. Site development can help integrate solar buildings into neighborhoods, and can maintain continuity with surrounding development. If carefully sited so that solar access is protected, vegetation also can help conceal a building's solar features, minimizing design contrasts with existing buildings.

ENVIRONMENTAL ASSESSMENTS

A simple and effective approach to protect solar access is to require that it be considered as part of any necessary environmental assessments which are part of the local development process. Site plans can be required to show the location and extent of shading by existing and proposed site features. Environmental impact reports can evaluate the effect of the development project on the solar access of neighboring parcels, as well as assess solar energy use on the project site.

Conclusion

Solar access planning can be considered a specialized category of energy planning. It is likely that as solar energy use increases, so too will solar access conflicts. But these conflicts can be mimimized by public and private actions.

The design implications of solar energy use are just now beginning to be explored. It is likely that solar technologies and solar access requirements will result in some visible change in the way we perceive our cities and neighborhoods, whether or not one believes that architecture is heading towards a new "solar aesthetic"[21] or that design implications will be minimal.[22] As cities have evolved throughout the centuries to meet new requirements of commerce, warfare, religion and social intercourse, so too can we anticipate a further evolution to meet new energy needs and technologies.

Notes

1. Central Naugatuck Valley Regional Planning Agency. *Overcoming Land Use Barriers to Solar Access.* Waterbury, CT.: August 1979, p. 102.

2. Thomas, Miller, and Robbins. *Overcoming Legal Uncertainties About Use of Solar Energy Systems.* Chicago: American Bar Foundation, 1978, p. 21.

3. This is because of the Earth's 22.5 degree tilt of its axis. In northern hemispheres, sunlight enters from the southern sky, while in the southern hemispheres, the sun is in the northern sky.

4. Hayes, Gail. *Solar Access Law.* Cambridge, MA: Ballinger, 1979, Parts II and III.

5. Johnson. *I Solar Law Reporter*, 110. "State Approaches to Solar Legislation: A Survey." Golden, CO: SERI, 1979.

6. Hayes. Op. Cit.

7. Ch. 1366, 1978 Cal. Stats.

8. Environmental Law Institute. *Solar Access and Land Use: State of the Law, 1977.* Rockville, MD: NSHCIC, 1978, pp. 6-8.

9. Ch. 169, 1977 N. M. Laws, Solar Rights Act; N. M. STAT ANNs. 70-8-2 et. seq. Interim Supp. 1976-1977.

10. Personal communication, Los Alamos City Planning Department, September, 1978.

11. Ch. 1154, 1978 Cal. Stats.

12. Personal communication, San Diego County Planning Department, September, 1979.

13. The states are California, Minnesota, Oregon, and Connecticut. See, Johnson, "State Approaches to Solar Legislation: A Survey," 1 *Solar Law Reporter* 117 (1979).

14. Personal communication, Central Naugatuck Valley Regional Planning Agency, September, 1979.

15. Testimony of Richard Stein in *A Forum on Solar Access.* Rockville, MD: NSHCIC, 1978, p. 19. The estimate is based on a site at 40-degrees north latitude on level terrain. It may be higher at lower latitudes and/or south sloping terrain, and less at more northern latitudes and/or north sloping terrain.

16. American Planning Association. *Solar Design Review: A Manual on Architectural Control and Solar Energy Use.* Rockville, MD: NSHCIC, forthcoming, Winter 1979.

17. American Planning Association. *Protecting Solar Access for New Residential Development: A Guidebook for Planning Officials.* Rockville, MD: NSHCIC, 1979, pp. 50-53.

18. Reported by Professor Ralph Knowles at *2nd Annual Open Workshops on Solar Technologies.* Washington, DC: DOE, September, 1979.

19. Ibid.

20. APA. *Protecting Solar Access.* p. 137.

21. Thayer, Robert Jr. "Bringing Solar Design Out of the Dark Ages," *Planning* (December, 1978).

22. Ritschard, Ronald. *Assessment of Solar Energy Within a Community: Summary of Three Community-Level Studies.* Washington, DC:DOE, September, 1979.

Legal and Institutional Barriers to Solar Energy: Some Early Demonstration Experiences

GERALD MARA AND DAVID ENGEL

Introduction

A new technology and political, social or institutional realities often conflict. In some societies this is due to a basic resistance on the part of traditional structures toward innovative practices, but in this country such conflicts are more likely caused by incongruities between a new way of doing
things and the structures that have grown up to accommodate its
predecessors. Among modern industrialized nations, American institutions
have perhaps been the most responsive to changing economic, commercial
and technological realities. This characteristic of American society has been
noted by such different cultural commentators as Franklin and Henry
Adams. Another acute observer of American life, de Toqueville, also com-
mented on how easily political questions tended to become legal questions.[1]
So, not surprisingly, both statute law and common law have generally
recognized, indeed they have supported and encouraged, the *responsive*
character of our institutions to economic and technological change.[2]

Historically, this legal responsiveness to changing technological realities
has been at best a mixed blessing for advocates of conservation and
renewable energy use. Most solar law experts can recite by heart portions of

the *Fountainebleu* opinion denying the applicability of the Doctrine of Ancient Lights to a developing urban economy.[3] Now, however, things may be turning full circle and institutional structures, including strictly *legal* ones, will have to become more accommodating to renewable and non-interruptible energy sources and the technologies which they support.

Conflicts between new technologies and existing institutions can be minimized if they are anticipated. Solar heating and cooling, indeed solar energy generally, has seen substantial anticipation of potential institutional or legal barriers. As a prudent, precautionary measure such anticipation is healthy. But carried too far, forecasting extensive legal barriers to solar before experience verifies such fears can result in involved and complicated mechanisms causing far more problems than they can solve.

In the early days of "solar law," predictions of significant and new legal barriers as well as proposed solutions to them were often articulated from perspectives that were too theoretical. This was particularly evident in the long and protracted debates over legal mechanisms to protect solar access.[4] Often these analyses recommended massive changes in existing property law[5] as well as federal or state takeover of prerogatives that traditionally belonged to local governments.[6] Too often, protection of solar access was seen as requiring clever and innovative legal distinctions whatever the practical consequences, rather than a sense of fairness or equity.[7] These narrow or restricted approaches to institutional or legal solar issues often ignored practical questions of implementation and unintentionally created "image" problems for the solar community.[8]

While these tendencies have not disappeared, they are less common now. In part this has been the result of increasing practical experience of this country's builders, local governments, utilities, lenders and consumers with solar applications. One of the chief avenues for this burgeoning experience was the Federal Solar Demonstration Program. The U.S. Department of Housing and Urban Development managed the residential portion of the demonstrations. One of that program's principal mandates had been to determine whether current institutional arrangements posed barriers to solar utilization. "Institutions" in this context refers not only to governmental structures but also to the financial or lending community, public utilities and other intermediaries necessary for the entrance of solar heating into the residential marketplace. The real world information derived from demonstration experiences is crucial for reasoned judgments about the authenticity of current or future institutional barriers to solar and for our capacity to deal with authentic problems effectively.

Evidence from the demonstrations suggests that the solar-institutional interface is mixed. A number of problems which were predicted to arise have not, while other unanticipated difficulties have been encountered. There are also certain areas (for example, solar's relationship with public utilities) where the existence or non-existence of a problem cannot yet be clearly demonstrated. In addition, the applicability of some of the experiences and findings to solar technologies other than heating and cooling is not clear and should be the subject of some future investigative efforts. This paper discusses the institutional experiences in the demonstration in more detail following a brief description of the program's structure.

The Residential Solar Program

The residential solar demonstration program was a part of the National Program for the Solar Heating and Cooling of Buildings created by Congress through Public Law 93-409. This program, one response to the Arab oil embargo of 1973, was designed in part to demonstrate the use of readily available solar technologies for heating and cooling in residential and commerical buildings.

Since the program's inception, HUD had funded the construction of over 700 projects (over 13,500 units) using solar energy for space and/or domestic water heating. Funding occurred through five competitive procurements (or cycles) as well as through a special Passive Design Competition.[9] Projects funded under cycles one through three (1975-7) are mostly completed and cycle four (1977-8) projects are under construction. Cycle five grants were awarded in early fall of 1979.

The technological mix of systems funded by the program has changed somewhat over the course of five cycles. In the first four cycles this was caused by the changing nature of the grant applications which reflected changing builder/domestic hot water systems with a preponderance of the applications in single-family residences. Cycles two through four saw proportionally more applications devoted to active domestic hot water systems in multi-family housing. The Passive Design Competition constituted a deliberate, large-scale foray into passive solar applications.[9] Cycle five projects continued that trend with grants awarded to developers of passive solar tract houses and to non-profit and community development groups retrofitting multi-family dwellings with passive solar techniques.

For all five cycles, HUD's role in the grant and construction processes had remained relatively constant. From the program's inception, HUD made the conscious decision not to involve itself any more than necessary in the market relationships among builders, consumers and institutional actors. Grantee builders were allowed to select their own solar manufacturers and solar sub-contractors. They made their own decisions on construction and marketing techniques. Although engineering resources were made available and technical concerns communicated to grantees, they were not *required* to make use of these resources or to act upon these concerns. Grantees also had to secure their own building and zoning permits, their own construction financing and their own utility hookups. While this arrangement at times had its drawbacks, this form of administration was essential to assess the existing state of the solar marketplace accurately.

Although HUD had consciously refrained from becoming *involved* in market relationships, it had conducted extensive *monitoring* of the experiences of program participants. This monitoring included extensive survey research into the experiences of participating market actors as well as other kinds of more technical information gathering.[10] HUD uses this information in a variety of ways, several of which are relevant to the question of institutional barriers to solar development. Most obviously the survey data and other results indicate where institutional impediments to solar do, *in fact*, exist in the residential marketplace. Second, technical data as well as builder and purchaser surveys also are used to help develop solar equipment and installation standards.[11] Finally, the information obtained through the program is used to educate builders, lenders, public officials, consumers and others about solar. From the beginning it was believed that only a strong information program could help to avoid or to remedy many of the institutional or other problems that could develop.

Although the experiences gained in the demonstration projects are the core of the program, HUD also has been conducting several supplementary activities to help insure reasonable institutional receptivity to solar heating and cooling systems. HUD funded a series of studies on ways to protect solar access in a variety of residential development contexts. *Solar Access Law* by Gail Boyer Hayes of the Environmental Law Institute[12] is a comprehensive document outlining the issues involved in and the available options for the reasonable protection of solar access in developed areas. Two guidebooks developed by the American Planning Association (APA), one for planning officials[13] and the other for private site planners,[14] were prepared to inform these audiences about the best mechanisms to protect current and future solar access in new developments through the use of conventional land use and planning tools. To remove some of the ambiguity that surrounds some of the financial predictions made on solar and to ease the traditional financial concerns of the housing industry, HUD also developed a computer program which can help builders and lenders assess the financial future of different solar applications.[15]

Program Experiences

Since there is no single "solar application," no single institutional response exists. Solar technologies can vary according to the application (space heating and hot water or hot water only) and the system type (active or passive). Buildings can range from single-family detached units to large multi-family structures. Obviously, the institutional issues for a domestic hot water system in a new single-family home in a new suburban development are very different from those affecting a passive space heating multi-family retrofit in the inner city. The different combinations of factors have to be borne in mind in any overall assessment of institutional barriers.

LAND USE AND SOLAR

Land use issues, particularly solar access, have been perhaps the most visible and most discussed of the potential institutional problems forecast for solar development. It is probably the only *strictly* legal problem associated with solar technologies in the sense that it offers a real potential for the involvement of courts and lawyers. In fact, those very few court cases which have already occurred have dealt with such issues.[16]

There are two potential problems in this area in all residential solar scenarios: (a) public regulations or private covenants which directly or effectively restrict or prohibit some portion of the solar installation and (b) subsequent shading of the solar collector area by other structures or vegetation (solar access).

Zoning, Prohibitive Ordinances and Restrictive Covenants—A very small percentage of our grantees have encountered problems in conforming their solar units to the current provisions of zoning codes.[17] Like all residences, houses with solar systems must meet height limitations and set-back and density requirements of all applicable zoning ordinances, but this does not appear to have caused many problems in the demonstrations. Most builders and zoning officials surveyed by HUD indicated that a flexible approach by both zoning authorities and builders could resolve nearly all potential problems.

There are no public ordinances which directly prohibit the installation of solar collectors. Owing to the current energy situation, it seems very unlikely that ordinances of this sort will be forthcoming in the future and, if they are, it is unlikely that they will be sustained in many courts.[18] More troublesome than public ordinances are the provisions of certain kinds of private covenants which, for example, prohibit the erection of roof structures or generally restrict the use of housing materials or designs to certain specified types. Provisions like this are generally attached to deeds and are transferred when the home is sold. Although there have been few instances where restrictive covenants have prohibited solar installations in demonstration homes,[19] a noticeable number of such instances have been reported from other sources. While these problems usually have occurred concerning active solar collectors, it is easy to visualize the same difficulties affecting passive applications featuring mass or water walls, greenhouses or clerestory windows.

Some communities have sought to balance the interests of architectural innovations and a dominant community style by requiring that proposed structural changes to a controlled property be approved by an architectural review board or jury. Attempts to accommodate solar within traditional housing styles are certainly laudable and the review board arrangement is by far preferable to outright prohibition. We also must realize that some communities do have a more legitimate stake than others in preserving an architectural *status quo*.[20] Nonetheless, given the current state of solar knowledge, such review boards ultimately can have a "chilling effect" on solar innovations. Extensive waiting periods and requirements for documentation and plan submissions could deter all but the most avid solar enthusiast.[2]

While the precise judicial disposition of suits challenging the validity of such covenants is uncertain, courts have generally been reluctant to be too daring in challenging this sort of private agreement.[22] A more certain remedy against the ill effects of private covenants is legislative.[23] Several states have introduced or drafted bills to void any restrictive covenants which would have the effect of prohibiting solar collectors.

In addition, the American Planning Association, as a follow-up to its planning and zoning guidebooks, is now conducting a study of the potential barriers to solar access posed by restrictive covenants and the potential use of zoning or land use controls as remedies.

Solar Access

Up to this point in the demonstrations very few instances of shading have been encountered, although this is hardly surprising given the mix of projects that have been funded. Many of our single-family solar houses have been located in fairly clear, large lots or among other (not necessarily solar) houses built by the same developer. In both cases the builder has substantial leeway to protect solar access through the use of familiar, conventional site planning tools. Many of our multi-family projects are domestic hot water retrofits with collectors mounted on the roofs of relatively tall buildings.

However, even though the shading of active collectors in the current demonstrations has not occurred, it is obvious that shading problems will affect certain types of solar applications in the future unless precautions or preparations are made by local authorities. In new residential developments which are low density, the shading of collectors by other buildings should not be a major concern at least for active space heating or domestic hot water systems. Most current zoning regulations restrict building height adequately to ensure access protection for active collectors in this kind of development. However, current zoning requirements may not be entirely sufficient to protect passive collectors from building shade even in new, low-density subdivisions. As the American Planning Association points out in its guidebook for planning officials, ensuring solar access protection for active solar energy systems usually requires only the protection of south *roofs*. But in order to secure access for passive systems, south *wall* protection is needed. Expanding protection from south roofs to south walls correspondingly expands restrictions on neighboring properties to the south.[24] In addition, both active and passive collectors for space heating are particularly vulnerable to serious shading from vegetation in new developments. While some vegetation control or vegetation planning occurs in newly developed areas, it generally is not adequate to protect solar collectors from eventual shading that could substantially damage performance.[25]

For these reasons, it seems that the best mechanisms for ensuring the widespread protection of solar access in new subdivisions are locally adopted planning or zoning regulations and private site plans that take solar access considerations into account *before* the subdivision is developed. The two guidebooks discussed earlier can be used on a wide scale right now to ensure that future developments will, at the very least, preserve options for including solar energy systems at some later date.

A few states have enacted legislation encouraging the inclusion of solar considerations in area or local land use plans.[26] However, even in the absence of enabling legislation most county or local planning authorities can now take steps to ensure solar access protection within the scope of their current authority. Private developers have many planning tools at hand to protect access to sunlight. The key is that they use them. In fact, from demonstration experiences, it seems apparent that the crucial actors for assuring solar access protection in new subdivisions will be builders and developers, not local zoning officials.

Special shading problems will occur in urban applications, active *or* passive. In the demonstration program many of these potential problems have not arisen because most of our inner city projects are domestic hot water retrofits on relatively tall structures. However, active and passive space heating applications in single-family residences are obviously vulnerable to shading by other buildings in urban settings.

It is certainly much more difficult to ensure solar access protection in urban or developed areas than in new subdivisions. In developed areas there are a number of potential obstructions to sunlight already existing and often localities have made substantial commitments to encouraging commercial or economic development. However, it is also important to realize that the extent of the potential problem will vary significantly within developed areas. For example, many large cities have areas zoned mostly residential or mixed residential and light commercial. Many inner city areas also have large redeveloping areas where older buildings are cleared away on a large scale to make room for new construction. In areas like these, the preservation of existing residential zones may be sufficient to afford reasonable solar access protection which is at the same time consistent with other development goals. For example, we have funded single-family projects in redeveloping areas of Philadelphia and Baltimore with no apparent shading problems.

The more difficult cases are those where there is a great deal of high-density commercial or multi-family residential development, those buffer areas between residential and heavy commercial zones and zones with highly varying mixed uses permitted. Realistically, most localities probably should try to solve potential access problems in these areas by balancing the probability of, and the community interest in, solar energy utilization with other development and land use goals.[27] Some planning departments have already attempted to make formal or informal judgments on these issues. For example, a study done in 1977 by the Los Angeles City Planning Department concluded that given anticipated levels and locations of solar utilization, the existing zoning scheme was adequate to allow reasonable solar access protection while not sacrificing other land use objectives.[28]

However, the resolution of these difficult cases will be done differently by different communities and some localities may wish to institute measures protecting solar access in buffer or mixed use zones. More importantly, in order to encourage the maximum reasonable use of solar in urban areas some assurances need to be provided to homeowners that existing *de facto* protections of solar access will not disappear in the future.[29] These considerations prompted HUD's funding of *Solar Access Law* by Gail Boyer Hayes of the Environmental Law Institute. Since no one strategy or mechanism to protect solar access is appropriate for all communities, this document presents a series of options from which a locality may choose that strategy which best fits its own needs. We hope that this document will clarify some of the legal and political issues involved in the protection of solar access in developed areas and allow local governments to make clear and reasoned choices among competing policy alternatives.

BUILDING CODES

Experiences—Uniformly, building codes have not posed serious problems in the construction of HUD demonstration homes. Although some builders have reported delays in the approval process, such delays were forecast in some early attempts to assess potential barriers to solar development.[30] However, in the HUD program at least, massive delays owing to, for example, the unavailability of materials' test results have not been encountered. Rather, most of the delays have been caused by code officials' unfamiliarity with solar equipment or their unfamiliarity with solar's mechanical interface with the conventional heating system. Although building inspectors may have been somewhat lenient because the construction is HUD-funded, evidence would suggest that codes will not be a major obstacle to future solar construction.

Of about 70 code jurisdictions recently surveyed in the program, very few required some design change specifically relating to solar before issuing code approval.[31] Most of these design changes were minor. In relatively few of the jurisdictions were additional site inspections required. However, the processing time for approval exceeded normal standards in about one-fifth of the cases because of the solar energy system.

Interviews with builders confirm that approval sometimes took longer. However, almost all of the builders who experienced delays cited the code officials' unfamiliarity with solar technology rather than the code itself as the reason. These delays will probably diminish as code officials get more experience. A small percentage of the officials surveyed said they thought delays would continue but, in fact, fewer code-related delays were reported by a recent sample of builders than in the first program surveys conducted in mid-1978.

It is encouraging that many of the code officials said they took extra time to issue a permit for solar construction precisely because they wanted to learn more about the solar energy system. To help respond to the need for solar information on the part of code officials some states, including California, Florida, Maryland and Massachusetts, have offered or will offer solar courses or training programs for building inspectors.[32]

One development that is occurring in some code jurisdictions and which will probably increase as solar construction becomes more widespread is the use of building permits to encourage solar utilization in private subdivisions. For example, Boulder, Colorado has a merit system for awarding building permits for residential construction, which is limited to 450 new units per year. Permits are awarded on a point system and builders who propose to use a solar energy system can earn as many as ten additional points.

Treatment of Solar in Existing Codes

Few code authorities have enacted building code provisions that apply exclusively to solar energy systems. Those few have concentrated in making sure that the work is done by a qualified person and on ensuring that solar installations pose no threat to health and safety.[33] Specific references to solar energy systems in state and local building codes will probably increase when Department of Energy-sponsored model solar building code language becomes available.[34]

It is important to remember that building codes only address issues or concerns that may affect health and safety. They are not concerned with the *performance* of any heating or mechanical system. Therefore, a number of important solar-specific technical concerns related to performance will never be relevant for the building inspector. Nonetheless, there are one or two areas *peculiar* to solar energy systems that are within the purview of the building official's concern with health and safety.

Active solar collector arrays are significant additions to the roof structure and therefore must meet all applicable wind, snow, hail and other roof load regulations. One or two new grant houses in the program had to add strength to roof trusses at the design stage in order to support additional collector weight. Because of requests of building officials, retrofit projects in particular may have problems in this regard since existing roof structures may be inadequate to support the additional weight of collectors. This may be a particular concern in *urban* retrofits, both single- and multi-family, although most of these would probably be hot water-only applications. While few of our inner city projects have encountered serious problems in this area, some officials in urban jurisdictions have suggested that inadequate roofing structures would be a significant barrier to widespread urban retrofits, at least in older inner city buildings or rehabilitation projects. Generally, such code provisions are not restrictions on solar development but, rather, are necessary safety precautions that aid solar utilization. We should probably add that, literally at least, no roofs have fallen in on any of our grant projects.

Another code area that specifically affects solar is the nearly universal requirement that any non-potable substance flowing through a plumbing system in a residence be completely separated from the potable water supply. Most codes which address this require double separation between potable and non-potable fluids and/or backflow prevention at cross-connection points. Of course, this code requirement was not adopted to deal specifically with solar energy use. But it does have significant impact on all active liquid systems which use non-potable fluids as, or in, the transfer media.[35]

There have been some recent fears expressed that codes will pose a problem for passive systems. However, it is still too early to say what if any elements of a passive solar energy system would be vulnerable to code problems. Some builders in the program have had delays in code approval caused by the absence of fire resistance rating for some material used in passive systems. And some grantees have expressed concern about building code provisions requiring that certain temperature levels be automatically attainable in all portions of the house at all times. More information on these questions should be available as Passive Competition and cycle five projects are monitored.

ADDITIONAL PROPERTY TAX ASSESSMENTS

Very few of the local tax assessors appraising HUD demonstration houses said that they had problems judging the value of the solar home. Most considered the solar energy system as an integral part of the house; thus, there seemed to be no significant disparities in methods used for property tax assessment for solar homes as compared with non-solar homes of similar size, type and neighborhood.[36]

At the beginning of the demonstration program, property taxes on solar equipment were expected to be disincentives to solar utilization. However, due to the fact that twenty-three states have passed some form of property tax exemption for houses using solar energy systems, the potential problem caused by upward adjustments of property taxes due to solar has not arisen on a wide scale.

AVAILABILITY OF FINANCING

The availability of credit is important at two stages of the housing transaction process, construction and sale. For speculative housing, construction financing is generally arranged by the builders, usually with an institution that has a history of making construction loans. Permanent financing (mortgage) arrangements are generally more flexible. They are often made individually by the purchaser with a conventional lending institution such as a savings and loan institution, a mutual savings bank or an independent mortgage company. In many larger developments the builder may offer prearranged permanent financing as part of his marketing package. One or two builders in the program have even provided permanent financing themselves.

Construction Financing

Apparently solar has caused few problems for grantee builders' obtaining construction financing. Where difficulties have been encountered, the cause generally cited was something other than the solar energy system. Grantee builders who did have difficulty obtaining financing said their problems were due to conservative lender attitudes or tight money in their own areas. Lenders who provided financing only after extensive checking usually claimed that there was insufficient information about the builder's previous track record or some other uncertainty about the builder's financial position. In general it appears that builders with good reputations and/or favorable past associations with the lending institution have had little difficulty obtaining construction financing in a favorable lending climate.

Permanent Financing[38]

It is quite difficult at this point to draw definitive conclusions from the grant program about the receptivity of permanent lenders to solar in the current market place. This is so because many permanent lenders in the program have not had to estimate the market value of the entire solar energy system.

Permanent lending institutions generally provide financing for a portion of the appraised value of a property. This portion, the loan-to-value ratio, is usually about 80 percent of the appraised value, with the remaining 20 percent provided by the borrower in the down payment. If the appraised value is significantly less than cost, either the seller must reduce the price of the property or the buyer (the borrower) must cover the additional difference between cost and value with a larger down-payment. However, the permanent lender will not need to make a judgment on the market value of a solar energy system if the borrower is not paying full price for the system or if he is prepared to provide a larger down payment than normal. In these cases, any subsidy to the buyer or additional down payment will usually cover the differences between cost and appraised value.

Some grantees in the program have passed part or all of the grant amount through to the purchaser. This kind of subsidization reduces the cost of the solar energy system for the purchaser and correspondingly reduces or eliminates the permanent lender's need to appraise the market values for the solar energy system in question,. About one-third of the permanent lenders interviewed by HUD said that the grant pass-through helped to bring the price of the solar unit into a conventional price range and thus reduced their risk.[39] At the same time, HUD is also finding that some of the purchasers of solar units are providing larger than normal down payments which also are serving to reduce permanent lenders' needs to judge the value of the solar system.

Within these rather broad limitations we can draw some conclusions from the experiences of program participants, however.

Appraisal Techniques

The techniques used to appraise the *value* of a solar energy system will obviously have a significant effect on the financing costs incurred by individual homeowners. As we indicated initially, there is considerable difficulty in drawing inferences about market-place realities in this area from the demonstration experiences. However, there are some definite trends which do stand out.

First, in the absence of data on resales there does not yet appear to be any uniform, agreed-upon method of appraising the market value of solar energy systems.[40] Most of the permanent lenders in the program make judgments about the value of the solar system that included *some* portion of the costs in the appraised value. However, no clear trends either in methods of appraisal used or in the portions of costs included in appraised values can be reported at this time.[41]

It is clear that lenders feel they need resale data in order to make fair appraisals of the value of solar energy systems. The history of what people are willing to pay for a given commodity is the best indicator of a commodity's value. However, resale data on a local or widespread scale are not yet available. There have been few resales in the program and their terms have been largely inconclusive. While there is slightly more resale data available from communities with relatively longstanding commitments to solar (Davis, California, for instance), there is some question about the degree to which these data are representative of expected trends in the rest of the nation.

In the absence of clear extensive resale histories, some private lenders as well as some federal financing programs have developed interim methods for determining the value of certain solar systems within certain cost or performance parameters.[42] There is no need to cover the details of such methods in an article of this nature. But the development of such procedures does indicate that lenders can and will devise means for appraising solar energy system value if sufficient markets for solar units exist in their areas.

It should be stressed that our experience with solar financing practices is focused almost exclusively on active applications. The experiences of passive homes constructed under the Passive Design Competition and cycle five will provide us with information on any particular problems of passive systems. However, for passive there *is* the potential problem of a lack of *clear* system cost data which can be used by the lender to determine market value. Since the passive system is integrated into the design and structure of the house, system costs must be determined as a percentage of total construction costs rather than as fixed costs for hardware and installation.

Finally there do not appear to have been any problems with general credit availability *peculiar* to solar homes. Homeowners in the program have not been charged higher mortgage rates or offered lower loan-to-value ratios because of the solar energy systems.

PUBLIC UTILITIES

The relationship of solar heating to public utilities is crucial for the future of the residential solar market. However, it is hard to predict this eventual interaction based on the experiences of a relatively small number of homes in the demonstrations. We have seen little evidence thus far supporting predictions of massive problems in this interface of solar with public utilities. However, many utilities are beginning to develop special nonstandard rates which may be applied to solar energy systems. It is therefore clear that we cannot treat the relatively few problem experiences to date as conclusive evidence that none will arise in the future.

Much of the literature on the relationship between solar energy systems and existing public utilities concentrates on two issues: the effect of current rate structures on cost savings provided by solar energy systems and the possibility of public utilities charging solar energy users special rates of service.[43]

Rate Structures

Rate structures adopted by public utilities presuppose a variety of technical and economic assumptions. Fortunately, this issue has been dealt with recently in enough contexts to make a discussion of the relevant concepts unnecessary here.[44] Simply put, rate structures can determine the degree to which energy savings are reflected in cost savings. Some rate structures, for example, the declining block, can have the effect of causing code savings to be proportionally lower than energy savings with the result that higher costs per unit of energy consumed are charged to customers who use fewer total energy units. We should be clear that this kind of effect would be visited on *any* customer who reduced total energy consumption and not simply on *solar* users.

Up to this point it does not appear that, in the demonstrations, solar customers are paying *substantially* more per unit of energy than conventional users in the same area. They *do*, however, appear to be paying slightly more—one or two cents per therm or between one and four tenths of a cent per KWH, for example. We should emphasize that this conclusion is highly speculative owing to the difficulties involved in comparing multiple utility bills. Differences in lifestyle, house structure, appliance type and use, and the different kinds of rates applicable in different service areas make broad analyses highly suspect. However, the phenomenon of slightly higher costs per energy unit for solar customers seems widespread enough, both geographically and across different fuel types, to justify further study.

Special Rates

Most of the builders interviewed by HUD were granted the standard rate for residential customers in their service areas.[45] However, the proportion of builders granted standard rates has dropped since the first interviews were conducted with program participants. In 1978, 91 percent of 30 builders said they were granted standard rates. But only 75 percent of 29 builders interviewed in 1979 received the standard residential rate charged in their service area. The number of builders being charged special rates has also increased from 1978 to 1979. However, only a few grantees or purchasers have complained about being charged discriminatory rates by utility companies. It also should be noted that one or two utilities have offered relatively lower rates to solar users because either they believed that the solar energy systems would contribute to reducing peak loads or they wanted to monitor the effect the solar energy system had on peak demand.

It is important to remember that in legal terms, rates are discriminatory only if they treat members of the same class of users differently. A number of utilities surveyed in the program believe that solar users may constitute a different class of users from the conventional residential customers in the same service areas. Many of these believe a special rate to be necessary because of the impact of solar utilization on peak loads and load factors. They claim that solar utilization would not reduce peak loads and therefore the need for maintaining peak generating capacity. But at the same time solar *would* reduce the load factor or total energy used. It is possible that special solar rates could be proposed and adopted under most current Public Service Commission guidelines if sufficient evidence were accumulated supporting such claims. This is an issue that needs further examination and one which probably should be addressed within a more widespread and systematic reconsideration of the effects of various rate structures on energy conservation.[46]

However, it is equally important that these general or theoretical kinds of inquiry be supplemented by a practical appreciation of the ways in which utility rates are applied and administered. For example, one grantee reported that the gas utility in his area would charge a different, higher rate to a solar customer if it were determined that the customer's load factor was 15 percent or less of the average load factor for that class of user. Again, this was not a specific solar rate, but a rate applicable to all limited use customers. On its face this requirement did not seem to be unreasonable. However, in practice the very presence of the solar system was taken to signify that the higher rate should be applied and moreover the higher rate included a minimum charge which was originally designed for commercial customers and was, as a result, very high. Clearly in this instance the separate rate for solar users *may* have been justified but its administration and its charges were applied in a way very harmful to solar. This indicates that the practical issues of implementation and billing deserve as much attention as theoretical analyses of the economics of pricing strategies.

Nearly all of our findings concerning public utilities stem from the experiences of projects using active systems. But it is likely that passive systems will have generally the same experiences.

CONSUMER PROTECTION

Experiences—Two or three years ago there were several forecasts of extensive problems with solar product and installation reliability.[47] A great deal of this concern was prompted by experiences in the New England Electric Solar Demonstration Program where faulty installation practices substantially degraded system performance on a wide scale.[48] The period during which this demonstration occurred probably constituted a low point for the solar industry. The number of manufacturers and installers had increased dramatically over a short period of time, and many inexperienced firms or individuals found themselves producing equipment or doing work without proper skills. There were some similarities in HUD's experiences in the demonstrations during that time period. A recent analysis of the occurrence of problems in the first three demonstration cycles showed an increase in most problem categories from cycle one to cycle two.

However, evidence suggests that the situation had improved. Continued monitoring of the New England Electric projects has apparently shown that after necessary installation repairs most systems performed much better in the second year of the program than they did initially.[49] Again, this parallels the HUD Demonstration experience. Reported hardware problems decreased from cycle two to cycle three and surveys of an admittedly small number of cycle three purchasers who have occupied solar units for six months or more shows a remarkable improvement in levels of consumer satisfaction over those reported in cycle two.

Warranties

In the demonstrations very active steps were taken to try to ensure that consumer safeguard mechanisms at least met current consumer protection standards in the conventional heating, ventilating and air conditioning industry. In the HUD Hot Water Initiative and in cycle four and after, major system components were required to be warranteed for five years by the manufacturer, with a one-year warranty on installation also mandated. Of course these warranties are not required to extend to non-federally assisted projects. But insofar as they are used as standards or benchmarks in the private market, they will provide the consumer with protection which will, in most cases, exceed the safeguards afforded by the conventional HVAC industry.

The extent of the burden imposed by these requirements on the solar industry is not yet clear but the issue certainly needs continuing attention.[50]

Current DOE efforts to determine, on the basis of actual experience, whether a federal warranty reinsurance program for the solar industry is needed should help to answer this question.[51]

Installation Quality

However, the best assurance for effective consumer protection is a competent, well-informed installation force capable of doing quality work. There have been serious and in some ways continuing problems with installation in the demonstrations.[52] This has not been due so much to any unusual features of solar equipment. Rather it appears that too often the installation of both solar and conventional heating systems is careless. However, these careless mistakes have had much more serious consequences for solar heating systems than for conventional ones. Solar energy systems must make more efficient use of the heat they obtain because heat is only available at certain times of the day.

Passive Issues

Consumer protection experiences and issues in the program so far have been concentrated on active systems. After more monitoring of some of the recently funded passive projects it will be clearer if passive solar energy presents distinctive consumer protection problems. Obviously, some of the areas which have caused consumer dissatisfaction in our active projects should not be problems in passive houses. There is little hardware involved in a passive system, and, usually, controls in a passive solar house are not mechanically complicated or subject to mechanical breakdown. However, since the passive solar heating system is incorporated within the design of the house itself, any problems with the heating system that do arise are likely to be much more serious and much less easily remedied.[53] A defective active collector can be replaced or removed whereas problems in passive collection, storage or distribution are much more likely to be ingrained in design defects of the structure.

Implications for Other Technologies

The use of active and passive solar heating and cooling systems can generally be integrated easily into existing institutional frameworks. Where potential problems or incongruities exist, means for resolving them are currently or potentially in place. If such means are applied flexibly on both sides, a resolution favorable to both the solar application and the institutional interest is usually possible. However, much of the congruity between solar heating and cooling and institutions stems from the relatively standard character of solar heating and cooling technologies. Generally, they involve only new configurations of familiar hardware and/or individual building design. All of the experiences we have been discussing are lot-by-lot applications rather than community or neighborhood-wide installations. It is not very difficult to see that institutional issues involved in the widespread use of wind energy systems, biomass facilities (excluding individual wood stoves) and solar electric applications are more problematic than those concerning solar heating and cooling precisely because these other applications involve the use of less familiar or less common equipment, and, often, on a neighborhood or community scale.

There are at least three general conclusions about future institutional problems with non-heating and cooling solar technologies which probably can be drawn from demonstration experiences. All of these are fairly apparent. First, it seems reasonably easy to predict where potential institutional problems with other solar technologies may arise, at least in broad brush. For example, there is certainly a potential problem involving the interface between public utilities and community-wide solar electric or wind facilities. Likewise, potential problems with zoning are probably far greater for solar electric, wind or biomass facilities than they are for heating and cooling technologies. The former solar applications raise more serious questions about adverse impacts on development patterns. Additionally they tend to blur the lines between residential and commercial zones. For example, a wind facility for a private residence could also be seen as a commercial or quasi-commercial application even if all of the energy generated were consumed on site. Finally, some of these applications may have environmental impacts which could concern local or state governments.

Our second conclusion is that there is no immediate reason to believe that any of these apparent problems are, in fact, non-problems. The reactions of many institutional actors surveyed have been cautious or skeptical even concerning heating and cooling applications alone. It is reasonable to assume that institutional concern will increase rather than diminish as less conventionally familiar solar applications are introduced.

Third, demonstration experience also suggests, however, that once sufficient local interest is perceived there is a general willingness on the part of institutions to deal fairly with solar applications. There is no reason to believe that this general tendency will disappear as other solar technologies evolve. However, in some areas the institutional receptivity to solar is certainly greater than in others. Institutions are probably most favorably disposed toward solar when there is some state or local policy support for the application in question and when favorable institutional experiences with that application are visible. In this regard, it would be difficult to overemphasize the importance of programs to encourage hands-on experiences with various solar technologies through demonstrations. The HUD demonstrations supported, reinforced and in some cases initiated this kind of community involvement with heating and cooling technologies.

Conclusion

It might be useful to reiterate some of our major conclusions and uncertainties as well as to outline some possible new directions in studying the relationship between institutions and residential solar applications.

The relatively small number of problems arising in the demonstrations in the areas of codes, the availability of construction financing and the effects of increased property taxes are probably very accurate representations of the current state of the private residential solar marketplace. Although the demonstration experiences with solar access are also reliable, additional problems could be posed by more *widespread* utilization of solar energy systems in certain areas. In addition, problems with consumer protection or consumer confidence should be no greater for active solar energy systems than for conventional HVAC systems once a solar industry infrastructure becomes more fully developed.

Two areas where real uncertainties still exist, however, concern the relationship between public utilities and widespread solar applications and the judgments made by permanent lenders about the market value of solar energy systems. Answers to these questions are equally uncertain for active and passive solar applications. This paper has outlined the reasons why these questions cannot be answered definitively as yet as well as what would be needed to do so. The HUD program will devote continued attention to these issues as its monitoring of demonstration sites progresses.

In the future, the HUD program will concentrate on institutional reactions to solar development in at least two other areas. First, it will examine passive projects to see how their experiences with institutions parallel or differ from those of the active demonstration homes. Second, the program will focus on the impacts of solar on areas where the energy and housing situations are most critical and where at the same time institutional concerns are most complex, for example, in our cities. These efforts, when added to the ongoing monitoring of projects already built and occupied, will provide a substantial body of empirical information necessary for reasoned and responsible solar policy decisions in the next decade.

Acknowledgments

The authors would like to thank Lance Hoch and Susan Perretta of the National Solar Heating & Cooling Information Center for their help and comments on earlier drafts of this paper.

Notes

1. de Toqueville, Alexis. *Democracy in America*. 2 vols. New York: Vintage, 1945. Vol. I, Chapter VI.

2. This is in no way to suggest that response has always favored narrowly defined business interests over competing social or political goals. Nor is it to suggest that various institutional responses to technological developments have always agreed. For example, a brief account of the Supreme Court's fluctuating reactions to the government regulation of business can be found in Henry J. Abraham. *Freedom and the Court*. New York: Oxford University Press, 1965. Chapter 2.

3. *Fountainebleu Hotel Corp. v. Forty-five Twenty-five, Inc.* 114 So. 2nd 357; (8) Fla. Supp. 74 (1959).

4. The amount of literature generated on this topic in barely a few years is staggering. For a reasonably complete bibliography see Hayes, Gail Boyer *Solar Access Law*. Cambridge, MA: Ballinger, 1979. pp. 290-293.

5. See White, D. Mary "The Allocation of Sunlight: Solar Rights and the Prior Appropriation Doctrine," *University of Colorado Law Review* 47 (1976), 421-447 and the criticism by Hayes, *supra*, pp. 187-192.

6. Although the traditional location of zoning authority in local government owes to an explicit delegation of that authority by the state, extensive state entrance into local zoning would certainly create something of a furor. For a very clear statement of the relationship between state and local zoning authorities in a solar context see the remarks of Johnson George. *A Forum on Solar Access*. Rockville, MD: National Solar Heating and Cooling Information Center, 1979. pp 38-39. A definitive treatment of zoning *per se* used extensively by Hayes, *supra*, is Anderson, M. Robert. *American Law of Zoning*. 2nd ed., 5 vols. Rochester, NY: Lawyers Cooperative Publishing Company, 1977.

7. See the comments by Engel, David. *Forum on Solar Access, supra*, pp. 1-5.

8. Both of these points are strongly made by Hayes, *supra*, p. 3.

9. For a summary of HUD's Passive Design Competition see *The First Passive Solar Home Awards*. Washington, D.C.: Government Printing Office, 1979.

10. For a full description of data gathered in the residential demonstration program see: Christopher, Patricia M. and Joan Krzewick. *Residential Solar Data Center: Data Resources and Reports*. Washington, D.C.: National Bureau of Standards, 1979.

11. Essentially two kinds of standards are being produced. The *Solar Supplement* to the *Intermediate Minimum Property Standards* details the structural and mechanical integration of the solar energy system with the house and the conventional heating system. The *Performance Criteria for Solar Heating and Cooling Systems in Residential Buildings* discusses what is needed to ensure adequate solar energy system performance. Both of these documents, are being prepared for the Residential Solar Demonstration Program by the Center for Building Technology, National Bureau of Standards (NBS). Interim versions of both documents are available from HUD or NBS.

12. Hayes, *supra.*

13. Jaffe, Martin and Duncan Erley. *Protecting Solar Access for Residential Development: A Guidebook for Planning Officials.* Washington, D.C.: Government Printing Office, 1979.

14. Erley, Duncan and Martin Jaffe. *Site Planning for Solar Access: A Guidebook for Residential Developers and Site Planners.* Washington, D.C.: Government Printing Office, 1979.

15. *RSVP: What It Can Do For You.* Rockville, MD: National Solar Heating and Cooling Information Center, 1978.

16. *Katz v. Zoning Board of Appeals, Kraye v. Old Orchard Association*, reported in 1 *Solar Law Reporter*, 495-506; *D'Aurio v. Board of Zoning Appeals*, 401 N.Y. State 2d 425 (1978). *Nicholas v. Gurther* reported in 1 *Solar Law Reporter*, 251-2.

17. Real Estate Research Corporation. *Working Papers on Marketing and Market Acceptance: Residential Solar Demonstration Program.* Washington, D.C.: U.S. Department of Housing and Urban Development, 1979.

18. At least in those states where some public policy commitment to solar has been made. See the opinion of the Court in *Katz v. Zoning Board of Appeals, supra*, note 16, which places heavy emphasis on New York State legislative commitments to solar in its decision to overrule the local zoning board's refusal to grant a variance in height limitation for an active solar collector.

19. Although there have been *some* problems encountered, see the comments of Martin Jaffe, cited by Wiley, John. "Private Land Use Controls as Barriers to Solar Development: The Need for State Legislation," 1 *Solar Law Reporter*. 283, note 9.

20. For example, the Beacon Hill area in Boston can probably make a stronger case for architectural controls than some newly developing subdivision.

21. Some communities, like Columbia, Maryland, have mitigated this somewhat by publishing guidelines for the incorporation of solar energy into the accepted architectural style.

22. At least in the absence of public policy to the contrary. See, for example, the opinion of the Court in *Nicholas v. Gurther*, Civil No. C-384239 reported in 1 *Solar Law Reporter*, 251-2. It should be noted, however, that the Court addressed only the question of the need to seek approval from the subdivision review board prior to the installation of a roof-mounted collector, not the question of the review board's treatment of that request.

23. cf. Wiley, *supra*, p. 300-303; *A Forum on Solar Access. supra* p. 5.

24. Jaffe and Erley, *supra*, pp. 23-4.

25. Jaffe and Erley, *supra*, pp. 99-110.

26. *State Solar Legislation.* Rockville, MD: National Solar Heating and Cooling Information Center, 1979.

27. For discussions of the ways in which solar access protection relates to and possibly competes with other land use or development goals see Jaffe and Erley, *supra*, pp. 40-43; *A Forum on Solar Access, supra*, pp. 1-5.

28. "Property Owners' Rights to Sunlight," City Plan Case No. 26110, Department of City Planning, City of Los Angeles, 1977. It should be noted that the report indicated that then-current zoning ordinances were adequate on an interim basis until a number of studies on the solar access question could be completed. For Los Angeles' most recent solar access activities see, 1 *Solar Law Reporter*, 259-260.

29. This point is strongly made by Hayes, *supra*, pp. 23-26; 57. There have been, however, some objections expressed in some quarters about the wisdom of creating institutional

remedies before the existence of a problem is demonstrated in a particular area. See Wallenstein, Arnold, ed. *Proceedings of a Workshop on Solar Access Legislation*. Cambridge, MA: Northeast Solar Energy Center, 1979. pp. 16-23.

30. Bezdek, Roger et. al. *Analysis of Policy Options for Accelerating Commercialization of Solar Heating and Cooling Systems*. Washington, D.C.: Energy Research and Development Administration. pp. 413-440. For a somewhat different perspective see Little, Arthur D. *Residential Solar Heating and Cooling Constraints and Incentives*. Washington, D.C.: U.S. Department of Housing and Urban Development, 1976. pp. 46-48.

31. *Working Papers on Marketing and Market Acceptance, supra.*

32. For a much fuller treatment of state information efforts see Ashworth, John et. al., *The Implementation of State Solar Incentives: A Preliminary Assessment*. Golden, CO: Solar Energy Research Institute, 1979, pp. 159-190.

33. Specific codes provisions have been adopted in Tampa, Florida and Houston, Texas. A number of localities and counties in the southwest have also adopted the *Solar Energy Code* developed by the International Assoc. of Plumbing and Mechanical Officials (IAPMO). See Solar Energy Code, Los Angeles, International Association of Plumbing and Mechanical Officials, 1978. For a critique of some of the code's provisions see Environmental Law Institute. *Legal Barriers to Solar Heating and Cooling of Buildings*. Washington, D.C.: Energy Research and Development Administration,1977. pp. 68-73.

34. DOE has contracted the Council of American Building Officials (CABO) to coordinate the development of model language appropriate to incorporate solar energy considerations into the three model codes which serve as bases for state or local building regulations. Participants in CABO's effort include representatives of the three major model code organizations (Building Official and Code Administrators International, International Conference of Building Officials, and the Southern Building Code Congress International), state and local governments and the building and solar industries as well as solar program staff from DOE, HUD and the National Bureau of Standards (NBS). Information from HUD's demonstrations is being used by CABO's committee as a partial basis for the model code's provisions. Draft copies of this document can currently be obtained by contacting Mr. Hank Wakabayashi, National Conference of States on Building Codes and Standards, 1970, Chain Bridge Road, McLean, Virginia.

35. HUD's current version of the *Intermediate Minimum Property Standards (IMPS) Solar Supplement* also requires that non-potable transfer fluid be separated from the home's domestic water supply by a double-walled heat exchanger. In general, the IMPS, which define the minimum structural quality acceptable for federally-financed housing, require that solar space and domestic water heating installed in such housing conform to all relevant local codes. See *Intermediate Minimum Property Standards Supplement, Solar Heating and Domestic Hot Water Systems*. Washington, D.C.: U.S. Department of Housing and Urban Development, 1977. S-615, 10.1.1.

36. *Working Papers on Marketing and Market Acceptance, supra.*

37. *State Solar Legislation, supra.*

38. For an extensive discussion of the financial issues involved in providing permanent financing for solar energy systems see Barret, David, Peter Epstein and Charles Haar. *Financing the Solar Home: Understanding and Improving Mortgage Market Receptivity to Energy Conservation and Housing Innovation*. Cambridge, MA: Regional and Urban Planning Implementation, Inc., 1976. A portion of this work has been reprinted as *Home Mortgage Lending and Solar Energy* by the National Solar Heating and Cooling Information Center. See Also Reiger, Arthur J. "Marketplace Realities and Solar Economics," SUN: *Mankind's Future Source of Energy*. London: Pergamon Press, 1978.

39. *Working Papers on Marketing and Market Acceptance, supra.*

40. Ibid.

41. For a more detailed analysis of issues connected with the appraisal of solar energy systems in homes see Reiger, Arthur J. "Appraising the Sun," a paper presented at the Society of Real Estate Appraisers National Convention, August, 1979.

42. Ibid.

43. For example, see Bezdek et. al., *supra*, pp. 148-162; Arthur D. Little, *supra*, pp. 48-50.

44. Feurstein, Randall. "Utility Rates and Solar Commercialization," 1 *Solar Law Reporter*. 305-368.

45. *Working Papers on Marketing and Market Acceptance, supra.*

46. Of course the broadest reconsideration mandated thus far is the Public Utilities Regulation Policies Act (P.L. 95-617), a part of the National Energy Act.

47. For example, "Solar Power: Solar Energy Devices Abound But Many are Useless or Inefficient," *Wall Street Journal* (April 28, 1977).

48. Smith, Robert O. and Associates, P.E. *Summary of Performance Problems of 100 New Residential Solar Water Heaters Installed by New England Electric Company Subsidiaries in 1976 and 1977.* Cambridge, MA: Arthur D. Little, 1978.

49. *Second Year of New England Electric Solar Water Heating Test Program.* March, 1979.

50. This is not just a solar industry issue. It should be noted that in the public meetings held as a part of the Solar Domestic Policy Review, many consumer spokespersons were also concerned with the effects of warranty requirements on prices. See, *The Great Adventure.* Washington, D.C.: U.S. Department of Energy, 1979. pp. 8, 23. Additional solar observations on the ultimate efficacy of warranties for ensuring consumer confidence can be found in *Solar Energy and Today's Consumer*, report by the Subcommittee on Oversights and Investigations of the House Committee on Interstate and Foreign Commerce, December 1978, pp. 27-34; *Commercializing Solar Heating: A National Strategy Needed*, report to the Congress by the Comptroller General of the United States, July 20, 1979, pp. 11-14.

51. Option II of the Domestic Policy Review Response Memorandum suggests why such a comprehensive warranty reinsurance program is unnecessary now. However, it also leaves the door open for some future federal role if private insurance proves to be an inadequate vehicle. See, *Domestic Policy Review of Solar Energy: A Response Memorandum to the President of the United States.* Washington, D.C.: U.S. Department of Energy, 1979. Attachment: pp. 21-22.

52. U.S. Department of Housing and Urban Development. *Building the Solar Home.* Washington, D.C.: U.S. Government Printing Office, 1978; Sparkes, R.H. and K. Raman. "Lessons Learned on Solar System Installation, Operation and Maintenance Problems from the HUD Solar Residential Demonstration Program," *Preconference Proceedings—Solar Heating and Cooling Systems Operational Results.* Golden, CO: Solar Energy Research Institute, 1978. pp. 253-258; Ward, Dan S. *Executive Summary: Solar Heating and Cooling Systems Operational Results Conference.* Golden, CO: Solar Energy Research Institute, 1979; Freeborne, William, Gerald Mara and Thomas Lent, "The Performance of Solar Energy Systems In The Residential Solar Demonstration Program," *Preconference Proceedings—Second Annual Solar Heating and Cooling Systems Operational Results Conference.* Golden, CO: Solar Energy Research Institute, 1979 (forthcoming).

53. *The First Passive Solar Home Awards. supra*, 190-192.

Federal Building Energy Performance Standards

GRANT P. THOMPSON

Introduction

MUCH to the despair of planners and utopians alike, Americans have a long and robust tradition concerning land use planning: they don't like it. However, this distrust of legal controls that seek to limit the kind of building a person may build on his own property has been gradually crumbling. Starting with cases upholding zoning by category of use, courts now embrace (though sometimes with residual distaste) even aesthetic and architectural restrictions.

This liberalizing trend in land use planning is well known and studied, but there is a less well known movement in another area of social control of building technology: the trend toward wider acceptance of the building code as a means of regulating various aspects of building design. Building codes now no longer are restricted to health and safety controls, but describe permitted heating systems, structural elements, window areas, ventilation rates, and a whole host of other design and construction features.

Building codes traditionally have regulated only those features of a building that were virtually invisible to the consumer. Whether soil pipes were made from cast iron or plastic pipe meant little to a homeowner contemplating a finished split-level near an express bus line and good neighborhood schools. Building codes played the role of silent guarantors of the basic soundness of construction. A building in compliance would not collapse, let us freeze, or stifle us with stale air and odors.

This paper describes what amounts to a new departure for building codes, the Federal Building Energy Performance Standards (BEPS). These energy efficiency standards for new buildings have the *potential* for changing the design of American homes and buildings in subtle fashions that are hard to predict but which *could* have profound effects on design. As I shall discuss at the end of this paper, there is a great range of uncertainty about the real effects of energy, energy prices and building regulations on land use and building design. Therefore, professionals engaged in the study of these issues must, I believe, be undogmatic in predictions of the future.

This paper has three parts. First, there is a short description of the pre-BEPS standards and codes that regulated energy consumption. Second, the BEPS are described and criticized. Finally, the paper speculates on some possible effects of a vigorous BEPS, adequately implemented.

Energy Standards for New Construction: Pre-BEPS

Regulation of energy use within buildings does not have a long history in this country. To be sure, a few examples of requirements for relatively minimal amounts of insulation can be found, but before 1970 there was little general interest in energy-efficient construction. The most wide-spread program was not mandatory, but was a set of voluntary guidelines adopted by electric utility companies interested in making the cost of operating electric heating competitive. Homes that complied with the guidelines were awarded a Gold Medallion. The first systematic, nationwide interest in energy conservation for buildings came as a reaction to the oil embargo of 1973-74. At that time, a voluntary group representing the heating, cooling and ventilating professions began the process of drawing up an energy efficiency code for new buildings. The group, the American Society of Heating, Refrigeration and Air-conditoning Engineers, Inc. (ASHRAE), intended to add to its growing list of standards covering subjects ranging from ventilation rates to humidity control. Because the standard was to be the ninetieth in the list of ones they had developed, it was called Standard 90P, the "P" standing for proposed.

ASHRAE and other similar voluntary professional societies were experienced in developing standards and had developed a process for making certain that all economically affected parties had an opportunity to comment on and offer revisions to a standard before it was issued in final form.

This so-called consensus process demands Herculean devotion from its participants who voluntarily attend meeting after meeting without pay or travel expenses, arguing over comments ranging from word changes to the most fundamental revisions. The consensus process, by its very nature, guarantees that any standard surviving the process will have two characteristics: it will not be unacceptably controversial and it will have had little input from any person who did not have some strong (usually economic) reason to donate a very large amount of time and effort to the process. ASHRAE standards are extremely influential. Their influence comes from the fact that they are technically sound, generally accepted by most directly affected interest groups, and usually cover highly technical subjects that no non-federal level of government would have the resources to regulate thoroughly and accurately. For this reason, many local or state laws and ordinances simply refer to a particular standard, thus giving it the force of law.

ASHRAE issued its standard in final form in 1975, and it was given a suffix indicating its vintage: Standard 90-75. In the meantime, Congress had reacted for the first time to the energy crisis by passing the Energy Policy and Conservation Act (P.L. 94-163, effective December 22, 1975). Section 362 of that Act required each state to develop energy conservation plans that included five mandatory provisions. One of those provisions was "mandatory thermal efficiency standards and insulation requirements for new and renovated buildings." The federal government seized upon ASHRAE Standard 90-75, declaring that any state that adopted the Standard or its equivalent would be deemed to have complied with the Energy Policy and Conservation Act, and thus be eligible for federal assistance. In order to assist states in using ASHRAE Standard 90-75 even more rapidly, the federal government funded another voluntary group, the National Conference of States on Building Codes and Standards, to change the Standard (a format that is unsuitable for enactment) into a code format, which local building code departments could apply. NCSBCS also developed, under contract to the federal government, a set of training courses for building code officials to familiarize them with the code based on ASHRAE Standard 90-75.

The ASHRAE standard is what is called a "component-performance" standard. This means that the builder is instructed to look at each element of a building (that is, the walls, the floor, the ceiling, the heating plant, and so on) and make certain that each one of those components had a certain minimum thermal integrity or performance. Any builder assembling a building made up of various elements, each of which had passed the Standard, would be guaranteed that the final building would be in compliance with the Standard.

This component-performance standard is relatively easy to administer, but various groups, spearheaded by the American Institute of Architects, argued that such a standard stifled innovation in building design and, in many cases, mandated construction practices that were actually wasteful of energy. After considerable efforts at persuasion, proponents of this viewpoint prevailed on Congress to mandate that states follow a quite different approach, one that looked at the *total* energy performance of a building. In the Energy Conservation and Production Act (P.L. 94-385, effective August 14, 1976) Congress required the Department of Energy to develop performance standards for new buildings. Section 303(9) of ECPA defined a performance standard as "an energy consumption goal or goals to be met without specification of the methods, materials, and processes to be employed in achieving that goal or goals, but including statements of the requirements, criteria and evaluation methods to be used, and any necessary commentary." The critics of the component performance standards had won a victory in the legislative arena. They had also set the Department of Energy on a long, technical, controversial and demanding course, whose end is not yet in sight. The next section of this paper describes development of BEPS, and where they stand at this time.

Development of BEPS and Critique of Their Current Form

The original legislation mandating development of BEPS gave the government three years—that is, until August 14, 1979—to develop the standards in their final form. The fact that BEPS were issued in *proposed* form on November 28, 1979 gives some clue of the actual schedule that has been followed. The following evaluation of BEPS is based on the Proposed Notice of Rulemaking (to be found in 44 *Federal Register* 68120). Changes are both desirable and likely in the final form of the rule. The present schedule calls for promulgation of the regulations in May, 1980. However, the Department is seeking to find more time to revise and rework parts of the rule. It may well be considerably later before final rules are available.

Under the original legislation, both the technical standard for BEPS and the implementation plan were to have been developed by the Department of Housing and Urban Development. Congress transferred the authority to develop the technical basis for the standard to the Department of Energy; however, it left the implementation development with HUD. It soon became clear that this arrangement was unworkable, since the standard development and the implementation plans are so closely allied to one another. Therefore, the two Departments entered into a Memorandum of

Understanding, delegating implementation to DOE. In a confusing development, HUD suddenly refused to renew the Memorandum, then just as suddenly agreed to renew it. The early development of both the BEPS standard and the implementation plan was carried out at HUD. This transfer and retransfer of authority has added enormously to the difficulty of developing a workable standard.

In the early stages of standard development, HUD chose the American Institute of Architects Research Corporation as the lead contractor for the standards. In broad outline, HUD and AIA/RC decided to look at what American builders were actually designing shortly after the oil embargo, then use the best of those actual designs as the new standard for all builders. This method of setting a standard demonstrated that HUD and AIA/RC had two strong views of energy conservation in buildings. First, it showed they believed it should be based on present technology in actual use. This view is contrary, in my opinion, to the intent of the legislation, which sought to use BEPS as a technology-forcing device, bringing new designs and new techniques into common use. Second, it showed that HUD and AIA/RC did not share the economists' view of energy conservation (that it is simply cost-minimization), but rather took the engineers' view (that it is plugging leaks as their existence becomes known to you).

Let us now turn to a somewhat more detailed description of the process AIA/RC followed in developing the standards. Although some of this story is now simply history, the data collected in the effort continues to exert a strong influence on the Department of Energy's views concerning what builders can actually do. AIA/RC began by surveying a large number of buildings designed during 1974-75. This period was picked both because designs were available, and because it was assumed that designers and engineers had by then taken new, higher energy prices into account. Enough buildings were selected so that a statistically significant sample was available for various building use categories and climate zones (defined on the basis of heating degree days only). AIA/RC used a sample size of 1,661 non-residential buildings. Data drawn from the plans of each of these 1,661 buildings were entered into a computer that estimated the amount of energy the building would consume, using a proprietary program (AXCESS) developed by the Edison Electric Institute. The computer output consisted of a figure showing how many British Thermal Units (BTUs) of energy each building would use per square foot per year.

The data generated by AXCESS permitted AIA/RC to prepare a matrix of American non-residential buildings, organized by climate zone, by building type, and by predicted energy consumption. (For example, by looking at the data books, it was possible to show a range of energy use for hotels located in climate zone 7). This large data collection effort formed the basis for standard setting.

In order to test how much further an average designer could improve on a design, the AIA/RC then selected a sample of about 10 percent of those buildings and asked the architect teams who designed the 161 buildings to attend a three-day training session on energy conservation in building design. Following this intensive session, each design team was asked to redesign its original building, but to do it within the original budget guidelines established by the client, with no additional use of active solar energy, and complying with any particular requests of the client, no matter what their energy consequences might be. The result of this Phase II redesign effort should give all of us renewed hope for the future of American education! Fully 80 percent of the redesigns were so good that if they had been categorized with the original 1,661 buildings, they would have fallen at or above the top fifth of that group as measured by energy efficiency.

In the case of residential buildings, the Department took a different approach, although it was likewise one based on technical improvements in the building stock. Using data collected for a different purpose by the National Association of Home Builders, the Department analyzed the energy consumption of these residences using a computerized version of the ASHRAE Modified Degree Day Method. Experienced designers were asked to develop prototype residences that were based on the median characteristics of the houses surveyed. These prototypes were then reanalyzed for energy consumption. Again, a technically-based methodology was used in order to set the energy performance standards.

Based on the analysis of the AIA/RC and the NAHB data, the Department selected energy budget figures that would have forced designers of all non-residential buildings subject to the BEPS to be as conscious of energy as were the better third-to-fifth of their colleagues. In the case of homes, builders would have been required to comply with the Thermal Performance Guidelines issued by the National Association of Home Builders. This form of the BEPS was released in an Advance Notice of Proposed Rulemaking at the end of 1978.

The criticism of these preliminary BEPS was immediate and harsh. Although there were many detailed criticisms of particular provisions, three important drawbacks were noted:

First, the standards were based simply on existing technology and based on buildings in which energy was not particularly singled out for special attention. Even in the case of the Phase II redesigns, the design teams were constrained by considerations that showed little sensitivity to use of new techniques, new machinery and new ways of persuading clients and designers to save energy.

Second, the standards were based on buildings that were designed almost immediately after the original oil embargo. The market had not had time to readjust to the higher prices, and many clients and architects believed that the crisis would soon be over with a return to lower prices. The standards that the Department was proposing to issue stated, in effect, that in the 1980s, American designers were required to design buildings only as well as many of their colleagues were already doing in 1974.

Third, and most fundamentally, the technically-based standards ignored the most basic question of energy consumption and conservation: what is the economic balance between the discounted present cost of using energy in the future and the capital cost of taking steps to avoid using that future energy. As the Ford Foundation sponsored energy study, *Energy: The Next Twenty Years*, states the case:

> We mean by conservation those energy-saving investments, operating decisions, and changes in the goods and services that we buy and use that save money over the life of energy-consuming products. Money can be saved by substituting intelligence, prudence, maintenance, better equipment, or different equipment for purchased energy; the substitution should be made up to the point where the cost of not using the energy is equal to the cost of the energy saved.

By ignoring the life cycle costs of buildings, the Department's strategy established standards that had no sound analytic basis. The House Report on H.R. 8650, an earlier version of the bill that was eventually passed establishing BEPS, made it clear that this economic basis was what Congress had in mind. The Report noted that the bill was designed to:

> introduce discipline in the construction process which will result in lower costs to the consumer and in higher quality buildings. The Committee recognizes that the construction of more energy-efficient buildings will result in higher development or initial costs under current design practices. However, the Committee received abundant evidence that the potential reductions in annual utility bills can offset the annual amortization costs of fairly substantial increases in front-end construction costs. . . . The Committee does not regard the higher capital costs involved in energy-efficient buildings to have any serious consequences with respect to the marketability of homes. . . ." H. R. Report No. 94-377, 94th Congress, in 1st Session, 3 (1975).

A technically-based standard is virtually impossible to revise intelligently as fuel prices rise, since the cost/benefit calculations that form the basis of such analysis are completely missing.

For whatever combination of reasons, the Department abandoned its original goal of promulgating final regulations in February, 1979. Instead, a major new research program was undertaken in order to put the BEPS on a sounder intellectual footing. The fruits of this further labor are now available and are analyzed in the following section considering the new format of the BEPS.

STANDARD FOR RESIDENTIAL BUILDINGS

For the revision of BEPS, the Department undertook a number of economic studies in order to determine the life-cycle costs of residential buildings. The preamble to the Notice of Proposed Rulemaking states that such a life-cycle analysis "permitted the use of well-defined economic criteria that have the potential of maximizing the net economic benefits to homeowners and to the Nation, as well as achieving maximum practicable energy conservation." In carrying out the life-cycle analysis, however, the Department constrained itself in a number of ways. It considered the use of energy conservation measures and techniques only if they were currently in common practice in the United States. Included were such conventional and timid measures as increased levels of insulation in the walls, ceilings and floors, and use of double and triple glazing. Similarly, no conservation measure that required any significant changes in behavior or level of amenity of the occupants was permitted. For calculation of costs and benefits, the Department used the Energy Information Administration's Series B Midterm Price Forecast (44 *Federal Register* 25369, April 30, 1979). The discount rate was set at 3 percent, corresponding to an interest rate 3 percent higher than the inflation rate. No doubt there will be wide and merited discussion concerning whether these parameters are correct, in view of the trend of price rises and the discount rates actually used by individuals in their own economic calculations.

What effects are the BEPS likely to have on real houses? Of course, in one sense, it is impossible to answer this question. By legislative design and purpose, the federal government is not to use these standards to dictate any particular architectural solution to meeting the standard. But in actual practice, the Department, from the first, recognized that small builders and designers would need assistance in understanding what kinds of buildings would be likely to pass an inspection based on BEPS. Therefore, the government intends to provide a number of "cookbook" solutions for use by designers. The HUD Minimum Property Standards will be revised so that builders complying with them will also comply automatically with BEPS. Instructions will be given concerning modifications that are necessary in ASHRAE Standard 90-75 in order to make buildings designed to meet it also meet BEPS.

Most helpful for persons trying to understand the effect of BEPS in the real world, the Notice of Proposed Rulemaking contains a sample list of measures that could be taken in the design of a single-family residence in order to comply with the Standards. Let us look at two examples:

> For a gas-heated home located in Chicago, Illinois, a designer could follow any of these three paths: (1) windows 15 percent of floor area distributed equally on the four walls, triple glazing, R-38 ceiling and R-19 wall insulation; or (2) windows redistributed so that south facing window area is increased by 75 percent, and east, west and north facing window area is decreased by 25 percent, double glazing, and R-38 ceiling and R-19 wall insulation; or (3) an active solar domestic hot water heating system supplying 60 percent of the hot water needs of the home, double glazing, and R-38 ceiling and R-11 wall insulation.
>
> For an electrically-heated home in Atlanta, Georgia, a designer could meet the standards in a number of ways, including by following any of these three packages: (1) windows 15 percent of the floor area distributed equally on the four walls, triple glazing, R-38 ceiling, R-19 wall, and R-11 floor insulation, heating supplied by a heat pump; or (2) windows redistributed so that south facing window area is increased by 80 percent, and east, west and north facing window area is decreased by 27 percent, double glazing, R-38 ceiling, R-19 wall, and R-11 floor insulation, heating supplied by a heat pump; or (3) an active solar domestic hot water heating system supplying 60 percent of the hot water needs of the home, double glazing, R-30 ceiling, R-19 wall, and R-11 floor insulation, heating supplied by electric resistance.

The careful reader will have noticed that although the heating loads vary enormously between Chicago and Atlanta (on base 65 F°, Chicago accumulates 6639 degree days, while Atlanta accumulates only 2961), the strategies that must be used to meet the BEPS are essentially identical. How can this be when the heating needs are so different? The explanation lies in the fuels the designer chose for heating: the Chicago home uses natural gas, while the Atlanta home uses electricity. The Department is thus taking into account more than just the energy use that registers on the customer's meter; it is subjecting this consumed energy to different weighting factors for each fuel type. What are these weighting factors? In effect, they are numbers assigned to each fuel type; the designer is required to multiply the amount of electricity, natural gas, or oil by the appropriate weighting factor before he adds up the number of BTUs per square foot per year the building uses.

How are these weighting factors derived? The weighting factor in the current version of the BEPS starts with the average price of fuel consumed. (Average prices of energy are based on the existing mix of old and new energy sources; replacement costs are the costs of new energy sources such as a new powerplant). Naturally, if the BEPS propose to use economic criteria for evaluating life-cycle costs, only the replacement or marginal cost of energy consumed is the proper measure of the value of energy consumed. Since the homeowner must make the choice between avoiding energy consumption (i.e., buying conservation) at marginal costs, only by considering the marginal costs of energy *not* consumed can the equation work fairly. As the Ford Foundation study notes, average prices

> are typically below the cost to the nation of replacing the energy consumed—that is, they are below the marginal cost of the energy. Analysis of the regulations based solely on prices paid by the consumer will therefore understate the value to the nation of more energy-efficient buildings.
>
> This failure to use the regulatory process for correcting deficiencies in the residential energy market is unfortunate because the housing market is almost a classic case in which intelligently conceived regulation has a place. Homebuyers do not generally think in terms of life-cycle costing. . . . A standard that . .took into account the benefits of energy conservation to both the consumer and to the nation, and that permitted exceptions in cases where direct regulation was inappropriate, would have a great deal to commend it.

The weighting factors used by DOE also include a premium for oil and natural gas, in order to press building designers away from using these fuels. Finally, the weighting factors were based on national averages, not on regional differences in fuel costs or availability. The weighting factors chosen by DOE are as follows: ˙

Building Type	Natural Gas	Oil	Electricity
Single-Family Residential	1.0	1.22	2.79
Commercial and Multi-family Residential	1.0	1.20	3.08

The effect of these weighting factors is to make it more "expensive" in any given energy budget to use electricity, somewhat less "expensive" to use oil, and least "expensive" to use natural gas. Solar energy and other renewable energy resources are "free" according to this scheme, so the use of such sources is highly encouraged. The other effect of the weighting factors, of course, is to announce in effect a fuels policy for the American building industry.

How Strict Are the BEPS?

Any detailed analysis of the BEPS for residences is certainly premature at this time. The Department based many of its decisions on Technical Support Documents (TSDs) that were not publicly released at the time the proposed rule was published in the *Federal Register*. These TSDs cover such crucial analytic topics as "Energy Budget Levels Selection," "Weighting Factors," "Economic Analysis," and "Passive and Active Solar Heating Analysis." Neither is it necessary to offer a detailed criticism of rules that may well be improved by the comment process. Nonetheless, one can legitimately look to the cast of mind of the Department as it selected these budget figures. In the selection of the energy budget figures for homes, the Department considered four levels: the level they chose, 10 percent tighter, 20 percent tighter, and 25-to-30 percent looser. Energy savings were 11 quads (summed over the 40 years from 1980 to 2020) for the alternative selected, but 16.5 quads for the tightest standard. Both the standard selected and the tightest standard were found to have approximately equal and favorable economic impacts on the nation and on the homeowner. The first costs of the alternative selected would be between $750 and $1,500 added to the base cost; for the tightest alternative, the additional first cost ranges from $1,500 to $3,000 (although the Department's analysis shows this first cost will tend to be smaller as new energy conservation technology is introduced to meet the tighter standard). Yet in spite of the additional energy saved, the benefits to homeowner and nation, and the relatively small additional first cost, DOE selected the less favorable alternative on the basis of "the difficulty of achieving those levels at the present time." This reasoning is hard to understand if BEPS are to be a technology forcing regulation.

Standards for Nonresidential Buildings

For nonresidential buildings, the Department was unable to conduct the kind of life-cycle analysis that it did on the single-family dwellings. Therefore, the proposed rules are based on the older, technically-based data base collected by AIA/RC. Using the Phase II buildings redesigned by their original architects following the three-day energy conservation course, the Department looked at three budget levels for such buildings. R_{30} means that

30 percent of all building redesigns for that building type achieved that level of design energy requirement or lower. DOE calls this "strict." R_{50} indicates that 50 percent met the figure; that is called "nominal." R_{70} means that 70 percent met that level of performance; this is called "lenient." (DOE's calculations reveal that for a large office building in Kansas City, these levels of performance translate into the following number of BTUs per square foot per year: R_{30} = 46 MBTU/sq. ft./year; R_{50} = 49 MBTU/sq. ft./year; and R_{70} = 51 MBTU/sq. ft./year.) Again, in each case, DOE found that "the net present value to the nation of the proposed Energy Budget Levels was greatest for the strict case and lowest for the lenient case. Thus, national economic benefits are greatest for the more strict levels." Likewise, DOE reveals that in a preliminary life-cycle study of a large office building, "there are designs that are economically beneficial at design energy requirement levels more stringent than those achieved by most of the redesigns in Phase II."

Nonetheless, DOE feared that designers would have difficulty in reaching strict levels, not because of costs or technical constraints, but merely because of "unfamiliarity of design professionals with energy-efficient design strategies and available technology." For this reason, DOE has selected the following levels:

Large and small office buildings: R_{30} ("strict").

Hospitals and multifamily low-rise residential buildings: R_{70} ("lenient").

All other commercial and multifamily residential buildings: R_{50} ("nominal").

Again, as a preliminary matter, it appears unwise to select standards on the basis that design professionals are unfamiliar with existing technology; a better strategy would be to set stricter standards and let the manufacturers, trade associations, continuing education course instructors, and the Federal Energy Extension Service educate the professionals to meet the new, higher standard.

SANCTIONS

Finally, in this description of BEPS, it is worth discussing how they will actually come to have the force of law at the state and local level. The building code professionals are conservative and clannish; from the beginning, there has been considerable distrust of the federal effort, and an active movement on the part of some states to have alternative energy conservation building codes in place in order to head off the federal BEPS when it finally emerged.

Unfortunately, Congress in the original legislation devised a Draconian remedy, one that is so excessive that it certainly would never be used. According to Section 305(c) of the Energy Conservation and Production Act, the President is to transmit the final BEPS regulations to Congress with a recommendation concerning their adoption. Congress then has ninety days in which to consider them. If both Houses pass a resolution approving the regulations, they become effective. Following that, any state that does not adopt BEPS or its equivalent, can lose all federal financial assistance for building. This includes "any form of loan, grant, guarantee, insurance, payment, rebate, study, subsidy, or any other form of direct or indirect federal assistance" and "any loan made or purchased by any bank, savings and loan association, or similar institute subject to regulation" by the federal government or insurance by a government agency. This sanction is equivalent to sending policemen out in patrol cars, equipped only with fragmentation bombs: cutting off all federal aid to the building industry is too extreme a penalty ever to be imposed by Congress on any state.

We may hope that in its consideration of the standards themselves, Congress will try to develop a more graduated set of incentives and penalties for states that refuse to adopt BEPS or its equivalent. Training grants for state and local officials, incentive payments to state building agencies, educational efforts for the national code groups and voluntary organizations, and partial withholding of federal benefits are a better array of carrots and sticks with which to equip DOE.

What Effects Will Follow from Adoption of BEPS?

Now let us turn to the final section of this paper and address the question whether a strong standard, aggressively enforced, will make much difference to the way we use land and build buildings.

There can be little doubt but that BEPS will have an enormous effect on the amount of energy we use in this country. Buildings that comply will use from between 30-to-50 percent *less* energy than their counterparts now do. If, as I believe possible and desirable, BEPS are tightened up to a truly cost-effective level, the amount of energy saved in new buildings could rise even further. Therefore, BEPS deserves to be supported for the effect it will have on our energy problems alone.

What effect will BEPS have on the way we use land? Here, we must be much more speculative, and I would like to suggest two different answers that are intended to be contradictory. I believe that there is virtually no way to predict in advance whether energy and higher energy prices will drive the shape of our cities or the appearance of our housing and other building stocks. I am truly agnostic, able to be persuaded in either direction. Let us look at two of the possible alternative futures.

FUTURE ONE

BEPS will have an enormous, though gradual, effect on design of our buildings and cities.

According to this scenario, BEPS will continue to be tightened down, as the price of oil and other energy sources rises and as conservation technology becomes more and more available. Gradually, over the next twenty or thirty years, the energy budget, the weighting factors, and the land use patterns that favor use of solar energy will make buildings more solar dependent, sloped, shaped and oriented toward catching the winter sun, rejecting the summer heat, and cooling with the use of natural ventilation. Likewise, BEPS will be sufficiently tight so that houses with common walls will make the energy budget easier to meet: row houses, apartment buildings and compact developments will be the norm, with the single-family house confined to those who are willing to invest a considerable amount of capital in conservation and solar energy.

Our office buildings will show less glass, will be more turned in upon themselves, looking into central atriums that serve as greenhouses, capturing solar heat and letting it pass through offices before it is wasted outside. More air-handling equipment, sensitively controlled by microprocesser chips, will shift heat and "coolth" around the building, from duplicating room to outside office, from south side to north side. Office buildings will also have to expose less of their bulk, and almost none of their windows, to the cool north side. On the south and west sides, windows will be small and shaded during the summer.

These trends, if they occur, could indeed make BEPS an important driving force in reshaping the face and the skeleton of American cities. The premium would be on buildings that conserved energy, that were placed to use the sun's heat and the wind's cooling, and that let natural light take the place of electricity as much as possible.

FUTURE TWO

Conservation technologies will develop so rapidly, so cheaply, and so unobtrusively that energy prices and energy regulation will have little or no effect on American building design or the cityscape.

An alternative view of the future looks at the many ways we have for coping with BEPS and the higher energy prices that drive them *without* making any substantial changes in our buildings. The strategies that the Department of Energy has proposed for meeting its own BEPS are not outlandish or design forcing. Consider again the strategies one can follow in a cold climate like Chicago and still comply with BEPS: a little more insulation, some triple glazing, and little restriction on window placement or size. Even in an electrically-heated home in Atlanta, the restrictions are relatively minor (the principal one being the substitution of a heat pump for electric resistance heating). The studies undertaken by the Princeton University Center for Environment and Energy in the planned unit development of Twin Rivers, New Jersey, have demonstrated how little work it takes in the interior of a standard townhouse to cut energy use by 30-to-50 percent. Checking insulation, filling cracks, and finding and blocking air paths to the attic have little or no effect on design, scale, or placement of a house. Thus, it is possible to increase the energy efficiency of houses so cheaply through simple conservation measures that they will exceed BEPS without the introduction of any solar technology at all and without changing the outside design of the house in any great detail.

Similarly, in office buildings, the simple shading devices, the use of more insulating glass, and the more careful selection and control of mechanical equipment can improve the energy performance of a building without affecting its appearance or function.

Thus, according to this view of the future, Americans will accommodate to higher energy prices by investing a rather small additional amount of capital in their houses and offices in order to preserve other amenities they value. The amount of extra capital required for an energy-efficient building is quite small and has a fast payback period. It may well be that Americans will make the kinds of accommodations to higher prices for heating and cooling as they have to higher prices for gasoline. Instead of abandoning their cars for public transportation, they have simply redesigned their cars for greater fuel efficiency. In exactly the same way, we may make simple, incremental changes in our buildings so that they continue to provide us adequate warmth and shelter yet satisfy the many other psychological demands we all make on our cities and on our dwellings.

Whichever future (or mixture of the two) turns out to occur in fact will not become obvious for a decade or more. What will become obvious is the energy and economic savings BEPS can help us capture. Other forces, such as economic growth, changes in personal lifestyle not related to energy, family size, and revitalization of inner cities, to name a few, will make it hard to judge, even in retrospect, the effect of energy-efficiency requirements on the cities of tomorrow.

BIBLIOGRAPHY

Energy and Land Use: A Bibliography

EDWARD DUENSING

---

100. Energy and the City—General

101. Albert, Joseph D., and H. Stan Banton. "Urban spatial adjustments resulting from rising energy costs." *Annals of Regional Science* 12(2):64–71, July 1978.

102. Bacon, Edmund N. "Energy: shaper of future living patterns." *AIA Journal*: 39–41, December 1973.

103. Burchell, Robert W. and David Listokin. *Future Land Use: Energy, Environmental and Legal Constraints*. New Brunswick, New Jersey: Rutgers University Center for Urban Policy Research, 1975.

104. Burby, Raymond J., III, and A. Fleming Bell, eds. *Energy and the Community*. Cambridge, Mass.: Ballinger, 1978.

105. Case, E. *Energy and the Urban Crisis*. Washington, D.C.: Joint Center for Political Studies, 1978.

106. Chibuk, John. *Energy and Urban Reform*. Ottowa, Ontario: Ministry of State for Urban Affairs, Information Resource Service, 1977.

107. "Cities as energy systems." *Building Systems Design* 73:9, February-March, 1976.

108. Dantzig, George B., and Thomas L. Saaty. *Compact City*. San Francisco, Calif.: W.H. Freeman and Co., 1973.

109. Dendrinos, D.S. "Energy costs, the transport network, and urban form." *Environment and Planning* (London) 11(6): 655–664, 1979.

110. Doernberg, A. *Comparative Analysis of Energy Use in Sweden and the United States.* Springfield, Va.: NTIS, 1975.

111. Downs, Anthony. "Squeezing Spread City." *New York Times Magazine* 38, 17 March 1974.

112. Dutt, Ashok. *Energy Resources and Urban Spatial Patterns.* Paper delivered at the 57th Annual Conference of the American Institute of Planners, Denver, Colo., 27–30, October 1974.

113. "Energy and land use: a statement by ULI-the Urban Land Institute." *Environmental Comment*: 1–7, September 1974.

114. "Energy and patterns of land use." *Journal of Architectural Education* 30(3): 62–66, February 1977.

115. "Energy and the city." *Lamp* 57:12–17, Summer 1975.

116. Energy Policy Project of the Ford Foundation, *A Time to Choose*: America's Energy Future, Cambridge, Mass.: Ballinger, 1974.

117. "Evaluating economics of energy consequences for land use decisions." *Journal of Urban Planning and Development* 102:105, August 1976.

118. Fels, Margaret Fulton, and Michael J. Munson. "Energy thrift in urban transportation: Options for the future." In Robert H. Williams, ed. *The Energy Conservation Papers.* Cambridge, Mass.: Ballinger, 1975.

119. Franklin, Herbert M. "Will the new consciousness of energy and environment create an imploding metropolis?" *AIA Journal*: 28–36, August 1974.

120. Goodale, Stephen G., and Phillip A. Leon. *The Energy/Fuel Shortage and Land Development Trends in the Richmond Metropolitan Area: A Survey.* Paper presented at the 57th Annual Conference of the American Institute of Planners, Denver, Colo., 27–30 October 1974.

121. Henderson, Lenneal J. "Energy urban policy, and socioeconomic development." *Urban League Review*: 34–40, Winter, 1978.

122. Hirst, E. *An Engineering-Economic Model of Residential Energy Use.* Springfield, Va.: NTIS, 1976.

123. Huffman, J. "Energy: the limit to growth." *Environmental Law* 5:1–28, Fall 1974.

124. Jarkewiz, Shirley A. "Stubby skyscraper: shorter and squatter office towers rise in some cities in an energy saving drive." *Wall Street Journal* 194:44, 12 September 1979.

125. Kaplan, G. "Cities: energy gluttons." *IEEE Spectrum* 13(7):73–74, July 1976.

126. Keyes, Dale L. *Urban Form and Energy Use.* Washington, D.C.: Urban Institute, 1975.

127. *Metropolitan Development and Energy Consumption.* Washington, D.C.: Urban Institute, 1976.

128. Keyes, Dale L., and George R. Peterson. *Urban Development and Energy Consumption.* Washington, D.C.: Urban Institute, 1977.

129. Knowles, Ralph L. *Energy and Form: An Ecological Approach to Urban Growth.* Cambridge, Mass.: The MIT Press, 1974.

130. Kruvant, William. *City Future—Energy Future.* Washington, D.C.: U.S. General Accounting Office, (n.d.).

131. Kydes, A.S. et al. *Regional Land Use and Energy Modeling.* Upton, N.Y.: Brookhaven National Laboratory, 1976.

132. McNamara, J. "Integrating energy development and land use management goals." *Natural Resource Lawyer* 11 (2): 325–41, 1978.

133. Meir, R. *The Design of Resource Conserving Cities.* Cambridge, Mass.: MIT Press, 1974.

134. National Energy Strategies Project. *Energy in America's Future: The Choices Before U.S.* (n.p.), (n.d.).

135. Reilly, William D., ed. *The Use of Land: A Citizen's Policy Guide to Urban Growth.* New York: Thomas Y. Crowell Co., Inc., 1973.

136. Roberts, James S. *Energy, Land Use, and Growth Management.* Washington, D.C.: Metropolitan Washington Council of Governments, 1975.

137. ———. *Compact Cities: A Neglected Way of Conserving Energy-Panel 1: Has Sprawl Been Tamed.* Testimony presented before U.S. House of Representatives, Committee on Banking Finance, and Urban Affairs, 11 December 1979.

138. Romanos, M.C. "Energy price effects on metropolitan spatial structure and form." *Environment and Planning* 10(1): 92–104, January 1978.

139. Spink, Frank H., Jr. *Regarding Energy and the City.* Testimony submitted to Committee of Banking. Finance and Urban Affairs, Subcommittee on the City, 14 September 1977.

140. Sternlieb, George et al. "Back to the central city: myths and realities." *Traffic Quarterly* 33(4): 617–136, October 1979.

141. Tanzler, Hans G., Jr. *Energy and the City* (n.p.), (n.d.).

142. Til, J. Von. "Spatial form and structure in a possible future: some implications of energy shortfall for urban planning." *Journal of the American Planning Association* 45(3): 318–29, July 1979.

143. U.S. Congress House. Committee on Banking, Finance, and Urban Affairs, Subcommittee on the City. *Energy and the City. Hearing . . . 95th Congress - 1st Session, September 14–16, 1977.* Washington, D.C.: U.S. GPO, 1977.

144. U.S. Congressional Budget Office. *Urban Transportation and Energy: The Potential Modes of Different Savings.* Washington, D.C.: U.S. GPO, 1977.

145. U.S. Congress Joint Economic Committee, Subcommittee on Energy. *Impact of the President's Energy Plan on the Northeast: Hearing . . . 95th Congress - 1st session, May 13, 1977*. Washington, D.C.: U.S. GPO, 1977.

146. U.S. Congress Senate. Committee on Government Operations. *Current Energy Shortages Oversight Series: Mayors Panel—Urban Impact—Hearing . . . 93rd Congress - 2nd session, March 6, 1974*. Washington, D.C.: U.S. GPO, 1974.

147. Vaughan, R.J. *Urban Impacts of Federal Policies. Vol. 2. Economic Development*. Santa Monica, Calif.: Rand Corporation, 1977.

148. Werth, Joel T. *Energy in the Cities Symposium*. American Planning Association, Planning Advisory Service, Report No. 349.

149. Windheim, Lee Stephen, and Rebecca R. Wodder. "Cities as energy systems." *Building Systems Design*: 9–30, February/March 1976.

150. Woodruff, A.M. *Urbanization and Land Use Controls*. (n.p.), (n.d.).

151. Zimmerman, Rae. *Energy Resource Management, Environmental Impacts and the Competition for Land*. Paper presented at 57th Annual Conference of the American Institute of Planners, Denver, Colo., 27–30, October 1974.

200. Land Use Measures to Limit Energy Consumption

201. Alaska. Division of Energy and Power Development. *Minimizing Consumption of Exhaustible Energy Resources through Community Planning: Final Report*. Anchorage, Ala.: The Division, 1977.

202. "Allocative and energetic implications of land use planning." *Environmental Lawyer* 5(3), Spring 1975.

203. American Society of Planning Officials. *Institutional Factors Influencing the Acceptance of Community Energy Systems and Energy-Efficient Community Design: Public Planning, Administration, and Regulation*. Argonne, Ill.: Argonne National Laboratory, 1976.

204. Bacon, Edmund N. "Energy and land use." *Urban Land*: 13–16, July/August 1973.

205. Baltimore Regional Planning Council. *The Effect of Urban Growth Alternatives on Travel, Air Quality, and Fuel Consumption*. Baltimore, Md.: The Council, 1977.

206. Bhagat, N., and H.G.M. Jones. *Simulation of Residential Energy Use and its Dependence on Land Use and Economic Parameters*. Upton, N.Y.: Brookhaven National Laboratory, 1974.

207. Booz, Allen, and Hamilton, Inc. *Interaction of Land Use Patterns and Residential Energy Conservation*. Washington, D.C.: Federal Energy Administration, 1976.

208. Brandywine Associates. *Land Use Configurations and the Utilization of Distributed Energy Technology: Final Report*. Springfield, Va.: NTIS, 1977.

209. *Brief Summary of the Land Use and Energy Utilization Project.* Upton, N.Y.: Brookhaven National Laboratory, 1975.

210. Brookhaven National Laborabory and the State University of New York. *Planner's Energy Workbook.* Springfield, Va.: NTIS, 1977.

211. Brookhaven National Laboratory and the State University of New York Land Use-Energy Utilization Project. *Land Use and Energy Utilization.* Springfield, Va.: NTIS, 1975.

212. "Can urban planning cut energy use?" *Environmental Action* 77(9): 3 + , 24 September 1977.

213. Carella, V.A. "Upon land use and urban planning." In R.A. Fazzolare, ed. *Energy Use Management, Vol. II.* Elmsford, N.Y.: Pergamon Press, Inc., 1977.

214. Carrol, T.O. *Land-Use Energy Simulation Model: A Computer Based Model for Exploring Land Use and Energy Relationships.* Upton, N.Y.: Brookhaven National Laboratory, 1977.

215. "City planners aid in energy conservation." *Land Research/Development* 22:98 + , March 1980.

216. Clark, James W. *Assessing the Relationships Between Urban Form and Travel Requirements: A Literature Review and Conceptual Framework.* Springfield, Va.: NTIS, 1970.

217. Clawson, Marion. *America's Land and Its Uses.* Baltimore, Md.: Johns Hopkins, 1972.

218. *Comprehensive Community Planning for Energy Management and Conservation: Developing and Applying a Coordinated Approach to Energy Related Community Development.* Springfield, Va.: NTIS, 1977.

219. Conklin and Rossant and Flack and Kubtz. *Reading the Energy Meter on Development.* Washington, D.C.: Federal Energy Administration, 1976.

220. Daniel, G.H. "Energy and land policies." *Journal of the Institute of Fuel* 49(400): 115–222, September 1976.

221. Darmstadter, Joel. *Conserving Energy, Prospects and Opportunities in the New York Region.* Baltimore, Md.: Johns Hopkins, 1975.

222. Denton, Jesse, ed. *Energy Conservation Through Effective Energy Utilization. Proceedings of the 1973 Engineering Foundation Conference, New England College, Henniken, New Hampshire, August 19–24, 1973.* Washington, D.C.: National Bureau of Standards, 1973.

223. Doe, M. Chiogioji. *International Energy Agency (IEA) Working Paper on Energy Conservation Research and Development Annual Report 1976–1977.* Springfield, Va.: NTIS, 1977.

224. Donovan, Hamester and Rattien, Inc. *Energy, The Environment and Land Use: Literature Review.* Washington, D.C.: Donovan, Hamester and Rattien, Inc., 1976.

225. Dowall, D.E. *Mapping the Land Use and Energy Policy Terrain: Land Use and Energy Conflicts.* Berkeley, Calif.: University of California, Institute of Urban and Regional Development, 1978.

226. Dumas, Lloyd J. *The Conservation Response.* Lexington, Mass.: D.C. Heath, 1976.

227. Edwards, Jerry L., and Joseph L. Schofer. *Relationships Between Energy Consumption in Transportation and Urban Spatial Structure.* Evanston, Ill.: Northwestern University, Department of Civil Engineering, 1974.

228. ———. *Relationships Between Transportation Energy Consumption and Urban Structure: Results of Simulation Studies.* Evanston, Ill.: Northwestern University, Department of Civil Engineering, 1975.

229. "Energy and land use: an instrument of U.S. conservation policy." *Energy Policy* 4:225, September 1976.

230. "Energy and land use. A statement by the ULI-the Urban Land Institute." *Environmental Comment*: 4–7, September 1974.

231. "Energy conservation and land development." *Environmental Comment*, July 1977 (entire issue).

232. "Energy conservation in human settlements." *Ekistics*, May 1978.

233. *Energy Conservation Training Institute.* Washington, D.C.: Conservation Foundation, (n.d.).

234. *The Energy Vista: Policy Perspectives on Energy Conservation through Land Use Management.* Cambridge, Mass.: Technology and Economics, Inc., 1976.

235. Fazzolare, R., and C.B. Smith, eds. *Energy Use Management.* Elmsford, N.Y.: Pergamon, 1978.

236. Ford Foundation, Energy Policy Project. *A Time to Choose: America's Energy Future.* Cambridge, Mass.: Ballinger, 1974.

237. Fraker, Harrison, and Elizabeth Schorske. *Energy Husbandry in Housing: An Analysis of the Development Process in a Residential Community.* Princeton, N.J.: Princeton University, Center for Environmental Studies, 1973.

238. Gibbons, J. *Conservation and Governance: Proceedings of a Symposium on Implications of Energy Conservation and Supply Alternatives.* East Brunswick, N.J.: Science Applications, Inc., 1978.

239. Gil, Ephraim. *Energy Efficient Planning: An Annotated Bibliography.* Chicago, Ill.: American Society of Planning Officials, 1976.

240. Gilbert, G. "Energy, urban form and transportation policy." *Transportation Research* 4–5: 267–76, October 1974.

241. Grew, Terry et al. "Fuel conservation and applied research." *Science*: 135–42, 14 April 1978.

242. Grot, Richard A., and Robert H. Socolow. *Energy Utilization in a Residential Community.* Cambridge, Mass.: (n.p.), (n.d.).

243. Hammond, Allen L. "Conservation of energy: the potential for more efficient use." *Science*, 5 December 1972.

244. Hardnett, James B., ed. *Alternative Energy Futures.* New York: Academic Press, 1976.

245. Harwood, Corbin Crews. "Planning for energy conservation." In *ECP* (Energy Conservation Project). (n.p.), (n.d.).

246. ———. *Using Land to Save Energy.* Cambridge, Mass.: Ballinger, 1977.

247. Hatsopoulos, G.N. et al. "Capital investment to save energy." *Harvard Business Review*: 111-22, March-April 1978.

248. Hayes, Denis. *Energy: The Case for Conservation.* Washington, D.C.: Worldwatch Institute, 1976.

249. Hirst, E. *Residential Energy Conservation Strategies.* Springfield, Va.: NTIS, 1967.

250. Hirst, E., and J. Carney. *Residential Energy Use to the Year 2000: Conservation and Economics.* Springfield, Va.: NTIS, 1977.

251. Hittman Associates, Inc. *Comprehensive Community Planning for Energy Management and Conservation.* Washington, D.C.: Energy Research and Development Administration, 1977.

252. Institute for Urban Studies, State University of New York at Stony Brook. *Planner's Energy Workbook: A Users Manual for Land Use and Energy Utilization.* Stony Brook, N.Y.: The Institute, 1976.

253. Jackson, C.I., ed. *Human Settlement on Energy: An Account of the ECE Seminar on the Impact of Energy Considerations on the Planning and Development of Human Settlements, Ottowa, Canada, October 3-14, 1977.* United Nations Economic Commission for Europe Series. Oxford, England: Pergamon Press, 1978.

254. Kaiser, Edward J. "Land use planning: The cornerstone of local environmental planning and control." In *An Anthology of Selected Readings for the National Conference on Managing the Environment.* Washington, D.C.: U.S. Environmental Protection Agency, 1972.

255. Keyes, Dale L. "Energy and land use: an instrument of U.S. conservation policy." *Energy Policy*: 225-36, September 1976.

256. McCoy, Hugh A., and Joseph F. Singer. "Energy growth management systems planning considerations." In *Community Energy Planning.* American Institute of Planners, Annual Meeting, New Orleans, La.: 28-30 September 1978.

257. McFarland, William F. *Energy Development and Land Use in Texas.* Springfield, Va.: NTIS, 1975.

258. Milstein, Jeffrey S. *Energy Conservation and True Behavior.* U.S. Department of Energy Paper, October 1977.

259. Myhra, David. *Energy Husbandry in the Homes Building Industry—A Look at Planned Unit Development.* (n.p.), (n.d.).

260. ———. *Saving Energy in the Residential Sector through Planning.* Paper delivered at the 57th Annual Conference of the American Institute of Planners, Denver, Colo: 27–30 October 1974.

261. National Petroleum Council. Committee on Energy Conservation. *Energy Conservation in the United States.* Washington, D.C.: (n.p.), 1974.

262. National Science Foundation Research Applied to National Needs (RANN) Program. *Summary Report of the Cornell Workshop on Energy and the Environment.* Washington, D.C.: U.S. GPO, 1972.

263. O'Donnell, Robert M., and James Parker. "Large-scale development: breeder for energy conservation." (n.p.), (n.d.).

264. Pauker, Guy J. *Can Land Use Management Reduce Energy Consumption for Transportation?* Santa Monica, Calif.: Rand Corp., 1974.

265. Penner, S.S., and L. Icerman. *Energy Demands, Resources, Impact, Technology and Policy.* Reading, Mass.: Addison-Wesley, 1974.

266. Peskin, R.C., and J.L. Schafer. "Transportation energy conservation through directed urban growth." In R.A. Fazzolare, ed. *Energy Use Management, Vol. II.* Elmsford, N.Y.: Pergamon Press, 1977.

267. Pikarsky, M. "Land use and transportation in an energy efficient society." In National Research Council, *Transportation and Land Development: Conference Proceedings Special Report 183, Commission on Socio-technical Systems,* National Academy of Sciences, 1978.

268. Portland Energy Conservation Project. *Energy and Land Use: Comprehensive Plan Working Paper no. 13.* Portland, Ore.: Skidmore, Owings and Merrill, 1976.

269. Pozzo, Robert J., and James Clark. *A Planner's Handbook on Energy (with Emphasis on Residential Uses).* Tallahassee, Fla.: State Energy Office, Department of Administration.

270. Priest, W.C., and M.R. Elgerman. *Energy Conservation through Land Use Management.* Springfield, Va.: NTIS, 1976.

271. Real Estate Research Corporation. *The Costs of Sprawl: Environmental and Economic Costs of Alternative Residential Development Patterns at the Urban Fringe.* Washington, D.C.: U.S. GPO, 1974.

272. Resource Planning Associates, Inc. *Comprehensive Community Planning for Energy Management and Conservation: Developing and Applying a Coordinated Approach to Energy Related Community Development.* Springfield, Va.: NTIS, 1977.

273. Resources for the Future. *Energy, Economic Growth and Environment.* Baltimore, Md.: Johns Hopkins, 1975.

274. Rice, Richard A. "System energy and future transportation." *Technology Review*: 37, January 1972.

275. Roberts, J. *A Preliminary Report for Community Systems: Advanced Technology-Mixed Energy System.* Springfield, Va.: NTIS, 1976.

276. Roberts, James S. "Energy and land use: analyses of alternative development patterns." *Environmental Comment*: 3–11, September 1975.

277. ———. *Energy, Land Use and Growth Policy: Implications for Metropolitan Washington.* Washington, D.C.: Metropolitan Washington Council of Governments, 1975.

278. Romanos, M.C. *Community Planning for Energy Conservation: A Scenario for ERDA's Community Design Research Development and Demonstration.* Springfield, Va.: NTIS, 1977.

279. Schipper, Lee, and A.J. Lichtenberg. *Efficient Energy Use and Well Being: The Swedish Example.* Berkeley, Calif.: University of California, Energy and Environment Division, Lawrence Berkeley Laboratory, 1976.

280. Schipper, Lee, and Joel Darmstadter. "The logic of energy conservation." *Technology Review*: 41–50, January 1978.

281. Schmalz, Anton B., ed. *Energy: Today's Choices, Tomorrow's Opportunities.* Washington, D.C.: World Future Society, 1974.

282. Schneide, Jerry B., and Joseph R. Beck. *Reducing the Travel Requirements of the American City: An Investigation of Alternative Urban Spatial Structures.* Seattle, Wash.: University of Washington, Department of Planning and Civil Engineering, 1973.

283. Socolow, Robert. "The coming age of energy conservation." *Annual Review of Energy* 2:239–89, 1977.

284. Taft, G.S. *Land-Use Guidance Strategies as a Means of Achieving National Energy Goals.* Unpublished Ph.D. dissertation, Purdue University, 1978.

285. Technology and Economics, Inc. *An Overview and Critical Evaluation of the Relationship between Land Use and Energy Conservation.* 3 vols. Washington, D.C.: U.S. Federal Energy Administration, 1976.

286. U.S. Congress. Joint Economic Committee. Subcommittee on Energy. *Energy Conservation: Hearings . . . 94th Congress, second session, February 2, 3, and 24, and April 13, 1976.* Washington, D.C.: U.S. GPO, 1977.

287. U.S. Congress, Senate. Committee on Commerce. *ERDA Energy Conservation Programs: Hearings . . . 94th Congress, second session, March 5, 1976.* Washington, D.C.: U.S. GPO, 1976.

288. U.S. National Bureau of Standards. *Technical Options in Energy Conservation on Buildings.* Washington, D.C.: U.S. GPO, 1973.

289. U.S. National Laboratory, Oak Ridge, Tenn. *Transportation Energy Conservation Data Book*. Springfield, Va.: NTIS, 1979.

290. Urban Land Institute. *Large Scale Development: Benefits, Constraints, and State and Local Policy Incentives*. Washington, D.C.: The Institute, 1977.

291. *Urban Planning As an Impediment to Energy Conservation: An Examination of Potential Conflicts Between Existing Planning Regulations and Energy Conserving Site Planning Alternatives*. Paper presented at Energy Management International Conference, Tuscon, Ariz.: 24–28 October 1977.

292. Watson, Donald, ed. *Energy Conservation Through Building Design*. New York, N.Y.: McGraw-Hill, 1979.

293. Webber, Andrew. "Energy prospects require new planning considerations in housing and community development fields." *Journal of Housing*: 253–55, May 1979.

294. Williams, Robert A., ed. *The Energy Conservation Papers*. Cambridge, Mass.: Ballinger, 1975.

300. Land Use Measures to Assure Energy Supply

301. *An Analysis of Federal Incentives Used to Stimulate Energy Production*. Richland, Wash.: Pacific Northwest Laboratory, 1978.

302. *Application of Solar Technology to Todays Energy Needs*. Washington, D.C.: June 1978.

303. Ballou, Stephen W. *Socio-Economic Aspects of Surface-Mining: Effects of Strip Mine Reclamation Procedures Upon Assessed Land Values*. Paper presented at National Coal Association. 4th Symposium on Surface Mining and Reclamation, Louisville, Ky.: 19–21 October 1976.

304. Behrman, Daniel. *Solar Energy: The Awakening Science*. Boston, Mass.: Little, Brown, 1976.

305. Belknap, Richard H., ed. *Energy: Future Alternatives and Risks*. Cambridge, Mass.: Ballinger, 1974.

306. Braunstein, H.M. et al., eds. *Environmental Health and Control Aspects of Coal Conversion: An Information Overview*. 2 Vols. Oak Ridge, Tenn.: Oak Ridge National Laboratory, 1977.

307. Carter, Anne P. *Energy and the Environment: A Structural Analysis*. Hanover, N.H.: University Press of New England, 1976.

308. Chamber of Commerce of the United States of America, Community and Regional Development Group. *Meeting National Energy Needs*. Washington, D.C.: The Chamber, 1972.

309. Dalstad, Norman L. et al. *Economic Impact of Coal Development in Western North Dakota*. Paper presented at the Fort Union Coal Field Symposium. Fargo, N.Dak.: North Dakota State University, 1975.

310. Ebbin, Stephen. *Citizen Groups and the Nuclear Power Controversy*. Cambridge, Mass.: MIT Press, 1974.

311. Edwards, R.G. et al. *Social, Economic, Environmental Impacts of Coal Gasification and Liquefication Plants*. Lexington, Ky.: University of Kentucky, Institute of Mining and Mining Research, 1975.

312. Ford Foundation. *A Time to Choose: America's Energy Future, Final Report*. Cambridge, Mass.: Ballinger, 1974.

313. Gilmore, John S., and Mary K. Duff. *Impacts of Intensive Oil Shale Development: Concepts and Remedies*. Denver, Colo.: University of Denver, Denver Research Institute, 1973.

314. Gordon, Howard, and Roy Meander. *Perspectives on the Energy Crisis*. Ann Arbor, Mich.: Ann Arbor Science, 1977.

315. Governor's Commission on Cogeneration. *Cogeneration: Its Benefits to New England*. Boston, Mass.: Commonwealth of Massachusetts, 1978.

316. Hagel, John. *Alternative Energy Strategies. Constraints and Opportunities*. New York, N.Y.: Praeger, 1976.

317. Hammond, Allen L. et al. *Energy and the Future*. Washington, D.C.: American Association for the Advancement of Science, 1973.

318. Hardesty, C.H., Jr. "New values of coal and the energy crisis." *West Virginia Law Review* 76:255 + , April 1974.

319. Harrah, Barbara K., and David Harrah. *Alternative Sources of Energy: A Bibliography*. Metuchen, N.J.: Scarecrow Press, 1975.

320. Hayes, Denis. "Energy and building in the 1980s: the solar prospect." (address) *Journal of Property Management* 43:36–42, January-February 1978.

321. Helman, G. "Energy: searching for substitutes." *Nations Business* 66(9): 78–80, 83–84, September 1978.

322. Huffman, R.L. *Coal Gasification and Its Alternatives*. Oklahoma City, Okla.: Cities Service Gas Company. 1977.

323. Issacs, Charles, S. "On-site cogeneration: energy conservation and energetic conservation." *Builings*: 87–90, September 1979.

324. Jackson, J.R., and W.S. Johnson. *Commercial Energy Use: A Disaggregation by Fuel, Building Type, and End Use*. Oak Ridge, Tenn.: Oak Ridge National Laboratory, 1978.

325. Kalter, R.J., and W.A. Vogely, eds. *Energy Supply and Government Policy*. Ithaca, N.Y.: Cornell Univeristy, 1976.

326. Landsberg, H.H. *Energy-The Next Twenty Years*. Cambridge, Mass.: Ballinger, 1979.

327. Lettes, R., and A. Ferrari. "Energy future: summary of recent studies." *Revue Generale Nucleaire* (2): 108–16, 1978.

328. Lovins, Amory B. *Soft Energy Paths: Towards a Durable Peace*. San Francisco, Calif.: Friends of the Earth, 1977.

329. Meadows, D., and J. Stanley Miller. "The transition to coal" *Technology Review* 75(1): 19–29, October/November, 1975.

330. Milora, S.J., and J.W. Tester. *Geothermal Energy as a Source of Electric Power.* Cambridge, Mass.: MIT Press, 1976.

331. Mink, Patsy, T. "Reclamation and rollcalls: the political struggle over stripmining." *Environmental Policy and Law* 2(4):176 + , December 1976.

332. Monasterio, F.J. "Ideology of the use of alternative sources of energy." In R. Fazzolara and C.H. Smith, eds. *Energy Use Management.* Elmsford, N.Y.: Pergamon, 1978.

333. Murphy, John J. *Energy and Public Policy - 1972.* New York, N.Y.: Conference Board, 1972.

334. National Coal Policy Project. "Summary and synthesis." In *Where We Agree: Report of the National Coal Policy Project.* Washington, D.C.: Automated Graphic Systems, 1978.

335. National Consumer Research Institute. *Proceedings of Conference on Cogeneration and Integrated Utility Systems.* Springfield, Va.: NTIS, 1977.

336. National Electric Reliability Council. *Fossil and Nuclear Fuel for Electric Utility Generation.* Springfield, Va.: NTIS, 1977.

337. National Petroleum Council. *U.S. Energy Outlook - Coal Availability.* Washington, D.C.: The Council, 1973.

338. National Research Council, Ad Hoc Panel on the Liquefacation of Coal. *Assessment of Technology for the Liquefaction of Coal.* Washington, D.C.: National Academy of Sciences, 1977.

339. National Research Council Solar Resource Group. *Domestic Potential of Solar and Other Renewable Energy Sources.* Washington, D.C.: National Academy of Sciences, 1979.

340. National Research Council, Committee on Literature Survey of Risks Associated with Nuclear Power. *Risks Associated with Nuclear Power: A Critical Review of the Literature.* Washington, D.C.: National Academy of Sciences, 1979.

341. National Research Council, Solar Research Committee on Nuclear and Alternative Energy Systems, Supply and Delivery Panel. *U.S. Energy Prospects to 2010.* Washington, D.C.: National Academy of Sciences, 1979.

342. Olsen, McKinley C. *Unacceptable Risk.* New York: Bantam, 1976.

343. Ramsay, William, and Phillip R. Reed. *Land Use and Nuclear Power Plants. Case Studies of Siting Problems.* Washington, D.C.: U.S. GPO, 1974.

344. Ridgeway, James, and Bettina Conner. *New Energy: Understanding the Crisis and a Guide to an Alternative Energy System.* Boston, Mass.: Beacon, 1975.

345. Rieck, T.A. "Economics of nuclear power." *Science* 201(18): 582–89, August 1978.

346. Rimberg, David. *Utilization of Waste Heat from Power Plants.* Park Ridge, N.J.: Data Corp., 1974.

347. "Solar energy and land use-problems and prospects especially in urban areas (seven articles)." *Environmental Comment*: 3–16, May 1978.

348. Steward, W. Herman, and James S. Cannon. *Energy Futures.* Cambridge, Mass.: Ballinger, 1977.

349. Stobaugh, R., and D. Yergin. *Energy Futures - Report of the Energy Project at the Harvard Business School.* New York, N.Y.: Random House, 1979.

350. Stroup, Richard L., and Verne House. *The Political Economy of Coal Gasification: Some Determinants of Demand for Western Coal.* Bozeman, Mon.: Montana State University, Agricultural Economics and Economics Department, 1975.

351. U.S. Atomic Energy Commission, Office of Information Services, Technical Information Services. *Coal Processing: Gasification, Liquefacation, Desulphurization: A Bibliogrpahy.* Springfield, Va.: NTIS, 1974.

352. U.S. Bureau of Mines. *Converting Organic Wastes to Oil: A Replenishable Energy Source.* Washington, D.C.: The Bureau, 1971.

353. U.S. Congress. Committee on Government Operations. *Nuclear Power Costs . . . 95th Congress, second session.* Washington, D.C.: U.S. GPO, 1978.

354. U.S. Congress. Office of Technology Assessment. *Application of Solar Technology to Today's Energy Needs.* Washington, D.C.: U.S. GPO, 1978.

355. U.S. Congress Office of Technology Assessment. *The Direct Use of Coal: Prospects and Problems of Production and Combustion.* Washington, D.C.: U.S. GPO, 1979.

356. U.S. Department of Energy. *An Assessment of National Consequences of Increased Coal Utilization: Executive Summary.* Washington, D.C.: The Department, 1979.

357. U.S. Department of Energy. *Report of the President's Domestic Policy Review of Solar Energy.* Washington, D.C.: The Department, 1979.

358. U.S. Federal Energy Administration. *Final Task Force Report on Coal Project Independence.* Washington, D.C.: U.S. GPO, 1974.

359. U.S. Interagency Task Force on Synthetic Fuels. *Recommendations for a Synthetic Fuels Commercialization Program: Report Submitted by Synfuels Interagency Task Force to the President's Energy Resources Council.* Washington, D.C.: U.S. GPO, 1975.

360. White, I.D.W., and D.L. Williams, eds. *Assessment of Geothermal Resources of the United States - 1975.* Washington, D.C.: U.S. GPO, 1975.

361. Williams, R.H. "Toward a fuel policy for industrial cogeneration." In R.A. Fazzolare, and C.B. Smith, eds. *International Conference on Energy Use Management*. Elmsford Park, N.Y.: Pergamon Press, 1979.

362. *World Energy Resources*. Guilford, U.K.: IPC Science and Technology Press, 1978.

363. Yannacone, Victor J. *Energy Crisis: Danger and Opportunity*. New York, N.Y.: Western, 1974.

400. Implementation of Energy-Sensitive Land Use Measures

401. "Access to sunlight: New Mexico's Solar Rights Act." *Natural Resources* 19:957–68, October 1979.

402. Adams, R.J. "Analysis of solar legislation - taxes and easements." *Land and Water Law Review* 14:393–413, 1979.

403. Aderman, Gary. "Energy standards for new buildings - an excercise in futility." *National Journal* 11: 1083–85, 30 June 1979.

404. Alfers, S.D. "Accommodation on preemption: state and federal control of private coal lands in Wyoming." *Land and Water Law Review* 12(1): 73–130, 1977.

405. Allen, Edward H. *Handbook of Energy Policy for Local Government*. Lexington, Mass.: Lexington, 1975.

406. "Allocation of sunlight: solar rights and the prior appropriation doctrine. Adaptation with title. Who owns the right to sunlight?" *University of Colorado Law Review* 47:421, Spring 1976.

407. American Planning Association: *Protecting Solar Access for New Residential Development: A Guidebook for Planning Officials*. Chicago, Ill.: American Planning Association, 1978.

408. American Society of Heating, Refrigerating and Air Conditioning Engineers. *Energy Conservation in New Building Design*. New York, N.Y.: The Society, 1975.

409. "Assuring legal access to solar energy: an overview with proposed legislation for the state of Nebraska." *Creighton Law Review* 12:567–627, Winter 1978–1979.

410. Bagge, Carl E. "One dimensional land use legislation: a threat to rational energy development." *Natural Resources Lawyer*. Winter 1973.

411. Becher, Ralph E., Jr. "Common law sun rights: an obstacle to solar heating and cooling." *Journal of Contemporary Law* 3:19+, 1976.

412. Bershon, David L. "Securing solar energy rights: easements, nuisance or zoning?" *Columbia Journal of Environmental Law* 3(1):112–52, Fall 1976.

413. Binns, P. *Turning Towards the Sun - Abstracts of State Legislature Enactments. 1974-1975*. Denver, Colo.: National Council of State Legislatures, 1976.

414. Building Officials and Code Administrators International. *Code for Energy Conservation in New Building Construction.* (n.p.), 1977.

415. Caldwell, L.K. "Energy crisis and environmental law: paradox of conflict and reinforcement." *New York Law Forum* 20:751–801, Spring 1975.

416. *The California Residential Energy Standards.* Springfield, Va.: NTIS, 1977.

417. "A California town is able to kill a watt in its war on waste; Davis learns to save energy with cyclists, solar power and severe building codes." *Wall St. Journal* 191:1 + , 17 May 1978.

418. City of Clarksburg/Inter-Technology Corp. *Comprehensive Community Plan for Energy Management at Clarksburg, West Virginia.* Springfield, Va.: NTIS, 1976.

419. Cluck, John M. "Local energy codes mandatory in 1979." *Alabama Municipal Journal:* 12–13 + , July 1978.

420. Collins, Eugene N. "Conservation of energy in Chattanooga: a study of one city's attempt to legislatively reduce energy consumption." *Municipal Attorney*, January 1975.

421. Conway, Nicholas, T., and Gregory L. Sumay. "Energy research and development: a partnership between federal and local government." *Public Administration Review:* 711–13, November-December 1977.

422. Council of State Governments. *State Response to the Energy Crisis.* Lexington, Ky.: The Council, 1974.

423. Deal, David T. "The Durham controversy; energy facility siting and the land use planning control process. *Natural Resource Lawyer* 8(3):437 + , 1975.

424. Doub, W.O. "Federal energy regulation - toward a better way." *ABA Journal* 60:920–23, August 1974.

425. Eisenhard, Robert M. *A Survey of State Legislation Relating to Solar Energy.* Washington, D.C.: National Bureau of Standards, 1976.

426. Eisenstadt, M.M. et al. "Proposed solar zoning ordinance." *Urban Law Annual* 15:211–50, 1978.

427. "Energy and the Law." *Hastings Law Journal* 28:1075–1273, May 1977.

428. "Energy and the law: a symposium." *Oregon Law Review* 54:503–679, 1975.

429. *Energy Conservation in New Building Design: An Impact Assessment of ASHRAE Standards 90-75.* Cambridge, Mass.: Arthur D. Little, Inc., 1976.

430. "Energy facility siting in North Dakota." *North Dakota Law Review* 53:703–23, Summer 1976.

431. Environmental Law Institute. *Solar Access and Land Use: State of the Law, 1977.* Rockville, Md.: NSHCIC, 1978.

432. *Environmental Planning and Siting of Nuclear Facilities: The Integration of Water, Air Coastal, and Comprehensive Planning into the Nuclear Siting Process.* Springfield, Va.: NTIS, 1977.

433. Erley, Duncan, and David Mosena. *Energy-Conserving Development Regulations: Current Practice.* Chicago, Ill.: APA, 1980.

434. Federal Housing Administration. *Minimum Property Standards for One and Two Housing Units.* Washington, D.C.: Department of Housing and Urban Development, 1971.

435. "Federal-state conflict in energy development: an illustration." *Denver Law Journal* 53:521–52, 1976.

436. Finlay, L.W., and M.S. McKnight. "Law of the sea: its impact on the international energy crisis." *Law and Policy in International Business* 6:639–76, Summer 1974.

437. Francis, D. "Energy conservation: guidelines for municipal programs." *Environmental Law* 8:131–71, Fall 1977.

438. Freeman, P.K. *States Response to the Energy Crisis: An Analysis of Innovation.* Milwaukee, Wis.: University of Wisconsin, 1978.

439. Goble, Dale D. "Solar rights: guaranteeing a place in the sun." *Oregon Law Review* 57:94+, 1977.

440. Hackman, Frank C. "Energy development and community growth: suggestions for elements of state policy toward boomtowns." *Utah Economics and Business Review* 39:1–8, January 1979.

441. Hammond, Jonathan et al. *A Strategy for Energy Conservation: Proposed Energy Conservation and Solar Utilization Ordinances for the City of Davis.* Davis, Calif.: (n.p.), 1974.

442. Hansen, Roger P. et al. *Legal and Regulatory Aspects of Land Reclamation.* Paper presented at Workshop on Reclamation of Western Surface Mined Lands, 7 +. Collins, Colo.: 1–3 March 1976.

443. Harrington, Winston. "Where do local governments fit into an energy conservation strategy." *Carolina Planning*: 43–52, Winter 1977.

444. Hawaii State Department of Planning and Economic Development. *State of Hawaii Energy Policies Plan.* Honolulu, Hi.: The Department, 1974.

445. Hayes, Gail Boyer. *Solar Access Law: Protecting Access to Sunlight for Solar Energy Systems.* Cambridge, Mass.: Ballinger, 1979.

446. Hiller, R.L. *Legal Aspects of Financing Solar-Heated Residential Structures. Final Report of Innovative Research Program Subtask, December 1977.* Springfield, Va.: NTIS, 1978.

447. Howard, J.J. "Energy crisis and its impact upon environmental law." *New York Law Forum* 20:711–30, Spring 1975.

448. *Industrial Siting Reform at the State Level.* Presented at Environmental Law Institute Conference on Energy and Public Lands, Utah: 23–26 August 1976.

449. Jaffe, Martin et al., *Protecting Solar Access for Residential Development: A Guidebook for Planning Officials.* Washington, D.C.: U.S. GPO, (n.d.).

450. Kapaloske, J. "Power plant siting on public lands: a proposal for resolving the environmental-developmental conflict." *Denver Law Journal* 56:233+, 1979.

451. Kauman, Lee, and Stuart L. Rehr. *Energy Conservation and Building Codes: The Legislative and Planning Process.* Boston, Mass.: Environmental Design and Research Center, 1977.

452. "Legislative approach to solar access-transferable development rights." *New England Law Review* 13:835-39, Spring 1978.

453. "Legislative response to solar access: a lesson for Michigan?" *Detroit College of Law Review*: 261-79, Summer 1979.

454. Little, Arthur D., and Company. *Residential Heating and Cooling: Constraints and Incentives.* Springfield, Va.: NTIS, 1976.

455. ———. *Technical Opportunities for Energy Conservation in Buildings through Improved Controls.* Springfield, Va.: NTIS, 1976.

456. McGregor, Gloria Shepart. "Davis, California implements energy building code." *Practicing Planner* 6(1):24-26+, February 1976.

457. Martin, Cinda, and David W. McKenna. *Energy and Local Government: A Report to the Cities and Counties of Texas.* Arlington, Tex.: University of Texas, Institute of Urban Studies, 1974.

458. Martin, L.H. "Many sides to government involvement in energy-use management." In R. Fazzolare, and C.B. Smith, eds. *Energy Use Management.* Elmsford Park, N.Y.: Pergamon, 1978.

459. Massachusetts. Department of Community Affairs. Energy Conservation Project. *Practical Steps Government Can take to Conserve Energy.* Springfield, Va.: NTIS, 1976.

460. Miller, A.S. et al. *Solar Access and Land Use: State of the Law, 1977.* Washington, D.C.: Environmental Law Institute, 1977.

461. Miller, A.S., and G.P. Thompson. *Legal Barriers to Solar Heating and Cooling of Buildings.* Springfield, Va.: NTIS, 1977.

462. Mills, J., and R.D. Woodson. "Energy policy: a test for federalism." *Arizona Law Review* 18:405-51, 1976.

463. Mills, J.L. *Energy: The Power of the States.* Gainsville, Fla.: University of Florida.

464. Morell, David, and Grace Singer. *State Legislatures and Energy Policy in the Northeast: Energy Facility Siting and Legislative Action.* Springfield, Va.: NTIS, 1978.

465. Moring, F. *Legal Issues in the Implementation of New Technology.* Washington, D.C.: Government Institute, Inc., 1976.

466. Nassikas, J. N. "Energy, the environment, and the administrative process." *Administrative Law Review* 26:165-90, Spring 1974.

467. National Conference of States on Building Codes and Standards. *Model

Code for Energy Conservation in New Building Construction. Springfield, Va.: NTIS, 1977.

468. National Institute of Building Sciences. *State's Energy Conservation Standards for Buildings.* Washington, D.C.: The Institute, 1978.

469. Olson, R.K. "Coal liquefaction: issues presented by a developing technology." *Tulsa Law Journal* 12:657–81, 1977.

470. *Overcoming Legal Uncertainties About the Use of Solar Energy Systems.* American Bar Foundation, 1977.

471. Peters, J. Douglas et al. "Durham, New Hampshire: a victim of home rule." *Ecology Law Quarterly* 5:53+, 1975.

472. "Progress toward solar building codes." *Solar Energy Research and Development Report*: 3, 1 May 1977.

473. Quinn, James T. "Implementation of energy consevation codes." *Construction Review*: 4–8, October 1978.

474. Real Estate Research Corporation. Public Affairs Counseling Division. *Local Government Approaches to Energy Conservation.* Washington, D.C.: U.S. GPO, 1979.

475. Revelle, Randy. *Staff Report on Energy Conservation in Seattle.* Wash.: (n.p.), (n.d.).

476. Robbins, R.L. "Building codes, land use controls, and other regulations to encourage solar energy use." In *Proceedings of Consumer Conference on Solar Energy Development*: 283–300, Albuquerque, N.M.: New Mexico Energy Resources Board, 1976.

477. Sawhill, John C. *Energy Conservation and Public Policy.* Englewood Cliffs, N.J.: Prentice-Hall, 1979.

478. Seely, D. et al. *Solar Energy Legal Bibliography.* Springfield, Va.: NTIS, 1979.

479. "Solar heating and cooling: state and municipal legal impediments and incentives." *Natural Resources Journal* 18(2): 313–24, April 1978.

480. Soloman, L.D., and F.H. Riesmeyer. "Development of alternative energy sources: a legal policy analysis." *Oklahoma Law Review* 30:319–53, Spring 1977.

481. Southeast Georgia Area Planning and Development Commission. *Energy Conservation and Management Guide for Small Municipalities and School Districts.* Washington, D.C.: U.S. Department of Housing and Urban Development, Office of Policy Development and Research, 1979.

482. Stang, D.P. "Energy legislation of the 93rd Congress—policy trends and implications for the future." *Oil and Gas Inst* 26:167–86, 1975.

483. Thomas, W.A., ed. *Proceedings of the Workshop on Solar Energy and Law.* Springfield, Va.: NTIS, 1975.

484. U.S. Congress. Senate. Committee on Government Operations. *Impact of the Energy Crisis on State and Local Governments*. Hearings, January, February 1974. Washington, D.C.: U.S. GPO, 1974.

485. U.S. Federal Energy Administration. *Project Independence Report*. Washington, D.C.: The Administration, 1976.

486. U.S. General Accounting Office. *National Standards Needed for Residential Energy Conservation*. Washington, D.C.: The Office, 1975.

487. U.S. National Bureau of Standards, Office of Building Standards and Code Services. *Building Energy Authority and Regulations Survey: State Activity*. Springfield, Va.: NTIS, 1976.

488. Watson, K.S. "Measuring and mitigating socio-economic environmental impacts of constructing energy projects: an emerging regulatory issue." *Natural Resources Law* 10:393–403, 1977.

489. White, H.D., and H.J. Barry III. "Energy development in the west: conflict and coordination of governmental decision making." *North Dakota Law Review* 52:451–528, Spring 1976.

490. Whitney, S.C. "Siting of energy facilities in the coastal zone - a critical regulatory hiatus." *William and Mary Law Review* 16:805–22, Summer 1975.